ADVANCES IN CELL AND MOLECULAR BIOLOGY OF MEMBRANES

Volume 1 • 1993

ENDOSOMES AND LYSOSOMES: A DYNAMIC RELATIONSHIP

ADVANCES IN CELL AND MOLECULAR BIOLOGY OF MEMBRANES

ENDOSOMES AND LYSOSOMES: A DYNAMIC RELATIONSHIP

Series Editor: ALAN M. TARTAKOFF
Institute of Pathology
Case Western Reserve University

Editors: BRIAN STORRIE
Department of Biochemistry and Nutrition
Virginia Polytechnic Institute and State University

ROBERT F. MURPHY
Department of Biological Sciences
Carnegie Mellon University

VOLUME 1 • 1993

JAI PRESS INC.

Greenwich, Connecticut *London, England*

55 Old Post Road, No. 2
Greenwich, Connecticut 06836

JAI PRESS LTD.
The Courtyard
28 High Street
Hampton Hill, Middlesex TW12 1PD
England

ISBN: 1-55938-362-3

Transferred to digital printing 2006

CONTENTS

LIST OF CONTRIBUTORS

Eliene Augenbraun — Department of Biological Sciences, Columbia University, New York, New York

Vytas A. Bankaitis — Department of Cell Biology, University of Alabama at Birmingham, Birmingham, Alabama

William A. Braell — Department of Biological Chemistry and Molecular Pharmacology, Harvard Medical School, Boston, Massachusetts

James A. Cardelli — Department of Microbiology and Immunology, Louisiana State University Medical Center, Shreveport, Louisiana

Penelope A. Colbaugh — The Molecular and Cell Biology Program, The University of Texas at Dallas, Richardson, Texas

Rockford K. Draper — The Molecular and Cell Biology Program, The University of Texas at Dallas, Richardson, Texas

William A. Dunn, Jr. — Department of Anatomy and Cell Biology, University of Florida, College of Medicine, Gainesville, Florida

H. Shelton Earp — UNC Lineberger Comprehensive Cancer Center, University of North Carolina at Chapel Hill, Chapel Hill, North Carolina

Bernard Hoflack
European Molecular Biology Laboratory
Cell Biology Programme
Heidelberg, Germany

Eric Holtzman
Department of Biological Sciences
Columbia University
New York, New York

William R. Huckle
UNC Lineberger Comprehensive
Cancer Center
University of North Carolina at Chapel Hill
Chapel Hill, North Carolina

Michael Koval
Department of Internal Medicine
Infectious Diseases Division
Washington University School of Medicine
St. Louis, Missouri

Peter Lobel
Center for Advanced Biotechnology
and Medicine and Department of
Pharmacology
Robert Wood Johnson Medical School
University of Medicine and
Dentistry of New Jersey
Piscataway, New Jersey

Robert W. Mason
Department of Biochemistry
Virginia Polytechnic Institute
and State University
Blacksburg, Virginia

Bryan K. McCune
Department of Pathology
Johns Hopkins University School
of Medicine and Frances Scott
Key Medical Center
Baltimore, Maryland

Robert F. Murphy
Department of Biological Sciences
Carnegie Mellon University
Pittsburgh, Pennsylvania

Michael G. Roth Department of Biochemistry
University of Texas Southwestern
Medical Center at Dallas
Dallas, Texas

Maria Santa-Hernandez Department of Biological Sciences
Columbia University
New York, New York

Henry B. Skinner Department of Cell Biology
University of Alabama at
Birmingham
Birmingham, Alabama

Robert St. Jules Department of Biological Sciences
Columbia University
New York, New York

Brian Storrie Department of Biochemistry and
Nutrition
Virginia Polytechnic Institute
and State University
Blacksburg, Virginia

Eric A. Whitters Department of Cell Biology
University of Alabama at
Birmingham
Birmingham, Alabama

Donna Wilcox Department of Cell Biology and
Physiology
Washington University School of Medicine
St. Louis, Missouri

INTRODUCTION TO THE SERIES

The remarkable vigor and central importance of cell biology result from the realization that emphasis on structure/function relations at the cellular and subcellular levels is essential for a rigorous and satisfactorily complete understanding.

Unlike many subdivisions of biomedical science, cell biology is not linked to any one methodology. It often emphasizes topological or topographic questions, and it is concerned with the structure, biogenesis, and turnover of macromolecular structures; however, there is no limit to the techniques and conceptual approaches which it brings to bear on these issues. Indeed, this has even been true since the term *cell biology* was first used. Certain scientists trace its origins back to ultrastructure and histology; others consider E.B. Wilson's, *The Cell in Development and Heredity*, to epitomize the foundations of cell biology; while others consider cell biology as an outgrowth of somatic cell genetics or the extension of biophysics to objects of increasingly large size. This varied and often changing identity makes it easier to specify what cell biology is not, rather than to describe it in a positive sense. It also suggests the rich intellectual mix which underlies today's successes in cell biology research.

This series aims to match the continuing evolution of cell biology—with particular emphasis on cell membranes—by treating coherent areas in multiauthored volumes. This approach allows a multifaceted coverage of topics which benefits from the unity of vision of the volume editor(s), but does not rely on one individual to synthesize an entire subject. Hopefully, the result will be more readable and intrinsically richer than lengthy review chapters which attempt to be all-encompassing.

Alan M. Tartakoff
Series Editor

PREFACE

Our intent in initiating this volume was to bring together a series of essays which would define our present understanding of the endosome and lysosome and their interrelationship. We deliberately encouraged our contributors to be speculative; to strive to put order to the "real" world of incomplete and sometimes conflicting data. Seeing science from the laboratory bench can often be like viewing an impressionistic painting from up close; a series of paint dabs with no apparent order. The contributors to this volume were asked to step back and leave the reader with a sense of the whole as well as the detail. To the extent that this has happened, the credit should go to the individual authors.

Our understanding of endosomes and lysosomes has undergone a molecular revolution over the last decade. Hence, we now know much about the molecular features required for internalization of an endocytic receptor, or the function of mannose 6-phosphate receptors in the transport of lysosomal enzymes. We can trace and follow the flow of molecules. In this volume current molecular knowledge concerning the function and relationship of endosomes and lysosomes is presented. Because of this vast increase in knowledge of molecules, we have realized that endosomes in particular are very ephemeral organelles. In fact, endosomes may well not be discrete entities but rather continuously changing and evolving in their molecular composition. The dynamic nature of the relationship between endosomes and lysosomes is the unifying focus of the genetic, biochemical, microscopic, and molecular biological approaches described in the chapters which follow.

In preparing this volume, we would like to acknowledge the continuous encouragement of Alan Tartakoff, senior editor for the series, whose suggestion led to this volume. We would also like to acknowledge the skillful work of JAI Press and the support and patience of our wives, Muriel and Cyndi.

Brian Storrie
Robert F. Murphy
Editors

MODELS OF ENDOSOME AND LYSOSOME TRAFFIC

Robert F. Murphy

Advances in Cell and Molecular Biology of Membranes
Volume 1, pages 1–17

ISBN: 1-55938-362-3

I. INTRODUCTION

The purpose of this chapter is to present current, often differing, concepts of how specific aspects of the endocytic system function. It is hoped that it will serve both as an introduction to the more detailed information in the following chapters and as a stimulus for further thought and experimentation about a fascinating but complex cell biological process. An artificial distinction is made here between *concepts* and *models*, with the latter term being reserved for proposals of how the complete endocytic system functions. The concepts described in the following sections constitute a set of parts from which models may be assembled; it is entirely possible that all of them may be embodied to a certain extent in the endocytic system of at least one cell type. Figure 1 illustrates the combination of a number of these concepts into a working model proposed by our laboratory. The discussion below will not consider the specialized phagocytic pathway for the uptake of large particles, but will consider only the internalization of soluble macromolecules.

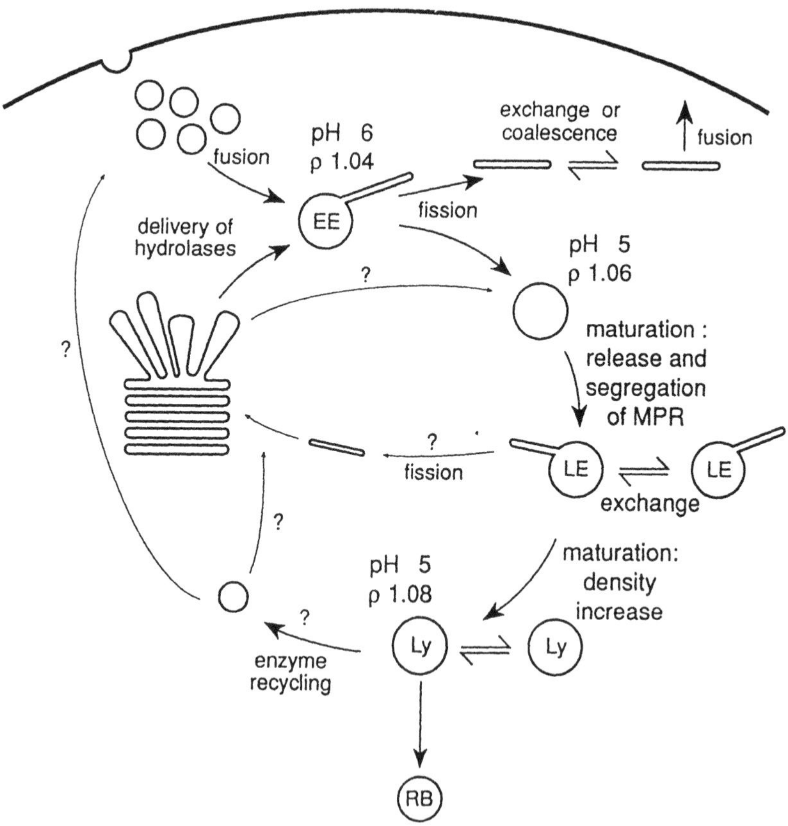

II. FORMATION OF PRIMARY ENDOCYTIC VESICLES

The formation of clathrin-coated pits and production of vesicles enriched in receptors from them has been described extensively (for reviews see references *8* and *47* and the chapter by Roth in this volume). At least two possible alternate pathways for internalization of membrane-impermeant material have also been described. The first is the so-called clathrin-independent pathway, which has received recent attention due to the ability in many cell types to inhibit clathrin-dependent uptake (e.g., by K^+ depletion or cytoplasmic acidification) without completely blocking uptake of fluid-phase markers (*28*). The extent to which this pathway intersects the clathrin-dependent pathway has not been fully evaluated. The second is the formation of caveollae and associated transmembrane transport

---

Figure 1. A working model for lysosome biogenesis by endosome maturation. The distinguishing feature of maturation models is the concept that early and late endosomal compartments are transient structures that are generated by fusion steps and consumed by fission and maturation steps. New endocytic vesicles are proposed to fuse with each other and/or previous early endosomes (**EE**) which are still fusion accessible. At least some lysosomal hydrolases are added at this step, presumably by fusion with vesicles derived from the Golgi complex. The resulting early endosomes undergo receptor-ligand dissociation due to the mildly acidic pH and membrane-fluid segregation due to the formation of tubules. Lysosomal hydrolases containing mannose 6-phosphate that are present in the early endosomes are expected to remain bound to the mannose 6-phosphate receptor (MPR) and be mainly excluded from the recycling tubules. The removal of these tubules by fission eventually gives rise to a late endosome (which initially is roughly spherical). Whether additional lysosomal enzymes are delivered directly to this compartment remains controversial. Reorganization of this endosome due to continued acidification leads to dissociation and segregation of hydrolases and MPRs. While some current evidence suggests that unoccupied MPRs recycle to the *trans*-Golgi network directly from this compartment, this question remains open. Late endosomes subsequently undergo an increase in density to produce dense lysosomes (**Ly**). At least some lysosomal hydrolases (and perhaps lysosomal membrane proteins) may recycle from this compartment, either back to the Golgi complex or directly to early endosomes. Dense lysosomes continue to increase in density (perhaps as a result of removal of membrane during enzyme recycling), giving rise to residual bodies (**RB**) or very dense lysosomes. Three sites of lateral exchange or coalescence of compartments (*14,61*) are included in the model: recycling tubules, late endosomes (**LE**), and dense lysosomes (**Ly**). Note that the cyclical nature of the model permits individual protein or lipid components to achieve their unique equilibrium distributions based on their affinity for the segregation machinery at each fission step. Note also that regulation of pH is envisaged to play two significant roles, controlling ligand-receptor dissociation in early and late endosomes and triggering the density increase leading to the formation of new dense lysosomes. The approximate density values (ρ) given are buoyant densities in Percoll. Reprinted from reference *46* with permission.

of small molecules (*2*). While measurements of probe uptake must consider the contribution from this possible pathway, it will not be discussed further since it does not transfer macromolecular material into other vesicular compartments. Collectively, vesicles arising from the plasma membrane via the clathrin-dependent and clathrin-independent pathways are termed primary endocytic vesicles (PEV).

III. GENERATION OF EARLY ENDOSOMES BY FUSION OF PRIMARY ENDOCYTIC VESICLES

Whether specific compartments of the endocytic pathway are stable or transient structures is a fundamental issue which concerns those proposing and supporting models of the endocytic system. This issue first arises when considering the transition from PEV to early endosomes, since early endosomes are proposed to form from PEV only when early endosomes are considered to be transient structures. As has been pointed out previously (*46*), evidence from studies of the *shibire* mutant of *Drosophila melanogaster* suggests that early endosomes are indeed transient. When formation of coated pits is inhibited by shifting to a non-permissive temperature, early endosomal tubules have been found to disappear, both in oocytes (*33,59*) and nephrocytes (*35*). This disappearance is reversible, as a return to the permissive temperature results in the resumption of coated pit endocytosis and reformation of early endosomes (*33,35*). The implication is that inability to generate new PEV leads to consumption of existing early endosomes. Of course, the possibility that the *shibire* mutation leads to disassembly of early endosomes and indirectly blocks coated pit formation must also be considered.

The fact that fusion events are clearly required in the early stages of endocytosis has provided a strong motivation for studying them using cell-free systems. The requirements for fusion of early endocytic compartments (PEV and/or early endosomes) have been studied by a number of investigators (see reference *27* and the chapter by Braell in this volume). Among the recent results from these studies is the observation that low molecular weight GTP-binding proteins, such as rab5 (*23*), are required for these fusion events.

An important point to be addressed in the concept of early endosome formation is whether components not found in PEV are required. If required, these components may be delivered by at least two means. First, proteins may become associated with the cytoplasmic face of the forming early endosome. An association/dissociation cycle is central to the current view of the catalytic function of the small GTP-binding proteins (e.g., rab proteins) and other "coat" proteins in Golgi traffic (for review see reference *37*). Association of cytoplasmic subunits may also play a role in activation of membrane pumps and channels (*49,66*). Second, vesicles derived from compartments within the cell may participate in the fusions giving rise to early endosomes. Among these may be vesicles containing proton pumps and hydrolases, either newly synthesized or not. The existence of vesicles (termed recycling lysosomes) carrying hydrolases recycled from previous lysosomes has been proposed (*45*) based on the failure of cycloheximide treatment to inhibit hydrolysis of an endocytosed cathepsin B substrate in early endosomes (*7*), but there is no direct evidence for their existence. Results from cell-free studies suggest that exposure of internalized material to cathepsin D requires a fusion event with characteristics similar to those of PEV/early endosome fusion (*41*). Either attachment of peripheral membrane proteins or fusion with internal vesicles may be invoked to explain the absence of fusion activity of vesicles derived from plasma membrane (*64*).

The possibility that membrane recycling (i.e., of transferrin and recycled receptors) begins prior to the complete formation of an early endosome must also be considered. Analysis of acidification kinetics (reviewed in reference *45*) suggests that segregation of transferrin from lysosomally directed ligands begins within minutes of internalization. In addition, examination of the colocalization of transferrin and clathrin indicates that transferrin recycling can occur without entry of transferrin into a classical tubulovesicular endosomal structure, but rather may take place from a coated endocytic vesicle (CEV) (*18*). The CEV may be considered as an alternative to the early endosome, a precursor to an early endosome, or as a product formed after recycling begins (e.g., from recycling tubules, see below). Whether fusion of PEV is required for formation of this early recycling structure, and whether the frequency of observation of CEV varies with cell type, are currently unclear. It is attractive to speculate that the early products of fusion between PEV may be competent to begin recycling, and that the balance between the number of CEV and early endosomes may be controlled by a balance between the rates of fusion of PEV and fission of CEV/early endosomes (a balance between the concentrations/amounts of components required for each). In line with this concept, overexpression of rab5 protein produces an increase in size of early endosomes while inhibition of normal rab5 function by transfection with a defective rab5 gene causes accumulation of small vesicles (presumably due to inhibition of their fusion with each other or early endosomes) (*9*).

The concept that the balance between rates of delivery/formation and removal/consumption controls compartment morphology has a precedent in models of Golgi traffic. Treatment of mammalian cells with brefeldin A leads to a coalescence of the endoplasmic reticulum and Golgi apparatus which is reversible upon drug removal (for review see reference (*34*). This is thought to result from disrupting the normally balanced traffic between the Golgi and the endoplasmic reticulum to favor the retrograde traffic to the endoplasmic reticulum. Analysis of Golgi morphology during brefeldin A treatment suggests that at least one of its effects is enhancing tubule formation while inhibiting assembly of coat proteins required for fission events. The observation that brefeldin A causes a similar coalescence of the *trans*-Golgi network with early endosomes (*38,63*) led to the generalization that organelle morphology and membrane traffic may be controlled by the balance between coat assembly and tubule formation (*34*).

IV. FUSION OF PRIMARY ENDOCYTIC VESICLES WITH A PREEXISTING EARLY ENDOSOME OR RETICULUM

Current data are also consistent with the concept that PEV fuse with preexisting early endosomes or an endosomal reticulum. The latter proposal (*31*) envisages a single interconnected system of tubules and varicosities to which incoming material may gain access via a single fusion event. The distinct types of compartments observed upon cell fractionation (or even by microscopy under conditions which favor disassembly of the reticulum) are viewed as specialized regions of a continuous whole. In contrast, many stable compartment models (*25*) propose that early endosomes are stable discrete compartments which accept material from PEV by fusion and generate endocytic carrier vesicles (ECV) for delivery of material to later compartments. While the cell-free systems for fusion of early compartments have examined fusion of PEV and early endosomes as a group (and therefore have not ruled out the possibility of fusion between PEV), the absence of fusion activity in vesicles derived from the cell surface (*64*) suggests that early endosomes rather than other PEV are the targets for fusion *in vivo*.

V. INITIATION AND REGULATION OF ACIDIFICATION

A central requirement for proper function of the endocytic system is appropriate acidification of endocytic compartments. Specifically, acidification via a vacuolar-type H^+-ATPase must be initiated after internalization of ligands and receptors and this acidification must be regulated to achieve the pH required for specific functions (and characteristic of particular compartments). Since PEV have been observed to lack H^+-ATPase activity (*20*), mechanisms such as those discussed above for addition of components not found in PEV may be required to initiate acidification. For example, the binding of a missing H^+-ATPase subunit to the cytoplasmic face of a forming early endosome may confer proton-pumping capacity (*49*). Alternatively, fusion with a compartment containing functional H^+-ATPases (either a putative transient structure, such as a recycling lysosome, or a putative stable structure, such as an early endosome) may initiate acidification of the endocytic content of a PEV.

Since in most cell types, early endosomes have a higher pH than late endosomes, or dense lysosomes (for reviews see references *43* and *45*), concepts have been developed for the regulation of pH within different endocytic compartments. Most of these focus on control of membrane potential as an indirect means of regulating H^+-ATPase activity, as this ATPase is electrogenic (*19*). Since the pH gradient which can be achieved by this ATPase will be limited by the potential across the membrane containing it, the generation of an interior-positive membrane potential by the action of a different cation pump would inhibit acidification and the dissipation of an interior-positive membrane potential by inward passage of anions or outward passage of cations would stimulate it. Thus, action of the Na^+, K^+-ATPase has been

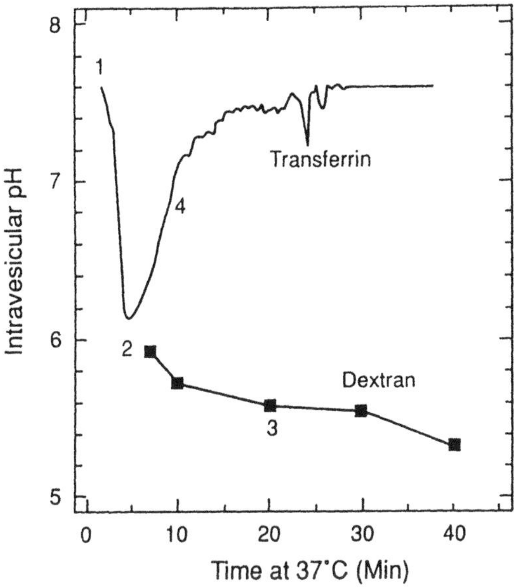

Molecule	1 PEV	2 Early Endosome	3 Lysosome	4 Recycling Tubule
		Acidification Phase / Compartment		
H^+-ATPase	Active	Active	Active	Inactive/Removed
Na^+,K^+-ATPase	Active	Active	Inactive/Removed	Active
Cl^--Channel	Inactive	Active	Active	?

Figure 2. Possible mechanisms for different phases of ligand acidification after endocytosis. The kinetics of acidification by 3T3 cells of a recycled ligand (transferrin, data from reference *55*) and a lysosomally-delivered probe (dextran, data from reference *52*) are shown. The lag before the onset of acidification (**1**) and the two phases of acidification (**2**,**3**) along the pathway to lysosomes are indicated. Recycled ligands follow the first two steps (**1**,**2**) but segregation in endosomes is followed by alkalinization during recycling (**4**). Also shown is a possible combination of activities by which the characteristic pH values of the compartments involved may be attained. The absence of acidification in PEV may also be due to H^+-ATPase being inactive or absent.

proposed to inhibit acidification in early endosomes (*10,21*) and action of Cl^- channels has been proposed to facilitate acidification in late endosomes and lysosomes (*6,60*). Any of the means by which the action of these components can be regulated, such as exclusion from/inclusion in forming compartments, covalent modification, or removal/addition of subunits, may be invoked to control acidification, and, potentially, subsequent events. For those cell types whose endosomes have a higher pH than lysosomes (e.g., fibroblasts), three stages in endocytic acidification may be envisaged (see Fig. 2):

1. Weak or no acidification: PEV contain internalized proton pump, chloride channel, and Na^+,K^+-ATPase (or acquire them by fusion with internal compartments). Initially, Na^+,K^+-ATPase is active and chloride channel is inactive, leading to maximal inhibition of acidification. This produces the lag between internalization and onset of acidification typically seen in kinetic measurements on intact cells (*54,55*). The possible absence of active proton pump in PEV may also explain this lag (*20*).
2. Intermediate acidification: Activation of chloride channels, such as by phosphorylation (*4,6*), permits the limited acidification typically seen in early endosomes (*43,45*) due to competing effects of the Cl^- and Na^+ fluxes on endosomal membrane potential (*22*).
3. Full acidification: Removal of Na^+, K^+-ATPase from early endosomes along with recycling receptors (or inactivation) then permits full acidification in late endosomes and lysosomes. This further acidification has been proposed to trigger eventual maturation into dense lysosomes (*45*).

VI. FISSION OF RECYCLING TUBULES

Recycling of receptors from material destined for lysosomes requires a fission event. Fission of a tubular element containing predominantly membrane components from a vesicular portion containing primarily lumenal content is most frequently envisaged as the event implied by the tubulovesicular structure typical of early endosomes. Fission of recycling tubules is either viewed as balanced by addition of membrane from PEV (if early endosomes are assumed to be stable) or eventually produces a late endosome depleted of recycled components (if early endosomes are assumed to be transient). The process of receptor segregation from non-recycled content has been reconstituted in cell-free systems (*48,62*), potentially allowing the biochemical requirements for the fission process to be determined.

In at least some cell types, segregation of transferrin is accompanied by an increase in vesicular pH (*55*), possibly due to enrichment of Na^+,K^+-ATPase in and/or exclusion of H^+-ATPase from recycling tubules. Incubation at temperatures below 19 °C has been observed to slow or block both the alkalinization of transferrin (*55*) and receptor recycling (*29,54*). The fact that these temperatures also block transfer of endocytosed material to dense lysosomes (*17*) has been suggested (*52*) as evidence that recycling of receptors (and presumably pH-regulatory elements) is required for generation of late endosomes and lysosomes, a notion more consistent with transient than stable compartment models. A similar linkage between inhibition of receptor recycling and inability to transfer endocytosed material to dense lysosomes has been observed in the temperature-sensitive, pleiotropic receptor recycling mutant TfT1.11 (*11*; see chapter by Colbaugh and Draper in this volume for review of endocytosis mutants).

VII. COALESCENCE OF RECYCLING TUBULES

Given the apparent rapidity with which tubules containing recycled receptors and transferrin are removed from early endosomes (implying a rapid rate of formation of discrete recycling tubules), one might expect that little mixing of separate cohorts of transferrin would be observed during recycling. However, pulses of transferrin separated by as much as 4 min can be found in the same compartment (*58*), indicating that coalescence of recycling tubules or fusion of recycling tubules with a common compartment allows mixing of material on its way back to the plasma membrane (*16,53,58*).

VIII. FISSION OF ENDOCYTIC CARRIER VESICLES FROM EARLY ENDOSOMES

In the stable compartment model described by Griffiths and Gruenberg (*25*), generation of endocytic carrier vesicles (ECV) carrying material to late endosomes is proposed to occur independently of the formation of recycling tubules. ECV are proposed to be large, spherical structures containing endocytosed material destined for lysosomes which pinch off from early endosomes (*26*). In this model, ECV are not proposed to become late endosomes, but rather to fuse with late endosomes to deliver their contents. They are predicted to have a protein composition distinct from that of early endosomes or late endosomes. While enrichment of certain proteins in preparations of early or late endosomes has been reported, no proteins have been proven to be unique to either compartment.

IX. MATURATION OF EARLY ENDOSOMES INTO LATE ENDOSOMES

An alternative to the concept of an ECV is that the the vesicular portion of an early endosome becomes a late endosome after removal of recycling tubules. As discussed above, the trigger for this maturation may be the further acidification permitted by removal of pH regulators (such as the Na^+,K^+-ATPase).

Support for the maturation of early endosomes has come from two lines of experiments on ligands processed by intact cells. Stoorvogel et al. (*58*) determined that endosomes gradually lose accessibility to incoming material and become enriched in mannose 6-phosphate receptors (MPR), suggesting a maturation process. Dunn and Maxfield (*15*) followed individual endosomes labeled with sequential pulses of LDL conjugated with two different fluorescent probes and found that the content of ligand in each early (recycling) endosome remained constant during a chase, implying that transport to late endosomes was essentially an all-or-none process (ruling out transport via small carrier vesicles).

While both require at least one fission event, theoretical differences exist between the generation of an ECV and maturation of an early endosome into a late endosome by removal of recycling tubules. For example, in the former concept, retention in early endosomes of components which neither recycle nor appear in late endosomes is permitted and recycling tubule removal need not be coupled to ECV formation. It should be noted that if the size of the proposed ECV in a stable compartment model approaches that of the vesicular portion of an early endosome, the practical difference between the stable and transient concepts diminishes (since little of the early endosome remains).

In addition to internal pH changes, the specific remodeling events associated with conversion of a portion of an early endosome into a late endosome may include changes in ion fluxes, association/dissociation of peripheral membrane proteins, degradation of contents, posttranslational modification of membrane or content proteins, invagination to generate a multivesicular morphology (such structures are often called multivesicular bodies (MVB) or multivesicular endosomes), and formation of an internal matrix. Lateral exchange with other late endosomes may also occur (see the chapter by Storrie in this volume).

X. FORMATION OF VARICOSITIES ALONG AN ENDOSOMAL RETICULUM

An alternative proposed means by which transport and processing events associated with the transition from early to late endosomes may occur is via the formation of varicosities along an endosomal reticulum (*31*). In addition, invaginations of the reticular membrane may generate vesicular inclusions enriched in membrane proteins to be degraded (e.g., EGF receptors); a collection of these inclusions in a varicosity would have the morphology characteristic of MVB. How MVB varicosities remain physically and functionally distinct from the remainder of the reticulum is not obvious; for example, a mechanism for generating pH differences between regions of a continuous reticulum must be developed to explain the differences in pH observed for ligands internalized for different times (see above). In addition, a number of studies have shown a decreased accessibility of internalized probes to previously internalized material destined for lysosomes (*52,53,57*), requiring that passage of contents in the reticulum be somehow directed and restricted. The possibility of transient disconnections between elements of the reticulum may be considered as a possible mechanism for reconciling these observations with the concept of a reticulum (*32*). Just as the transient and stable compartment concepts converge as the size of the portion of the compartment remaining stable diminishes, the discrete compartment and reticulum concepts converge as the frequency of fission and fusion events rises.

XI. TEMPERATURE EFFECTS ON APPEARANCE OF ENDOCYTOSED MATERIAL IN DENSE LYSOSOMES

Incubation of cells or tissues with probes at temperatures below 20 °C permits internalization but dramatically reduces degradation and appearance in dense lysosomes (*17,39*). This phenomenon has been used frequently in experimental protocols and must be accounted for in models of the endocytic system. The most frequently invoked explanation (*30*) is that low temperatures inhibit transfer to/fusion with hydrolase-containing compartments (either dense lysosomes themselves or compartments required for the generation of dense lysosomes). An alternative is that low temperatures inhibit receptor recycling and removal of pH regulators and thereby retain endocytosed material primarily in low density, high pH compartments (*45*; see above). This question has been addressed in a careful series of experiments on perfused liver by Hubbard and co-workers (*12,50*) that used two approaches to determine whether a full complement of hydrolases is present in early endosomes. The first, using endocytosed fluorogenic substrates, revealed that while acid phosphatase and aryl sulfatase are present at the plasma membrane and in early endosomes, exposure of endocytosed material to active β-hexosaminidase does not occur in early endosomes and is inhibited at 16 °C (*12*). The possibility that β-hexosaminidase is present in early endosomes but not active (e.g., due to pH, steric inhibition, or presence of inhibitors) was not ruled out. The second, monitoring specific proteolytic cleavages of EGF, indicated that only one cleavage of EGF (by a cathepsin B-like enzyme) occurred at 16 °C while further cleavages did not occur. To determine whether early endosomes contained the enzymes necessary for these cleavages but were inactive, fractions from livers incubated for 10 minutes at 37 °C or 60 minutes at 16 °C were incubated *in vitro* at pH 5 at 37 °C. Some further proteolysis of EGF was observed but complete conversion of the singly-cleaved EGF to the shorter forms did not occur (implying the absence of the required enzymes in the EGF-containing compartments). Since full proteolysis was not observed *in vitro* for any labeling condition, (i.e., even for fractions from livers incubated for 60 minutes at 37 °C), the failure to observe further proteolysis *in vitro* when EGF was predominantly in early compartments may be due to suboptimal *in vitro* assay conditions rather than absence of the appropriate enzymes. The possibility also remains that low temperatures partially inhibit the proposed delivery of hydrolases to early endosomes (e.g., via putative recycling lysosomes). However, it would appear more likely that exposure of endocytosed material to the full complement of active hydrolases requires an event (such as fusion with a compartment containing "late" hydrolases) which does not occur at low temperatures.

XII. RELEASE OF LYSOSOMAL ENZYMES FROM MANNOSE 6-PHOSPHATE RECEPTORS (MPRs)

The primary role of MPRs is thought to be in carrying newly synthesized lysosomal enzymes to lysosomes (see reference *36* and the chapter by Hoflack and Lobel in this volume). Dissociation of lysosomal hydrolases from MPRs requires a pH below 5.5 (*24*) while dissociation of cargo from recycling plasma membrane receptors can occur above pH 6 [e.g., α-macroglobulin from its receptor (*40*) and iron from the transferrin/transferrin receptor complex (*5,56*)]. Thus, lysosomal hydrolases are likely to remain bound to MPRs in early endosomes and be released only in late endosomes (*45,65*). This implies a second round of receptor segregation and tubule fission in late endosomes which functions to recycle MPRs. For transient compartment models, one must therefore account for the change in surface area to volume ratio required for the transition from a roughly spherical to a tubulovesicular structure. One possible explanation is that the increase in density characteristic of the transition of ligand from early endosomes to late endosomes and dense lysosomes results from a loss of water (see below) which creates excess membrane permitting tubule formation.

One possible mechanism to explain the observation that different hydrolases appear to be delivered to different extents to early and late endosomes (see above) is based on results indicating that the two forms of the MPR play different roles in enzyme targeting and secretion. Chao et al. (*13*) have shown that secretion of lysosomal enzymes is mediated by the 46-kDa (cation-dependent) MPR but not by the 300-kDa (cation-independent) MPR. Newly synthesized (or preexisting) hydrolases appear to have different affinities for the two receptors, and thus the 46-kDa MPR may deliver enzymes to early endosomes (from which they may then be "secreted" during recycling to the plasma membrane) while the 300-kDa MPR may deliver enzymes primarily to late endosomes.

Two other explanations for differences in levels of hydrolases in early and late compartments may be considered. Enzymes may be delivered to lysosomes indirectly by export to the plasma membrane of membraneous precursor forms and subsequent internalization (see reference *36* and chapter by Mason in this volume). In addition, a "prohydrolase receptor" which binds L- and D-procathepsins at pH 5 but not at pH 6 to 7 (*42*) may function to carry some enzymes from acidic portions of the Golgi (*3*) to early endosomes or from late endosomes to dense lysosomes.

XIII. FUSION OF LATE ENDOSOMES WITH DENSE LYSOSOMES

The classical model for delivery of endocytosed material to dense lysosomes is via fusion of some prior compartment [i.e., an entire late endosome (transient compartment models) or portions thereof (stable compartment models)] with a preexisting dense (secondary) lysosome. This late fusion event was explicit in both the vesicle shuttle and maturation models described by Helenius et al. (*30*). The origin of the preexisting dense lysosomes (and the means by which they become dense) are often unspecified in models postulating endosome-lysosome fusion. Cell-free conditions which result in fusion of early endocytic compartments (PEV and endosomes) with dense lysosomes have been described (*1,44*); under at least some conditions, fusion is required for delivery of endocytosed probe to compartments with the buoyant density of lysosomes.

XIV. MATURATION OF LATE ENDOSOMES INTO LYSOSOMES

The realization that early and late endosomes contain lysosomal hydrolases made possible models in which late endosomes mature into lysosomes without a requirement for fusion with a pre-existing dense lysosome. The absence of this requirement is possible in both transient and stable compartment models. A prediction in both cases is that isolated endosomes should be capable of giving rise to dense lysosomes in the absence of other dense lysosomes (in the latter case, only the portion which buds from the late endosome is proposed to undergo this maturation). This phenomenon was observed in a cell-free system requiring ATP and elevated temperature (*51*). A similar increase in buoyant density of a fraction of endocytosed fluid-phase marker was observed in a cell-free system for reconstituting receptor segregation (*62*).

XV. REUTILIZATION OF LYSOSOMAL ENZYMES

The long half-life of lysosomal enzymes implies that these enzymes are utilized for degradation of more than a single bolus of endocytosed material. Thus, it is important to consider means by which lysosomal enzymes may be reutilized. The simplest such means is the proposed fusion of secondary lysosomes with late endosomes (or portions thereof) after MPR recycling. A related alternative, suggested by the fact that late endosomes have significant amounts of lysosomal enzymes, MPRs and lysosomal membrane proteins, is the possibility that secondary lysosomes fuse with late endosomes prior to MPR recycling.

Delivery of enzymes from secondary lysosomes to even earlier compartments in the lysosomal biogenesis pathway is supported by the observation that cycloheximide treatment for as long as 8 hours does not affect the kinetics of hydrolysis of a cathepsin B substrate (*7*), hydrolysis which begins within minutes of addition of substrate and is therefore likely to occur in early endosomes. As discussed above, this led to the proposal that separate vesicles (speculatively termed recycling lysosomes) carrying lysosomal enzymes to be reused (perhaps only those enzymes which retain mannose 6-phosphate markers or other targeting signals) may be generated from dense lysosomes (*45*). Thus, two more controversial delivery sites for enzymes from secondary lysosomes, the Golgi/*trans*-Golgi network and early endosomes, must be considered. In either case, enzymes which had not yet completely lost the mannose 6-phosphate marker would be able to bind to MPR to be retargeted to lysosomes.

XVI. CONCLUSIONS

This chapter has summarized many aspects of current models of endocytic processing. It is hoped that one point that will be taken from this discussion is how concepts that are diametrically opposed may often be seen as equivalent in practice given appropriate values of free parameters (e.g., size of shuttle vesicles). While a particular conceptual framework is helpful for presentation and interpretation of experimental data, it is desirable to explicitly consider alternatives that are also consistent with those data.

ACKNOWLEDGMENTS

I thank Drs. Frederick R. Maxfield and Michael Forgac for helpful discussions and Sandra A. Brockman for critical reading of the manuscript. Work from my laboratory on some of the subjects discussed here has been supported by the National Institutes of Health (grant GM32508) and by the National Science Foundation (grant DCB8903657).

REFERENCES

1. Altstiel, L. and Branton, D. (1983). Fusion of coated vesicles with lysosomes: Measurement with a fluorescence assay. *Cell* 32: 921–929.
2. Anderson, R.G.W., Kamen, B.A., Rothberg, K.G., and Lacey, S.W. (1992). Potocytosis: sequestration and transport of small molecules by caveolae. *Science* 255: 410–411.
3. Anderson, R.G.W. and Pathak, R.K. (1985). Vesicles and cisternae in the trans Golgi apparatus of human fibroblasts are acidic compartments. *Cell* 40: 635–643.
4. Bae, H.-R. and Verkman, A.S. (1990). Protein kinase A regulates chloride conductance in endocytic vesicles from proximal tubules. *Nature* 348: 637–639.
5. Bali, P.K., Zak, O. and Aisen, P. (1991). A new role for the transferrin receptor in the release of iron from transferrin. *Biochem.* 30: 324–328.
6. Barasch, J., Gershon, M.D., Nunez, E.A., Tamir, H., and Al–Awqati, Q. (1988). Thyrotropin induces the acidification of the secretory granules of parafollicular cells by increasing the chloride conductance of the granular membrane. *J. Cell Biol.* 107: 2137–2147.
7. Bowser, R. and Murphy, R.F. (1990). Kinetics of hydrolysis of endocytosed substrates by mammalian cultured cells: Early introduction of lysosomal enzymes into the endocytic pathway. *J. Cell. Phys.* 143: 110–117.
8. Brodsky, F.M. (1988). Living with clathrin: its role in intracellular membrane traffic. *Science* 242: 1396–1402.
9. Bucci, C., Parton, R.G., Mather, I., Stunnenberg, H., Simons, K., and Zerial, M. (1992). The small GTPase rab 5 functions as a regulatory factor in the early endocytic pathway. *Cell* 70: 715–728.
10. Cain, C.C., Sipe, D.M., and Murphy, R.F. (1989). Regulation of endocytic pH by the Na^+,K^+-ATPase in living cells. *Proc. Natl. Acad. Sci. USA* 86: 544–548.

11. Cain, C.C., Wilson, R.B., and Murphy, R.F. (1991). Isolation by fluorescence-activated cell sorting of Chinese hamster ovary cell lines with pleiotropic, temperature-conditional defects in receptor recycling. *J. Biol. Chem.* 266: 11746–11752.
12. Casciola-Rosen, L.A.F. and Hubbard, A.L. (1991). Hydrolases in intracellular compartments of rat liver cells: Evidence for selective activation and/or delivery. *J. Biol. Chem.* 266: 4341–4347.
13. Chao, H.H.-J., Waheed, A., Pohlmann, R., Hille, A., and von Figura, K. (1990). Mannose 6-phosphate receptor dependent secretion of lysosomal enzymes. *EMBO J.* 11: 3507–3513.
14. Deng, Y. and Storrie, B. (1988). Animal cell lysosomes rapidly exchange membrane proteins. *Proc. Natl. Acad. Sci. USA* 85: 3860–3864.
15. Dunn, K.W. and Maxfield, F.R. (1992). Delivery of ligands from sorting endosomes to late endosomes occurs by maturation of sorting endosomes. *J. Cell Biol.* 117: 301–310.
16. Dunn, K.W., McGraw, T.E., and Maxfield, F.R. (1989). Iterative fractionation of recycling receptors from lysosomally destined ligands in an early endosome. *J. Cell Biol.* 109: 3303–3314.
17. Dunn, W.A., Hubbard, A.L., and Aronson, Jr., N.N. (1980). Low temperature selectively inhibits fusion between pinocytic vesicles and lysosomes during heterophagy of ^{125}I-asialofetuin by the perfused rat liver. *J. Biol. Chem.* 255: 5971–5978.
18. Eskelinen, S., Kok, J.W., Sormunen, R., and Hoekstra, D. (1991). Coated endosomal vesicles: sorting and recycling compartment for transferrin in BHK cells. *Eur. J. Cell Biol.* 56: 210–222.
19. Forgac, M. (1989). Structure and function of vacuolar class of ATP-driven proton pumps. *Physiol. Rev.* 69: 765–795.
20. Fuchs, R., Ellinger, A., Pavelka, M., Peterlik, M., and Mellman, I. (1987). Endocytic coated vesicles do not exhibit ATP-dependent acidification *in vitro*. *J. Cell Biol.* 105: 91a (abstract).
21. Fuchs, R., Schmid, S., and Mellman, I. (1989). A possible role for the Na^+,K^+-ATPase in regulating ATP-dependent endosome acidification. *Proc. Natl. Acad. Sci. USA* 86: 539–543.
22. Gaete, V., Nunez, M.T., and Glass, J. (1991). Cl^- , Na^+, and H^+ fluxes during the acidification of rabbit reticulocyte endocytic vesicles. *J. Bioenerget. Biomemb.* 23: 147–160.
23. Gorvel, J.-P., Chavrier, P., Zerial, M., and Gruenberg, J. (1991). Rab5 controls early endosome fusion *in vitro*. *Cell* 64: 915–925.
24. Gonzalez-Noriega, A., Grubb, J.H., Talkad, V., and Sly, W.S. (1980). Chloroquine inhibits lysosomal enzyme pinocytosis and enhances lysosomal enzyme secretion by impairing receptor recycling. *J. Cell Biol.* 85: 839–852.
25. Griffiths, G. and Gruenberg, J. (1991). The arguments for pre-exlsting early and late endosomes. *Trends Cell Biol.* 1: 5–9.
26. Gruenberg, J., Griffiths, G., and Howell, K.E. (1989). Characterization of the early endosome and putative endocytic carrier vesicles *in vivo* and with an assay of vesicle fusion *in vitro*. *J. Cell Biol.* 108: 1301–1316.
27. Gruenberg, J. and Howell, K.E. (1989). Membrane traffic in endocytosis: Insights from cell-free assays. *Annu. Rev. Cell Biol.* 5: 453–481.
28. Hansen, S.H., Sandvig, K., and van Deurs, B. (1991). The preendosomal compartment comprises distinct coated and noncoated endocytic vesicle populations. *J. Cell Biol.* 113: 731–741.
29. Harding, C.V. and Unanue, E.R. (1990). Low-temperature inhibition of antigen processing and iron uptake from transferrin: deficits in endosome functions at 18 °C. *Eur. J. Immunol.* 20: 323–329
30. Helenius, A., Mellman, I., Wall, D., and Hubbard, A. (1983). Endosomes. *Trends Biochem. Sci.* 8: 245–249.
31. Hopkins, C.R., Gibson, A., Shipman, M., and Miller, K. (1990). Movement of internalized ligand-receptor complexes along a continuous endosomal reticulum. *Nature* 346: 335–339.
32. Hopkins, C.R., and Miller, K.M. (1991). New insights into the endocytic pathway. In: *Current Topics in Endocytosis: Proceedings of the Second European Workshop on Endocytosis* (Courtoy, P.J., Ed.), NATO ASI Series, Vol. H 62, Springer-Verlag, Berlin.
33. Kessell, I., Holst, B.D., and Roth, T.F. (1989). Membranous intermediates in endocytosis are labile, as shown in a temperature-sensitive mutant. *Proc. Natl. Acad. Sci. USA* 86: 4968–4972.

34. Klausner, R.D., Donaldson, J.G., and Lippincott-Schwartz, J. (1992). Brefeldin A: Insights into the control of membrane traffic and organelle structure. *J. Cell Biol.* 116: 1071–1080.
35. Koenig, J.H. and Ikeda, K. (1990). Transformational process of the endosomal compartment in nephrocytes of *Drosophila melanogaster. Cell Tissue Res.* 262: 233–244.
36. Kornfeld, S., and Mellman, I. (1989). The biogenesis of lysosomes. *Annu. Rev. Cell Biol.* 5: 483–525.
37. Kreis, T.E (1992). Regulation of vesicular and tubular membrane traffic of the Golgi complex by coat proteins. *Curr. Op. Cell Biol.* 4: 609–615.
38. Lippincott-Schwartz, J., Yuan, L., Tipper, C., Amherdt, M., Orci, L., and Klausner, R.D. (1991). Brefeldin A's effects on endosomes, lysosomes, and the TGN suggest a general mechanism for regulating organelle structure and membrane traffic. *Cell* 67: 601–616.
39. Marsh, M., Bolzau, E., and Helenius, A. (1983). Penetration of semliki forest virus from acidic prelysosomal vacuoles. *Cell* 32: 931–940.
40. Maxfield, F.R. (1982). Weak bases and ionophores rapidly and reversibly raise the pH of endocytic vesicles in cultured mouse fibroblasts. *J. Cell Biol.* 95: 676–681.
41. Mayorga, L.S., Diaz, R., and Stahl, P.D. (1989). Reconstitution of endosomal proteolysis in a cell-free system. *J. Biol. Chem.* 264: 5392–53999.
42. McIntyre, G.F. and Erickson, A.H. (1991). Procathepsins L and D are membrane-bound in acidic microsomal vesicles. *J. Biol. Chem.* 266: 15438–15445
43. Mellman, I., Fuchs, R., and Helenius, A. (1986). Acidification of the endocytic and exocytic pathways. *Annu. Rev. Biochem.* 55: 663–700.
44. Mullock, B.M., Branch, W.J., van Schaik, M., Gilbert, L.K., and Luzio, J.P. (1989). Reconstitution of an endosome-lysosome interaction in a cell-free system. *J. Cell Biol.* 108: 2093–2099.
45. Murphy, R.F. (1988). Processing of endocytosed material. *Adv. Cell Biol.* 2: 159–180
46. Murphy, R.F. (1991). Maturation models for endosome and lysosome biogenesis. *Trends Cell Biol.* 1: 77–82.
47. Pearse, B.M.F (1990). Clathrin, adaptors, and sorting. *Annu. Rev. Cell Biol.* 6: 151–171.
48. Podbilewicz, P., and Mellman, I. (1990). ATP and cytosol requirements for transferrin recycling in intact and disrupted MDCK cells. *EMBO J.* 9: 3477–3488.
49. Puopolo, K. and Forgac, M. (1990). Functional reassembly of the coated vesicle proton pump. *J. Biol. Chem.* 265: 14836–14841.
50. Renfrew, C.A. and Hubbard, A.L. (1991). Sequential processing of epidermal growth factor in early and late endosomes of rat liver. *J. Biol. Chem.* 266: 4348–4356.
51. Roederer, M., Barry, J.R., Wilson, R.B., and Murphy, R.F. (1990). Endosomes can undergo an ATP-dependent density increase in the absence of dense lysosomes. *Eur. J. Cell Biol.* 51: 229–234
52. Roederer, M., Bowser, R., and Murphy, R.F. (1987). Kinetics and temperature dependence of exposure of endocytosed material to proteolytic enzymes and low pH: evidence for a maturation model for the formation of lysosomes. *J. Cell. Physiol.* 131: 200–209.
53. Salzman, N.H., and Maxfield, F.R. (1989). Fusion accessibility of endocytic compartments along the recycling and lysosomal endocytic pathways in intact cells. *J. Cell Biol.* 109: 2097–2104.
54. Sipe, D.M., Jesurum, A. and Murphy, R.F. (1991). Absence of Na^+,K^+-ATPase regulation of endosomal acidification in K562 erythroleukemia cells. *J. Biol. Chem.* 266: 3469–3474.
55. Sipe, D.M., and Murphy, R.F. (1987). High resolution kinetics of transferrin acidification in BALB/c 3T3 cells: exposure to pH 6 followed by temperature-sensitive alkalinization during recycling. *Proc. Natl. Acad. Sci. USA* 84: 7119–7123.
56. Sipe, D.M. and Murphy, R.F. (1991). Binding to cellular receptors results in increased iron release from transferrin at mildly acidic pH. *J. Biol. Chem.* 266: 8002–8007.
57. Stoorvogel, W., Geuze, H.J., Griffith, J.M., Schwartz, A.L., and Strous, G.J. (1989). Relations between the intracellular pathways of the receptors for transferrin, asialoglycoprotein, and mannose 6-phosphate in human hepatoma cells. *J. Cell Biol.* 108: 2137–2148.

58. Stoorvogel, W., Strous, G.J., Geuze, H.J., Oorschot, V., and Schwartz, A.L. (1991). Late endosomes derive from early endosomes by maturation. *Cell* 65: 417–427.
59. Tsuruhara, T., Koenig, J.H., and Ikeda, K. (1990). Synchronized endocytosis studied in the oocyte of a temperature-sensitive mutant of *Drosophila melanogaster. Cell Tissue Res* 256: 199–207.
60. Van Dyke, R.W. (1988). Proton pump-generated electrochemical gradients in rat liver multivesicular bodies. *J. Biol. Chem.* 263: 2603–2611.
61. Ward, D.M., Hackenyos, D.P., and Kaplan, J. (1990). Fusion of sequentially internalized vesicles in alveolar macrophages. *J. Cell Biol.* 110: 1013–1022.
62. Wessling-Resnick, M. and Braell, W.A. (1990). The sorting and segregation mechanism of the endocytic pathway is functional in a cell-free system. *J. Biol. Chem.* 265: 690–699.
63. Wood, S.A., Park, J.E., and Brown, W.J. (1991). Brefeldin A causes a microtubule-mediated fusion of the trans-Golgi network and early endosomes. *Cell* 67: 591–600.
64. Woodman., P.G. and Warren, G. (1988). Fusion between vesicles from the pathway of receptor-mediated endocytosis in a cell-free system. *Eur. J. Biochem.* 173: 101–108.
65. Yamashiro, D.J. and Maxfield, F.R (1988). Regulation of endocytic processes by pH. *Trends Pharmacol. Sci.* 9: 190–193.
66. Zhang, J., Myers, M., and Forgac, M. (1992). Characterization of the V_0 domain of the coated vesicle H^+-ATPase. *J. Biol. Chem.* 265: 9773–9778.

ENDOCYTIC RECEPTORS

Michael G. Roth

ABSTRACT

The plasma membrane of animal cells contains proteins designed for the transport of extracellular ligands into the cell. These endocytic receptors are internalized by specialized structures of the plasma membrane, the coated pits, and after internaliza-

Advances in Cell and Molecular Biology of Membranes
Volume 1, pages 19–50

ISBN: 1-55938-362-3

tion usually return rapidly to the cell surface to complete another cycle of internalization. For each internalization cycle, there are two sorting events that separate endocytic receptors from other proteins. The first occurs at the coated pits, which internalize a variety of proteins in addition to endocytic receptors, but exclude others. The second event occurs in organelles of the endocytic pathway where recycling receptors are sorted from proteins destined for lysosomes. Several short amino acid motifs have been identified that are necessary and sufficient for allowing proteins to be internalized through coated pits, although there is no general agreement yet on the identity of the protein(s) that bind these sequences. For the second sorting event, it is now clear that the separation of receptors from their ligands usually depends upon a lower binding affinity at the acidic pH of the endosome, causing ligands to partition into the fluid phase for transport to lysosomes. Most internalized membrane recycles to the plasma membrane, but the mechanism by which membrane proteins are sorted in the endosome into recycling and lysosomally directed populations is currently unknown. A model for how this might occur is proposed.

I. INTRODUCTION: ENDOCYTIC FUNCTIONS OF THE PLASMA MEMBRANE

Endocytosis in eukaryotic cells serves many purposes. Cells internalize plasma membrane as part of the change in surface to volume that accompanies changes in shape. They constantly renew the plasma membrane by internalizing and degrading surface proteins and they internalize certain other proteins as a part of a response to external stimuli, such as the binding of hormones by their receptors. At the same time, they selectively and continually internalize special plasma membrane receptors for the purpose of delivering various ligands from the extracellular environment to lysosomes. Although there are multiple pathways for the internalization of cell surface proteins, for many cell types the major pathway appears to be through specialized domains of the plasma membrane, the clathrin coated pits (*29,30,44,69*).

Cell surface proteins can be divided into five classes according to their ability to interact with coated pits. Certain proteins are essentially immobile, apparently anchored to the cell cytoskeleton, and do not enter coated pits. Other proteins, such as the influenza virus hemagglutinin expressed from a cDNA, are highly mobile (*23,59*) but enter coated pits 50 to 100 times more slowly than the bulk internalization of membrane (*56,59,88*). A third class of proteins is internalized at a rate comparable to the uptake of lipid bilayer and appear to lack the capacity for significant concentration in coated pits, but are not excluded (*9,83,107*). Finally, there are two types of receptors that are internalized 5- to 40-fold faster than proteins that do not concentrate in coated pits. These receptors differ in the degree to which binding of ligand promotes interaction with coated pits. The best studied of these proteins, from the point of view of their transport within the endocytic pathway, do not require ligand for rapid internalization. Some members of this latter class of receptors are now known to contain specific cytoplasmic sequences that interact in some way with elements of the clathrin coat. For the purpose of the following discussion, such proteins with constitutively high rates of internalization will be referred to as "endocytic receptors" since their main function is to deliver ligand inside the cell. The other type of receptor, typified by the receptors for polypeptide hormones and growth factors, requires the binding of ligand to generate a high-affinity interaction with coated pits. This interaction also involves recognition of cytoplasmic residues of the receptor, but exact sequences responsible have not been identified.

After internalization, the subsequent fate of proteins depends upon a sorting event in the early, or recycling endosome, where proteins destined for lysosomes are segregated from those that recycle to the cell surface (*101*) or are transported to other destinations. The separation of ligands from receptors and the intracellular pathways followed by receptors and ligands after this sorting event are reasonably well understood now and have been the subject of reviews both in this volume and others (*29,44*). However, the molecular details regulating the entry of endocytic receptors into, and their return from, the endocytic pathway are incompletely understood, even for the best studied of cell surface receptors, such as those for low-density lipoproteins (LDL) or transferrin. The purpose of this chapter is to discuss what we currently know about the traffic control of endocytic receptors internalized through coated pits. Although the focus is on proteins that function primarily to deliver ligand into the endocytic pathway, important contributions from the study of receptors that function primarily for transmembrane signaling are mentioned. In the past decade, the ability to clone, express, and specifically mutate the genes for endocytic receptors has resulted in a large increase in our knowledge of the structure of those proteins and a considerable advance in understanding which features are important for internalization. A summary of the current data on the nature of internalization "signals" is presented, along with models for the interaction of receptors and coated pits. Much less developed is our understanding of features important for sorting proteins in the endosome into those that recycle to the plasma membrane and those that continue on to lysosomes. Some observations relevant to this subject are also discussed.

II. ENDOCYTIC PATHWAYS

It is important to remember that endocytosis through clathrin-coated pits is not the only route for internalization of plasma membrane proteins. A non-clathrin-mediated pathway has been described (*93–95*) that is distinct from phagocytosis and probably is distinct from the internalization of membrane that occurs as pseudopodia or membrane ruffles fold back and are internalized on motile, or hormone-stimulated cells (although these latter two processes are quantitatively important under certain conditions). This non-clathrin pathway can be distinguished from the better known clathrin-dependent pathway through its insensitivity to treatments that prevent proper clathrin lattice formation, such as hypertonic media (*17,38*), potassium depletion (*55*), or acidification of cytosol (*39,95*). The nonclathrin pathway is, however, inhibited by cytochalasin D, whereas clathrin-dependent internalization is not (*94*), and internalization by this pathway is slower. Little is known about the selectivity of the clathrin-independent pathway. Although under normal circumstances internalization of receptors with high affinity for clathrin pits occurs almost entirely by the clathrin-mediated pathway (*69*), in cells such as Chinese hamster ovary (CHO) where the volume of non-clathrin-mediated internalization is significant, treatments that inhibit the function of clathrin coats do not entirely block receptor internalization. In the same way, mutations of an endocytic receptor that block most internalization in human fibroblasts, such as conversion of tyrosine 807 to cysteine in the LDL receptor, allow 15% of wild-type internalization in CHO cells (*19*). Thus, it is possible that for endocytic receptors expressed at levels that do not exceed the capacity of coated pits to bind them, exclusion from the nonclathrin pathway is merely due to their much higher affinity for clathrin coats. One would predict, then, that overexpression of receptors in cells transfected with cloned genes could have quite complex effects, depending upon the cell type. In a cell with a significant volume of non-clathrin internalization, proteins kept out of coated pits might be internalized with lower efficiency through another pathway (*66*). The reader should be aware that most studies of the effects of mutating endocytic receptors known to be internalized through coated pits have not considered the possibility that mutant receptors with reduced affinity for coated pits might be internalized by a different pathway. The effect of such pathway switching would be to reduce the apparent affect of mutations that diminish clathrin-mediated internalization, producing an underestimate of the decrease in affinity of receptors for coated pits.

III. THE STRUCTURE OF ENDOCYTIC RECEPTORS

Receptors that constitutively concentrate in coated pits for the purpose of delivering ligand into cells are transmembrane glycoproteins that span the lipid bilayer once. Although no crystal structure has yet been solved for any of these receptors, we have some idea of their shape from biochemical tests, such as sensitivity to proteases, and from analyses of their primary sequences by secondary structure prediction algorithms. In general, they appear to be elongated proteins with globular-ligand binding domains held away from the lipid bilayer by a stalk that often is heavily glycosylated. Many receptors are oligomeric. Although for some receptors it has been difficult to establish the native oligomeric state of the protein, the trend has been to discover that proteins thought to be monomeric are, in fact, homooligomers that are difficult to isolate intact. An example of this is the LDL receptor, which was long thought to be a monomer but has been demonstrated to be a dimer (*38,105*). Thus, proteins listed in Table 1 as monomeric might well be multimers in the cell.

Table 1. Characteristics of Common Endocytic and Hormone Receptors

Receptor			*Domain size*			*Posttranslational modifications*		
	species	*type*	*Ecto*	*Tm*	*Cyto*	*oligomeric state*		*Ref.*
TfR	human	II	671	28	61	dimer	a,d,e	*1,45,46,71*
LDLR	human	I	767	22	50	dimer	a,b,d,e	*51,52,112*
M-6-P/IGF-IIR	human	II	2269	23	164	monomer	a,d	*65,90a*
Mannose R	human	I	1348	28	45	monomer	a	*103*
mIgG	human	I	400-500	26	28	dimer	a	*49,109*
ASGPR								
H1/R1	human/	II	223	21	40	multimer	a,d	*35,99,100,73a*
H2/R2	rat		222	21	58			
hepatic lectin	chicken	II	207	23	24	hexamer	a	*20*
pIgR	rabbit	I	629	23	103	monomer	a,d	*12,79*
EGFR	human	I	621	23	542	dimer[1]	a,c,d	*104a*
IR	human	I	929	22	403	dimer	a,c,d	22

Notes: Tfr, transferrin receptor. LDLR, low density lipoprotein receptor. M-6-P/IGF-IIR, mannose 6-phosphate/insulin-like growth factor-II receptor. mIgG, membrane-bound IgG heavy chain. ASGPR, asialoglycoprotein receptor. pIgR, polyimmunoglobulin receptor. EGFR, epidermal growth factor receptor. IR, insulin receptor.

[1]Oligomeric state after binding ligand.

Domain size is given in amino acid residues: Ecto, external domain. Tm, transmembrane domain. Cyto, cytoplasmic domain. Proteins listed as monomers may really have an undefined oligomeric state. The insulin receptor is historically called a heterotetramer, but is really a homodimer of polypeptides that are each cleaved intracellularly into two disulfide bonded subunits. Through the membrane and inside the cell, the insulin receptor is a dimer.

Posttranslational modification codes are as follows: **a**, N-linked glycosylation. **b**, O-linked glycosylation. **c**, tyrosine phosphorylation. **d**, serine/threonine phosphorylation. **e**, palmitylation.

The cytoplasmic domains of endocytic receptors are small, varying from a few to over 100 amino acids, as one might expect for proteins that do not carry cytoplasmic enzymatic functions. Either the carboxy-terminus or amino-terminus can be in the cytoplasm. For all receptors that have been investigated, deletion of the cytoplasmic domain severely inhibits internalization (*58,64,74,78,86,89*). The cytoplasmic domains of some, but not all receptors receive posttranslational modifications such as phosphorylation on serine, threonine, or tyrosine and the palmitylation of cysteines. Phosphorylation has been shown to be unnecessary for internalization of the LDL receptor (*18*), the transferrin receptor (*89*) and the asialoglycoprotein receptor (*27*). At least one type, and perhaps two types of cytoplasmic internalization feature, often containing a tyrosine residue, have been identified so far (*14,15,19,48,52,73*). A number of different approaches including the use of antibodies specific for phosphotyrosine have failed to detect even transient phosphorylation of the tyrosines in these signals. However, phosphoryla-tion may be important in some cell types for sorting events other than internalization. An internal pool of asialoglycoprotein receptors that does not participate in endocytosis appears to be stably phosphorylated on serine whereas the asialoglycoproteins cycling between the plasma membrane and endosomes are transiently phosphorylated (*102*). Phosphorylation of a specific cytoplasmic serine is necessary to sort the polyimmunoglobulin receptor into the transcytotic pathway (*12*). LDL receptors in bovine adrenal cells are extensively phosphorylated on serine 833 (*50*) by a casein kinase II-like protein that has an activator subunit that binds to the cytoplasmic domain of the receptor (*51*). The function of this phosphorylation is not known. An attractive, but untested hypothesis is that in certain cell types phosphorylation might actually inhibit internalization, either by promoting sequestration of certain receptors as observed for the asialoglycoprotein receptor, or perhaps by inducing conformational changes in the cytoplasmic domain.

For EGF receptors and insulin receptors, the effect of phosphorylation on internalization is less certain. Autophosphorylation of the cytoplasmic domains of these receptors occurs rapidly after hormone binding and probably accompanies conformational changes of the proteins, including aggregation or oligomerization (*96,97*), that immediately precede rapid internalization. Under certain conditions of high levels of expression, mutant EGF receptors defective in autophosphorylation were shown to be internalized (*40,63*). However, another study reported a loss of high-affinity binding for coated pits in kinase-negative mutants, but continued low-affinity internalization by an undefined pathway that has the capacity to internalize a significant number of receptors (*66*). An inhibition of hormone-induced internalization of insulin receptors with defective kinase activity has been reported (*70*). It remains to be seen whether phosphorylation of these receptors results in rather long-range conformational changes that affect internalization, or whether in some instances a phosphorylated residue is included in the feature that interacts with proteins of the endocytic apparatus.

Another modification of cytoplasmic domains, the palmitylation of cysteines, may also influence internalization in some cases but is unlikely to be part of an internalization signal. Two different groups have removed the palmitylation site in the transferrin receptor and report unchanged (*46*) or improved (*2*) internalization. Removal of the palmitylation sites in wild-type HA or in a mutant HA competent for internalization, HA-Y543, has no effect on the ability of HA to be excluded from coated pits and a minor effect on the rate of internalization of the mutant (H. Naim and M. Roth, unpublished data). Thus, to summarize, there are no convincing data that a posttranslational modification forms part of a recognition feature for internalization. However, for some proteins such modification may contribute to the conformation of the cytoplasmic domain and in that way influence the functioning of the polypeptide sequence that directly interacts with proteins of the clathrin coat.

IV. INTERACTION OF RECEPTORS WITH COATED PITS

Historically there have been two different ideas of how cell surface receptors could be distinguished from other transmembrane proteins and sorted into coated pits. One view, based upon studies of the behavior of EGF receptors at the cell surface after binding ligand, is that formation of microaggregates by endocytic receptors is sufficient for internalization (*32,96,113*). A second concept, originating from consideration of the molecular structure of the clathrin coat, is that the cytoplasmic sequences of receptors contain a feature that is bound by a protein associated with the cytoplasmic coat (*30*). These ideas are not exclusive, since microaggregation could be due to cross-linking of receptors by cytoplasmic proteins to form a protein complex in the cytoplasm that can subsequently bind to other proteins of the clathrin coat. Conversely, for certain receptors, microaggregation, perhaps caused by conformational changes after the binding of ligand, might form a cytoplasmic feature for binding to clathrin coats that was cryptic or incomplete in those receptors in their disaggregated state. However, in many cell types aggregation is not in itself sufficient to cause internalization of proteins through coated pits. Cross-linking of proteins normally excluded from coated pits, such as HA in fibroblasts (*56*) or MHC class I antigens in resting T cells (*84*), with multiple layers of antibody does not cause internalization. Mutant HAs have been made that form microaggregates after assembling into trimeric proteins but nevertheless are not internalized (*114*). The main consequence of aggregation that has been observed so far is an inhibition of recycling from the endosome to the plasma membrane after internalization (*43,75*).

A. Endocytic Receptors with a Discrete, Tyrosine-Containing Internalization Signal

The case for a cytoplasmic feature important for internalization is now quite strong. Although the cytoplasmic sequences of receptors do not conserve a readily recognizable linear sequence, many receptors share a small cytoplasmic feature containing a tyrosine or phenylalanine that is necessary and sufficient for internalization. This has been determined through experiments employing site-directed mutagenesis of a number of endocytic receptors. In addition, it has been possible to create internalization features in the cytoplasmic domains of two proteins that lack them, HA (*52,56,114*) and glycophorin (*52*). Figures 1 through 6 show the mutations that have been made in the cytoplasmic domains of several proteins and how they affected endocytosis.

Progress towards identifying a cytoplasmic feature important for internalization began with the discovery that internalization defective human LDL receptors had mutations in the cytoplasmic domain, including one in which a single tyrosine at position 807 had been changed to cysteine (*58*). Subsequent work employing site-directed mutagenesis indicated that only a few residues in the neighboring sequence containing tyrosine 807 were important for internalization and that the tyrosine could be replaced by other aromatic amino acids (*19*). These experiments suggested that either an aromatic residue at position 807 was part of a discrete feature recognized by the endocytic apparatus, or that residue contributed importantly to the structure of an internalization feature located elsewhere. Evidence for the first interpretation came when it was demonstrated that introduction of a tyrosine at only one of three nearby positions in the cytoplasmic domain of the influenza hemagglutinin converted that protein from one excluded from coated pits to one that was efficiently internalized (*56*). Recently sequences containing aromatic residues important for internalization have been identified in a number of receptors. Based on this recent work, three slightly different sequence motifs have been suggested for a consensus internalization signal (*10,15,48,52*). For different proteins, the assays used to measure the effects of mutations on internalization differ in some respects, primarily in whether they measure only changes in the interaction of a protein with coated pits, or include a number sorting events, any or all of which could be changed. The results of mutagenesis experiments on receptors that have been studied in detail will be discussed separately below, followed by a proposal for a consensus internalization signal that incorporates the current data.

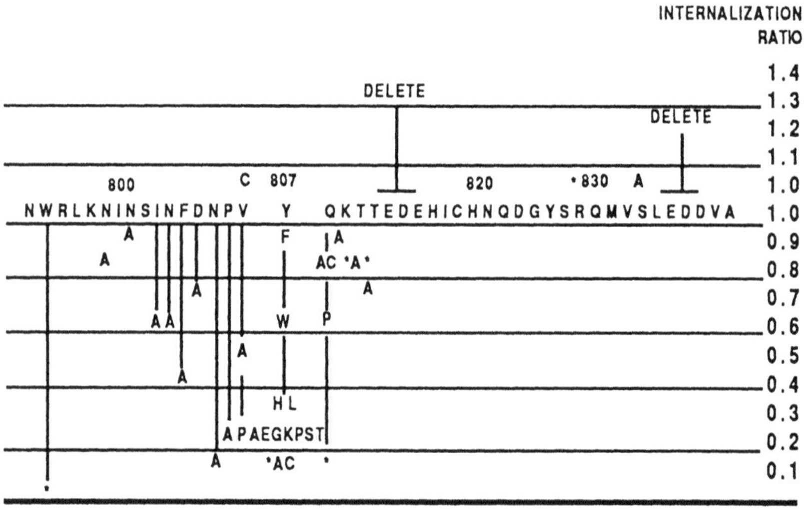

Figure 1. The effect of mutations in the cytoplasmic domain of the LDL receptor on internalization. The internalization ratio is computed by dividing the internalization index of the mutant by the internalization index of the wild-type receptor. The internalization index is computed as described in the text. Numbers above the wild-type sequence designate positions in the entire LDLR sequence. N794 is the first cytoplasmic residue. Mutations are designated by the single letter amino acid code placed above or below the wild-type sequence a distance corresponding to the change in the internalization ratio. * Denotes a stop codon.

LDL Receptor

Measurement of the rates of endocytosis of low-density lipoproteins by a series of mutant LDL receptors (*14,18,19*) indicates that in the full-length receptor the amino acid sequence necessary for internalization in fibroblasts requires an aromatic residue at position 807, with tryptophan functioning significantly worse than tyrosine or phenylalanine (*19*). At least two amino acids to the carboxy-terminal side, and one to the amino terminal side are also important. Chen and colleagues, investigating the effects of a series of mutations made in the background of a receptor truncated at amino acid 811, determined that the amino acids FXNPXY are the critical residues for internalization (*14*). These residues are conserved in LDL receptors of six species, and, if isoleucine is allowed to replace the first phenylalanine, in the large LDL-related proteins of two species (*14,37*). By NMR, Bansal and Gierasch (*5*) have recently determined that nonapeptides which have the sequence of the LDL receptor surrounding the important tyrosine form a reverse turn involving the residues NPXY. Peptides with sequences that work poorly for internalization also are less able to form this reverse turn, suggesting that structure is a critically important part of the internalization signal. If the internalization indices (see below) of the mutants studied by Chen are compared to that of the wild-type receptor, a more extensive sequence of eight residues to the amino terminal side of the tyrosine also appears to contribute to the internalization feature (Fig. 1). Thus, in the LDL receptor a sequence of at least 10 residues including the tyrosine is important for endocytosis. This does not mean that all of these residues are interacting with other proteins of the endocytic machinery. It is likely that most of these residues are important for their contribution to the presentation of the feature recognized by proteins of coated pits. In addition, because of the way that internalization of mutant LDL receptors was measured, some of the effects of these mutations might not be on internalization at all, but on a subsequent step in the transport of the receptor.

The studies of mutant LDL receptors compared the internalization index for each mutant, a ratio calculated by adding the amount of ^{125}I-LDL inside the cell and the amount of monoiodotyrosine released into the medium and dividing by the amount of ^{125}I-LDL that could be released from the cell surface. This figure involves the sum of two sorting events, the interaction with coated pits leading to internalization, and sorting in the endosome for receptor recycling. A reduction in either of these rates would produce a decrease in the ratio. Thus, the internalization index is a valid measure of receptor affinity for coated pits only if the recycling event is shown to be unaffected by a mutation, or if the internalization event is always rate-limiting. For the LDL receptor the exact step of endocytosis affected by most mutations has not been reported. For the mutation of tyrosine 807 to cysteine, however, electron microscopy indicates that the defect is in the interaction with coated pits (*3*).

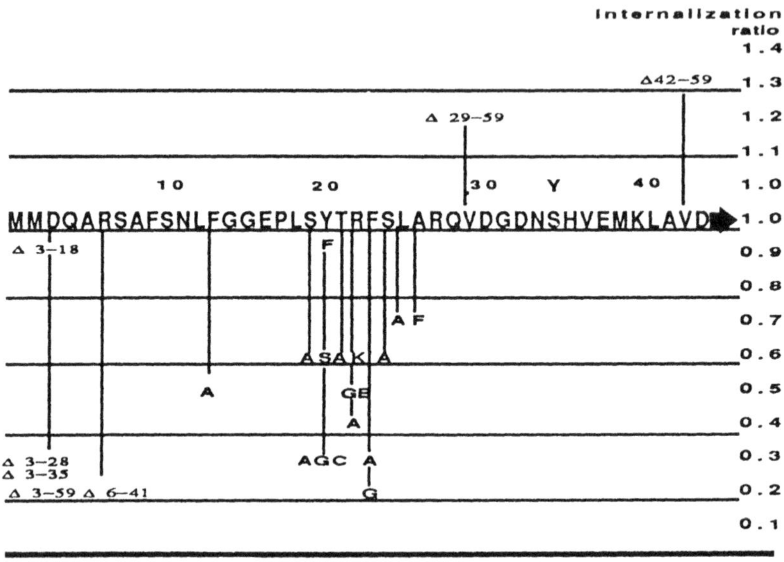

Figure 2. The effect of mutations in the cytoplasmic domain of the transferrin receptor on internalization. The internalization ratio is the initial rate of internalization of a mutant receptor divided by the initial rate of the wild-type receptor. Deleted sequences are designated by Δ followed by the position of amino acids removed. The identity and effect of single amino acid changes are shown as in Figure 1. The numbers above the wild-type sequence correspond to the positions of amino acids in the transferrin receptor sequence, with the N-terminal methionine at position 1. The arrow indicates that the cytoplasmic sequence continues.

Transferrin Receptor

The transferrin receptor, in contrast to the LDL receptor, has the N-terminus in the cytoplasm. Residues of the cytoplasmic domain of this receptor that are important for endocytosis have been identified (Fig. 2). Sequence requirements for internalization of the human transferrin receptor have been investigated in Chinese hamster ovary (CHO) and chicken embryonic fibroblasts, and internalization of the hamster receptor has been investigated in CHO cells. In CHO cells deletion of the cytoplasmic sequences of the full-length receptor reduces internalization 10-fold (*47*) and mutation of tyr 20 decreases the rate from three- to five-fold, depending upon the residue chosen as a replacement (*47,72,73*). Mutation of phenylalanine 24 also causes a three-fold reduction in internalization (*15,73*). Essentially wild-type rates of internalization were observed for the human receptor lacking residues 3 to 18 and 29 to 59 (*47*). Analysis of residues 19 to 28 by alanine-scanning mutagenesis of this deleted receptor determined that the most important residues of the internalization signal were tyrosine 20, arginine 23, and phenylalanine 24. A search of the data base of known protein structures revealed that sequences related to the tetrapeptide YXRF or the NPXY sequence from the LDL receptor often occurred in tight turns, with the aromatic residues at the first and or last positions of the turn, suggesting that this structure forms an important part of an internalization sequence found in two different receptors. Although the sequence YXRF has not yet been tested by NMR, as mentioned above, the LDL tetrapeptide is quite likely to actually be in a tight turn (*5*). An attractive aspect of an internalization signal that incorporates a tight turn as a central element is that structure depends mainly on local interactions and, thus, is relatively autonomous. However, mutation of phenylalanine 12 outside the region sufficient for internalization of the transferrin receptor caused a significant reduction in the internalization rate (*73*). Thus, in their native structural context, internalization signals may depend in complex ways upon other residues for their proper presentation.

An additional complexity with regard to the transferrin receptor, thus far unexplained, is that the hamster receptor has been discovered to differ from the human in the cytoplasmic domain by a single cysteine at the critically important position 20 (*1*). Two groups have measured the internalization of this receptor in CHO cells with very different results, reporting either a twofold faster rate compared to the human receptor in CHO cells (*73*), or a 25% slower rate (1). Three groups have investigated the effect of mutations of the internalization signal of the transferrin receptor on the rate of recycling, and all agree that decreasing the efficiency of the internalization signal does not decrease the rate of return from endosome to the plasma membrane.

Mannose 6-Phosphate Receptors (MPRs)

Two receptors have been identified that bind to mannose 6-phosphate (M6P) residues and deliver proteins bearing that ligand to lysosomes. One, the M6P/IGF-II receptor, is a 275-kDa protein that also binds insulin-like growth factor-II (see the chapter by Hoflack and Lobel in this volume). The intracellular traffic followed by this protein is more complex than for most endocytic receptors, since a major activity of the receptor is to bind lysosomal enzymes in the *trans*-Golgi network and direct them to the lysosome by an intracellular route. Thus, this protein travels continually between the Golgi, endosome, and cell surface. However, at the plasma membrane the M6P/IGF-II receptor acts as an efficient endocytic receptor and constitutively enters and is internalized by clathrin-coated pits. The other receptor, the 46-kDa cation-dependent receptor also mediates sorting of lysosmal enzymes at the Golgi and also cycles to and from the plasma membrane. This smaller protein binds its ligand poorly at the cell surface at neutral pH (*108*), but is nevertheless efficiently internalized through coated pits. Residues important for internalization of both of these receptors in mouse L cell fibroblasts have been identified (*10,48*). In the case of the large receptor, the internalization signal is particularly well identified (Fig. 3).

For the M6P/IGF-II receptor, the most important sequence contributing to its internalization is YXYSKV beginning 24 residues to the carboxy-terminal side of

Cytoplasmic sequence	Internalization ratio
10 20 30 40	
KKERREMV MSRLTNCCRRSA NVSYKYSKV NKEEEADENET⇨	1.00
•••••••• • ••••••••••• •••F•F••••KEEEADENET⇨	1.08
KKERRE*	0.09
KKERREMV MSRLTNCCRRSA *	0.05
KKERREMV MSRLTNCCRRSA NVSYKYSK*	0.16
•••••••• • •••••••••••NVSYKYSKV*	1.05
•••••••• • •••••••••••NVSYKYSKVN*	0.88
•••••••• • •••••••••••NVSYKYSKVNKEEEADENET*	0.99
•••••••• • ••••••••••• •••A•A••••KEEEADENET⇨	0.25
•••••••• • ••••••••••• •••••A••••KEEEADENET⇨	0.77
•••••••• • ••••••••••• ••A•A•••••KEEEADENET⇨	1.00
•••••••• • ••••••••••• •••A••••••*	0.72
•••••••• • ••••••••••• •••••A••• *	0.24
•••••••• • ••••••••••• •••A••A•••*	0.75
•••••••• • ••••••••••• •••A•••A••*	0.49
•••••••• • ••••••••••• •••A••••A•*	0.18
•••••••• • ••••••••••• •••A•••••A*	1.53
•••••••• • ••••••••••• •••A••A•AA*	0.19
•••••••• • ••••••••••• •••A•••R••*	1.23
•••••••• • ••••••••••• •••A•••H••*	0.89
•••••••• • ••••••••••• ••••AA••••*	0.06
•••••••• • ••••••••••• •••••A•A••*	0.16
•••••••• • •••••••••AAAAA•••••••*	2.74

Figure 3. The effect of mutations in the cytoplasmic domain of the M6P/IGF-II receptor. The numbers above the wild-type sequence indicate positions in the domain relative to the transmembrane domain. The internalization ratio for this protein is computed as is that of the LDL receptor in Figure 1. • indicates that a residue in that position matches the wild-type; ⇒ indicates that the sequence continues; * indicates a stop codon at that position.

the transmembrane sequences (*10*). The most important residues within this sequence are the second tyrosine and the terminal valine. When both tyrosines were replaced by alanines in the full-length sequence, the internalization index (a measure of both internalization and recycling rates) dropped to 25% of normal. This is compared to 9% for a truncated protein with just six cytoplasmic amino acids (*64*). Phenylalanines can fully replace the tyrosines in the M6P/IGF-II receptor internalization sequence, and efficiency of the signal is actually increased by replacing lysine with either arginine or histidine (Fig. 3).

The small CD-MPR is internalized rapidly with a half-time of internalization of less than two minutes (*48*). When the 67 amino acid cytoplasmic domain is truncated to the first seven residues, the receptor is internalized at the same rate as the bulk of membrane proteins. Mutation of all the aromatic residues of the cytoplasmic domain to alanine has the same effect as the truncation. Aromatic residues at two locations separated by 26 amino acids are both required for the full internalization rate of the receptor. Mutation of phenylalanines at positions 13 and

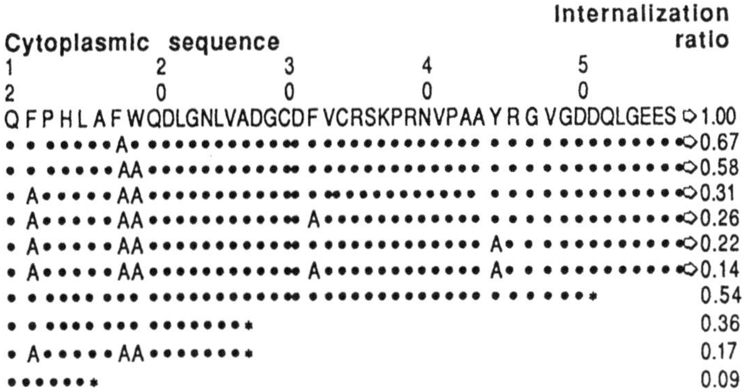

Figure 4. Effect of mutations in the cytoplasmic domain on internalization of the small M6P receptor. The internalization ratio is the initial rate of internalization of the mutant divided by the internalization rate of the wild-type receptor. The numbers above the sequence indicate position in the chain relative to the last hydrophobic membrane spanning residue. • indicates that a residue in that position matches the wild-type; ⇒ indicates that the sequence continues; * indicates a stop codon at that position.

18 to alanine dropped the internalization rate by three- to fourfold, and mutation of tyrosine 45 decreased internalization almost twofold. It was suggested by Johnson and colleagues that the two internalization signals in this receptor involve six and four amino acids, respectively. However, since very few mutations around the important aromatic residues were made, it is not yet clear exactly which sequences are involved in internalization.

HA-Y543

Identification of a discrete internalization signal requires not only that a sequence be shown necessary for internalization of a proteins, but also that it is shown to be sufficient for internalization. To accomplish the latter, it is necessary to transfer the suspected signal to a reporter protein that by itself does not have the capacity to enter coated pits. A good reporter protein for internalization through coated pits is the influenza virus HA, which is internalized approximately 40-fold slower than the bulk internalization of membrane through coated pits in CV-1 African green monkey fibroblasts (*88*). The cytoplasmic domain of HA has been replaced with that of the LDL receptor or the polyimmunoglobulin receptor and the resulting chimeric proteins observed to be internalized through coated pits and to recycle to the plasma membrane (S.-P. Shia and M. G. Roth, unpublished results). In polarized Madin-Darby canine kidney epithelial cells, the chimeric HA/pIgR protein also was targeted to the basolateral membrane (*80*) as is the pIgR itself. In an attempt to identify sequences both necessary and sufficient for the internalization of HA, several mutations were made in the short cytoplasmic sequence of HA, and two

	1									10			Internalization ratio
HA wt	C	S	N	G	S	L	Q	C	R	I	C	I	0.02
	•	•	Y	•	•	•	•	•	•	•	•	•	0.02
	•	•	•	•	•	•	•	•	•	Y	•	•	0.02
	•	•	•	•	•	•	•	•	•	•	Y	•	0.02
	•	•	•	•	•	•	•	F	•	•	•	•	0.02
	•	•	•	•	•	•	•	W	•	•	•	•	0.02
HA-Y543	C	S	N	G	S	L	Q	Y	R	I	C	I	1.00
	•	•	•	•	•	•	•	Y	L	•	•	•	0.12
	•	•	•	•	•	•	L	Y	S	•	•	•	0.20
	•	•	•	•	•	•	E	Y	•	•	•	•	0.36
	•	•	•	•	A	•	•	Y	•	•	•	•	0.42
	•	•	•	•	•	•	•	Y	W	•	•	•	0.42
	D	•	•	•	•	•	L	Y	S	•	•	•	0.42
	•	•	•	•	•	•	•	Y	S	•	•	•	0.50
	•	•	•	•	•	•	•	Y	V	•	•	•	0.48
	•	•	•	•	•	•	•	Y	A	•	•	•	0.46
	•	•	•	•	•	•	•	Y	•	•	*		0.50
	•	•	•	•	•	•	E	Y	S	•	•	•	0.58
	•	•	•	•	•	•	•	Y	•	•	S	•	0.70
	•	•	•	•	•	•	•	Y	E	•	•	•	0.70
	•	•	•	•	•	•	L	Y	•	•	•	•	0.85
	•	•	•	•	•	•	•	Y	•	Y	•	•	1.00
	D	•	•	•	•	•	•	Y	•	•	•	•	1.20

Figure 5. Effects of mutations of the cytoplasmic domain of HA on internalization. The internalization ratio is the rate of internalization of mutant or wild-type HAs divided by the rate of internalization of HA-Y543. Numbers above the wild-type sequence indicate position relative to the hydrophobic transmembrane domain. • indicates that a residue in that position matches the wild-type; * indicates a stop codon at that position.

were identified that allowed the mutant proteins to efficiently enter coated pits (*56,114*).

The cytoplasmic domain of HA is predicted by a hydropathy algorithm (*53*) to be the 12 amino acids preceding the carboxy terminus, residues 536 to 548. Change of cysteine 543 to tyrosine allows HA to be internalized through coated pits, but tyrosines at three other positions do not. In an investigation of the role of nearby residues for the internalization of HA-Y543, Ktistakis and colleagues (*52*) introduced a number of second and third site mutations into the HA-Y543 sequence (Fig. 5). These mutations were guided by a comparison of sequences surrounding cytoplasmic tyrosines of proteins known or strongly suspected of containing an internalization signal dependent upon tyrosine. Near the tyrosine, usually on the side of the sequence towards the transmembrane sequence, a statistically significant preference was noticed for residues that break regular structure, and, at certain positions on both sides of the tyrosine, for polar or basic residues (acidic residues showed no preference). The results of mutagenesis of the cytoplasmic domain of

	10										20			Internalization ratio
HA +8	C	R	I	C	I	Y	D	Y	K	S	F	Y	N	1.00
	•	•	•	•	•	S	•	Y	•	•	•	S	•	1.00
	•	•	•	•	•	S	•	Y	•	•	A	S	•	0.08
	•	•	•	•	•	Y	•	S	•	•	F	•	•	0.02
	•	•	•	•	•	Y	•	F	•	•	Y	•	•	0.02
	•	•	•	•	•	Y	•	Y	•	•	*			0.02

Figure 6. Effects of mutations of the cytoplasmic domain on internalization of HA +8. The internalization rates of mutants in the HA +8 background are expressed as a fraction of that of HA +8. Numbers above the HA +8 sequence indicate position relative to the transmembrane domain. • indicates that a residue in that position matches the wild-type. * indicates a stop codon at that position.

HA-Y543 indicated that secondary structure played an important role in the function of the critically important tyrosine, since mutations that reduced the rate of internalization of HA-Y543 could be compensated by other second-site mutations. Using the two parameters of the presence of turn-promoting residues, such as proline, glycine, serine, and aspartate, and basic or polar residues at defined positions, Ktistakis et al. (*52*) were able to successfully predict where an internalization signal could be introduced into the cytoplasmic domain of glycophorin, another protein that lacks cytoplasmic tyrosines and is excluded from coated pits.

A different mutant of HA was also found to be internalized through coated pits (Fig. 6). This protein, HA+8, has an extension of eight amino acids caused by a frame-shift mutation introduced at the HA termination codon. Of these eight amino acids, three are tyrosines, and one is phenylalanine. This sequence resembles the internalization signal in the M6P/IGF-II receptor quite closely, although at the time that the mutation was made, no sequence data for that receptor was available. The intracellular traffic of the HA+8 protein is quite remarkable. The surface concentration of this protein is very low because the protein is internalized quite rapidly (at 50%/min) and does not detectably recycle, but is transported efficiently to lysosomes and degraded. Additional mutation of the HA+8 cytoplasmic sequence reveals that the internalization signal is YKSF, with the tyrosine being absolutely necessary and the phenylalanine important for internalization. Thus, in HA+8 the signal runs from N-terminal to C-terminal, using the most important residue as the point of reference. In this respect, HA+8 resembles both the transferrin receptor and the M6P/IGF-II receptor. HA-Y543, in contrast, has a signal oriented from C-terminal to N-terminal, as does the LDL receptor. As will be discussed below, an interesting question is whether or not these are the same or different signals.

Other Receptors and Lysosomal Membrane Proteins

For a number of other receptors, mutagenesis experiments have identified sequences in the proteins that are necessary for internalization, and in certain cases the requirement for a tyrosine has been established. In the case of the polyimmunoglobulin receptor, there are two cytoplasmic tyrosines and both function for internalization (*8,80*). One, at position 668, is a good fit for the consensus proposed by Ktistakis et al. (*52*) and lacks a second large hydrophobic residue. The second tyrosine, at position 734, is in a sequence with other hydrophobic residues nearby, and in particular there is a phenylalanine at position 737. Of the two signals, the one including tyrosine 734 is the more important, since a protein containing only that tyrosine is internalized threefold faster than a protein containing only tyrosine 668 (*80*). No other amino acids near the two tyrosines have been mutated, so the exact sequence requirements for the signals are unknown.

Several lysosomal proteins have been shown to contain tyrosine-dependent internalization signals in their cytoplasmic domains that allow them to be efficiently internalized by coated pits (*34,85,111*). A peptide with the sequences known to be important for internalization of lysosomal acid phosphatase has been shown by NMR to form a tight turn similar to that observed for the internalization sequence of the LDL receptor (*5*) and predicted for the transferrin receptor (*15*). Thus, it is likely that lysosomal membrane proteins contain sorting information identical to that of endocytic receptors. Lysosomal membrane proteins are thought, based primarily on kinetic arguments, to be sorted at the *trans*-Golgi network to the late endosome, presumably following the pathway identified morphologically for the MPR/IGF-II receptor (*28,31,33*). Mutation of the unique cytoplasmic tyrosine of several of these proteins causes them to accumulate at the cell surface. This has led to the proposal that the tyrosine is part of a lysosomal targeting signal (*34,111*), an interesting idea since clathrin-coated membrane is observed at the *trans*-Golgi network. However, if the lysosomal membrane proteins and endocytic receptors both contain a signal that can be recognized at the *trans*-Golgi network for sorting to endosomes by an intracellular route, the signal on endocytic receptors must be inactive at that location, or endocytic receptors must follow the same route. The data for passage of newly synthesized lysosomal membrane proteins directly from the Golgi to late endosomes are not extensive. Most of the assays for surface appearance of these proteins have measured the steady-state concentration in the membrane, which would be vanishingly small if these proteins are internalized as efficiently as is the HA+8 mutant, and like that protein, do not return to the plasma membrane. Overexpression of wild-type lysosomal membrane proteins does, in fact, result in a detectable surface population capable of internalization through clathrin-coated pits and low levels of lysosomal proteins have been detected on the cell surface of some cell types (*7,25,62,68,111*). Thus, it is possible that some lysosomal membrane proteins normally travel first to the cell surface before entering the endocytic pathway and reaching lysosomes.

A role for cytoplasmic sequences has been demonstrated for the internalization of several other receptors for which the exact sequences involved have not been identified. In the insulin receptor, deletion of cytoplasmic amino acids located near the transmembrane sequences reduces the internalization rate to approximately 25% of wild-type (*4,104*). Included in this segment are two tyrosines, one at position 960 in the context NPXY that has been identified as the core of the internalization feature of the LDL receptor. Mutation of this tyrosine to phenylalanine decreases the internalization rate by only a third (*4*). Phosphorylation of this tyrosine appears to be required for some of the biological effects of insulin binding to the receptor (*110*). If this tyrosine is really part of a recognition signal in the insulin receptor, it will be interesting to learn how phosphorylating the tyrosine affects internalization.

An important endocytic receptor of the liver is the mammalian lectin asialoglycoprotein receptor, ASGPR. This receptor is a hetero-oligomer of two subunits H1 and H2 that assemble into a complex in the endoplasmic reticulum (*35,99*). H1, but not H2, can be transported to the cell surface where it is endocytosed and recycles (*26*). Both subunits are oriented with the N-terminus in the cytoplasm, as is the transferrin receptor. H1 contains a single tyrosine as the fifth residue from the N-terminus, and mutation of this residue reduces the internalization rate by a factor of 4 (*24*). This residual internalization is, however, through coated pits.

B. Consensus Features of the Signal

It is now clear that many endocytic receptors contain a small cytoplasmic sequence sufficient and necessary for internalization and that these sequences have very little conservation of primary sequence between receptors, other than containing a tyrosine, or less often, a phenylalanine (Fig. 7). This degeneracy of sequence means that the chemical properties and spatial organization of these sequences, and perhaps their location in the cytoplasm, must be responsible for conferring specificity. Recently several proposals for internalization signals have been made that share certain common elements. Ktistakis and colleagues (*52*) have proposed that the central elements of a consensus internalization feature are a loop exposed at the cell surface that contains a critically important tyrosine located to one side and important, hydrogen-bond forming residues present in defined places on either side of the tyrosine. According to this model a region of eight to 10 amino acids is sufficient for the signal with the tyrosine positioned in the portion of the sequences farthest from the membrane, placing it no closer than six residues from transmembrane sequences. No particular structure for the loop is proposed by this model, however, a feature of this proposal is that the polarity of the polypeptide chain does not influence where the tyrosine is located. Collawn and co-workers (*15*), based upon their work with the transferrin receptor and a comparison of cytoplasmic sequences of other receptors, propose that internalization signals contain two aromatic residues at positions i and i + 3 of a β-turn with residues capable of forming hydrogen bonds in the intervening i + 1 and i + 2 positions. They propose that the tyrosine can be no closer than seven residues from membrane-spanning hydrophobic sequences. Johnson et al. (*48*) based upon identification of aromatic residues important for endocytosis at two separate locations of the CI-M6P receptor cytoplasmic domain, suggest that there may be two different signals, one involving six amino acids with large hydrophobic residues at positions 1, 3, and 6, and a second

4-amino acid sequence with an aromatic residue at position 1 and another large hydrophobic residue at position 4. They also propose that for greatest activity, one of the two residues in positions 2 and 3 should be positively charged.

It is possible that some of the differences in the signals proposed reflects the fact that there really are several independent signals that are recognized by different cellular proteins. Alternatively, different sequence motifs may be bound by the same cellular protein, but with very different affinities, separating various receptors into a hierarchy in terms of ability to compete for incorporation into coated pits. According to this concept, there would be one "best" or archetypal sequence from which other sequences differ by being to some degree degenerate. Thus, the differences between the internalization signals in the transferrin receptor and, for example, the artificial protein HA-Y543, would simply be that HA-Y543 is incomplete, lacking an aromatic residue at position 540. In an important recent experiment, McGraw and colleagues (*73*) have changed serine 34 to tyrosine in the cytoplasmic domain of a transferrin receptor that lacks the critically important tyrosine 20. This mutant transferrin receptor is internalized as efficiently as is the wild-type protein, but lacks a neighboring large hydrophobic residue, at either the second or fourth position from the tyrosine. The high rate of internalization of this protein indicates that the internalization signal created is at least as complete as is the wild-type signal. McGraw and colleagues remark that the sequence surrounding tyrosine 34 would fit the consensus proposed by Ktistakis et al. (*52*) provided that turn forming residues are allowed to be located on either side of the tyrosine and not just on the side toward the membrane. It is also of interest that the novel signal generated in this mutant absolutely requires a tyrosine, just as does the HA-Y543 mutant.

Figure 7 presents cytoplasmic sequences aligned N- to C-terminal from proteins in which the presence of an internalization signal has been confirmed by mutagenesis. Regions shown to be important for internalization are boxed, and residues at positions where mutations result in a loss of 75% of the internalization rate are outlined. From this data, it is obvious that for efficient endocytosis the position of the critically important aromatic residue is not fixed in the region necessary for internalization, nor is there an absolute requirement for a large hydrophobic residue in the fourth position from the tyrosine or phenylalanine. And since the pIg receptor is internalized at one of the highest rates yet measured, efficient internalization does not require a basic residue in one of the two positions to the carboxy-terminal side of the tyrosine or phenylalanine. It is not yet possible to distinguish between the existence of one type of signal, or more than one. Additional mutagenesis of the type proposed by Collawn et al. (*15*) probing the effect of amino acid substitutions chosen to change secondary structure independently of charge or hydrophilicity would help to determine whether such apparently different primary sequences actually arrive at a common three-dimensional shape. Analysis of the solution structures of peptides with amino acids sequences known to be important for internalization *in vivo* would more directly determine if identified internalization sequences can form similar structures. However, to determine the number of

Protein	Sequence N-terminal to C-terminal								
TfR (tyr20)	G	E	P	L	S	Y	T	R	F
TfR (tyr34)*	V	D	G	D	N	Y	H	V	E
LDL	F	D	N	P	V	Y	Q	K	T
M6P/IGF-IIR	N	V	S	Y	K	Y	S	K	V
CI-M6PR (F18)	F	P	H	L	A	F	W	Q	D
CI-M6PR (Y45)	N	V	P	A	A	Y	R	G	V
HA-Y543*	N	G	S	L	Q	Y	R	I	C
HA+8*	I	C	I	Y	D	Y	K	S	F
ASGP H1		M	T	K	E	Y	Q	D	L
LAP	A	Q	P	P	G	Y	R	H	V
glyc-Y106*	K	S	P	S	D	Y	K	P	L
pIgR (tyr668)	V	S	I	G	S	Y	R	T	D
pIgR (tyr734)	E	A	D	L	A	Y	S	A	F
LAMP-1	R	S	H	A	G	Y	Q	T	I

Figure 7. Cytoplasmic sequences demonstrated to be important for internalization through coated pits. The amino acid sequences near tyrosine that have been shown to be important for endocytosis are shown for 10 proteins. Sequences with boxes have been shown by mutagenesis to be important for internalization. Positions where mutagenesis causes greater than 75% reduction in the internalization rate are designated by an outlined letter. Proteins designated * contain internalization signals created *de novo* by site-directed mutagenesis.

independent binding sites for endocytic receptors in coated pits, ultimately the protein or proteins binding these signals must be identified.

C. Receptors for Which the Internalization Signal is Not Known

There are a number of important cell surface proteins that are internalized quite efficiently under certain circumstances, or in certain cell types, but for which the internalization signals have not been identified. These include polypeptide hormone receptors, including those with cytoplasmic tyrosine kinase activities, receptors of the immune system including the histocompatibility antigens and Fc receptors, and proteins that span the bilayer multiple times, such as the adrenergic receptors. For each of these classes of proteins, deletion experiments have demonstrated that the cytoplasmic domain is necessary for internalization. Thus, these proteins presumably enter coated pits through a cytoplasmic interaction with elements of the coat, as do the endocytic receptors for which recognition sequences have been identified. A complication for study of some of these proteins, however, is that their internalization rates and destinations after internalization often change dramatically after they bind ligand. In most cases these changes occur rapidly after transmembrane signalling events have occurred, either due to signaling from the receptor itself, or in response to an independent signaling system. The relationship between this signaling, posttranslational modifications of the receptor, such as phosphorylation, and internalization is not precisely known and could be very complicated.

An example of this complication occurs with the EGF receptor, which exhibits high-affinity association with coated pits rapidly after binding ligand, and subsequently is sorted to lysosomes. Mutation of the ATP-binding site inhibits the intrinsic kinase activity of the receptor and also blocks ligand induced, high-affinity binding to coated pits (*66*) and receptor downregulation (*13,40*). However, the internalization of the unphosphorylated receptor continues, but now with low-affinity binding to coated pits and efficient recycling to the plasma membrane (*41,66*). A sequence towards the C-terminus of the EGF receptor that is not involved in kinase activity has been identified by deletion analysis as critical for both high-, and low-affinity internalization (*13*). Thus, autophosphorylation of the receptor is likely to cause conformational changes necessary for internalization. These changes might expose a recognition sequence for internalization (the region required for internalization contains the sequence NFYRAL in a position predicted to form a turn) or might allow a cytoplasmic protein containing an internalization sequence to bind to the receptor. Other than for the EGF and the insulin receptors, sequences important for internalization of receptors with tyrosine kinase activity have not been identified.

Another important endocytic receptor is the mannose receptor of macrophages and hepatic sinusoidal cells (see Figure 1). This receptor contains a C-terminal cytoplasmic domain of 45 amino acids with a unique tyrosine in the sequence FENTLYF, which resembles that containing Y734 of the polyimmunoglobulin receptor (*103*). Mutagenesis of this sequence has not been reported.

A number of receptors of the immune system can be endocytosed through coated pits, at least under certain circumstances. These include Class I and II major histocompatibility antigens (*16,84,87*), Fc receptors (*74*), and surface Ig in B cells (*109*). For Fc receptors and Class I antigens, the cytoplasmic domain has been shown to be necessary for internalization and regions of that domain required have been identified through mutagenesis, but the exact sequences responsible are not yet known (*11,74,106*). For Class II proteins, a recent report suggests that the α and β subunits of these proteins may lack internalization signals, since they are not internalized in transfected L cells (*91*). Class II proteins are internalized in certain murine B cells, suggesting that either they associate with another protein that allows their internalization, or they receive some as yet unidentified modification that uncovers a cryptic internalization signal. In the Fc receptor, deletion of cytoplasmic sequences decreases the internalization rate by only 50% (*74*) which is much less than observed for similar types of mutations of other receptors. An insertion in the cytoplasmic sequence of the lymphocyte receptor FcRII-B1 decreases internalization slightly more than removing the entire cytoplasmic domain. Near the insertion point in the cytoplasmic domain a region of 10 amino acids including a tyrosine is highly conserved among Fc receptors for which sequence data is available. The tyrosine occurs in the sequence context ITYSLL (*60*), which fits one of the proposed consensus patterns (*48*). However, mutation of the tyrosine in this sequence does not alter the internalization rate (I. Mellman, personal communication). Current data suggest that Fc receptors interact with coated pits through a different type of recognition feature than those identified to date. Perhaps Fc receptors bind to an accessory protein that contains an internalization signal.

D. Internalization by Clathrin-Coated Pits in the Absence of a Tyrosine-Containing Signal

In addition to endocytic receptors, which concentrate in coated pits and internalize at rates much faster than the 1 to 2% of plasma membrane taken up through coated pits each minute, there is another class of proteins that are constitutively internalized at rates typically near the rate of bulk internalization of membrane. Included in this class are the Fc receptors (*74*), fibronectin receptor (*9*), and CD4 protein of lymphocytes (*83*). In the case of CD4 expressed in transfected fibroblasts, deletion of the cytoplasmic domain has only a modest effect on internalization, similar to the results with the FcRII-B1 Fc receptor (*74,83*). Little work has been done to investigate the mechanism of internalization of this class of proteins, in part because it is difficult to demonstrate clustering in pits, and thus difficult to establish the internalization pathway by morphology. However, Verrey and colleagues have reported interesting experiments in which the cytoplasmic domain of the chicken hepatic lectin (CHL) was replaced with sequences from the Na/K ATPase or globin (*107*). These chimeric molecules were internalized 10-fold more slowly than the CHL, but at rates that suggested some concentration in coated pits. This was confirmed by electron microscopy. An inspection of the internalization rates of endocytic receptors in which internalization signals have been deleted reveals a similar phenomenon (Figs. 1–5). Even in the absence of the cytoplasmic domain, these proteins are internalized at rates 10-fold higher than that of highly mobile proteins like HA (*23*) or Thy-1 (*59*).

Chimeric HA proteins have been constructed in which the transmembrane domain of HA have been replaced by the homologous domains of either the VSV G protein or the Herpes simplex I gC protein (*88*). In both cases these chimeric proteins were internalized through coated pits at rates of 1 to 3%/min, suggesting that some feature of the HA transmembrane domain is responsible for preventing internalization (J. Lazarovits and M. Roth, unpublished results). Another possibility is that the foreign transmembrane domains alter the structure of the HA cytoplasmic domain in some way that allows an interaction with components of the clathrin coat. This possibility is supported by the observation that HAs with foreign transmembrane domains have demonstrable changes in the external domains (*57*).

If there were only three options for cell surface proteins—high-affinity binding and clustering in coated pits, low-affinity binding, and no binding due to immobility imparted by anchoring to the cytoskeleton—then all internalization data could be explained by two parameters, receptor mobility and relative affinity for coated pits. However, the existence of mobile proteins not internalized and the class of slowly internalized proteins suggests that there are at least two separate interactions of proteins with coated pits.

E. Proteins that Bind to Internalization Motifs

Of the proteins known to be present in clathrin-coated vesicles, the clathrin-associated proteins have long been suspected of mediating the association of the coat with cytoplasmic domains of endocytic receptors. Two experiments have been performed that suggest that this might be the case. Pearse has reported that purified clathrin associated proteins will bind to columns containing sequences of the cytoplasmic domains of receptors, and that this binding can be competed with peptides containing internalization motifs, but not with a peptide with the sequence of the HA (*81*). Pearse has named these proteins "adaptins". The clathrin-associated proteins are sticky and difficult to work with *in vitro*, and other laboratories have had difficulty reproducing Pearse's result using slightly different protocols. Beltzer and Speiss (*6*) have used the asialoglycoprotein receptor to probe clathrin-coated proteins that have been transferred to nitrocellulose and report that the receptor specifically binds to one of the 100-kDa clathrin-associated proteins (*6*). In contrast, Anderson and colleagues have established a different *in vitro* model for coated pit formation in which coated vesicle fractions can be bound to isolated plasma membranes from which clathrin and associated proteins have been stripped (*67,77*). Treatment of these membranes with elastase inactivates them for binding adaptins (R. G. W. Anderson, personal communication). This result would not be expected if adaptins required only the cytoplasmic domains of receptors for binding, since not all cytoplasmic domains are elastase-sensitive. Thus, at the moment, the question of exactly which proteins bind to receptor cytoplasmic domains is still actively pursued by those working in this field.

V. MODELS FOR COATED PIT LOCALIZATION AND COATED PIT FORMATION

Coated pits assemble and bud off to form vesicles that rapidly uncoat so that the coat components can return to the plasma membrane to form new coats. Endocytic receptors diffuse in the plane of the plasma membrane, concentrate around and in coated pits and are internalized and delivered to endosomes to be rapidly recycled to the plasma membrane. Do the proteins of the clathrin coat need receptors in the membrane to nucleate the polymerization of the coat? Do receptors bind to empty sites in preformed coated membrane? Two studies of the transferrin receptor have suggested that overexpression of the receptor leads to increased polymerization of clathrin on the cell surface (*45,76*). These findings would support a hypothesis in which the nucleation point for clathrin lattice formation is the binding of a coat component to the cytoplasmic domains of receptors. In such a model, receptors could also diffuse to the edges of pre-existing coats where they would be incorporated into the growing lattice and fixed in place (*82*). In support of this concept, binding *in vitro* of a 100- kDa clathrin-associated protein to peptides with sequences from receptor tails has been reported, although not with the sequence from the transferrin receptor (*81*). However, no similar increase in clathrin-coated membrane has been observed in cells overexpressing the LDL receptor. Using a technique similar to that of Miller et al. (*76*), Sanan and Anderson (*92*) have observed areas of high concentration of clathrin on membranes from fibroblasts lacking LDL receptors, suggesting that overexpression of receptors may not be required for recruiting large amounts of clathrin to certain areas of the plasma membrane.

Using a different approach, Fire and colleagues compared the lateral diffusion of the wild-type HA and a mutant, Tyr543, that is internalized through coated pits (*23*). Rather than detecting a difference in the mobile fraction of these proteins, they found that the entire mobile population of Tyr543 diffused threefold slower, and this difference was lost when cells were incubated in conditions that disrupt clathrin lattice structure, and was decreased when the temperature was lowered to 10 °C. The simplest interpretation of these data are that the Tyr543 binds reversibly to coated pits and this binding shows the temperature-dependence characteristic for a hydrophobic interaction. These results are not easily explained by a model for coated pit assembly in which receptors are added to the margins of coated pits as the membrane attachment point for additional coat components, unless the coat is bound to the membrane by other interactions after polymerization. However, similar measurements are needed for several endocytic receptors before the implications of dynamic binding of receptors for coated pit assembly are clear.

VI. THE RELATION BETWEEN INTERNALIZATION AND RECYCLING

The interaction with coated pits is but one of several sorting events for endocytic receptors. After internalization, most endocytic receptors are sorted in the early endosome and recycled to the plasma membrane. Fluid content of the early endosome, including released ligand, is preferentially sorted to lysosomes and most membrane returns to the cell surface. Whether one postulates a sorting event in which each endosome is separated into membrane domains with large surface area and low volume (tubules) and one with large volume and less surface area (central vacuole) (*61*), or postulates repeated budding of small vesicles preferentially carrying away receptors from a larger endosome retaining most of the fluid volume (*21*), the problem is the same; how are *membrane* proteins destined for lysosomes separated from those that recycle?

Fc receptors when bound by Fab′ fragments recycle to the cell surface but when bound by polyvalent antibodies are routed to lysosomes (*75*). Similar results have been reported for the transferrin receptor (*43*). Receptors such as the EGF receptor or insulin receptor, when bound to their monovalent ligand, are preferentially sorted to lysosomes. Sorting of EGF receptors to lysosomes appears related to the concentration of EGF-receptor complexes in the endosome and not the concentration of constitutively internalized unoccupied receptor (*54*). One interpretation of this observation is that some sort of self-interaction, or formation of higher order oligomers, might be necessary for sorting to lysosomes. If this is the case, the weak relation between rate of lateral diffusion in the membrane and size (*36*) suggests that aggregates would need to be quite large if sorting is due primarily to an inability to move rapidly into recycling membrane domains.

An HA mutant has been constructed (HA+8, see Figure 6) that is internalized at 50%/min but recycles little, if at all, and is rapidly degraded. This protein can be isolated from cells extracted with octylglucoside as aggregated, well-folded, trimeric HAs, whereas the Tyr543 mutant HA that is internalized and recycled has never been observed to form aggregates. These aggregates are remarkable in that they form early in the biosynthetic pathway, delaying, but not preventing, HA+8 from reaching the Golgi. From there the kinetics of transport are normal and during their brief stay at the plasma membrane the proteins can be cleaved into HA1 and HA2 subunits by trypsin in the extracellular medium. This latter event demonstrates that the trypsin-sensitive site of HA, located 4 nm from the membrane, is accessible to a protein the size of trypsin. This would not be expected if the external domains of HA+8 proteins were tightly packed in the aggregate. A second mutation of HA+8 that changes two tyrosines to serines has no effect on the rate of internalization or degradation, but prevents isolation of the aggregated protein. Thus, HA+8 aggregates are sensitive to small changes in the cytoplasmic domain and the formation of aggregates through high-affinity interactions is not required for sorting this protein to the lysosomal pathway in the early endosome.

When the high-affinity internalization signal is mutated in the transferrin receptor, recycling of the slowly internalized proteins continues normally (*1,72*). Thus, possession of a high-affinity internalization signal is not a requirement for efficient recycling. However, when the tyrosine signal in HAY-543 is mutated, the proteins that are slowly internalized also show increases in the fraction of the proteins that are internal at steady-state, indicating that recycling of those proteins is slowed or the fraction that recycles is decreased (*52*). Chimeric HAs that have the cytoplasmic domain of the pIgR (*80*) or of the LDL receptor, (*98*) show the same pattern. Mutation of the tyrosine in the internalization signal reduces the rate of endocytosis much more than the rate of degradation or the fraction of the protein that is internal at steady-state. Wild-type HA, which is highly mobile in the cell membrane and is internalized very slowly, is degraded at the same rate as it is internalized, indicating that little, if any, recycles. These observations do not necessarily conflict with the observation that transferrin receptors with defective internalization signals nevertheless recycle normally. It is possible that proteins that lack high-affinity binding to coated pits are incorporated into the pits through another interaction that also functions at the endosome to sort them to the lysosomal pathway. The transferrin receptor sequences either would not participate in this sort of interaction, or would do so reversibly and the receptor would efficiently recycle. In this view, proteins that do not participate in a second-sorting event in the endosome would return to the cell surface with the bulk of the membrane.

VII. A MODEL FOR RECEPTOR INTERACTION WITH COATED PITS AND RECYCLING

The coated pit serves the dual function of reserving space for the efficient internalization of endocytic receptors while also internalizing other cell surface proteins destined for lysosomes during the continual renewal of plasma membrane. During this process other proteins are essentially excluded from the pit. The observations that proteins apparently lacking internalization signals are nevertheless incorporated into coated pits creates a problem for explaining how a highly mobile protein, such as HA, is excluded. The exclusion of mobile proteins from the coat indicates that proteins that are internalized slowly through coated pits are not simply occupying free space, but must interact in some way with elements of the clathrin coat.

A working hypothesis of how this might be accomplished is as follows. The cytoplasmic sequences identified so far as necessary and sufficient for internalization of endocytic receptors serve as "clustering" signals by which proteins of the coat bind the receptors dynamically, concentrating them in the vicinity of the coated pit. However, retention in the pit requires a second event that does not necessarily require the clustering signal. Proteins that lack the clustering signal, but that can participate in the second event, diffuse into coated membrane and remain. The degree to which they appear to cluster is controlled by the amount of space not taken by receptors possessing clustering signals and by the density of packing of proteins into the coat, which might be greater than the average density of these proteins elsewhere in the plasma membrane. Thus, internalization of this class of protein would not necessarily be at the rate of bulk internalization of membrane by the coat and they could, under some circumstances, exhibit a modest degree of clustering. For receptors with high-affinity signals, the second event could be movement from the edge towards the center of the coat, where the proteins would be surrounded by high affinity binding sites and might simply exchange sites with other receptors and be held in place. For proteins lacking such signals, the second interaction is likely to involve either the transmembrane domains, or some malfolding of the cytoplasmic domains, resulting in an interaction that allows the protein to stay in the coat during budding. This second, "packing" interaction would be the means for selecting "damaged" proteins from the cell surface pool for internalization. If this is true, damage ought to be invisible to the cell if it is limited to the external domain of proteins (a point that can be tested with a protein such as HA for which assays for proper folding are available). Since in the endosome-"damaged" proteins must subsequently be sorted away from the recycling endocytic receptors, it is attractive to argue that the interaction allowing "damaged" proteins to stay in the clathrin coat is also the event used to sort them to lysosomes. This may occur if the cytoplasmic domain of a damaged protein exposes sequences that attract a chaperon-like "garbage collector" protein. The hypothetical "garbage collector" would be multivalent, binding weakly to proteins of the clathrin coat and more strongly to other garbage collectors that had bound damaged proteins. Thus, the packing interaction responsible for holding damaged proteins in the coat long enough for internalization would be produced by reversible, weak interactions among the "garbage collectors." After delivery to the endosome, endocytic receptors (and anything else) free of garbage collectors would rapidly enter recycling membrane. Garbage collectors still bound to cytoplasmic sequences at the endosome would self-associate, just as they had in the clathrin coat, to form a sort of matrix in the cytoplasm that would retard diffusion of damaged proteins into recycling membranes. Like the cytoplasmic matrix proteins of enveloped viruses, this matrix of "garbage collectors," perhaps with help of other cytosolic proteins, might be capable of extending into, or even budding into the central vacuole of the endosome, contributing to the internal membranes observed in multivesicular endosomes.

It has been difficult to directly demonstrate the association of adapter proteins to the known target sequences in receptor tails, or of cytosolic chaperon-like proteins with their presumed targets, probably because the binding is both relatively non-specific and weak (*81,90*). However, a prediction of the proposal just described is that there should be a chaperon, or "heat-shock"-like protein located on the cytosolic face of endosomal membranes. This protein should also be on the inside of invaginations, or of central vesicles, found in multivesicular endosomes, where it could be detected by immunocytochemistry. A second prediction is that such a protein should be found as a component of clathrin coated pits and vesicles, although if it bound weakly, it might be apparent mainly in crude preparations and resemble a cytosolic contaminant.

ACKNOWLEDGMENTS

The author is grateful to a number of colleagues who supplied copies of manuscripts prior to their publication. Work from the author's laboratory was supported by grants from the NIH and from the American Heart Association. M.G.R. is an Established Investigator of the AHA.

REFERENCES

1. Alvarez, E., Girones, N., and Davis, R.J. (1990). A point mutation in the cytoplasmic domain of the transferrin receptor inhibits endocytosis. *Biochem. J.* 267: 31–35.
2. Alvarez, E., Girones, N., and Davis, R.J. (1990). Inhibition of the receptor-mediated endocytosis of diferric transferrin is associated with the covalent modification of the transferrin receptor with palmitic acid. *J. Biol. Chem.* 265: 16644–16655.

3. Anderson R.G.W., Goldstein, J.L., and Brown, M.S. (1977). A mutation that impairs the ability of lipoprotein receptors to localize in coated pits on the cell surface of human fibroblasts. *Nature* 270: 695–699.
4. Backer, J.M., Kahn, C.R., Cahill, D.A., Ullrich, A., and White, M.F. (1990). Receptor-mediated internalization of insulin requires a 12-amino acid sequence in the juxtamembrane region of the insulin receptor b-subunit. *J. Biol. Chem.* 265: 16450–16454.
5. Bansal A. and Gierasch, L. (1991). The NPXY internalization signal of the LDL receptor adopts a reverse turn conformation. *Cell* 67: 1195–1201.
6. Beltzer, J.P. and Speiss, M. (1990). The cytoplasmic tail of the asialoglycoprotein receptor binds to a plasma membrane clathrin-coated vesicle adaptin. *J. Cell Biol.* 111: 202a.
7. Braun M., Waheed, A., and von Figura, K. (1989). Lysosomal acid phosphatase is transported to lysosomes via the cell surface. *EMBO J.* 8: 3633–3640.
8. Breitfeld, P.P., Casanova, J.E., McKinnon, W.C., and Mostov, K.E. (1990). Deletions in the cytoplasmic domain of the polyimmunoglobulin receptor differentially affect endocytotic rate and postendocytotic traffic. *J. Biol. Chem.* 265: 13750–13757.
9. Bretscher, M.S. (1989). Endocytosis and recycling of the fibronectin receptor in CHO cells. *EMBO J.* 8: 1341–1348.
10. Canfield, W.M., Johnson, K.F., Ye, R.D., Gregory, W., and Kornfeld, S. (1991). Localization of the signal for rapid internalization of the bovine cation-independent mannose 6-phosphate/insulin-like growth factor-II receptor to amino acids 24-29 of the cytoplasmic tail. *J. Biol. Chem.* 266: 5682–5688.
11. Capps, G.G., Van Kampen, M., Ward, C.L., and Zuniga, M.C. (1989). Endocytosis of the Class I major histocompatibility antigen via a phorbol myristate acetate-inducible pathway is a cell-specific phenomenon and requires a cytoplasmic domain. *J. Cell Biol.* 108: 1317–1329.
12. Casanova, J.E., Breitfeld, P.P., Ross, S.A., and Mostov, K.E. (1990). Phosphorylation of the polymeric immunoglobulin receptor required for its efficient transcytosis. *Science* 248: 742–745.
13. Chen, W., Lazar, C., Lund, K., Welsh, J., Chang, C.-P., Walton, G., Der, C., Wiley, H.S., Gill, G.N., and Rosenfeld, M.G. (1989). Functional independence of the epidermal growth factor receptor from a domain required for ligand-induced internalization and calcium regulation. *Cell* 59: 33–43.
14. Chen, W.J., Goldstein, J.L., and Brown, M.S. (1990). NPXY, a sequence often found in cytoplasmic tails, is required for coated pit-mediated internalization of the low density lipoprotein receptor. *J. Biol. Chem.* 265: 3116–3123.
15. Collawn, J.F., Stangel, M., Kuhn, L.A., Esekigwu, V., Jing, S., Trowbridge, I.S., and Tainer, J.A. (1990). Transferrin receptor internalization sequence YXRT implicates a tight turn as the structural recognition motif for endocytosis. *Cell* 63: 1061–1072.
16. Dasgupta, J.D., Watkins, S., Slayter, H., and Yunis, E.J. (1988). Receptor-like nature of class I HLA: endocytosis via coated pits. *J. Immunol.* 141: 2577–2580.
17. Daukus, G. and Zigmond, S.H. (1985). Inhibition of receptor-mediated but not fluid-phase endocytosis in polymorphonuclear leucocytes. *J. Cell Biol.* 101: 1673–1679.
18. Davis, C.G., Lehrman, M.A., Russell, D.W., Anderson, R.G.W., Brown, M.S., and Goldstein, J.L. (1986). The J.D. mutation in familial hypercholesterolemia: amino acid substitution in cytoplasmic domain impedes internalization of LDL receptors. *Cell* 45: 15–24.
19. Davis, C.G., van Driel, I.R., Russell, D.W., Brown, M.S., and Goldstein, J.L. (1987). The low-density lipoprotein receptor: identification of amino acids in cytoplasmic domain required for rapid endocytosis. *J. Biol. Chem.* 262: 4075–4082.
20. Drickamer, K. (1981). Complete amino acid sequence of a membrane receptor for glycoproteins. *J. Biol. Chem.* 256: 5827–5839.
21. Dunn, K.W., McGraw, T.E., and Maxfield, F.R. (1989). Iterative fractionation of recycling receptors from lysosomally distined ligands in an early sorting endosome. *J. Cell Biol.* 109: 3303–3314.

22. Ebina, Y., Ellis, L., Jarnagin, K., Edery, M., Graf, L., Clausner, E., Ou, J.-H., Masiarz, F., Kan, X.W., Goldfine, I.D., Roth, R.A., and Rutter, N.J. (1985). The human insulin receptor cDNA: the structural basis for hormone activated transmembrane signalling. *Cell* 40: 747–758.
23. Fire, E., Zwart, D., Roth, M.G., and Henis, Y.I. Comparisons of lateral mobility of the influenza virus hemagglutinin and an internalization-competent mutant: evidence that binding to pits is dynamic. *J. Cell Biol.* 115: 1585–1594.
24. Fuhrer, C., Geffen, I., and Spiess, M. (1991). Endocytosis of the ASGP receptor H1 is reduced by mutation of tyrosine-5 but still occurs via coated pits. *J. Cell Biol.* 114: 423–432.
25. Furuno, K., Yano, S., Akasaki, K., Tanaka, Y., Yamaguchi, Y., Tsuji, H., Himeno, M., and Kato, K. (1989). Biochemical analysis of movement of a major lysosomal membrane glycoprotein in the endocytic membrane system. *J. Biochem.* 106: 717–722.
26. Geffen, I., Wessels, H.P., Roth, J., Shia, M.A., and Spiess, M. (1989). Endocytosis and recycling of subunit H1 of the asialoglycoprotein receptor is independent of oligomerization with H2. *EMBO J.* 8: 2855–2862.
27. Geffen, I., Fuhrer, C., and Spiess, M. (1991). Endocytosis of the asialoglycoprotein receptor is independent of cytoplasmic serine residues. *Proc. Natl. Acad. Sci. USA* 88: 8425–8429.
28. Geuze, H.J., Stoorvogel, W., Strous, G.J., Slot, J.W., Bleekemolen, J.E., and Mellman, I. (1988). Sorting of mannose 6-phosphate receptors and lysosomal membrane proteins in endocytic vesicles. *J. Cell Biol.* 107: 2491–2501.
29. Goldstein, J.L., Brown, M.S., Anderson, R.G.W., Russell, D.W., and Schneider, W.J. (1985). Receptor-mediated endocytosis. *Ann. Rev. Cell Biol.* 1: 1–39.
30. Goldstein, J.L., Anderson, R.G.W., and Brown, M.S. (1979). Coated pits, coated vesicles and receptor-mediated endocytosis. *Nature* 279: 679–685.
31. Green, S.A., Zimmer, K.-P., Griffiths, G., and Mellman, I. (1987). Kinetics of intracellular transport and sorting of lysosomal membrane and plasma membrane proteins. *J. Cell Biol.* 105: 1227–1240.
32. Gregory, H., Taylor, C.C., and Hopkins, C.R. (1982). Leutinizing hormone release from associated pituitary cells by dimerization of occupied LHRH receptors. *Nature* 300: 269–271.
33. Griffiths, G., Hoflack, B., Simons, K., Mellman, I., and Kornfeld, S. (1988). The mannose 6-phosphate receptor and the biogenesis of lysosomes. *Cell* 52: 329–341.
34. Harter C., Hull, M., Webster, P., and Mellman, I. (1990). Differential requirements in the cytoplasmic tail of lgp120 for transport to lysosomes and internalization. *J. Cell Biol.* 111: 66a.
35. Henis, Y., Katzir, I.Z., Shia, M.A., and Lodish, H.F. (1990). Oligomeric structure of the human asialoglycoprotein receptor: nature and stoichiometry of mutual complexes containing H1 and H2 polypeptides assessed by fluorescence photobleaching recovery. *J. Cell Biol.* 111: 1409–1418.
36. Henis, Y. (1990). Lateral and rotational diffusion in biological membranes. In: *Handbook of Biomembranes*; *Mammalian Membranes*; *Structure and Function*, Vol. 2. (Shinitzky, M., Ed.). VCH Publishers, Weinheim.
37. Herz, J., Hamann, U., Rogne, S., Myklebost, O., Gausepohl, H., and Stanely, K.K. (1988). Surface location and high affinity for calcium of a 500-kDa liver membrane protein closely related to the LDL-receptor suggest a physiological role as lipoprotein receptor. *EMBO J.* 7: 4119–4127.
38. Heuser, J.E. and Anderson, R.G.W. (1989). Hypertonic media inhibit receptor-mediated endocytosis by blocking clathrin-coated pit formation. *J. Cell Biol.* 108: 389–400.
39. Heuser, J. (1989). Effects of cytoplasmic acidification on clathrin lattice morphology. *J. Cell Biol.* 108: 401–412.
40. Honeggar, A.M., Dull, T.J., Felder, S., Van Obberghen, E., Bellot, F., Szapary, D., Schmidt, A., Ullrich, A., and Schlessinger, J. (1987). Point mutation at the ATP binding site of EGF receptor abolishes protein-tyrosine kinase activity and alters cellular routing. *Cell* 51: 199–209.

41. Honeggar, A.M., Schmidt, A., Ullrich, A., and Schlessinger, J. (1990). Separate endocytic pathways of kinase-defective and -active EGF receptor mutants expressed in same cells. *J. Cell Biol.* 110: 1541–1548.
42. Hopkins, C.R., Gibson, A., Shipman, M., and Miller, K. (1990). Movement of internalized ligand-receptor complexes along a continuous endosomal reticulum. *Nature* 346: 335–339.
43. Hopkins, C.R. and Trowbridge, I.S. (1983). Internalization and processing of transferrin and the transferrin receptor in human carcinoma A431 cells. *J. Cell Biol.* 97: 508–521.
44. Hubbard, A.L. (1989). Endocytosis. *Curr. Opin. Cell Biol.* 1: 675–683.
45. Iacoppetta, B.J., Rothenberger, S., and Kuhn, L.C. (1988). A role for the cytoplasmic domain in transferrin receptor sorting and coated pit formation during endocytosis. *Cell* 54: 485–489.
46. Jing, S. and Trowbridge, I.S. (1990). Nonacylated human transferrin receptors are rapidly internalized and mediate iron uptake. *J. Biol. Chem.* 265: 11555–11559.
47. Jing, S., Spencer, T., Miller, K., Hopkins, C., and Trowbridge, I.S. (1990). Role of the human transferrin receptor cytoplasmic domain in endocytosis: localization of a specific signal sequence for internalization. *J. Cell Biol.* 109: 283–294.
48. Johnson, K., Chan, W., and Kornfeld, S. (1991). Cation-dependent mannose 6-phosphate receptor contains two internalization signals in its cytoplasmic domain. *Proc. Natl. Acad. Sci. USA* 87: 10010–10014.
49. Kataoka-Yamawaki, Y., Nakai, S., Miyata, T., and Honjo, T. (1982). Nucleotide sequences of gene segments encoding membrane domains of immunoglobulin gamma chains. *Proc. Natl. Acad. Sci. USA* 79: 2623–2627.
50. Kishimoto A., Brown, M.S., Slaughter, C.A., and Goldstein, J.L. (1987). Phosphorylation of serine 833 in cytoplasmic domain of low density lipoprotein receptor by a high molecular weight enzyme resembling casien kinase II. *J. Biol. Chem.* 262: 1344–1351.
51. Kishimoto A., Goldstein, J.L., and Brown, M.S. (1987). Purification of catalytic subunit of low density lipoprotein receptor kinase and identification of heat-stable activator protein. *J. Biol. Chem.* 262: 9367–9373.
52. Ktistakis, N., Thomas, D., and Roth, M.G. (1990). Characteristics of the tyrosine recognition signal for internalization of transmembrane surface glycoproteins. *J. Cell Biol.* 111: 1393–1407.
53. Kyte, J. and Doolittle, R.F. (1982). A simple method for displaying the hydropathic character of a protein. *J. Mol. Biol.* 157: 105–132.
54. Lai, W.H., Cameron, P.H., Wada, I., Doherty II, J.-J., Kay, D.G., Posner, B.I., and Bergeron, J.J.M. (1989). Ligand-mediated internalization, recycling, and downregulation of the epidermal growth factor receptor *in vivo*. *J. Cell Biol.* 109: 2741–2749.
55. Larkin, J.M., Brown, M.S., Goldstein, J.L., and Anderson, R.G.W. (1983). Depletion of intracellular potassium arrests coated pit formation and receptor-mediated endocytosis. *Cell* 33: 273–285.
56. Lazarovits, J. and Roth, M.G. (1988). A single amino acid change in the cytoplasmic domain allows the influenza virus hemagglutinin to be endocytosed through coated pits. *Cell* 53: 743–752.
57. Lazarovits, J., Shia, S.-P., Ktistakis, N., Lee, M.S., Bird, C., and Roth, M.G. (1990). The effects of foreign transmembrane domains on the biosynthesis of the influenza virus hemagglutinin. *J. Biol. Chem.* 265: 4760–4767.
58. Lehrman, M.A., Goldstein, J.L., Brown, M.S., Russell, D.W., and Schneider, W.J. (1985). Internalization-defective LDL receptors produced by genes with nonsense and frameshift mutations that truncate the cytoplasmic domain. *Cell* 41: 735–743.
59. Lemanski, P., Fatemi, S.H., Gorican, B., Meyale, S., Rossero, R., and Tartakoff, A. (1990). Dynamics and longevity of the glycolipid-anchored membrane protein Thy-1. *J. Cell Biol.* 110: 1525–1532.
60. Lewis, V.A., Koch, T., Plutner, H., and Mellman, I. (1986). A complementary DNA clone for a macrophage-lymphocyte Fc receptor. *Nature* 324: 372–375.

61. Linderman, J.L. and Lauffenberger, D.A. (1988). Analysis of intracellular receptor/ligand sorting in endosomes. *J. Theor. Biol.* 132: 203–245.
62. Lippincott-Schwartz, J. and Fambrough, D.M. (1987). Cycling of the integral membrane protein LEP100, between plasma membrane and lysosomes: kinetic and morphological analysis. *Cell* 49: 669–677.
63. Livneh, E., Reiss, N., Berent, E., Ullrich, A., and Schlessinger, J. (1987). An insertional mutant of epidermal growth factor receptor allows dissection of diverse receptor functions. *EMBO J.* 6: 2669–2676.
64. Lobel, P., Fujimoto, K., Ye, D., Griffiths, G. and Kornfeld, S. (1989). Mutations in the cytoplasmic domain of the 275-kDa mannose 6-phosphate receptor differentially alter lysosomal sorting and endocytosis. *Cell* 57: 787–796.
65. Lobel, P., Dahms, N.M., Breitmeyer, J., Chirgwin, J.M., and Kornfeld, S. (1987). Cloning of the bovine 215-kDa cation-independent mannose 6-phosphate receptor. *Proc. Natl. Acad. Sci. USA* 84: 2233–2237.
66. Lund, K.A., Opresko, L.K., Starbuck, C., Walsh, B.J., and Wiley, H.S. (1990). Quantitative analysis of the endocytic system involved in hormone-induced receptor internalization. *J. Cell Biol.* 265: 15713–15723.
67. Mahaffey, D.T., Moore, M.S., Brodsky, F.M., and Anderson, R.G.W. (1989). Coat proteins isolated from clathrin coated vesicles can assemble into coated pits. *J. Cell Biol.* 108: 1615–1624.
68. Mane, S.M., Marzella, L., Bainton, D.F., Holt, V.K., Cha, Y., Hildreth, J.E.K., and August, J.T. (1989). Purification and characterization of human lysosomal membrane glycoproteins. *Arch. Biochem. Biophys.* 268: 360–378.
69. Marsh, M. and Helenius, A. (1980). Adsorptive endocytosis of Semliki Forest virus. *J. Mol. Biol.* 142: 439–454.
70. McClain, D.A., Maegawa, H., Lee, J., Dull, T.J., Ulrich, A., and Olefsky, J.M. (1987). A mutant insulin receptor with defective tyrosine kinase displays no biologic activity and does not undergo endocytosis. *J. Biol. Chem.* 262: 14663–14671.
71. McClelland, A., Kuhn, L.C., and Ruddle, F.H. (1984). The human transferrin receptor gene: genomic organization, and the complete primary structure of the receptor deduced from a cDNA sequence. *Cell* 39: 267–274.
72. McGraw, T.E. and Maxfield, F.R. (1990). Human transferrin receptor internalization is partially dependent upon an aromatic amino acid on the cytoplasmic domain. *Cell Regulation* 1: 369–377.
73. McGraw, T.E., Pytowski, B., Arzt, J., and Ferrone, C. (1991). Mutagenesis of the human transferrin receptor: two cytoplasmic phenylalanines are required for efficient internalization and a second-site mutation is capable of reverting an internalization-defective phenotype. *J. Cell Biol.* 112: 853–861.
73a. McPhaul, M. and Berg, P. (1987). Identification and characterization of cDNA clones encoding two homologous proteins that are part of the asialoglycoprotein receptor. *Mol. Cell Biol.* 7: 1841–1847.
74. Meittinen, H.M., Rose, J.K., and Mellman, I. (1989). Fc receptor isoforms exhibit distinct abilities for coated pit localization as a result of cytoplasmic domain heterogeneity. *Cell* 58: 317–327.
75. Mellman, I. and Plutner, H. (1984). Internalization and degradation of macrophage Fc receptors bound to polyvalent immune complexes. *J. Cell Biol.* 98: 1170–1177.
76. Miller K., Shipman, M., Trowbridge, I.S., and Hopkins, C.R. (1991). Transferrin receptors promote the formation of clathrin lattices. *Cell* 65: 621–632.
77. Moore, M.S., Mahaffey, D.T., Brodsky, F.M., and Anderson, R.G.W. (1987). Assembly of clathrin-coated pits into purified plasma membranes. *Science* 236: 558–563.
78. Mostov, K.E., de Bruyn Kops, A., and Deitcher, D.L. (1986). Deletion of the cytoplasmic domain of the polymeric immunoglobulin receptor prevents basolateral localization and endocytosis. *Cell* 47: 359–364.

79. Mostov, K.E., Friedlander, M. and Blobel, G. (1984). The receptor for *trans*-epithelial transport of IgA and IgM contains multiple immunoglobulin-like domains. *Nature* 308: 37–43.
80. Okamoto, C., Shia, S.-P., Bird, C., Mostov, K.E., and Roth, M.G. (1991). The cytoplasmic domain of the polymeric immunoglobulin receptor contains two internalization signals that are distinct from its basolateral sorting signal. *J. Biol. Chem.* 267: 9925–9932.
81. Pearse, B.M.F. (1988). Receptors compete for adaptors found in plasma membrane coated pits. *EMBO J.* 7: 3331–3336.
82. Pearse, B.M.F. and Crowther, R.A. (1987). Structure and assembly of coated vesicles. *Ann. Rev. Biophys. Chem.* 16: 49–68.
83. Pelchen-Mathews, A., Armes, J.E., and Marsh, M. (1989). Internalization and recycling of CD4 transfected into Hela and NIH3T3 cells. *EMBO J.* 8: 3641–3649.
84. Pernis, B. (1985). Internalization of lymphocyte membrane components. *Immunology Today* 6: 45–49.
85. Peters, C., Braun, M., Weber, B., Wendland, M., Schmidt, B., Pohlmann, R., Waheed, A., and von Figura, K. (1990). Targeting of a lysosomal membrane protein: a tyrosine-containing endocytosis signal in the cytoplasmic tail of lysosomal acid phosphatase is necessary and sufficient for targeting to lysosomes. *EMBO J.* 9: 3497–3506.
86. Prywes R., Livneh, E., Ullrich, A., and Schlesinger, J. (1986). Mutations in the cytoplasmic domain of EGF receptor affect EGF binding and receptor internalization. *EMBO J.* 5: 2179–2190.
87. Reid, P.A. and Watts, C. (1990). Cycling of cell-surface MHC glycoproteins through primaquine-sensitve intracellular compartments. *Nature* 346: 655–657.
88. Roth, M.G., Doyle, C., Sambrook, J., and Gething, M.-J. (1986). Heterologous transmembrane and cytoplasmic domains direct functional chimeric influenza virus hemagglutinins into the endocytic pathway. *J. Cell Biol.* 102: 1271–1283.
89. Rothenberg, S., Iacopetta, B.S., and Kuhn, L.C. (1987). Endocytosis of the transferrin receptor requires the cytoplasmic domain but not its phosphorylation site. *Cell* 49: 423–431.
90. Rothman, J.E. (1989). Polypeptide chain binding proteins: catalysis of protein folding and related processes in cells. *Cell* 59: 591–601.
90a. Sahagian, G.G. and Neufeld, E.F. (1983). Biosynthesis and turnover of the mannose 6-phosphate receptor in cultured chinese hamster ovary cells. *J. Biol. Chem.* 258: 7121–7128.
91. Salamero, J., Humbert, M., Cosson, P., and Davous, J. (1990). Mouse B lymphocyte specific endocytosis and recycling of MHC class II molecules. *EMBO J.* 9: 3489–3496.
92. Sanan, D. and Anderson, R.G.W. (1991). Simultaneous visualization of LDL receptor distribution and clathrin lattices on membranes torn from the upper surface of cultured cells. *J. Histochem. Cytochem.* 39: 1017–1024.
93. Sandvig, K., Sundan, A. and Olsnes, S. (1985). Effect of potassium depletion of cells on their sensitivity to diphtheria toxin and pseudomonas toxin. *J. Cell Physiol.* 124: 54–60.
94. Sandvig, K. and van Deurs, B. (1990). Selective modulation of the endocytic uptake of ricin and fluid phase markers without alteration in transferrin endocytosis. *J. Biol. Chem.* 265: 6382–6388.
95. Sandvig, K., Olsnes, S., Petersen, O.W., and van Deurs, B. (1987). Acidification of the cytosol inhibits endocytosis from coated pits. *J. Cell Biol.* 105: 679–689.
96. Schelsinger, J. (1980). The mechanism and role of hormone-induced clustering of membrane receptors. *Trends Biochem. Sci.* 5: 210–214.
97. Schlesinger, J. (1986). Allosteric regulation of the epidermal growth factor receptor kinase. *J. Cell Biol.* 103: 2067–2072.
98. Shia, S.-P., Lazarovits, J., Ktistakis, N., and Roth, M.G. (1989). Chimeric glycoproteins can enter coated pits in the absence of a "tyrosine internalization signal": evidence for alternative pathways of endocytosis. *J. Cell Biol.* 109: 140a.
99. Speiss, M. (1990). The asialoglycoprotein receptor: a model for endocytic transport receptors. *Biochemistry* 29: 1009–1018.

100. Speiss, M., Schwartz, A.L., and Lodish, H.F. (1985). Sequence of human asialoglycoprotein receptor cDNA. *J. Biol. Chem.* 260: 1979–1982.

101. Stoorvogel, W., Geuze, H.J., and Strous, G.J. (1987). Sorting of endocytosed transferrin and asialoglycoprotein occurs immediately after internalization in HepG2 cells. *J. Cell Biol.* 104: 1261–1268.

102. Stoorvogel, W., Schwartz, A.L., Strous, G.J., and Fallon, R.J. (1991). A pool of intracellular phosphorylated asialoglycoprotein receptors which is not involved in endocytosis. *J. Biol. Chem.* 266: 5438–5444.

103. Taylor, M.E., Conary, J.T., Lennartz, M.A., Stahl, P.D., and Drickamer, K. (1990). Primary structure of the mannose receptor contains multiple motifs resembling carbohydrate-recognition domains. *J. Biol. Chem.* 265: 12156–12162.

104. Thiess, R.S., Webster, N.J., and McClain, D.A. (1990). A domain of the insulin receptor required for endocytosis in rat fibroblasts. *J. Biol. Chem.* 265: 10132–10137.

104a. Ullrich, A., Coussens, L., Hayflick, J.S., Dull, T.J., Gray, A., Tam, A.W., Lee, J., Yarden, Y., Libermann, T.A., Schlessinger, J., Downward, J., Mayes, E.L.V., Whittle, N., Waterfield, M.D., and Seeburg, P.H. (1984). Human epidermal growth factor receptor cDNA sequence and aberrant expression of the amplified gene in A431 epidermoid carcinoma cells. *Nature* 309: 418–425.

105. van Driel, I.R., Davis, C.G., Goldstein, J.L., and Brown, M.S. (1987). Self-association of the low density lipoprotein receptor mediated by the cytoplasmic domain. *J. Biol. Chem.* 262: 16127–16134.

106. Vega, M.A. and Strominger, J.L. (1989). Constitutive endocytosis of HLA class I antigens requires a specific portion of the intracytoplasmic tail that shares structural features with other endocytosed molecules. *Proc. Natl. Acad. Sci. USA* 86: 2688–2692.

107. Verrey, F., Gilbert, T., Mellow, T., Proulx, G., and Drickamer, K. (1990). Endocytosis via coated pits mediated by glycoprotein receptor in which the cytoplasmic tail is replaced by unrelated sequences. *Cell Regulation* 1: 471–486.

108. Watanabe, H., Grubb, J.H., and Sly, W.S. (1990). The overexpressed human 46-kDa mannose 6-phosphate receptor mediates endocytosis and sorting of b-glucuronidase. *Proc. Natl. Acad. Sci. USA* 87: 8036–8040.

109. Watts, C. and Davidson, H.W. (1988). Endocytosis and recycling of specific antigen by human B-cell lines. *EMBO J.* 7: 1937–1945.

110. White, M.F., Livingston, J.N., Backer, J.M., Lauris, V., Dull, T.J., Ullrich, A., and Kahn, C.R. (1988). Mutations of the insulin receptor at tyrosine 960 inhibits signal transmission but does not affect its tyrosine kinase activity. *Cell* 54: 641–649.

111. Williams, M.A. and Fukuda, M. (1990). Accumulation of membrane glycoproteins in lysosomes requires a tyrosine residue at a particular position in the cytoplasmic tail. *J. Cell Biol.* 111: 955–966.

112. Yamamoto, T., Davis, C.G., Brown, M.S., Schneider, W.J., Casey, M.L., Goldstein, J.L., and Russell, D.W. (1984). The human LDL receptor: a cystein-rich protein with multiple alu sequences in its mRNA. *Cell* 39: 27–38.

113. Zidovetski, R., Yarden, Y., Schlessinger, J., and Tovin, T.M. (1981). Rotational diffusion of epidermal growth factor complexed to cell surface receptors reflects rapid microaggregation and endocytosis of occupied receptors. *Proc. Natl. Acad. Sci. USA* 78: 6981–6985.

114. Zwart, D., Lazarovits, J., and Roth, M.G. A mutant influenza virus hemagglutinin that aggregates rapidly after synthesis is competent for exocytosis but does not recycle after endocytosis. *J. Biol. Chem.* (Submitted).

FUNCTIONS OF THE MANNOSE 6-PHOSPHATE RECEPTORS

Bernard Hoflack and Peter Lobel

Advances in Cell and Molecular Biology of Membranes
Volume 1, pages 51–80

ISBN: 1-55938-362-3

ABSTRACT

In higher eukaryotic cells, the mannose 6-phosphate receptors (MPRs) are essential components of the targeting system that delivers lysosomal enzymes to lysosomes. Two related MPRs have been characterized in many mammalian cells. The first MPR has a dual function in binding mannose 6-phosphate (M6P) containing ligands and insulin-like growth factor II (IGF-II). Expression of the gene for this receptor is imprinted and is developmentally regulated. The second MPR only binds M6P-containing ligands. Why cells require two MPRs that show overlapping targeting functions is an important unresolved question. The MPRs are distributed over several intracellular compartments. Receptors in the *trans*-Golgi network mediate sorting of lysosomal enzymes from secretory proteins. Receptors on the plasma membrane function in endocytosis of extracellular ligands. The bulk of the MPRs reside in early and late endosomes, which are the compartments where the receptors release their ligands. The MPRs are in rapid equilibrium between these different intracellular compartments. While the precise routes taken by the MPRs within the cell are still a matter of debate, it is clear that some crucial steps in their routing require specific determinants contained in their cytoplasmic domains. The cytoplasmic determinants of both receptors interact with specific cellular sorting components, some of which involve formation of clathrin-coated vesicles.

I. INTRODUCTION

Lysosomes are acidic high-density organelles which function in the catabolism of macromolecules. They are responsible for the degradation of both extracellular molecules internalized by endocytosis and intracellular molecules that are targeted by the autophagic pathway. To carry out their function, lysosomes contain over 50 different hydrolases (proteases, glycosidases, nucleases, phosphatases, lipases) that function in concert to degrade substrates into their basic components. Lysosomal storage disorders illustrate the importance of these functions; a deficiency in a single lysosomal enzyme, or a dysfunction in the targeting of these proteins, can drastically affect the fitness of cells and organisms (*100*). Thus, targeting of lysosomal enzymes is of interest for understanding both fundamental problems in cell biology and mechanisms underlying several human diseases.

In multicellular organisms, most of the lysosomal hydrolases are synthesized as soluble proteins and are transported to lysosomes by the mannose 6-phosphate (M6P)-dependent pathway. However, some proteins such as acid phosphatase or β-glucocerebrosidase are synthesized as membrane proteins and are targeted by a M6P-independent pathway (*140*). Unicellular organisms, such as *Dictyostelium discoideum*, *Acanthamibae*, or yeasts, also target vacuolar hydrolases by a M6P-independent pathway (see chapter by Cardelli), though some of these organisms are able to phosphorylate soluble acid hydrolases efficiently. The function of the M6P modification in organisms that do not utilize the M6P-dependent targeting system has not been established. In contrast, the basic properties of the M6P-dependent pathway in mammalian cells is well understood.

In mammalian cells, lysosomal enzymes, and secretory proteins contain signal sequences that direct their insertion into the lumen of the rough endoplasmic reticulum. These proteins are cotranslationally glycosylated by transfer of a pre-formed high mannose oligosaccharide onto selected asparagine residues (for review see *72a*). The fates of lysosomal enzymes and secretory proteins differ in a post-ER–pre-Golgi compartment. Here, the lysosomal enzymes are selectively recognized by a phosphotransferase that transfers N-acetylglucosamine-1-phosphate onto the 6 hydroxyl position of mannose residues on the α-1,6 branch of their high mannose oligosaccharides (*78*). Although the phosphotransferase has not yet been directly localized, it appears that it is also present in the *cis*-Golgi, where it phosphorylates mannose residues on the α-1,3 branch after the oligosaccharide is trimmed by a Golgi α-mannosidase (*79*). An N-acetylglucosamine-1-phosphodiester α-N-acetylglucosaminidase removes the N-acetylglucosamine to generate the M6P monoester. M6P serves as a transient recognition marker that allows the lysosomal enzymes to bind to high-affinity M6P receptors (MPRs) in the *trans*-face of the Golgi apparatus also referred to as the *trans*-Golgi network (TGN). A major function of the MPRs is to divert the lysosomal enzymes from the secretory pathway to the endocytic pathway, prior to their delivery to the lysosomes. The MPRs and their bound ligands are packaged into clathrin-coated vesicles which bud from the TGN and subsequently fuse with endosomes. After acid-dependent release of the lysosomal enzymes, the MPRs return to the TGN while the lysosomal enzymes are transported to lysosomes where they loose their M6-P recognition marker. Thus, some key steps in the targeting process are the phosphorylation of lysosomal enzymes by the phosphotransferase, the recognition of the phosphomannosyl marker of the lysosomal enzymes by the MPRs, and the selective packaging of the MPRs into transport vesicles.

During the past decade, much has been learned about the lysosomal enzymes and their receptors. A complete summary of the early work can be found in the excellent reviews of A. Robbins (*117*), S. Kornfeld and I. Mellman (*73*) and K. von Figura (*140*). In this chapter, we will summarize recent findings on the biochemistry and the cell biology of the lysosomal enzyme receptors and current views on how they may function in the cell.

II. PHOSPHORYLATION OF LYSOSOMAL ENZYMES

Studies on fibroblasts from patients with I-cell disease (mucolipidosis II) indicate that a single enzyme functions to phosphorylate lysosomal enzymes. I-cell fibroblasts accumulate undigested substrates in their lysosomes due to a pleiotropic deficiency of many lysosomal enzymes. The molecular basis for this defect is a lack of functional phosphotransferase activity (*61,139*). While the underlying chromosomal lesion in I-cell disease has not been identified, it is likely that a single defective gene is responsible for the lack of activity. Fibroblasts from patients with pseudo-Hurler polydystrophy (mucolipidosis III) have diminished but still detectable phosphotransferase activity. Heterocaryons of I-cell and pseudo-Hurler fibroblasts, from different patients, indicate that there are at least two complementation groups (*9,65a*), suggesting that the phosphotransferase contains multiple subunits (*76*). Little is known regarding the structure of the phosphotransferase, which is present in cells in low amounts and has not yet been purified.

An intriguing problem is how the phosphotransferase can selectively recognize many different lysosomal enzymes. To date, the primary sequence of twenty different proteins that are the targets of the phosphotransferase have been determined. These proteins do not contain any obvious consensus sequence. Recent studies have made some progress towards identifying the phosphotransferase recognition determinants on the lysosomal cathepsin D (*4,4a*). Cathepsin D and pepsinogen (a non-glycosylated secretory protein) are 45% identical in sequence and share structural motifs which are conserved among several members of the aspartyl protease family. Chimeras between cathepsin D and a glycosylated form of pepsinogen were expressed in *Xenopus* oocytes. Glycopepsinogen is not phosphorylated and therefore secreted. When discontinuous sequences of cathepsin D were substituted onto glycopepsinogen, the chimera was efficiently phosphorylated. Analysis of the location of the transplanted residues on the three dimensional structure of pepsinogen and a computer-generated model of cathepsin D revealed that the sequences came together to form a recognition patch that covered ≈15% of the surface of the molecule. While this study is an important step towards identifying the recognition determinants on cathepsin D, structural determination of other lysosomal enzymes will be required in order to decipher the common structural features that are shared by substrates of the phosphotransferase.

III. TWO MANNOSE 6-PHOSPHATE RECEPTORS

The identity of the M6P recognition marker on lysosomal enzymes was identified by Kaplan and collaborators in 1977 (*70*). Subsequent uptake and binding studies demonstrate that, unlike most endocytic receptors, the M6P-binding protein did not require divalent cations (*47,120*). Shortly afterwards, a M6P receptor (MPR) was isolated by affinity chromatography using β-galactosidase (a phosphorylated lysosomal enzyme) coupled to a matrix (*123*). This receptor was a large (>200 kDa) integral membrane glycoprotein that was predominantly localized in intracellular membranes. Cell lines were identified that lacked the large MPR (*48*). These cells secreted most of their newly synthesized lysosomal enzymes, but still retained some hydrolases. While fibroblasts of patients with I-cell disease secrete large amounts of their lysosomal enzymes as a consequence of the phosphorylation defect, other tissues such as liver or brain appeared to have normal levels of lysosomal enzymes (*105,143a*). These observations led to the search for an alternate, M6P-independent targeting pathway for lysosomal enzymes in mammalian cells. In the search for a putative receptor, P388 D1 cells which lack the previously characterized MPR were found to bind β-galactosidase in the presence of divalent cations and, unexpectedly, in a M6P-inhibitable manner (*64*). This led to the isolation of a second, smaller (≈46 kDa) integral membrane protein that was also predominantly present in intracellular membranes (*65*). Subsequent work revealed that both MPRs are present in most cell types and as discussed below, may have different functions in targeting lysosomal enzymes. In addition, cloning of the large MPR (*86,104*) and the insulin-like growth factor II receptor (*91,96*) led to a surprising finding: the large MPR is a multifunctional protein, binding both IGF-II and phosphorylated lysosomal enzymes. Based on their different binding properties, the two receptors have been assigned a plethora of names [(large MPR: cation-independent M6PR; M6P/IGF-IIR; IGF-II/M6PR; 215-, 275-, or 300-kDa M6PR; and MPR1) (small MPR: cation-dependent MPR, 43- or 46-kDa M6PR, and MPR2)]. However, since one can find exceptions in the different characteristics of the MPRs, and for the sake of simplicity, we will refer to the two lysosomal enzyme receptors as the *large* and the *small* MPRs.

A. Structure of the MPRs

Hydrodynamic studies indicate that purified, detergent-solubilized large MPR is a monomer (*109*). The cloning of the cDNAs for the large MPR from bovine (*86*), human (*96,104*) and rat sources (*91*) has allowed insights into the structure of the receptors. These cDNAs code for proteins of ≈2500 amino acids which are ~80% identical. The large MPR has a calculated protein subunit molecular weight of 270 kDa. Sequence analysis, combined with proteolysis experiments of the receptor in membranes, indicates that the bovine large MPR consists of a 44-residue amino terminal signal sequence, a 2269-residue extracytoplasmic domain, a single 23-residue transmembrane region and a 163-residue carboxyl-terminal cytoplasmic domain. The extracytoplasmic domain has a highly repetitive structure consisting of 15 contiguous units that have an average length of 147 amino acids. When the repeating units are aligned, numerous sequence identities are seen with the percentage of identical residues ranging from 16 to 38%. In addition, there are numerous regions of conservatively substituted amino acids and a characteristic spacing of cysteine residues. One repeat contains a 43-residue insertion that is related to the type II repeat of fibronectin. Similar sequences are found in coagulation factor XII, type IV collagenase, and two bovine seminal plasma proteins. The biological significance of this insertion is still unknown. The extracytoplasmic domain contains 19 potential N-linked glycosylation sites, some of which are utilized. Therefore, the estimated molecular weight of this receptor is close to 300 kDa.

The small MPR is a dimer in the membrane, and either a dimer, trimer, or tetramer in solution (*33,131*). In detergents, formation of tetramers is favored by neutral pH, high receptor concentration, and presence of M6P-containing ligands (*143,144*). The dimeric and tetrameric forms of the detergent solubilized small MPR undergo rapid dissociation in acidic solutions. The sequence of the cDNA for the small MPR from bovine (*34*), human (*114*), and murine (*89,90*) sources has been determined. These sequences encode proteins that are 90% identical. The mature small MPR has a calculated molecular weight of 28 kDa. Sequence analysis, combined with proteolysis experiments, indicates that the small MPR consists of a 28-residue amino-terminal signal sequence, a 159-residue extracytoplasmic domain, a single 25-residue transmembrane region, and a 67-residue carboxyl-terminal cytoplasmic domain. Comparison of the sequences of the large and small receptors indicates that they are related. The entire extracytoplasmic domain of the small MPR is similar to each of the repeating units of the large MPR, with sequence identities ranging from 14 to 28%. In addition, the six cysteines of the small receptor extracytoplasmic domain have a similar spacing to the cysteine residues (six or eight per repeat) of the large receptor. The cysteines of the small receptor form disulfide bonds whose connectivity has been indirectly determined by site-directed mutagenesis (*147*). Disruption of any of these disulfide bonds results in loss of binding activity, as expected for structurally important residues. The similarity of the predicted disulfide bonding patterns of the small and large MPRs, and their sequence identities, suggest that the extracytoplasmic domains arose from a common ancestral gene. Surprisingly, while the two receptors have similar intracellular distributions and routing, there are no obvious sequence similarities between their transmembrane or cytoplasmic domains.

The small MPR is heavily glycosylated, with carbohydrate contributing significantly to its apparent molecular weight on SDS-gels of 46 kDa. The extracytoplasmic domain contains five potential N-linked glycosylation sites, four of which are utilized (*33,82,148*). The human small MPR expressed in BHK cells and receptor-purified from bovine testis have two complex-type and two high mannose-type oligosaccharides (*82,148*). The oligosaccharide processing can vary in different cells, which may explain the heterogeneity observed in molecular weights of small MPRs isolated from different sources. For example, 45- and 41-kDa forms of the small MPR isolated from bovine testis have identical protein sequence but contain oligosaccharides which differ in their terminal glycosylation (*82*). Enzymatic deglycosylation (*65*) and site-directed mutagenesis of the different glycosylation sites (*148*) indicate that the N-linked oligosaccharides are important to maintain a high stability conformation and therefore the binding activity of the small MPR. It is still unknown whether the oligosaccharides of the large MPR also contribute to its binding activity.

B. Ligand Binding Properties

Interaction with Phosphorylated Ligands

Newly synthesized lysosomal enzymes contain high mannose oligosaccharides with phosphomonoesters and phosphodiesters, as well as complex type oligosaccharides. Both MPRs bind high mannose oligosaccharides containing phosphomonoesters with high affinity (*46,63,138*). They interact only weakly with high mannose oligosaccharides-containing phosphodiesters and do not bind non-phosphorylated high mannose oligosaccharides or complex-type sugar chains.

Equilibrium dialysis binding studies of purified detergent-solubilized large MPR with M6P demonstrate that the receptor binds 2 moles of M6P per mole of receptor monomer, with an affinity of 7 μM (*135*). The receptor binds 1 mole of oligosaccharide with two phosphomonoesters per mole of receptor, with an affinity of 2 nM. Lysosomal enzymes typically bind to the large receptor with nanomolar affinities (*135* and references within). Most lysosomal enzymes have multiple oligosaccharides, each of which can contain up to two phosphates. Thus, the simplest interpretation is that the large MPR has two independent M6P binding sites, and that the two binding sites function in concert to bind lysosomal enzymes with high affinity.

Equilibrium dialysis binding studies of the detergent-solubilized small MPR with M6P demonstrate that the receptor binds 1 mole of M6P per mole of monomeric subunit, with an affinity of 8 μM (*136*). The small MPR binds 1 mole of divalent phosphorylated oligosaccharide per mole of dimer, with an affinity of 200 nM. Thus, each dimer has two functional M6P binding sites. The small receptor in membranes binds the lysosomal enzyme β-glucuronidase with an affinity of 0.5 to 4 nM (*90,145*). The difference in affinity between the divalent oligosaccharide and β-glucuronidase may be the consequence of β-glucuronidase being able to bridge the two binding sites of the dimer more easily than the oligosaccharide. If so, one might expect that different lysosomal enzymes that have different spatial orientations of their phosphorylated oligosaccharides may bind with markedly different affinities to the small receptor. However, no preference in the binding specificity of the small receptor for different phosphorylated lysosomal enzymes has been demonstrated to date.

Although the two MPRs are related proteins, they exhibit slightly different binding properties. In contrast to the small MPR, the large MPR interacts with the methyl-covered M6P residues contained in oligosaccharides of lysosomal enzymes from *Dictyostelium discoideum* (*138*). Unlike the large receptor, the small receptor requires divalent cations for optimal ligand binding (*64,136*). These different properties can be used to purify the two proteins (*65*). Another interesting difference is their sensitivity to the variations of pH. Both MPRs bind their phosphorylated ligand at a slightly acidic pH (≈pH 6–6.5) and release it at an acidic pH (pH 5) (*47,63,135,136*). This is consistent with their function in transport of ligands to acidified endosomes. However, in contrast to the large MPR, the small MPR has low affinity for phosphorylated ligands at the neutral pH of the extracellular medium (*135,136*). This may explain why the small MPR functions extremely poorly in endocytosis of phosphorylated ligands (*132*), although it is clearly present at the cell surface and is rapidly internalized (*44,69,130*).

The M6P recognition marker is present on oligosaccharides of some nonlysosomal proteins. Secreted proteins containing this modification include transforming growth factor-β (TGF-β) precursor, proliferin, and thyroglobulin. TGF-β is involved in wound healing, growth, and differentiation (*115*). Activation of latent

TGF-β1 and TGF-β2 precursors requires binding to the large MPR which may mediate transport to acidic compartments where processing occurs (*83a*). Alternatively, the receptor may function by immobilizing latent TGF-β precursor on the cell surface where it is activated by plasmin and plasminogen activator (*39*). Proliferin (also called mitogen regulated protein) is a prolactin-related glycoprotein that is secreted during pregnancy and may regulate growth and/or differentiation of specific cells in the mouse fetus or in maternal tissues (*80*). This protein binds to membranes in a M6P-dependent manner. Thyroglobulin, the precursor of thyroid hormones, also bears M6P residues on some of its oligosaccharides (*62*). It has been postulated that the large MPR mediates transport of thyroglobulin from the follicle lumen to lysosomes where it is processed to produce the active hormones T3 and T4 (*125*). In addition to these soluble proteins, the epidermal growth factor receptor isolated from A-431 cells contains M6P (*134*).

Interaction with IGF-II

Mammalian large MPRs bind both M6P-containing ligands and the nonphosphorylated ligand IGF-II while the small receptor does not bind IGF-II (for review see *35,122*). The large MPR binds one mole of IGF-II per mole of receptor monomer, with an affinity of 0.2 nM (*137*). The IGF-II and M6P binding sites are distinct, as saturating concentrations of M6P do not inhibit IGF-II binding, and the receptor can bind some phosphorylated substrates and IGF-II simultaneously (*137,12*). In some cases, phosphorylated lysosomal enzymes and IGF-II can inhibit each others binding. This has been investigated by measuring the effect of IGF-II on the binding of radiolabeled phosphorylated lysosomal enzymes to purified receptor and to cultured cells, and by measuring the effect of phosphorylated lysosomal enzymes on the binding of radiolabeled IGF-II (*74,74a*). M6P prevents the inhibitory effects of the phosphorylated lysosomal enzymes on IGF-II binding. The simplest interpretation of these results is that the two binding sites are distinct, but close enough to each other so that IGF-II and certain high molecular weight lysosomal enzymes obstruct each other's binding.

IGF-II is a non-glycosylated polypeptide that is structurally related to insulin. IGF-II has many effects on cultured cells, and is present in high levels during development (reviewed in *36*), but its physiological function has not been established. Three possible functions for binding of IGF-II to the large receptor can be postulated. First, IGF-II binding to the large MPR may result in signal transduction. This has been difficult to determine since IGF-II also binds to the IGF-I receptor, which is a tyrosine kinase. While many of the documented actions of IGF-II are probably mediated through the IGF-I receptor, several reports suggest that IGF-II mediates a response through the large MPR (reviewed in *32,122*). Signal transduction may be mediated by a guanine nucleotide binding protein (G-protein) coupled pathway involving a Gi-2α subunit (*101*). This is unusual, as trimeric G proteins typically interact with the family of receptors that have seven transmembrane regions. However, IGF-II stimulates hydrolysis of GTP in a defined system of liposome-reconstituted and purified large MPR and G-protein (*98*). The G protein interacts with a 14 amino acid segment of the large MPR cytoplasmic domain that is similar to mastoparan, a peptide which activates G proteins, mimicking the action of G-protein-coupled receptors (*103*). IGF-II binding to large MPR also triggers the formation of inositol triphosphate by activation of phospholipase C (*119*). Second, IGF-II by steric hindrance may modulate the trafficking of select lysosomal enzymes. This could be important in those processes where it is necessary to secrete lysosomal enzymes such as tissue remodeling in growth and development. However, while IGF-II does inhibit endocytosis of certain lysosomal enzymes (*74*), incubation of human fibroblasts with IGF-II (*14*) or overexpression of mature IGF-II in NIH-3T3 cells (*11b*) does not increase the secretion of newly synthesized lysosomal enzymes. Finally, the large receptor may function in clearance of IGF-II. The receptor binds IGF-II at the cell surface, resulting in the internalization, delivery to lysosome, and degradation of this ligand (*102*). In addition, a soluble form of the receptor could modulate the action of IGF-II by functioning as an IGF-II binding protein (*21,48a,92*). The soluble large MPR is likely to be a breakdown product of the membrane-bound receptor: the major turnover pathway of the large MPR in tissue culture cells is through release into the medium (*27*) although it is still unclear where the processing occurs. Fetal serum contains high levels of IGF-II, half of which is bound to circulating large MPR (*48a*). Regardless of whether the large MPR is directly involved in signal transduction, the receptor is likely to have an important role in IGF-II physiology. Just as acetylcholine esterase is essential to inactivate acetylcholine at the muscle endplate to prevent overstimulation of acetylcholine receptors, clearance of IGF-II by the large MPR should be important in the IGF-II signaling pathway.

Ligand Binding Domains

As the small receptor is an oligomer and the large receptor contains repeating units of sequence, a fundamental question is what constitutes a functional M6P binding domain? Studies performed on the small MPR give important clues to this question. Evidence that each small MPR monomer can fold into an independent binding unit has been obtained by demonstrating that the truncated form of the small MPR, when expressed in *Xenopus* oocytes, behaves as a soluble monomer and is capable of binding M6P (*33*). In some cases, the soluble small receptor can dimerize (*146*), but this does not rule out the ability of the monomer to bind ligand. Studies using chemical modification and site-directed mutagenesis indicate that His 131 and Arg 137 of the human small MPR are essential for binding (*131,146a*). Binding studies suggest that two repeating units of the large MPR contain functional M6P binding sites. Sequence comparisons do not reveal that two repeats of the large receptor contain residues analogous to His 131 and Arg 137, or linear sequences that are strikingly similar to the small receptor. Thus, it is not obvious which repeating units contain M6P binding activity. Activity blotting on proteolytic fragments of the large MPR separated by SDS-Gels demonstrate that two regions containing the repeating units 1–3 and 7–11 contain functional M6P binding sites, and a recombinant protein consisting of domains 1–3 binds M6P (*149*). Further work is necessary to identify the minimal M6P binding units.

One fascinating question is why does the large receptor contain fifteen repeating units? Two of the repeats are likely to be M6P binding domains. The IGF-II binding site is likely to involve only one or a few of the remaining repeating units, as chicken and *Xenopus* contain a large MPR that binds M6P but does not interact with IGF-II (*20,26*). It is possible that most of the repeats function as spacers to allow flexibility so that the two M6P binding domains can have a range of spatial orientations. This may allow the receptor to bind different lysosomal enzymes with high affinity, each of which may have multiple oligosaccharides with different conformations. If this were the case, one might expect that the repeats that function as spacers would have diverged more rapidly between different species than the IGF-II and M6P binding domains. However, comparison of the bovine and human large MPRs reveal that all the repeats have similar levels of homology (70–90% identical). One intriguing possibility is that some of the domains function to bind other ligands that have not yet been identified. A single structural motif for binding many different classes of ligands is not unprecedented, as different immunoglobulins have similar structures but bind many different types of molecules.

IV. FUNCTION OF MPRS IN LYSOSOMAL ENZYME TRANSPORT

The reasons why mammalian cells require two lysosomal enzyme binding proteins are still elusive. The availability of the corresponding cDNA clones have provided valuable tools that have shed some light on the function of each MPR in lysosomal enzyme transport.

The large MPR has a major role in intracellular retention of newly synthesized lysosomal enzymes. The evidence comes from studies performed on cell lines lacking this receptor. These cultured cells secrete a large proportion of the newly synthesized lysosomal enzymes (*48*) while the residual targeting is mediated by the small MPR (*65,132*). Expression of the large MPR in these cells partially corrects the hypersecretion of newly synthesized lysosomal enzymes, reducing secretion from ≈75 to 30% (*75,87*). Conversely, incubation of normal cells with anti-large MPR antisera resulting in depletion (or inactivation) of the preexisting intracellular pool of this MPR, induces a hypersecretion of newly synthesized lysosomal enzymes (*141*). The large MPR also mediates endocytosis of exogenous phosphorylated ligands and IGF-II.

The function of the small MPR in lysosomal enzyme transport is far from clear. Up to now, our understanding of its function has been limited by the lack of cell lines devoid of this receptor. A series of experiments performed in different cell systems suggest that the small MPR is involved in both intracellular retention and secretion of lysosomal enzymes. Early studies on cells possessing only the small

MPR suggested that this MPR is involved in partial intracellular retention of newly synthesized lysosomal enzymes; large MPR-negative cells retain 25 to 40% of their newly synthesized lysosomal enzymes (*48*), and incubation of these cells with anti-small MPR antibodies results in a higher secretion of newly synthesized enzymes (*132*). Moreover, extremely high expression of the small MPR in these cell lines slightly reduces the hypersecretion of lysosomal enzymes (*145*). Taken together, these observations demonstrate the role of the small MPR in targeting of lysosomal enzymes to lysosomes. However, in cells containing both MPRs, the small MPR appears to have only a minor role in intracellular targeting of lysosomal enzymes. Thus, immunodepletion of the small MPR performed on normal cells does not induce a higher secretion of lysosomal enzymes (*132*). This shows that in these cells, the large MPR can take over the function of the small MPR and demonstrates that the large MPR is responsible for most of the intracellular targeting of lysosomal enzymes. This would be consistent with the higher affinity of the large MPR for phosphorylated ligands and would indicate that the MPRs probably compete for ligand binding in the *trans*-Golgi network (TGN). More recent studies have suggested that the small MPR may also function in secretion of lysosomal enzymes (*23*). Massive overexpression of the human small MPR in BHK and mouse L cells (which express both receptors) produces a higher secretion of lysosomal enzymes. One possibility is that the small MPR-mediated secretion of lysosomal enzymes reflects inefficient sorting of this receptor in the TGN and thus, some receptor molecules follow the constitutive secretory pathway. As mentioned previously, the conditions found in the extracellular medium (ions and pH) would most likely induce rapid dissociation of a phosphorylated ligand bound to the small MPR. Alternatively, the small MPR-dependent secretion may reflect the direct targeting of the small MPR from the TGN to a compartment such as an early endosome which is connected to the plasma membrane via a recycling pathway. If some small MPR molecules are recycled to the cell surface, these may release their bound ligands. Alternatively, the small receptor may release its ligand in the early endosome, and the free ligand could be regurgitated into the medium. Mapping the precise pathways followed by the two MPRs within the various intracellular compartments will facilitate a better understanding of their function.

V. MODELS FOR MPR TRAFFICKING

Several models have been proposed for targeting of lysosomal enzymes (for review see *73,117,142*). Neufeld and co-workers initially proposed a model where lysosomal enzymes were secreted and subsequently routed to lysosomes by receptor-mediated endocytosis. This model was based on early observations that different lysosomal enzymes secreted by normal fibroblasts can be taken up by mutant fibroblasts lacking the enzymes. Alternatively, it was proposed by von Figura and colleagues that lysosomal enzymes could travel to the cell surface as receptor-bound ligands and be reinternalized. After the characterization of the M6P recognition marker, it was shown that incubation of cells with concentrations of M6P that prevent uptake of lysosomal enzymes and induce rapid dissociation of MPR-bound enzymes has little effect on intracellular retention of hydrolases. These results demonstrated that most of the lysosomal enzymes were diverted intracellularly from the bulk flow secretory pathway for subsequent transport to lysosomes. Early biochemical (*124*) and morphological (*150*) studies indicated that the large receptor was not present in dense lysosomes, suggesting that lysosomal enzymes are discharged into an acidic prelysosomal compartment prior to reaching the lysosome. It is now clear that the prelysosomal compartment is part of the endocytic pathway; the large MPR is detected by electron microscopy in structures accessible to markers internalized by fluid-phase endocytosis and colocalizes in part with endocytic receptors (*16,49–51,56*). The endocytic pathway consists of four main structures which can be characterized by their large MPR and lysosomal membrane glycoprotein (e.g., lgp 120) content as well as by their accessibility to endocytic

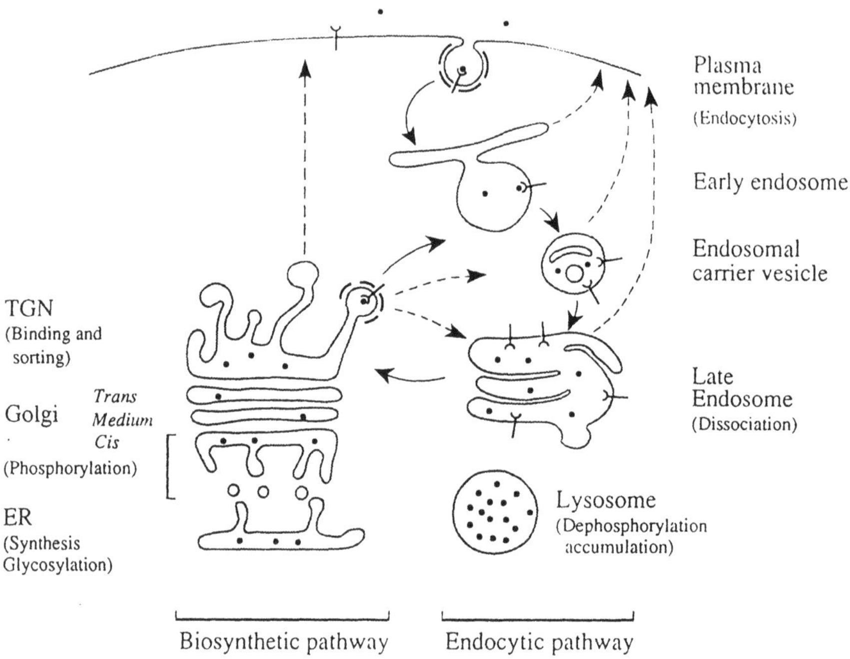

Figure 1. Model of lysosomal enzyme targeting (See text for details). After synthesis and acquisition of the M6P recognition marker, lysosomal enzymes (black dots) bind to MPRs (Y) in the TGN. The receptor-lysosomal enzyme complexes leave the TGN in clathrin-coated vesicles which travel to and fuse with endosomes. After dissociation of the bound ligand (most likely in late endosomes), the MPRs recycle back to the TGN while lysosomal enzymes travel to lysosomes. Cell surface MPRs are also internalized via clathrin-coated vesicles and can recycle back to either the plasma membrane or the TGN. The arrows show the pathways followed by the MPRs, with dashed lines indicating steps which have not been demonstrated experimentally.

markers (for review see *59,73*). Endocytic markers sequentially appear in early endosomes, in endosomal carrier vesicles (multivesicular body-like structures), then in late endosomes (also referred to as the prelysosomal compartment), and finally accumulate in lysosomes. While early and late endosomes have different protein compositions and acidification properties (*126*), it is still unclear whether they represent two distinct functional compartments connected by endosomal carrier vesicles or whether late endosomes and ultimately lysosomes originate from early endosomes by a maturation process (for review see *55,99*). We will summarize the lines of evidence which support the model for transport of lysosomal enzymes shown in Figure 1.

A. Distribution of Markers

Newly synthesized lysosomal enzymes are often difficult to detect by morphological means. In addition, most cell fractionation techniques do not allow clean resolution of Golgi and different populations of endosomes. Also, until recently, there has been a lack of good markers for the different compartments. This complicates biochemical studies directed at following transport of newly synthesized lysosomal enzymes. Therefore, in order to dissect the routes leading to lysosomes, many groups have studied the distribution of the large MPR by electron microscopy. It is now well documented that the large MPR is present in the Golgi (predominantly in the TGN), on the plasma membrane, and in several populations of endosomes, but not in dense lysosomes (*16,49–51,56,150*). In several cell types such as NRK and Madin–Dardy canine kidney (MDCK) cells (*57,106*), most of the intracellular large MPR is detected in late endosomes and small amounts are found over the TGN and early endosomes. In others cell types such as rat liver or HS4 cells, the large MPR is more prevalent in the TGN and is also present in early and late endocytic structures (*50,51*). While these different results are difficult to interpret in terms of transport between Golgi and endosomes, they do suggest that the rate-limiting steps in the cycling of the large MPR may vary from one cell type to another, leading to the preferential accumulation of this receptor in different compartments at steady state. Due to the lack of potent antibodies to the small MPR, much less is known about the distribution of this receptor. However, the available data suggest that both MPRs have similar intracellular distributions (*10,93*). The MPRs are in rapid equilibrium between these different compartments, the entire intracellular pool of the MPRs being rapidly accessible to exogenously added antibodies (*132,141*).

In polarized cells, trafficking of lysosomal enzymes appears to be restricted to one cellular domain according to the specialization of the cell type. For example, while the bulk of the large MPR is found in late endosomes of MDCK cells (*106*), this receptor is detected only on the basolateral plasma membrane and in early endosomes of the basolateral domain (*116*). In contrast, the large MPR is found on the apical membrane of osteoclasts (*6*) and appears to be involved in secretion of lysosomal enzymes into the acidic apical extracellular lacuna. Similarly, the large

MPR is detected on the apical membrane of the thyroid follicle cells and has been proposed to function in the reuptake of thyroglobulin stored in the follicle lumen for subsequent processing in protease-rich compartments (*125*).

B. Transport To and From the Golgi

The tubulo-vesicular, *trans*-face of the Golgi apparatus, now referred to as the *trans*-Golgi reticulum or *trans*-Golgi network (TGN) (reviewed in *58)*, is probably the site where newly synthesized lysosomal enzymes encounter MPRs and are diverted from the secretory pathway. Although the large MPR is occasionally detected in *cis*-Golgi-like elements in a few cell types (*15*), other morphological studies show that the receptor is more frequently detected in the TGN than in the rest of the Golgi stacks (*16,49*). This is in agreement with several lines of biochemical evidence. First, lysosomal enzymes occasionally contain a few complex-type oligosaccharides, some of which are sialylated (*45*). Since the sialyltransferase has been localized to the TGN (for review see *121*), this indicates that at least a few lysosomal enzymes travel through this compartment. Furthermore, kinetic analysis shows that both MPRs rapidly travel from the cell surface to a sialyltransferase-containing compartment, but do not efficiently travel to mannosidase- or galactosyltransferase-rich compartments (*44,68*). The latter two enzymes are commonly used as markers of the *cis*- and *trans*-Golgi, respectively. Studies performed on brefeldin A-treated cells are consistent with this scheme. In many cell types, this drug inhibits transport of proteins from the endoplasmic reticulum (ER) to the Golgi (reviewed in *108*). This block is accompanied by a disappearance of the typical organization of the Golgi stacks and a concomitant appearance of the Golgi markers in the ER (*84*). However, the distribution of TGN markers is apparently not affected (*42,85*), nor is resialylation of cell surface large MPR (*24*). Several biochemical investigations conclude that the cycling of the MPRs is constitutive (*13,44,110*) and does not require triggering by ligand binding.

The TGN contains clathrin-coated buds, and morphologically distinct clathrin-coated vesicles are frequently observed nearby. Several groups have observed by electron microscopy that the buds and vesicles contain large MPR and lysosomal enzymes (*16,49,50*). The MPRs, as well as bound lysosomal enzymes, can be detected in purified clathrin-coated vesicles (*18*). However, in most cases, these preparations contain both exocytic and endocytic clathrin-coated vesicles. Lemanski and collaborators have fractionated exocytic from endocytic clathrin-coated vesicles using density shift protocols and have found that the exocytic vesicles contain the precursor form of cathepsin D (*81*). Taken together, these observations indicate that lysosomal enzymes bound to the large MPR leave the TGN in clathrin-coated vesicles.

The efficiency of lysosomal enzyme sorting in the TGN is still unclear. Early studies have shown that low amounts (5–10%) of newly synthesized lysosomal enzymes can be recovered in the medium of cultured cells. This secretion is usually interpreted as overloading of the MPRs in the TGN with the excess lysosomal enzymes following the constitutive secretory pathway. There is no direct evidence that this interpretation is correct. Even in cells overexpressing either of the MPRs, detectable secretion still occurs (*23,75,145*). This hypothesis may be reconsidered in light of more recent findings indicating that the small MPR may be involved in secretion of lysosomal enzymes (*23*), and that some newly synthesized lysosomal enzymes can be diverted from the TGN to recycling early endosomes (*88*).

A cell-free system has been used to examine the requirements for the *in vitro* recycling of the large MPR to the TGN from endosomes. Return to the TGN occurs via an energy-dependent vesicular transport mechanism and requires donor membranes, acceptor membranes, and cytosol (*53*). Clathrin is not essential to the process (*43*). GTP hydrolysis is required for transport (*54*). It is thought that different small GTPases and associated proteins are involved in multiple steps to ensure specificity and vectorial transport between different cellular compartments (reviewed in *11a*). It is not known if the GTPases, guanine-nucleotide releasing factors, or GTPase-activating proteins interact directly with the MPRs.

C. Delivery of Newly Synthesized Lysosomal Enzymes to Endosomes

Electron microscopic studies suggest that late endosomes are the delivery site of the newly synthesized lysosomal enzymes. They usually contain high amounts of MPRs, significantly high levels of lysosomal enzymes, and their low pH is compatible with the release of bound lysosomal enzymes (*56*). However, there is no direct evidence that this really occurs and it is possible that late endosomes represent kinetic intermediates in the recycling of receptors and in transport of lysosomal enzymes to lysosomes. Biochemical studies have documented that internalized molecules can be rapidly inactivated or degraded (*22,41*), suggesting that some proteolysis may occur in early endosomes. More recent biochemical and morphological studies on NRK cells have shown that newly synthesized lysosomal enzymes are present both in early and late endosomes (*88*). Their distribution in the endocytic pathway was not affected when MPR-mediated endocytosis was blocked by incubating cells with M6P. However, they could not be detected in early endosomes after inhibition of protein synthesis. The simplest interpretation of these observations is that some newly synthesized lysosomal enzymes are directly delivered from the TGN to early endosomes. The early endosome may correspond to the first convergence point of the biosynthetic and endocytic pathways. A pathway connecting the TGN and early recycling endosomes could also explain the presence of MPRs on the cell surface without the need to postulate direct transit of MPRs from the TGN to the plasma membrane. Morphological studies using brefeldin A would be consistent with this notion: treatment of cells with this drug also results in the formation of a mixed TGN/early endosomal compartment (*85a,151*).

D. Endocytosis of the MPRs

Small amounts of both MPRs (5–10%) are present on the surface of cells. The large MPR mediates endocytosis of both IGF-II and exogenous phosphorylated lysosomal enzymes. Cell surface small MPR is inefficient in endocytosis of phosphorylated ligands due to its low affinity in this neutral pH environment (*132*). It is likely that the MPRs follow the classical endocytic pathway upon internalization (i.e., clathrin-coated vesicles, early and then late endosomes) and are recycled to the TGN from the late endosomes. Therefore, late endosomes may be the last station along the endocytic pathway where membrane proteins are retrieved to escape degradation in lysosomes. Phosphorylated lysosomal enzymes internalized by the large MPR dissociate more slowly (half-life ≈ 11 min) than does insulin endocytosed by the insulin receptor (half-life < 5 min) (*11*). This could indicate that dissociation of ligand from the large MPR does not occur completely in the mildly acidic early endosomes and that dissociation is completed when the complex reaches the more acidic late endosomes. This is consistent with the binding properties of the large MPR, which requires pH < 5.8 to release bound ligands (*64,123,135,136*).

VI. SORTING SIGNALS AND TRANSPORT VESICLES

To be routed efficiently, it is likely that both large and small MPRs contain multiple signals that mediate their departure from and arrival at the Golgi, plasma membrane, and endosomes. The elemental steps in the transit between any two compartments include segregation of the receptors from other membrane components, formation of receptor-rich carrier vesicles, and transit of the carrier vesicles to, recognition of, and fusion with the target membrane. Some of these steps undoubtedly involve interactions of the receptors with other cellular proteins. Progress has been made towards identifying signals on the receptors involved in targeting and in identifying proteins that interact with these signals.

A. Targeting Signals

Experiments using transfection of normal and mutant receptors into cultured cells have begun to reveal functional determinants on the receptors that are involved in targeting. Mouse L(Rec-) cells that lack endogenous large MPR secrete ≈70% of their newly synthesized cathepsin D and are unable to endocytose extracellular lysosomal enzymes. L(Rec-) cell lines expressing the normal bovine large MPR secrete ≈25% of their cathepsin D and endocytose lysosomal enzymes efficiently (*87*). Cells that express mutant receptors with 123- or 74-residue cytoplasmic domains (deletion of the 40 or 89 C-terminal amino acids of the 163-residue cytoplasmic domain) secrete ≈50% but endocytose extracellular lysosomal enzymes as efficiently as cells expressing the full-length receptor. This demonstrates that intracellular targeting can be perturbed without affecting endocytosis, and suggests that these two aspects of receptor trafficking utilize different signals. Truncated large MPRs with 7- or 20-residue cytoplasmic domains, or with the two tyrosines at positions 24 and 26 mutated to alanines, were impaired in both endocytosis and intracellular targeting. These receptors accumulated on the cell surface, presumably because they were unable to enter clathrin-coated pits. Extensive site-directed mutagenesis has revealed that the determinants necessary for efficient endocytosis are contained within a linear sequence of six amino acids (Tyr 24-Lys-Tyr-Ser-Lys-Val 29) (*19*). Tyr 26 and Val 29 were shown to be critical residues while Tyr 24 and Lys 28 also contributed to the internalization signal. The tyrosines could be effectively substituted by phenylalanines, indicating that aromaticity and not tyrosine phosphorylation is required for rapid internalization.

Studies on L(Rec-) cells overexpressing the small MPR reveal that this receptor requires two independent discontinuous cytoplasmic sequences for rapid internalization from the cell surface (*69*). The more potent determinant is located with the sequence Phe 13-Pro-His-Leu-Ala-Phe 18 of the small receptor's 67 residue cytoplasmic domain. In this sequence, Phe 18 is critical while Phe 13 is less important. The second, independent sorting signal of the small receptor contains Tyr 45, but has not been analyzed in detail.

The internalization signals found in both MPR cytoplasmic domains share similarities with internalization signals found in cytoplasmic domains of other receptors (see chapter by M. Roth). The internalization sequences of different receptors can be exchanged and function efficiently in the context of a heterologous protein. For instance, normal endocytosis of the human transferrin receptor, a type II (cytoplasmic amino terminus) transmembrane protein , is still observed when its internalization signal is replaced by sequences from type I (cytoplasmic carboxyl terminus) and type II membrane proteins (large MPR, low-density lipoprotein receptor, poly-Ig receptor, and asialoglycoprotein receptor) (*29*). Computer modeling and secondary structure predictions performed on internalization signals from several different receptors suggests that the critical residues are in regions with tight turn conformations (*28,29*). Thus, a three-dimensional structural motif rather than a defined amino acid sequence appears to be the crucial determinant for efficient endocytosis.

While the cytoplasmic tail of the large MPR is necessary for receptor routing, other portions of the receptor may also contain sorting signals. One study has demonstrated that a chimeric protein that contains the EGF receptor extracytoplasmic and transmembrane regions fused to the large MPR cytoplasmic region has a predominantly cell-surface distribution, similar to that of the normal EGF receptor (*40*). The normal large MPR has a predominantly intracellular distribution. This may indicate that the extracytoplasmic or transmembrane domains of the large MPR contains an intracellular retention signal.

B. Interaction with Assembly Proteins

Departure of the MPRs from either the TGN or the plasma membrane involves the formation of clathrin-coated vesicles. Isolated vesicles contain two major sets of proteins, clathrin heavy and light chains forming triskelions and assembly protein (AP) complexes (reviewed in *71,97,113*) Two types of assembly proteins (also called adaptors) are commonly found in many cell types. They are referred as to AP1 and AP2 (*71*) or HA I and HA II (*112*). The HA I assembly proteins are a tetrameric complex consisting of two 100- to 116-kDa proteins (γ and β′ subunits) associated with a 47- and 20-kDa protein. HA II consists of two 100- to 115-kDa proteins (α and β subunits) associated with a 50- and 17-kDa protein (*1*). cDNAs encoding several of these proteins have been cloned from different sources. Fluorescence microscopy indicates that HA I and HA II have different intracellular localizations and therefore appear to be involved in two different transport processes (*1,118*). HA I is detected in the Golgi area and is most likely present in Golgi-derived clathrin-coated vesicles. HA II is found on the plasma membrane and in plasma membrane derived clathrin-coated vesicles. Both clathrin and assembly proteins are required for efficient budding of transport vesicles; receptor-mediated endocytosis can be blocked both *in vivo* and *in vitro* by antibodies against clathrin or the γ subunit of HA II (*25,43,83,127*).

The assembly proteins may function to recruit receptors into and mediate formation of clathrin-coated vesicles. Assembly proteins catalyze formation of clathrin cages *in vitro* (*112*) and are located between clathrin and the interior of the clathrin cages (*113*). *In vitro* reconstitution studies indicate that the β or β′ subunit mediates interaction with clathrin (*2,128*). The assembly proteins may catalyze recruitment of receptors into clathrin-coated pits (or buds) by interaction with receptor cytoplasmic domains (*52,111*). Purified HA II binds with micromolar affinity to immobilized cytoplasmic tail fusion proteins of the large MPR and the low-density lipoprotein (LDL) receptor. HA II does not bind to the corresponding two fusion proteins that lack the tyrosine internalization signals. Binding of the Golgi-specific HA I is restricted to the large MPR and does not require tyrosine residues. Using ligand blotting with an asialoglycoprotein receptor fusion protein, it has been proposed that the β subunit of the HA II complex mediates the association (*8*). In these *in vitro* studies, the affinity of receptor cytoplasmic domains for APs are rather low. However, HA II can bind with high affinity to isolated plasma membranes (*83*). This may indicate that oligomerization or aggregation of receptors is important to generate high affinity interactions with APs. Alternatively, additional proteins may be required to catalyze the interaction or to restrict the APs to different compartments. The receptors are likely to be rate limiting for formation of the clathrin network, as light and electron microscopic studies have reported that the number of clathrin-coated vesicles (*67*) or flat clathrin lattices (*95*) were increased in cells overexpressing the human transferrin receptor.

C. Modification of MPRs Cytoplasmic Tails

Early observations of Sahagian and Neufeld (*124*) have shown that the large MPR is phosphorylated predominantly on serine residues. Turnover of phosphate groups is much more rapid than turnover of the receptor (half-lives of 2.5 versus 16 hr), suggesting that phosphorylation and dephosphorylation mechanisms may be important in cycling. Several studies have reported that the state of phosphorylation of the large receptor is regulated. For example, insulin stimulation of adipocytes results in recruitment of large MPR to the cell surface. The plasma membrane associated receptor is less phosphorylated than the intracellular receptor (*30*). Plasma membrane could be separated into two fractions (*31*). One fraction, enriched in clathrin-heavy chain, contains large MPR that is more highly phosphorylated on serine residues than the receptor in a fraction devoid of clathrin. A calcium- and calmodulin-independent kinase (*107*) and a casein kinase II (*7*) have been detected in isolated clathrin-coated vesicles. The casein kinase II is capable of phosphorylating clathrin β-light chain *in vitro*. More recently, it has been reported that this casein kinase II is tightly associated with the Golgi-specific HA I complex and most likely represents the 47-kDa subunit (*94*). The kinase catalyzes the *in vitro* phosphorylation of serine residues located in two casein kinase sites on the cytoplasmic domain of the large receptor, and these sites are phosphorylated *in vivo* (*94*). Future studies on the *in vitro* interactions of phosphorylated and dephosphorylated large MPR cytoplasmic domain with HA I and clathrin, and on trafficking of receptor mutants *in vivo*, should reveal the role of the HA I associated casein kinase II in receptor routing.

VII. GENETICS OF THE MPRs

A. Genomic Organization

The human and murine small MPR genes have been cloned and characterized (*72,89*). Comparison of the 10 to 12-kb genes with their cDNAs indicate that they contain seven exons. The extracytoplasmic domain is contained in four exons, the transmembrane region in two exons, and the cytoplasmic domain in two exons. The promoter regions of the genes contain structural elements characteristic of constitutively expressed house-keeping genes. The genes map to human chromosome 12 (*114*) and mouse chromosome 6 (*89*). The mouse genome also contains a highly unusual processed-type pseudogene that resembles the small MPR cDNA. These intronless pseudogenes are believed to arise from integration of a cDNA generated by reverse transcription of a mRNA. As they normally do not contain functional promoters, they are not expressed and, without selective pressure, mutate over evolutionary time. Conceptual translation of the small MPR pseudogene indicates that it encodes a truncated protein whose extracytoplasmic domain contains three cysteines. Thus, it is unlikely to encode a stable, functional protein. The pseudogene is unusual in that it differs from the cDNA in two other aspects. First, it contains an extended 5′ region that is homologous to the promoter region of the normal gene. Second, it contains a "minicistron"-like insertion which is also present on the normal gene, but not in any cDNA isolated to date. In the normal gene, this sequence is flanked by splice consensus sequences and may represent an infrequently used exon that would introduce a short open reading frame and termination codon upstream to the initiator methionine of the small receptor. The presence of the 5′-extension and the insertion suggests that in some cases, expression of the mouse small MPR gene is regulated by alternate splicing and by use of additional upstream promoter elements.

The human and mouse large MPR genes map to chromosomes 6 and 17, respectively (*77*). The ≈100-kb large MPR mouse gene has been partially characterized (G. Szebenyi and P. Rotwein, personal communication). The organization of many of the exons encoding extracytoplasmic sequence is similar to that of the small receptor, strongly suggesting that both receptors arose from a common ancestral gene. Comparison among individual exons of the large and small receptor genes may give important clues concerning functional domains of the receptors.

B. Tissue Distribution and Developmental Expression

Early binding studies demonstrated that the large MPR is present in all major tissues. While most established cell lines express this receptor, there are some exceptions (*48*). All cell lines and tissues examined to date express the small receptor. While both receptors are ubiquitous, their relative expression varies considerably in different tissues and cell lines (*148a*). Although the promotor of the large MPR gene has not been characterized yet, it appears that the expression of this MPR is highly regulated. In mouse and rat development, expression of the receptor is extremely high during the early fetal life, the large MPR being one of the major proteins in several fetal tissues (*3,129*). There is a parallel increase in M6P binding sites in the maternal liver (*80*). However, its expression declines markedly in late gestation and in the early postnatal period. It is interesting to note that, during mouse development, IGF-II mRNAs are present at high levels prenatally and decline postnatally, paralleling expression of the large MPR (*17*). Similarly, there is a coordinate increase in IGF-II mRNA and in cell surface expression of the large MPR during the terminal myogenic differentiation of the mouse skeletal muscle-derived C2 cells (*133*). During the same period, expression of the small MPR was unchanged, indicating that the two MPRs are differentially regulated.

C. Imprinting of the Large MPR Gene

An exciting finding in terms of potential function of the large MPR has come from genetic studies demonstrating that the large MPR gene and the IGF-II gene are imprinted (reviewed in *60*). The large MPR message is expressed from the maternally inherited allele with little expression contributed by the paternal allele (*5*). IGF-II is imprinted in the opposite manner (*37*). Recent evidence suggests that deletion of the large MPR is closely related, if not identical, to the T-associated maternal effect (Tme), a mutation that causes fetal mice to die at day 15 of gestation. This indicates that the receptor probably is an essential gene. In mice, disruption of both alleles of the IGF-II gene by homologous recombination results in abnormally small but viable and fertile mice (*38*). If the growth promoting effects of IGF-II were exerted by binding to the large MPR, one might expect that Tme embryos that lack receptor would be small. However, the embryos are normal size at the time of death. The opposite imprinting of the large MPR gene and IGF-II has been interpreted as a balance between the reproductive demand of the father (IGF-II increasing the growth of his offspring) and the survival of the mother for subsequent reproduction (the large MPR functioning to clear IGF-II) (*60*). Why do Tme mice die at embryonic day 15? There are at least three possibilities.

1. Death is due to mistargeting of M6P-containing proteins. This seems unlikely as in humans, patients with I-cell disease do not phosphorylate lysosomal enzymes but do survive into early childhood.
2. Death may be a result of some other essential function of the receptor such as binding other classes of ligands that have not yet been identified.
3. Death may be due to impaired clearance of IGF-II. This last possibility is now experimentally testable. If this is the case, mating of Tme females and transgenic IGF-II negative males should produce progeny that do not express large MPR and IGF-II, and are viable.

VIII. CONCLUSION

During the last 10 years, considerable progress has been made to extend our knowledge of the targeting system which allows hydrolases to reach lysosomes. Structural features crucial for the acquisition of the mannose 6-phosphate marker have been identified on cathepsin D. Studies are still needed to establish that these structural features are also shared by other lysosomal enzymes. Also, cloning and characterization of the phosphotransferase will provide an important missing piece of the puzzle.

The primary amino acid sequences of the mannose 6-phosphate receptors (MPRs) have revealed interesting features of their structural organization. A more detailed dissection of the large MPR luminal domain is necessary to determine the identity of the ligand binding domains. Expression of functional binding units of the large MPR as well as crystallographic studies will contribute to understanding ligand binding and acid-induced dissociation at the molecular level.

A fascinating aspect of the cell biology of the MPRs concerns the mechanisms involved in intracellular trafficking. Although determinants for rapid endocytosis of the MPRs have been identified in their cytoplasmic domains, the determinants critical for other steps of routing remain to be characterized. While site-directed mutagenesis studies will be instrumental in identifying important residues on the primary sequence, determination of the three-dimensional structure of the MPR cytoplasmic domains is required for the in-depth understanding of the sorting signals. How these signals interact with other cellular proteins is also an important question. Assembly proteins of clathrin-coated vesicles play a critical role in segregation of receptors from other membrane components and departure from the Golgi and plasma membrane. The molecular events leading to binding and dissociation of the MPR cytoplasmic domains and assembly proteins are largely unknown. Assembly proteins are now biochemically well defined and several of their subunits have been cloned. Their precise functions in MPR sorting can be now investigated. In addition, there are likely to be many other cellular proteins that must be identified and characterized to understand receptor routing.

Most important, we still do not understand why there are two distinct MPRs. Are there yet undiscovered specificities for the MPRs towards lysosomal enzymes, with both receptor being necessary for efficient targeting of hydrolases to lysosomes? Do they release their ligands in different compartments? Why does the large receptor bind IGF-II, and does this influence lysosomal enzyme targeting? To date, clear answers to these questions have been elusive, as there are no cell lines that lack the small receptor. Disruption of the small MPR gene by homologous recombination may provide new systems to address these questions. Similarly, genetic studies with organisms that lack the receptors and their ligands are likely to be instrumental in determining the biological functions of the two MPRs. We can guess that many ongoing studies will answer some of these interesting questions in the near future and will provide, as in the past, unexpected findings.

ACKNOWLEDGMENTS

We would like to thank members of our laboratories and our colleagues for stimulating discussions. Our collaborative research was supported by NATO (Grant 900226).

REFERENCES

1. Ahle, S., Mann, A., Eichelsbacher, U., and Ungewickell, E. (1988). Structural relationships between clathrin assembly proteins from the Golgi and the plasma membrane. *EMBO J.* 7: 919–929.
2. Ahle, S. and Ungewickell, E. (1989). Identification of a clathrin-binding subunit in the HA2 adaptor protein complex. *J. Biol. Chem.* 264: 20089–20093.
3. Ballesteros, M., Scott, C.D., and Baxter, R.C. (1990). Developmental regulation of the insulin like growth factor II/mannose 6-phosphate receptor mRNA in the rat. *Biochem. Biophys. Res. Comm.* 172: 775–779.
4. Baranski, T.J., Faust, P.L., and Kornfeld, S. (1990). Generation of a lysosomal enzyme targeting signal in the secretory protein pepsinogen. *Cell* 63: 281–291.

4a. Baranski, T.J., Koelsch, G., Hartsuck, J.A., and Kornfeld, S. (1991). Mapping and molecular modeling of a recognition domain for lysosomal enzyme targeting. *J. Biol. Chem.* 266: 23365–23372.
5. Barlow, D.P., Stöger, R., Herrmann, B.G., Saito, K., and Schweifer, N. (1991). The mouse insulin-like growth factor type 2 receptor is imprinted and closely related to the Tme locus. *Nature* 349: 84–87.
6. Baron, R., Neff, L., Brown, W., Courtoy, P.J., Louvard, D., and Farquhar, M.G. (1988). Polarized secretion of lysosomal enzymes: codistribution of the cation-independent mannose 6-phosphate receptor and lysosomal enzymes along the osteoclast exocytic pathway. *J. Cell Biol.* 106: 1863–1872.
7. Bar-Zvi, D. and Branton, D. (1986). Clathrin-coated vesicles contain two protein kinase activities. *J. Biol. Chem.* 261: 9614–9621.
8. Beltzer, J.P. and Spiess, M. (1991). *In vitro* binding of the asialoglycoprotein receptor to the β adaptin of the plasma membrane. *EMBO J.* 10: 3735–3742.
9. Ben-Yoseph, K., Pack, B.A., Mitchell, D.A., Elwell, D.G., Potier, M., Melançon, S.B., and Madler, H.L. (1986). Characterization of the mutant N-acetylglucosaminyl-phosphotransferase in I-cell disease and pseudo-hurler polydystrophy. Complementation analysis and kinetic studies. *Enzyme* 35: 106–116.
10. Bleekemolen, J.E., Stein, M., von Figura, K., Slot, J.W., and Geuze, H.J. (1988). The two mannose 6-phosphate receptors have almost identical subcellular distributions in U937 monocytes. *Eur. J. Cell Biol.* 47: 366–372.
11. Borden, L.A., Einstein, R., Gabel, C.A., and Maxfield, F.R. (1990). Acidification-dependent dissociation of endocytosed insulin precedes that of endocytosed proteins bearing the mannose 6-phosphate recognition marker. *J. Biol. Chem.* 265: 8497–8504.
11a. Bourne, H.R., Sanders, D.A., and McCormick, F. (1990). The GTPase superfamily: A conserved switch for diverse cell functions. *Nature* 348: 125–132.
11b. Braulke, T., Bresciani, R., Buergisser, D.M., and von Figura, K. (1991). Insulin-like growth factor II overexpression does not affect sorting of lysosomal enzymes in NIH-3T3 cells. *Biochem. Biophys. Res. Commun.* 179: 108–115.
12. Braulke, T., Causin, C., Waheed, A., Jurghans, U., Hasilik, A., Maly, P., Humbel, R.E., and von Figura, K. (1988). Mannose 6-phosphate/insulin like growth factor II receptor: distinct binding sites for mannose 6-phosphate and insulin-like growth factor II. *Biochem. Biophys. Res. Commun.* 150: 1287–1293.
13. Braulke, T., Gartung, C., Hasilik, A., and von Figura, K. (1987). Is movement of mannose 6-phosphate-specific receptor triggered by binding of lysosomal enzymes? *J. Cell Biol.* 104: 1735–1742.
14. Braulke, T., Tippmer, S., Chao, H.J., and von Figura, K. (1990). Insulin-like growth factor I and II stimulate endocytosis but do not affect sorting of lysosomal enzymes in human fibroblasts. *J. Biol. Chem.* 265: 6650–6655.
15. Brown, W.J. and Farquhar, M.G. (1984). The mannose 6-phosphate receptor for lysosomal enzymes is concentrated in *cis*-Golgi cisternae. *Cell* 36: 295–307.
16. Brown, W.J., Goodhouse, J., and Farquhar, M.G. (1986). Mannose 6-phosphate receptors for lysosomal enzymes cycle between the Golgi complex and endosomes. *J. Cell Biol.* 103: 1235–1247.
17. Brown, A.L., Graham, D.E., Nissley, S.P., Hill, D.J., Strain, A.J. and Rechler, M.M. (1986). Developmental regulation of the insulin-like growth factor II mRNA in different rat tissues. *J. Biol. Chem.* 261: 13144–13150.
18. Campbell, C.H., Fine, R.E., Squicciarini, J., and Rome, L.H. (1983). Coated vesicles from rat liver and calf brain contain cryptic mannose 6-phosphate receptors. *J. Biol. Chem.* 258: 2628–2633.
19. Canfield, W.M., Johnson, K.F., Ye, R.D., Gregory, W., and Kornfeld, S. (1991). Localization of the signal for rapid internalization of the bovine cation-independent mannose 6-phosphate/

insulin-like growth factor II receptor to amino acids 24–29 of the cytoplasmic tail. *J. Biol. Chem.* 266: 5682–5688.

20. Canfield, W.M. and Kornfeld, S. (1989). The chicken liver cation-independent mannose 6-phosphate receptor lacks the high-affinity binding site for the insulin-like growth factor II. *J. Biol. Chem.* 264: 7100–7103.
21. Causin, C., Waheed, A., Braulke, T., Junghans, U., Maly, P., Humbel, R.E., and von Figura, K. (1988). Mannose 6-phosphate/insulin-like growth factor II-binding proteins in human serum and urine: their relation to the mannose 6-phosphate/insulin-like growth factor II receptor. *Biochem. J.* 252: 795–799.
22. Casciola-Rosen, L.A. and Hubbard, A.L. (1991). Hydrolases in intracellular compartments of rat liver cells. *J. Biol. Chem.* 266: 4341–4347.
23. Chao, H.H-J., Waheed, A., Pohlmann, R., Hille, A. and von Figura, K. (1990). Mannose 6-phosphate receptor-dependent secretion of lysosomal enzymes. *EMBO J.* 9: 3507–3513.
24. Chege, N.W. and Pfeffer, S.R. (1990). Compartmentation of the Golgi complex: Brefeldin A distinguishes *trans*-Golgi cisternae from the *trans*-Golgi network. *J. Cell. Biol.* 111: 893–900.
25. Chin, D.J., Straubinger, R.M., Acton, S., Näthke, I., and Brodski, F. (1989). 100-kDa polypeptides in peripheral clathrin-coated vesicles are required for receptor endocytosis. *Proc. Natl. Acad. Sci. USA* 86: 9289–9293.
26. Clairmont, K.B. and Czech, M.P. (1989). Chicken and *Xenopus* mannose 6-phosphate receptors fail to bind insulin-like growth factor II. *J. Biol. Chem.* 264: 16390–16392.
27. Clairmont, K.B. and Czech, M.P. (1991). Extracellular release as the major degradative pathway of the insulin-like growth factor II/mannose 6-phosphate receptor. *J. Biol. Chem.* 266: 12131–12134.
28. Collawn, J.F., Stangel, M., Kuhn, L.A., Esekogwu, V., Jing, S., Trowbridge, I.S., and Tainer, J.A. (1990). Transferrin receptor internalization sequence YXRF implicates a tight turn as the structural recognition motif for endocytosis. *Cell* 63: 1061–1072.
29. Collawn, J.F., Kuhn, L.A., Sue Liu, L-F., Tainer, J.A., and Trowbridge, I.S. (1991). Transplanted LDL and mannose 6-phosphate receptor internalization signals promote high-efficiency endocytosis of the transferrin receptor. *EMBO J.* 10: 3247–3253.
30. Corvera, S. and Czech, M.P. (1985). Mechanism of insulin action on membrane protein recycling: a selective decrease in the phosphorylation state of the insulin-like growth factor II receptors in the cell surface membrane. *Proc. Natl. Acad. Sci. USA* 82: 7314–7318.
31. Corvera, S., Folander, K., Clairmont, K.B., and Czech, M.P. (1988). A highly phosphorylated subpopulation of insulin-like growth factor II/mannose 6-phosphate receptors is concentrated in a clathrin-enriched plasma membrane fraction. *Proc. Natl. Acad. Sci. USA* 85: 7567–7571.
32. Czech, M.P. (1989). Signal transmission by the insulin-like growth factors. *Cell* 59: 235–238.
33. Dahms, N.M. and Kornfeld, S. (1989). The cation-dependent mannose 6-phosphate receptor: Structural requirements for mannose 6-phosphate binding and oligomerization. *J. Biol. Chem.* 264: 11458–11467.
34. Dahms, N.M., Lobel, P., Breitmeyer, J., Chirgwin, J.M., and Kornfeld, S. (1987). 46-kDa mannose 6-phosphate receptor: Cloning, expression and homology to the 215-kDa mannose 6-phosphate receptor. *Cell* 50: 181–192.
35. Dahms, N.M., Lobel, P., and Kornfeld, S. (1989). Mannose 6-phosphate receptors and lysosomal enzyme targeting. *J. Biol. Chem.* 264: 12115–12118.
36. Daughaday, W.H. and Rotwein, P. (1989). Insulin-like growth factors I and II. Peptide, messenger ribonucleic acid and gene structures, serum and tissue concentrations. *Endocrine Reviews* 10: 68–91.
37. De Chiara, T.M., Efstratiadis, A., and Robertson, E.J. (1990). A growth-deficiency phenotype in heterozygous mice carrying an insulin-like growth factor gene disrupted by targeting. *Nature* 345: 78–80.
38. De Chiara, T.M., Robertson, E.J., and Efstratiadis, A. (1991). Parental imprinting of the mouse insulin-like growth factor II gene. *Cell* 64: 849–859.

39. Dennis, P.A. and Rifkin, D.B. (1991). Cellular activation of latent transforming growth factor-β requires binding to the cation-independent mannose 6-phosphate/insulin-like growth factor type II receptor. *Proc. Natl. Acad. Sci. USA* 88: 580–584.
40. Dintzis, S.M. and Pfeffer, S.R. (1990). The mannose 6-phosphate receptor cytoplasmic domain is not sufficient to alter the cellular distribution of a chimeric EGF receptor. *EMBO J.* 9: 77–84.
41. Doherty, J.J., Kay, D.G., Lai, W.H., Posner, B.I., and Bergeron, J.J. (1990). Selective degradation of insulin within rat endosomes. *J. Cell Biol.* 110: 35-42.
42. Dohms, R., Russ, W.G., and Yewdell, J.W. (1989). Brefeldin A redistributes resident and itinerant Golgi proteins to the endoplasmic reticulum. *J. Cell Biol.* 109: 61–72.
43. Draper, R.K., Goda, Y., Brodski, F.M., and Pfeffer, S.R. (1990). Antibodies to clathrin inhibits endocytosis but not recycling to the *trans*-Golgi network. *Science* 248: 1539–1541.
44. Duncan, J.R. and Kornfeld, S. (1988). Intracellular movement of the two mannose 6-phosphate receptors: return to the Golgi apparatus. *J. Cell Biol.* 106: 617–628.
45. Fedde, K.N. and Sly, W.S. (1985). Ricin-binding properties of acid hydrolases from isolated lysosomes implies prior processing by terminal transferases of the *trans*-Golgi apparatus. *Biochem. Biophys. Res. Comm.* 133: 614–620.
46. Fischer, H.D., Creek, K.E., and Sly, W.S. (1982). Binding of oligosaccharides to immobilized phosphomannosyl receptors. *J. Biol. Chem.* 257: 9938–9943.
47. Fischer H.D., Gonzalez-Noriega, A., and Sly, W.S. (1980). β-Glucuronidase binding to human fibroblast membrane receptors. *J. Biol. Chem.* 255: 5069–5074.
48. Gabel, C.A., Goldberg, D.E., and Kornfeld, S. (1983). Identification and characterization of cells deficient in the mannose 6-phosphate receptor: evidence for an alternate pathway for lysosomal enzyme targeting. *Proc. Natl. Acad. Sci. USA* 80: 775–779.
48a. Gelato, M.C., Rutherford, C., Stark, R.I., and Daniel, S.S. (1989). The insulin-like growth factor II/mannose-6-phosphate receptor is present in fetal and maternal sheep serum. *Endocrinology* 124: 2935–2943.
49. Geuze, H.J., Slot, J.W., Strous, G.J., Hasilik, A., and von Figura, K. (1985). Possible pathways for lysosomal enzyme delivery. *J. Cell Biol.* 101: 2253–2262.
50. Geuze, H.J., Slot, J.W., Strous, G.J., Peppard, J., von Figura, K., Hasilik, A., and Schwartz, A. (1984). Intracellular receptor sorting during endocytosis: comparative immunoelectron microscopy of multiple receptors in Rat liver. *Cell* 37: 195–204.
51. Geuze, H.J., Stoorvogel, W., Strous, G.J., Slot, W.J., Bleekemolen, J.E., and Mellman, I. (1988). Sorting of mannose 6-phosphate receptors and lysosomal membrane proteins in endocytic vesicles. *J. Cell Biol.* 107: 2491–2501.
52. Glickman, J.N., Conibear, E., and Pearse, B.M.F. (1989). Specificity of binding of clathrin adaptors to signals on the mannose 6-phosphate/insulin-like growth factor II receptor. *EMBO J.* 8: 1041–1047.
53. Goda, Y. and Pfeffer, S.R. (1988). Selective recycling of the mannose 6-phosphate/IGF II receptor to the *trans*-Golgi network *in vitro*. *Cell* 55: 309–320.
54. Goda, Y. and Pfeffer, S.R. (1991). Identification of a novel, N-ethylmaleimide-sensitive cytosolic factor required for vesicular transport from endosomes to the *trans*-Golgi network *in vitro*. *J. Cell Biol.* 112: 823–831.
55. Griffiths, G. and Gruenberg, J. (1991). The arguments for pre-existing early and late endosomes. *Trends in Cell Biol.* 1: 5–9.
56. Griffiths, G., Hoflack, B., Simons, K., Mellman, I., and Kornfeld, S. (1988). The mannose 6-phosphate receptor and the biogenis of lysosomes. *Cell* 52: 329–341.
57. Griffiths, G., Matteoni, R., Bach, R., and Hoflack, B. (1990). Characterization of the cation-independent mannose 6-phosphate receptor enriched prelysosomal compartment in NRK cells. *J. Cell Sci.* 95: 441–461.
58. Griffiths, G. and Simons, K. (1986). The *trans*-Golgi network: sorting at the exit site of the Golgi complex. *Science* 234: 430–443.

59. Gruenberg, J. and Howell, K.E. (1989). Membrane traffic in endocytosis: insights from cell-free assays. *Ann. Rev. Cell Biol.* 5: 453–481.
60. Haig, D. and Graham, C. (1991). Genomic imprinting and the strange case of the insulin-like growth factor II receptor. *Cell* 64: 1045–1046.
61. Hasilik, A., Waheed, A., and von Figura, K. (1981). Enzymatic phosphorylation of lysosomal enzymes: absence of activity in I-cell fibroblasts. *Biochem. Biophys. Res. Comm.* 98: 761–767.
62. Herzog, V., Neumüller, W., and Holzmann, B. (1987). Thyroglobulin, the major and obligatory exportable protein of thyroid follicle cells, carries the lysosomal recognition marker mannose 6-phosphate. *EMBO J.* 6: 555–560.
63. Hoflack, B., Fujimoto, K., and Kornfeld, S. (1987). The interaction of phosphorylated oligosaccharides and lysosomal enzymes with bovine liver cation-dependent mannose 6-phosphate receptor. *J. Biol. Chem.* 262: 123–129.
64. Hoflack B. and Kornfeld, S. (1985). Lysosomal enzyme binding to mouse P388D1 macrophage membranes lacking the 215-kDa mannose 6-phosphate receptor: evidence for the existence of a second mannose 6-phosphate receptor. *Proc. Natl. Acad. Sci. USA* 82: 4428–4432.
65. Hoflack B. and Kornfeld, S. (1985). Purification and characterization of a cation-dependent mannose 6-phosphate receptor from murine P388D1 macrophages and bovine liver. *J. Biol. Chem.* 260: 12008–12014.
65a. Honey, N.K., Mueller, O.T., Little, L.E., Miller, A.L., and Schow, T.B. (1982). Mucolipidosis II and III. The genetic relationships between two disorders of lysosomal enzyme biosynthesis. *Proc. Natl. Acad. Sci. USA* 79: 7420–7424.
66. Hubbard, A.L. (1989). Endocytosis. *Current Opinion in Cell Biology* 1: 675–683.
67. Iacopetta, B.J., Rothenberger, S., and Kühn, L.C. (1988). A role for the cytoplasmic domain in transferrin receptor sorting and coated pit formation during endocytosis. *Cell* 54: 485–489.
68. Jin, M., Sahagian, G.G., and Snider, M.D. (1989). Transport of surface mannose 6-phosphate receptor to the Golgi complex in cultured human cells. *J. Biol. Chem.* 264: 7675–7680.
69. Johnson, K.F., Chan, W., and Kornfeld, S. (1990). Cation-dependent mannose 6-phosphate receptor contains two internalization signals in its cytoplasmic domain. *Proc. Natl. Acad. Sci. USA* 87: 10010–10014.
70. Kaplan, A., Achord, D.T., and Sly, W.S. (1977). Phosphohexosyl components of a lysosomal enzyme are recognized by pinocytosis receptors on human fibroblasts. *Proc. Natl. Acad. Sci. USA* 74: 2026–2030.
71. Keen, J.H. (1990). Clathrin and associated assembly and disassembly proteins. *Ann. Rev. Biochem.* 99: 415–438.
72. Klier, H.J., von Figura, K., and Pohlmann, R. (1991). Isolation and analysis of the human 46-kDa mannose 6-phosphate receptor gene. *Eur. J. Biochem* 197: 23–28.
72a. Kornfeld, R. and Kornfeld, S. (1985). Assembly of Asparagine-linked oligosaccharides. *Ann. Rev. Biochem.* 54: 631–664.
73. Kornfeld, S. and Mellman, I. (1989). The biogenesis of lysosomes. *Ann. Rev. Cell Biol.* 5: 483–525.
74. Kiess, W., Thomas, C.L., Greenstein, L.A., Lee, L., Sklar, M.M., Rechler, M.M., Sahagian, G.G., and S. P. Nissley (1989). Insulin-like growth factor II (IGF II) inhibits both the cellular uptake of β-galactosidase and the binding of β-galactosidase to purified IGF II/mannose 6-phosphate receptor. *J. Biol. Chem.* 264: 4710–4714.
74a. Kiess, W., Thomas, C.L., Sklar, M.M., and Nissley, S.P. (1990). Beta-galactosidase decreases the binding affinity of the insulin-like-growth factor-II/mannose 6-phosphate receptor for insulin-like-growth-factor II. *Eur. J. Biochem.* 190: 71–77.
75. Kyle, J.W., Nolan, C.M., Oshima, A., and Sly, W.S. (1988). Expression of the human cation-independent mannose 6-phosphate receptor cDNA in receptor-negative Mouse P388D1 cells following gene transfer. *J. Biol. Chem.* 263, 16230–16235.

76. Lang, L., Takahashi, T., Tang, J., and Kornfeld, S. (1985). Lysosomal enzyme phosphorylation in human fibroblasts: kinetic parameters offer a biochemical rationale for two distinct defects in the N-acetylglucosaminyl l-phosphotransferase. *J. Clin. Invest.* 76: 2191–2195.
77. Laureys, G., Barton, D.E., Ullrich, A., and Francke, U. (1988). Chromosomal mapping of the gene for the type II insulin-like growth factor/cation-independent mannose 6-phosphate receptor in man and mouse. *Genomics* 3, 224–229.
78. Lazzarino, D.A. and Gabel, C.A. (1988). Biosynthesis of the mannose 6-phosphate recognition marker in transport-impaired mouse lymphoma cells. *J. Biol. Chem.* 263: 10118–10126.
79. Lazzarino, D.A. and Gabel, C.A. (1989). Mannose processing is an important determinant in the assembly of phosphorylated high mannose-type oligosaccharides. *J. Biol. Chem.* 264: 5015–5023.
80. Lee, S.J. and Nathans, D. (1988). Proliferin secreted by cultured cells binds to mannose 6-phosphate receptors. *J. Biol. Chem.* 263: 3521–3527.
81. Lemanski, P., Hasilik, A., von Figura, K., Helmy, S., Fishman, J., Fine, R.E., Kedersha, N.L., and Rome, L.H. (1987). Lysosomal enzyme precursors in coated vesicles derived from the exocytic and endocytic pathways. *J. Cell Biol.* 104: 1743–1748.
82. Li, M. and Jourdian, G.W. (1991). Isolation and characterization of the two glycosylation isoforms of the low molecular weight mannose 6-phosphate receptor from bovine testis. *J. Biol. Chem.* 266: 17621–17630.
83. Lin, H.C., Moore, M.C., Sanan, D.A., and Anderson, R.G.W. (1991). Reconstitution of clathrin-coated pit budding from plasma membranes. *J. Cell Biol.* 114: 1029–1036.
83a. Lioubin, M.N., Madisen, L., Marquardt, H., Roth, R., Kovacina, K.S., and Purchio, A.F. (1991). Characterization of latent recombinant TGF-beta 2 produced by Chinese hamster ovary cells. *J. Cell Biochem.* 45: 112–121.
84. Lippincott-Schwartz, J., Donaldson, J.G., Schweizer, A., Berger, E.G., Hauri, H.P., Yuan, L.C., and Klausner, R.D. (1990). Microtubule-dependent retrograde transport of proteins into the ER in the presence of Brefeldin A suggests an ER recycling pathway. *Cell* 60: 821–836.
85. Lippincott-Schwartz, J., Yuan, L.C., Bonifacino, J.S., and Klausner, R.D. (1989). Rapid redistribution of Golgi proteins into the ER in cells treated with Brefeldin A: evidence for membrane recycling from Golgi to ER. *Cell* 56: 801–813.
85a. Lippincott-Schwarz, J., Yuan, L., Tipper, C., Amherdt, M., Orci, L., and Klausner, R. (1991). Brefeldin A's effects on endosomes, lysosomes and the TGN suggest a general mechanism for regulating organelle structure and membrane traffic. *Cell* 67: 601–616.
86. Lobel, P., Dahms, N.M., and Kornfeld, S. (1988). Cloning and sequence analysis of the cation-independent mannose 6-phosphate receptor. *J. Biol. Chem.* 263: 2563–2570.
87. Lobel, P., Fujimoto, K., Ye, R.D., Griffiths, G., and Kornfeld, S. (1989). Mutations in the cytoplasmic domain of the 275-kDa mannose 6-phosphate receptor differentially alter lysosomal enzyme sorting and endocytosis. *Cell* 57: 787–796.
88. Ludwig, T., Griffiths, G., and Hoflack, B. (1991). Distribution of newly synthesized lysosomal enzymes in the endocytic pathway of NRK cells. *J. Cell Biol.* 115: 1561–1572.
89. Ludwig, T., Rüther, U., Metzer, R., Copeland, N.G., Jenkins, N.A., Lobel, P., and Hoflack, B. (1992). Gene and pseudogen of the mouse cation-dependent mannose 6-phosphate receptor: genomic organization and chromosomal localization. *J. Biol. Chem.* 267: 12211–12219.
90. Ma, Z., Grubb, J.H., and Sly, W.S. (1991). Cloning, sequencing and functional characterization of the murine 46-kDa mannose 6-phosphate receptor. *J. Biol. Chem.* 266: 10589–10595.
91. MacDonald, R.G., Pfeffer, S., Coussens, L., Tepper, M.A., Brocklebank, C.M., Mole, J.E., Anderson, J.K., Chen, E., Czech, M.P., and Ullrich, A. (1988). A single receptor binds both insulin-like growth factor II and mannose 6-phosphate. *Science* 239: 1134–1137.
92. MacDonald, R.G., Tepper, M.A., Clairmont, K.B., Perregaux, S.B., and Czech, M.P. (1989). Serum form of the rat insulin-like growth factor II/mannose 6-phosphate receptor is truncated in the carboxyl-terminal domain. *J. Biol. Chem.* 264: 3256–3261.

93. Matovcik, L.M., Goodhouse, J., and Farquhar, M.G. (1990). The recycling itinerary of the 46-kDa mannose 6-phosphate receptor-Golgi to late endosomes-coincides with that of the 215-kDa M6PR. *Eur. J. Cell. Biol.* 53: 203–211.
94. Méresse, S., Ludwig, T., Frank, R., and Hoflack, B. (1990). Phosphorylation of the cytoplasmic domain of the bovine cation-independent mannose 6-phosphate receptor. *J. Biol. Chem.* 265: 18883–18842.
95. Miller, K., Shipman, M., Trowbridge, I.S., and Hopkins, C.R. (1991). Transferrin receptors promote the formation of clathrin lattices. *Cell* 65: 621–632.
96. Morgan, D.O., Edman, J.C., Standring, D.N., Fried, V.A., Smith, M.C., Roth, R.A., and Rutter, W.J. (1987). Insulin-like growth factor II receptor as a multifunctional binding protein. *Nature* 329: 301–307.
97. Morris, S.A., Ahle, S., and Ungewickell, E. (1989). Clathrin-coated vesicles. *Current Opinion in Cell Biology* 1: 684–690.
98. Murayama, Y., Okamoto, T., Ogata, E., Asano, T., Iiri, T., Katada, T., Ui, M., Grubb, J.H., Sly, W.S., and Nishimoto, I. (1990). Distinctive regulation of the functional linkage between the human cation-independent mannose 6-phosphate receptor and GTP-binding proteins by insulin-like growth factor II and mannose 6-phosphate. *J. Biol. Chem.* 265: 17456–17462.
99. Murphy, R.F. (1991). Maturation models for endosome and lysosome biogenesis. *Trends in Cell Biol.* 1: 77–82.
100. Neufeld, E. (1991). Lysosomal storage diseases. *Ann. Rev. Biochem.* 60: 257–280.
101. Nishimoto, I., Murayama, Y., Katada, T., Ui, M., and Ogata, E. (1989). Possible direct linkage of the insulin-like growth factor-II receptor with guanine nucleotide-binding proteins. *J. Biol. Chem.* 264: 14029–14038.
102. Oka, Y., Rosek, L.M., and Czech, M.P. (1985). Direct demonstration of rapid insulin-like growth factor II receptor internalization and recycling in rat adipocytes. *J. Biol. Chem.* 260: 9435–9442.
103. Okamoto, T., Katada, T., Murayama, Y., Ui, M., Ogata, E., and Nishimoto, I. (1990). A simple structure encodes G-protein-activating function of the IGF II/mannose 6-phosphate receptor. *Cell* 62: 709–717.
104. Oshima, A., Nolan, C.M., Kyle, J.W., Grubb, J.H., and Sly, W.S. (1988). The human cation-independent mannose 6-phosphate receptor: cloning and sequence of the full-length cDNA and expression of functional receptor in Cos cells. *J. Biol. Chem.* 263: 2553–2562.
105. Owada, M. and Neufeld, E.F. (1982). Is there a mechanism for introducing acid hydrolases into liver lysosomes that is independent of mannose 6-phosphate recognition? Evidence from I-cell disease. *Biochem. Biophys. Res. Comm.* 105: 814–820.
106. Parton, R.G., Prydz, K., Bomsel, M., Simons, K., and Griffiths, G. (1989). Meeting of the apical and basolateral endocytic pathways of the Madin-Dardy canine kidney cell in late endosomes. *J. Cell Biol.* 109: 3259–3272.
107. Pauloin, A., Bernier, I., and Jolles, P. (1982). Presence of cyclic nucleotide-Ca-independent kinase in bovine brain coated vesicles. *Nature* 298: 574–576.
108. Pelham, H.R.B. (1991). Multiple targets for Brefeldin A. *Cell* 67: 449–451.
109. Perdue, J.F., Chan, J.K., Thibault, C., Radaj, P., Mills, B., and Daughaday, W.H. (1983). The biochemical characterization of the detergent-solubilized insulin-like growth factor II receptors from rat placenta. *J. Biol. Chem.* 258: 7800–7811.
110. Pfeffer, S.R. (1987). The endosomal concentration of a mannose 6-phosphate receptor is unchanged in the absence of ligand synthesis. *J. Cell Biol.* 105: 229–234.
111. Pearse, B.M.F. (1988). Receptors compete for adaptors found in plasma membrane coated pits. *EMBO J.* 7: 3331–3336.
112. Pearse, B.M. and Robinson, M.S. (1984). Purification and properties of 100-kDa proteins from coated vesicles and their reconstitution with clathrin. *EMBO J.* 3: 1951–1957.
113. Pearse, B.M. and M.S. Robinson. (1990). Clathrin, adaptors and sorting. *Ann. Rev. Cell. Biol.* 6: 151–171.
114. Pohlmann, R., Nagel, G., Schmidt, B., Stein, M., Lorkowski, G., Krentler, C., Cully, J., Meyer, H.E., Grzeschik, K.H., Mersmann, G., Hasilik, A., and von Figura, K. (1987). Cloning of a cDNA

encoding the human cation-dependent mannose 6-phosphate-specific receptor. *Proc. Natl. Acad. Sci. USA* 84: 5575–5579.

115. Purchio, A.F., Cooper, J.A., Brunner, A.M., Lioubin, M.N., Gentry, L.E., Kovacina, K.S., Roth, R.A., and Marquardt, H. (1988). Identification of mannose 6-phosphate in two asparagine-linked sugar chains of the recombinant Transforming Growth Factor-β1 precursor. *J. Biol. Chem.* 263: 14211–14215.
116. Prydz, K., Brändli, A., Bomsel, M., and Simons, K. (1990). Surface distribution of the mannose 6-phosphate receptors in epithelial Madin-Darby canine kidney cells. *J. Biol. Chem.* 265: 12629–12635.
117. Robbins, A.R. (1988). In: Protein Transfer and Organelle Biogenesis, pp. 463–520. Academic Press, Orlando, FL.
118. Robinson, M.S. (1987). 100-kDa coated vesicle proteins: molecular heterogeneity and intracellular distribution studied with monoclonal antibodies. *J. Cell Biol.* 104: 887–895.
119. Rogers, S.A. and Hammerman, M.R. (1989). Mannose 6-phosphate potentiates insulin-like growth factor II-stimulated inositol triphosphate production in proximal tubular basolateral membranes. *J. Biol. Chem.* 264: 4273–4276.
120. Rome, L.H., Weissmann, B., and Neufeld, E.F. (1979). Direct demonstration of binding of a lysosomal enzyme, α-iduronidase, to receptors on cultured fibroblasts. *Proc. Natl. Acad. Sci. USA* 76: 2331–2334.
121. Roth, J. (1987). Subcellular organization of glycosylation in mammalian cells. *Biochem. Biophys. Acta* 906: 405–436.
122. Roth, R.A. (1988). Structure of the receptor for insulin-like growth factor II: the puzzle amplified. *Science* 239: 1269–1271.
123. Sahagian, G.G., Distler, J., and Jourdian, G.W. (1981). Characterization of a membrane-associated receptor from bovine liver that binds phosphomannosyl residues of bovine testicular β-galactosidase. *Proc. Natl. Acad. Sci. USA* 78: 4289–4293.
124. Sahagian, G.G. and Neufeld, E.F. (1983). Biosynthesis and turnover of the mannose 6-phosphate receptor in cultured Chinese hamster ovary cells. *J. Biol. Chem.* 258: 7121–7128.
125. Scheel, G. and Herzog, V. (1989). Mannose 6-phosphate receptor in porcine thyroid follicle cells. Localization and possible implication for the intracellular transport of thyroglobulin. *Eur. J. Cell Biol.* 49: 140–148.
126. Schmid, S., Fuchs, R., Male, P., and Mellman, I. (1988). Two distinct subpopulations of endosomes involved in membrane recycling and transport to lysosomes. *Cell* 52: 73–83.
127. Schmid, S.L. and Smythe, S. (199). Stage-specific assays for coated-pit formation and coated vesicle budding *in vitro*. *J. Cell Biol.* 114: 1017–1028.
128. Schröder, S. and Ungewickell, E. (1991). Subunit interaction and function of clathrin-coated vesicles adaptors from the Golgi and the plasma membrane. *J. Biol. Chem.* 266: 7910–7918.
129. Sklar, M., Kiess, M.W., Thomas, C.L., and Nissley, P.S. (1989). Developmental expression of the tissue insulin-like growth factor II/mannose 6-phosphate receptor in the rat measurement by quantitative immunoblotting. *J. Biol. Chem.* 264: 16733–16738.
130. Stein, M., Braulke, T., Krentler, C., Hasilik, A., and von Figura, K. (1987). 46-kDa mannose 6-phosphate-specific receptor: biosynthesis, processing, subcellular location and topology. *Biol. Chem. Hoppe-Seyler.* 368: 937–947.
131. Stein, M., Meyer, H.E., Hasilik, A., and von Figura, K. (1987). 46-kDa mannose 6-phosphate-specific receptor: Purification, subunit composition, chemical modification. *Biol. Chem. Hoppe-Seyler.* 368: 927–936.
132. Stein, M., Zijderhand-Bleekemolen, J.E., Geuze, H., Hasilik, A., and von Figura, K. (1987). M_r 46,000 mannose 6-phosphate specific receptor: its role in targeting of lysosomal enzymes. *EMBO J.* 6: 2677–2681.
133. Szebenyi, G. and Rotwein, P. (1991). Differential regulation of the mannose 6-phosphate receptor and their ligands during myogenic development of C2 cells. *J. Biol. Chem.* 266: 5534–5539.
134. Todderud, G. and Carpenter, G. (1988). Presence of mannose phosphate on the epidermal growth factor receptor in A-431 cells. *J. Biol. Chem.* 263: 17893–17896.

135. Tong, P.Y., Gregory, W., and Kornfeld, S. (1989). Ligand interactions of the cation-independent mannose 6-phosphate receptor: the stoichiometry of mannose 6-phosphate binding. *J. Biol. Chem.* 264: 7962–7969.
136. Tong, P.Y. and Kornfeld, S. (1989). Ligand interactions of the cation-dependent mannose 6-phosphate receptor. Comparison with the cation-independent mannose 6-phosphate receptor. *J. Biol. Chem.* 264: 7970–7975.
137. Tong, P.Y., Tollefsen, S.E., and Kornfeld, S. (1988). The cation-independent mannose 6-phosphate receptor binds insulin-like growth factor II. *J. Biol. Chem.* 263: 2585–2588.
138. Varki, A. and Kornfeld, S. (1983). The spectrum of anionic oligosaccharides released by *endo*-β-N-acetylglucosaminidase from glycoproteins: structural studies and interactions with the phosphomannosyl receptor. *J. Biol. Chem.* 258: 2808–2818.
139. Varki, A., Reitman, M.L., and Kornfeld, S. (1981). Identification of a variant of mucolipidosis III (pseudo-Hurler polydystrophy): a catalytically active N-acetylglucosaminylphosphotransferase that fails to phosphorylate lysosomal enzymes. *Proc. Natl. Acad. Sci. USA* 78: 7773–7777.
140. von Figura, K. (1991). Molecular recognition and targeting of lysosomal proteins. *Current Opinion in Cell Biology* 3: 642–646.
141. von Figura, K., Gieselmann, V., and Hasilik, A. (1984). Antibody to mannose 6-phosphate specific receptor induces receptor deficiency in human fibroblasts. *EMBO J.* 3: 1281–1286.
142. von Figura, K. and Hasilik, A. (1986). Lysosomal enzymes and their receptors. *Ann. Rev. Biochem.* 55: 167–193.
143. Waheed, A., Hille, A., Jurghans, U., and von Figura, K. (1990). Quaternary structure of the M_r 46,000 mannose 6-phosphate specific receptor: effect of ligand, pH, and receptor concentration on the equilibrium between dimeric and tetrameric receptor forms. *Biochemistry* 29: 2449–2455.
143a. Waheed, A., Pohlmann, R., Hasilik, A., von Figura, K., van Elsen, A., and Leroy, J. (1982). Deficiency of UDP-N-acetylglucosamine:lysosomal enzyme N-acetylglucosamine-1-phosphotransferase in organs of I-cell patients. *Biochem. Biophys. Res. Commun.* 105: 1052–1058.
144. Waheed, A. and von Figura, K. (1990). Rapid equilibrium between the monomeric, dimeric and tetrameric forms of the 46-kDa mannose 6-phosphate receptor at 37°C. Possible relation to the function of the receptor. *Eur. J. Biochem.* 193: 47–54.
145. Watanabe, H., Grubb, J.H., and Sly, W.S. (1990). The overexpressed human 46-kDa mannose 6-phosphate receptor mediates endocytosis and sorting of β-glucuronidase. *Proc. Natl. Acad. Sci. USA* 87: 8036–8040.
146. Wendland, M., Hille, A., Nagel, G., Waheed, A., von Figura, K., and Pohlmann, R. (1989). Synthesis of a truncated M_r 46,000 mannose 6-phosphate receptor that is secreted and retains ligand binding. *Biochem. J.* 260: 201–206.
147. Wendland, M., von Figura, K., and Pohlmann, R. (1991). Mutational analysis of disulfide bridges in the M_r 46,000 mannose 6-phosphate receptor. Localization and role for ligand binding. *J. Biol. Chem.* 266: 7132–7136.
148. Wendland, M., Waheed, A., Schmidt, B., Hille, A., Nagel, G., von Figura, K., and Pohlmann, R. (1991). Glycosylation of the M_r 46,000 mannose 6-phosphate receptor: Effect on ligand binding, stability, and conformation. *J. Biol. Chem.* 266: 4598–4604.
148a. Wenk, J., Hille, A., and von Figura, K. (1991). Quantitation of M_r 46,000 and M_r 300,000 mannose 6-phosphate receptors in human cells and tissues. *Biochem. Int.* 23: 723–731.
149. Westlund, B., Dahms, N.M., and Kornfeld, S. (1991). The bovine mannose 6-phosphate/insulin-like growth factor II receptor: localization of mannose 6-phosphate binding sites to domains 1-3 and 7-11 of the extracytoplasmic region. *J. Biol. Chem.* 266: 23233–23239.
150. Willingham, M., Pastan, I.H., and Sahagian, G. (1983). Ultrastructural immunocytochemical localization of the phosphomannosyl receptor in Chinese Hamster Ovary (CHO) cells. *J. Histochem. Cytochem.* 31: 1–11.
151. Wood, S.A., Park, J.E., and Brown, W.J. (1991). Brefeldin A causes a microtubule-mediated fusion of the *trans*-Golgi network and early endosomes. *Cell* 67: 591–600.

CHEMISTRY OF LYSOSOMAL CYSTEINE PROTEINASES

Robert W. Mason and Donna Wilcox

Advances in Cell and Molecular Biology of Membranes
Volume 1, pages 81–116

ISBN: 1-55938-362-3

ABSTRACT

The major mammalian lysosomal cysteine proteinases identified to date are cathepsins B, H, L, and S. They are soluble enzymes and chiefly are packaged into lysosomes by the mannose 6-phosphate dependent pathway. Each enzyme is from a different gene product and can be differentiated by catalytic criteria, although the differences between cathepsins L and S are rather subtle. The major physiological roles for these enzymes are undoubtedly the hydrolysis of extracellular proteins taken up into the cell by endocytosis and hydrolysis of intracellular proteins taken up by autophagy. There is, however, considerable circumstantial evidence that these enzymes play a pathological role in degenerative diseases and may act outside the cell. The major problems that need to be overcome by secreted enzymes, are a means of activation of the secreted zymogens and a way to avoid inhibition by extracellular inhibitors. Current research is aimed at elucidating the mechanism by which this occurs.

I. INTRODUCTION

Proteases (or peptidases) are defined as enzymes that hydrolyze peptide bonds. They can be divided into two main groups consisting of enzymes that cleave peptide bonds located at the N- or C-termini of proteins (exopeptidases) and enzymes that cleave peptide bonds within proteins (endopeptidases or proteinases). The exopeptidases are named according to their substrate specificity, as is standard for enzyme nomenclature (*100,101*). Thus amino-peptidases hydrolyze amino acids sequentially from the N-terminus, and carboxypeptidases hydrolyze amino-acids sequentially from the C-terminus of proteins. Enzymes that sequentially remove dipeptides from the C-terminus are termed peptidyl dipeptidases, and those that sequentially remove dipeptides from the N-terminus are called dipeptidyl peptidases. Although this terminology has been adopted by scientists in the protease field, it might have been less ambiguous to the nonexpert if terms such as amino-dipeptidase and carboxy-dipeptidase had been adopted, as such names

Table 1. The Nomenclature of the Cathepsins

Cathepsin	*Catalytic type*	*New name ?*
A	serine	carboxypeptidase A
B	cysteine	
C	cysteine	dipeptidyl peptidase I
D	aspartic	
E	aspartic	
F	?	?
G	serine	
H	cysteine	
I	cysteine	cathepsin H?
J	cysteine	cathepsin B?
K	cysteine	cathepsin B?
L	cysteine	
M	cysteine	cathepsin B?
N	cysteine	cathepsin S/L?
P	cysteine	cathepsin B
R	serine	
S	cysteine	
T	cysteine	cathepsin L

Notes: This table lists all of the cathepsins that have been named to date. The new name indicates either what has become accepted as a better name for a given enzyme (no ?), or what enzyme we believe is most likely responsible for the reported activity. A ? indicates that the new name proposed may be controversial. For more details of individual enzymes, see the text.

clearly state which end of the protein that the dipeptides are removed from. Such terminology has occasionally been used (*101*). Some of the exopeptidases exhibit a preference for particular amino acids and thus they can be named according to the preferred substrate (e.g. leucyl aminopeptidase). Lysosomes contain a range of exopeptidases including lysosomal carboxypeptidase A and dipeptidyl peptidase 1 which were previously named cathepsins A and C, respectively (see *101* and Table 1). (For further discussion on the nomenclature of exopeptidases, see reference *100*). The exopeptidases have generally been considered to be responsible for the later stages in protein turnover, and thus have been less well studied than the endopeptidases. However, after limited proteolysis by the endopeptidases, these enzymes may be responsible for most of the peptide bonds hydrolyzed in the cell.

Using the same nomenclature, all of the endopeptidases would be classified as the same type of enzymes, as they all hydrolyze internal peptide bonds. These are further classified according to kinetic mechanism (*17*). There are four known catalytic mechanisms, involving aspartic, cysteine, or serine side chains, or a bound metal ion (usually zinc). A fifth category of enzymes is used when either the catalytic mechanism is unknown or the criteria for classifications are unclear. Aspartic proteinases are usually classified as enzymes that can be inhibited by pepstatin, a microbial peptidyl inhibitor of pepsin; cysteine proteinases as enzymes that can be inhibited by L-3-carboxy-2,3-*trans*-epoxypropionyl-leucylamido-(4-guanidino)butane (E-64); serine proteinases as enzymes that can be inhibited by diisopropylfluorophosphate (DFP) or dichloroisocoumarin, and metallo-proteinases by chelators such as EDTA, 1,10-phenanthroline, or phosphoramidon. These inhibitors will not necessarily inhibit all enzymes of each class, but are the safest to use to be sure of the correct classification. The actual names given to the proteinases are generally randomly chosen, such as papain, trypsin, collagenase, neutrophil elastase, and cathepsin B. Such terminology is generally uninformative or even misleading, but unfortunately historical precedence and uncertainty as to the physiological substrates for many of these enzymes may well confound efforts to solve these problems.

II. THE DISCOVERY OF LYSOSOMAL PROTEINASES

Cathepsin B was identified as a proteinase from tissue extracts long before the discovery of lysosomes, being given the name cathepsin II in 1941 (*51*). The name cathepsin B was not used until 1951 (*144*). Cathepsin D was also detected in the late 1930s, but was not named until 1960 (*121*). Cathepsin D is an aspartic proteinase, having two essential aspartic acid residues that are required for peptide bond hydrolysis. Cathepsin B is a completely different type of proteinase, having an essential cysteine (hence its classification as a cysteine proteinase), which is involved in the nucleophilic attack of the peptide bond.

As mentioned in the preceding section, the name cathepsin is unfortunately not very informative since it also includes cathepsin G, a serine proteinase from azurophil granules, and cathepsin E, an aspartic proteinase found in the stomach. An early definition of a cathepsin as an intracellular proteinase also seems inappropriate and uninformative as there are now known to be many different intracellular proteinases (not all of them being called cathepsins), and also many of the cathepsins may act outside cells (Table 1). The majority of the enzymes that have been given the name cathepsin are however cysteine proteinases that are related to cathepsin B and papain by sequence and catalytic mechanism. The best characterized of these are cathepsins B, H, L, and S.

Since the discovery of lysosomes, it has become clear that these organelles contain a large number of hydrolases that are capable of degrading proteins and lipids down to their monomeric units for reutilization in biosynthetic processes. The lysosome is therefore an organelle that is responsible for the final degradation of components that have been taken up from either the extracellular environment or from within the cell by endocytosis or autophagy, respectively. Molecules that have been taken up into lysosomes are already removed from their functional location and therefore hydrolysis should be considered to be a recycling process rather than an inactivation process. There are over 50 different hydrolases in the mammalian lysosome. They share a number of common features that distinguish them from other cellular proteins. Lower eucaryotes such as yeast, plants, slime molds, and protozoa also contain lysosome-like organelles, but appear to have different packaging mechanisms and probably different functions. This review will focus on the mammalian lysosomal cysteine proteinases as models of lysosomal enzymes and discuss their possible functions in lysosomes and other cellular compartments.

III. THE CATALYTIC MECHANISM OF CYSTEINE PROTEINASES

The lysosomal cysteine proteinases belong to a family of sequence-related proteins from very diverse species, such as plants (*56*), microorganisms (*115*), and animals (*72*). The most studied cysteine proteinase, to date, is the plant proteinase papain (EC 3.4.22.2), which is found in the latex of the unripe papaya fruit (*Carica papaya*).

All cysteine proteinases have an essential cysteine in their active site (Cys 25 in papain). The role of the thiol group of cysteine in the reaction mechanism was first demonstrated by Finkle and Smith (*48*). A histidine residue has also been shown to be involved in catalysis. It is involved in the formation of an ion pair between the thiol group of Cys 25 and the imidazole group of His 159 resulting in the formation of a thiolate ion (S^-) (*118*). The thiolate ion acts as a nucleophile, which attacks the peptide bond, resulting in the formation of an acylated intermediate (*88*, Fig. 1). The histidine side chain also probably donates a proton to the substrate to facilitate the cleavage step during catalysis.

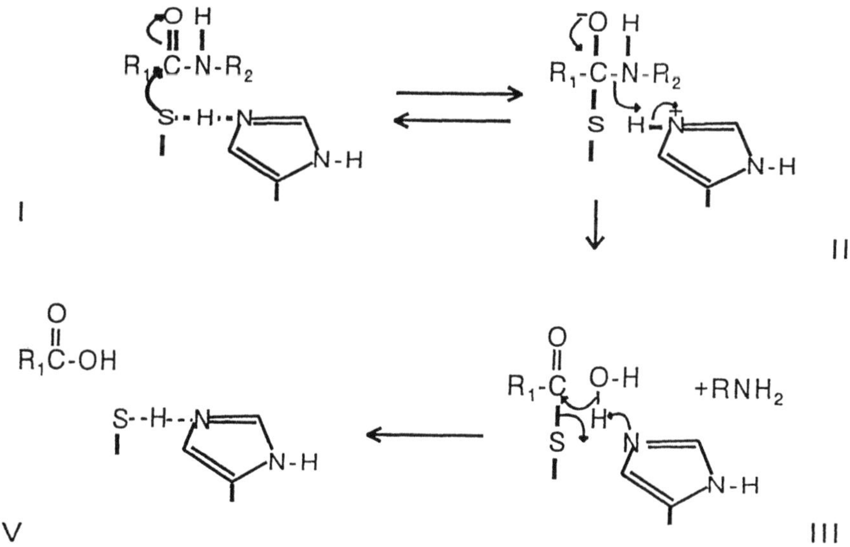

Figure 1. The mechanism of action of cysteine proteinases. **I**. The thiolate ion of the active-site cysteine attacks the carbonyl group of the scissile bond. **II**. Cleavage is catalyzed by the transfer of a hydrogen from the histidine to the amino group. **III**. The acyl group is hydrolyzed by a water molecule. **IV**. The enzyme is regenerated.

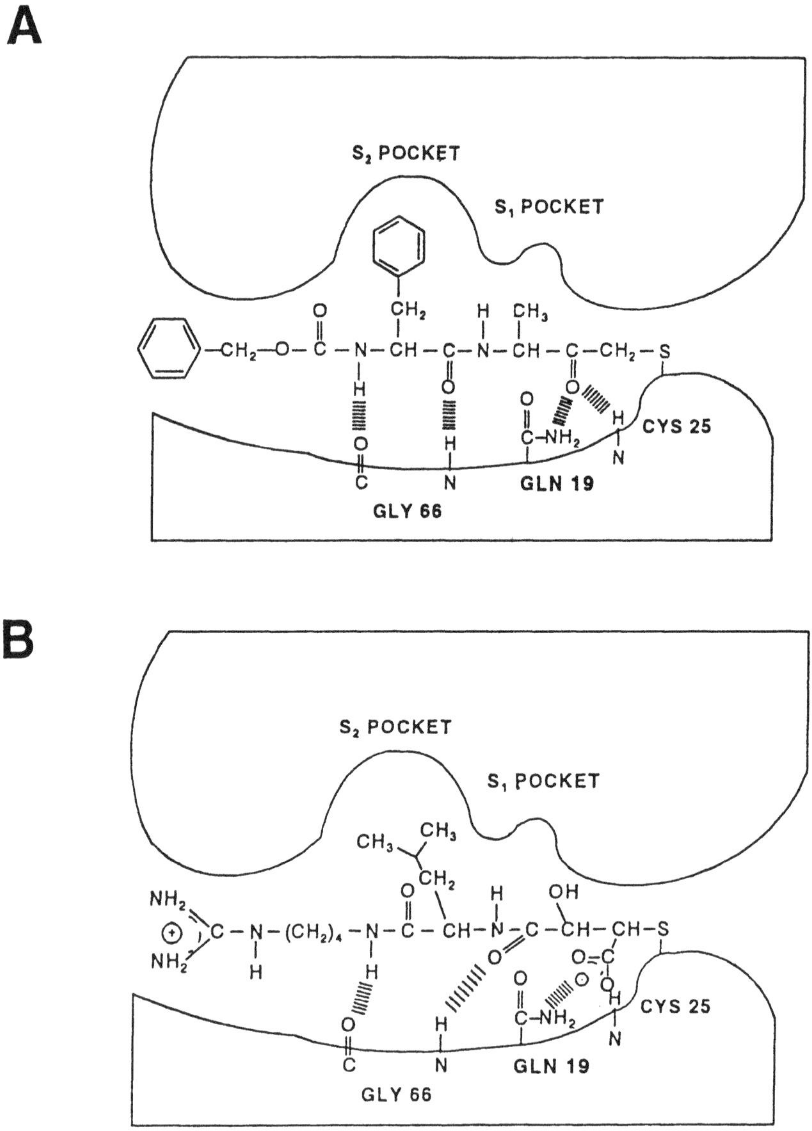

Figure 2. The mode of binding of Z-Phe-Ala-CH2Cl and E-64 to Papain. This figure shows models of how two inhibitors appear to bind to papain, as determined from X-ray crystallographic data. Important features to note are anti-parallel binding of the phenylalanine backbone of the Z-Phe-Ala-CH2Cl to the glycine backbone of the enzyme compared to the parallel binding of the backbone of leucine in E-64 to the same residue of the enzyme. This figure is based on that given in reference 151.

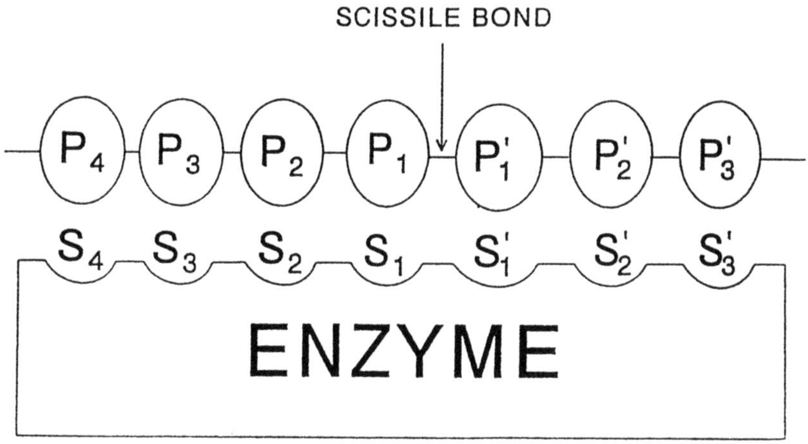

Figure 3. Model for the binding of substrates to proteinases. This figure is based on that proposed by Schecter and Berger (*130*) for the binding of substrates to papain. P_1, P_2, P_3 and P_4 refer to amino acids on the N-terminal side of the scissile bond in the substrate, and P'_1, P'_2, P'_3, and P'_4, refer to amino acids on the C-terminal side of the scissile bond. S and S′refer to sites in the enzyme that bind the corresponding residues.

The catalytic site of papain has been defined by the use of X-ray crystallographic data (*41,71*). The crystal structure of actinidin (*9*) has also been determined, but to date the crystal structures of the lysosomal cysteine proteinases have not been elucidated. Rat cathepsin B has been crystallized and therefore structural information may soon become available (*84*). It had been proposed, based on studies of the catalytic mechanism, that the His 159 and Cys 25 are in close proximity, and the X-ray data confirmed this. The structure of papain bound to an inhibitor, Z-Phe-Ala-CH_2Cl, suggests that the NH_2- group of a glutamine residue at position 19 of papain and the -NH- group of the active site cysteine are hydrogen bonded to the carbonyl group of the alanine of the inhibitor, suggesting that this "oxyanion hole" serves to bind and coordinate the carbonyl group of the bond to be hydrolyzed in the substrate (Fig. 2). This binding site would also stabilize the intermediate shown in Figure 1. The conservation of this amino acid in all of the cysteine proteinases supports this suggestion.

Papain has been shown to consist of two domains of equal size and the active site is formed by a cleft between the two domains. Amino acids from both domains are involved in the binding of substrates. A model for the papain substrate binding site was proposed by Schechter and Berger (*130*). This has been widely accepted as a convenient model for the study of most proteinase-substrate interactions. This model has been applied to the lysosomal cysteine proteinases and has been used to define the interaction of substrates and inhibitors. The model proposes that papain consists of up to seven subsites, a subsite being defined as the portion of the enzyme which reacts with one side chain of the substrate molecule. This is illustrated in Figure 3. The enzyme subsites are denoted by S_n and the peptide side chains which interact with the enzyme's subsites being denoted by P_n.

IV. THE PROPERTIES OF THE LYSOSOMAL CYSTEINE PROTEINASES

A. Cathepsin B

Purification

Cathepsin B is found in a wide range of tissues and has been purified from several sources, including rabbit liver, chicken liver, bovine spleen, sheep thyroid, and human liver (reviewed in *72*). Purified human and rat cathepsin B do not bind to concanavalin A-Sepharose, but they do contain a small amount of carbohydrate (*72,143*). Cathepsin B is unstable at alkaline or neutral pH, which has become accepted as a general feature (but see cathepsins S and H) of lysosomal cysteine proteinases (*13*). However, the enzyme is active at neutral pH against synthetic substrates, exhibiting activity at pH 8.0 (*106*). The purification of cathepsin B has been improved by affinity chromatography using a reversible inhibitor, the semicarbazone of Gly-Phe-glycinal, linked to Sepharose 4B. The eluted material was 80 to 100% active (*125*). The major advantages of this technique are that only active forms of the enzyme bind and that the enzyme is released by covalent binding to a disulfide, which renders the enzyme inactive, preventing potential autolysis. The enzyme is easily reactivated by the addition of reducing agents prior to use.

Specificity

The specificity of cysteine proteinases has largely been determined by activity against synthetic substrates and inhibitors. The advantage of such compounds over protein substrates is that kinetic data is easier to obtain to enable meaningful comparisons. "Preferred cleavage sites" in protein substrates such as the oxidized B-chain of insulin (a favorite polypeptide for many groups) can indicate some preferences, but these depend on the secondary and tertiary structures of the proteins as well as the combination of a range of amino acids that could bind to several potential subsites in the enzymes. Such studies have shown a general preference for hydrophobic amino acids in P_2 and P_3 (*103*), but have not given definitive answers to the specificity of cathepsin B. This is perhaps not surprising, because if cathepsin B has seven subsites, as proposed for papain, then the relative contributions of so many subsites would require a study with 20^7 different peptides! Studies with di- and tripeptide substrates and inhibitors therefore seem more feasible.

The characterization of cathepsin B was greatly facilitated by the development of the synthetic substrate, Bz-Arg-NH_2 (*51*). This was known to be a substrate for the serine proteinase, trypsin, but this enzyme is not active at acidic pH and is not activated by reducing agents. A more convenient substrate now available is Z-Phe-Arg-NHMec, which was also originally developed as a substrate for a serine proteinase, kallikrein (*14*). This is however also a substrate for cathepsins L and S, so a more selective substrate, Z-Arg-Arg-NHMec, is now used (*16*). It must be

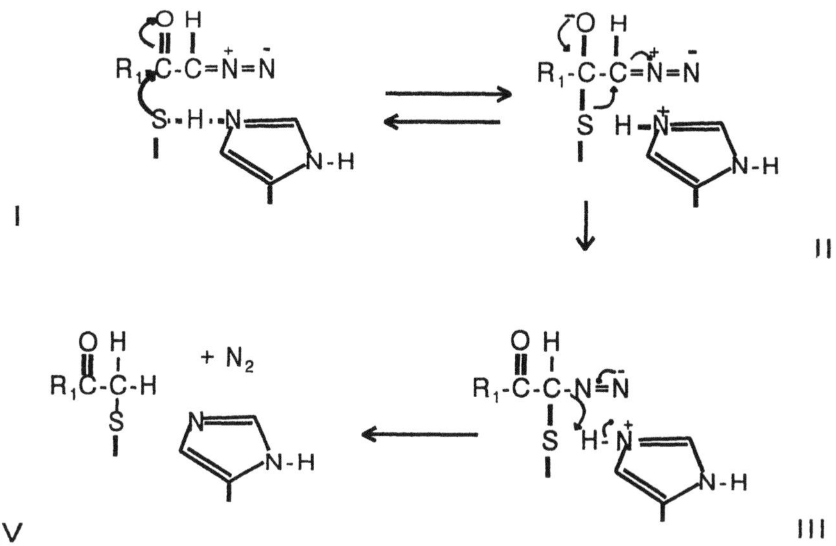

Figure 4. Mechanism for the binding of diazomethane inhibitors to cysteine proteinases. The mode of binding diazomethane (and chloromethane) inhibitors to cysteine proteinases is similar to that seen for the binding of substrates (Fig. 1). Thus the initial attack by the thiolate ion is the same (I). However the methylene group is not hydrolyzable and the second stage appears to involve a rearrangement that results in the transfer of the bond from the carbonyl group to the methylene group (II). The histidine side chain then donates a hydrogen, resulting in the loss of nitrogen gas and the enzyme is irreversibly alkylated (III, IV). The reason that these inhibitors have proven to be rather specific for cysteine proteinases, may be due to these enzymes being more flexible around the active-site cysteine due to the lower contribution to binding of amino acid side chains in P_1.

remembered that none of these substrates are sufficiently specific to be used to unequivocally identify cathepsin B in crude tissue extracts. Although these are the commonly used substrates for cathepsin B, they do not truly reflect the specificity of this enzyme, but rather are the most conveniently available substrates.

A systematic exploration of the primary specificity site of cathepsin B has been performed using diazomethane inhibitors of the type Z-Phe-X-CHN_2 (*135*). Such compounds bind to cysteine proteinases in a substrate-like manner, but alkylate the active-site cysteine and are not hydrolyzed (Fig. 4). They found that compounds containing tryptophan and nitro-phenylalanine in the position denoted by "X" were poorer inhibitors, indicating that cathepsin B could not accommodate bulky side chains in P_1. Interestingly, *O*-benzyl threonine was a very good inhibitor of cathepsin B. It was proposed that this might be due to a more remote binding site for the bulky benzyl group that could not be reached by the nitro-phenylalanyl side chain (*135*). It can therefore be concluded that the S_1 binding site of cathepsin B is restricted to less bulky amino acids, but that there is another pocket nearby that can occupy more bulky groups. Arginine fits into this category and both chloromethanes and diazomethanes containing this amino acid in P_1 are good inhibitors of cathepsin B. However, the positive charge on the arginine does not seem to contribute significantly to the binding of cathepsin B as the uncharged nitroarginine containing chloromethane is as effective an inhibitor as unmodified arginine containing chloromethane (*135*). We have found that inhibitors that contain iodo-tyrosine in P_1 do not bind to cathepsin B *in vitro* or *in vivo* (*155*), further supporting the observation that cathepsin B discriminates against amino acids with bulky side chains in P_1.

By contrast, the S_2 subsite seems to prefer bulky side chains. Tyrosine and phenylalanine in P_2 give good inhibitors of cathepsin B and addition of a bulky iodine group to the tyrosine of Z-Tyr-Ala-CHN_2 actually gives a 10-fold better inhibitor of this enzyme (*31*).

The other major type of inhibitors used to determine the specificity of cysteine proteinases are the peptidyl epoxides. These proved of limited use in distinguishing individual enzymes in earlier studies, and were thought to bind in the S′ subsites of the enzymes. The determination of the crystal structure of a papain/E-64 complex showed that in fact E-64 binds in the S_2 and S_3 subsites of papain (*151*, Fig. 2). In contrast to Z-Phe-Ala-CH_2Cl which binds anti-parallel to glycine residue 66 in papain, E-64 binds in a parallel fashion. The breaking of the epoxide ring leaves a hydroxyl group in the position that would be occupied by P_1 in other substrates and inhibitors, and the free carboxyl group on E-64 seems to bind to the oxyanion hole that is occupied by the carbonyl group of substrates to be cleaved or the ketone group of the chloro- and diazomethane inhibitors (see Fig. 2). This might explain the rather general nature of E-64 as a cysteine proteinase inhibitor, having no bulky groups to hinder binding in P_1, and a small hydrophobic group (leucine) in P_2. Since the crystal structure of the epoxide inhibitor with papain became known, it was apparent that a rational basis of design could be made to devise selective inhibitors of cathepsin B. Inhibitors that have the carboxyl group of a derivative of E-64 esterified, are equally as effective as inhibitors of cathepsin B as those that are not esterified (*107*). Even bulky groups such as hexyl amide and phenyl amide are accepted. Where this group binds is unknown, but it would appear likely that it would bind near to the active-site cysteine if the carbonyl group still binds in the oxyanion hole as seen for the non-esterified inhibitors. It is unlikely that it would bind in the S_1 subsite, as we know from the above that cathepsin B does not tolerate bulky groups here. Indeed, this site may be the same site that the extended bulky group in P_1 of the diazomethane groups occupy, as mentioned earlier. A model of the active site of cathepsin B based on these observations is shown in Figure 5a.

Epoxide inhibitors containing proline in P_2 and P_3 were good inhibitors of cathepsin B, which is in marked contrast to results found for a diazomethane inhibitor with proline in P_2 which was a very poor inhibitor of cathepsin B (*107,153*). The conformation of proline residues are severely restricted, and thus as the backbone peptide chain of epoxide inhibitors bind in the opposite direction to those of the ketone inhibitors, it is quite likely that prolines in analogous positions

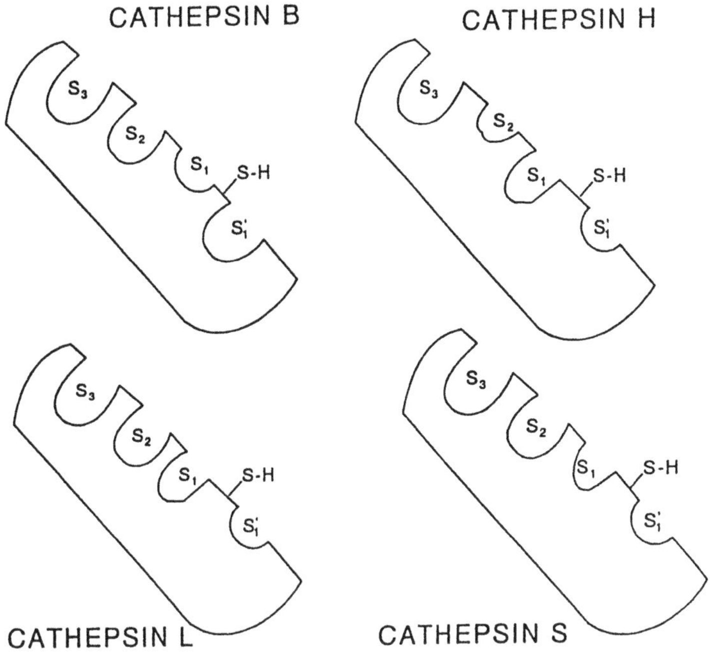

Figure 5. Comparisons between the active sites of lysosomal cysteine proteinases. These models are based on the substrate and inhibitor specificities of cathepsins B, H, L, and S. The distinguishing feature of cathepsin B is a small S_1 pocket but a larger pocket near to the active-site cysteine that may be in S_1' as drawn. Unlike the other proteinases, cathepsin H has only a small S_2 pocket. Cathepsin L is unique in having large pockets in S_1, S_2 and S_3. Cathepsin S is like cathepsin L, but has slightly more restricted S_1 and probably S_2 pockets. For more details see the text.

will not be in the same orientation and, hence, probably not contact the enzyme in the same subsites. The amide and carbonyl groups of the leucine of E-64 do not seem to be hydrogen bonded to papain, while those of phenylalanine in Z-Phe-Ala-CHN_2 are, so proline substituted in this P_2 position would result in loss of binding for the chloromethane, but not for E-64 (*151*) (see Fig. 2). Synthetic substrates with proline in P_2 are not good substrates for cathepsin B (*23*). It would therefore seem that analogues of E-64 will be unable to predict accurately the specificity of cysteine proteinases for substrates.

In addition to its activity as an endopeptidase, cathepsin B has a peptidyl dipeptidase activity, that is, it sequentially removes dipeptides from the C-termini of proteins. This has been shown for the degradation of glucagon (*6*) and fructose-1,6-bis-phosphate aldolase (*22*). It has been suggested that porcine cathepsin B degrades proteins by this exopeptidase activity only (*141*). This is probably not true, as the cleavage of and trapping by α_2-macroglobulin by porcine cathepsin B has been shown (*92*); a mechanism that requires the cleavage of an internal peptide bond of α_2-macroglobulin, providing evidence that at least for one protein, porcine cathepsin B acts as an endopeptidase. Therefore cathepsin B probably degrades proteins by both endo- and exopeptidase action. The relative contribution of these two activities to the degradation of proteins by cathepsin B *in vivo* remains unclear.

B. Cathepsin L

Purification

Cathepsin L (EC 3.4.22.15) was first purified from rat liver lysosomes (*37,73,147*). It was identified by its hydrolysis of the general proteinase substrate, azocasein, and its failure to hydrolyze Bz-Arg-NH_2. This distinguished it from the then better characterized cysteine proteinase, cathepsin B. Subsequently, it has been purified from several other species, including rabbit liver (*93*), human liver (*94*), and chicken liver (*42*). The purification procedure from lysosomes involved lysis, followed by gel filtration and cation exchange chromatography (*73*). Cathepsin L demonstrates anomalous tight binding to cation exchange resins, thus enabling its separation from both cathepsins B and H. The purification from whole tissue extracts involved a critical pH 4.2 treatment to release active enzyme (*94*). The major reason for the need for a special extraction technique is probably due to the fact that cathepsin L, unlike cathepsin B, binds very tightly to most of the known protein inhibitors that occur in mammalian tissues, and that the low pH disrupts the enzyme-inhibitor complexes by an (as yet) uncharacterized mechanism. Naturally occurring inhibitors are probably the major contributing factors that have hampered the purification or even identification of many cellular proteinases. Another problem in purification is the heterogeneity of the lysosomal enzymes, most probably due to the extensive posttranslational modifications that they undergo. Thus cathepsin L from rat and man consists of multiple pI forms (*73,94*) and only partially binds to concanavalin A-Sepharose (*74,94*).

Specificity

The specificity of bond cleavage by cathepsin L has been investigated using several protein substrates. This demonstrated that hydrophobic residues in P_2 and P_3 and polar residues in P_1 may increase susceptibility to cleavage by cathepsin L (*146*).

The development of specific synthetic substrates for cathepsin L has proved difficult, due to hydrolysis of these by cathepsin B (*16*). Z-Phe-Arg-NHMec has been used to assay cathepsin L, however, even though it is hydrolyzed by cathepsin B. Cathepsin B can be accurately quantified using Z-Arg-Arg-NHMec, and hence the Z-Phe-Arg-NHMec activity attributable to cathepsin L can be calculated. The kinetic parameters for the hydrolysis of this substrate differ for cathepsins B and L, with the K_M values being above 200 μM and approximately 2 μM respectively.

Irreversible inhibition by diazomethane inhibitors has shown that cathepsin L, in contrast to cathepsin B, can accommodate large hydrophobic groups in P_1

(*31,77*). Thus an active site directed inhibitor, Z-Leu-Leu-[^{125}I]Tyr-CHN_2 can discriminate between cathepsins B and L in living cells (*155*). Cathepsin L can also accommodate large bulky groups in P_2. Like cathepsin B, iodo-tyrosine in P_2 gives a better inhibitor of cathepsin L than the non-iodinated inhibitor (*31*). In a re-examination of the ability of cathepsin L to degrade protein substrates, it was also found that cathepsin L prefers bulky side chains in P_3 (*79*). The ability of cathepsin L to accommodate large groups in P_1, P_2, and P_3, are probably responsible for the broad specificity of this enzyme. Analogues of E-64 with the carboxyl group (that binds in the oxyanion hole of papain) esterified, were poor inhibitors of cathepsin L, suggesting that the site around the active-site cysteine is more restricted than that of cathepsin B (*107*). This has led to the development of a very specific inhibitor of cathepsin B.

This information allows a model of the architecture of the active-site of cathepsin L to be compared with that of cathepsin B (see Fig. 5). Amino acids in P_1 discriminate between cathepsins B and L with inhibitors, and therefore it could be anticipated that a specific substrate could be designed for cathepsin L. Such a substrate has not yet been found, and may be difficult to find due to the proximity of the proposed S_1 and S'_1 subsites. If the bulky group attached to the epoxide inhibitor actually binds in an alternative S_1 subsite rather than S'_1, then a substrate may bind differently to the inhibitors and be able to utilize this alternative pocket and be hydrolyzed by both cathepsins B and L. It should be noted that there is a difference between acylation of the carbonyl group of a substrate and the acylation by the methyl group of the diazomethanes (see Figs. 1 and 3). Further structural and kinetic data are required to resolve this problem.

C. Cathepsin H

Purification

Cathepsin H has been purified from several tissues and is found in a range of species (reviewed in *72*). The purification of human cathepsin H involved autolysis of tissue homogenates at pH 4.2, followed by ion exchange, gel filtration, hydroxyapatite chromatography and finally concanavalin A Sepharose (*131*). This showed that cathepsin H was a single polypeptide of M_r 28,000 and a glycoprotein.

Specificity

Cathepsin H has both aminopeptidase and endopeptidase activity. Its endopeptidase action is limited, but it is able to degrade azocasein by this mechanism (*131*). The aminopeptidase action has been demonstrated by its ability to degrade synthetic substrates with a free amino terminus, the current commonly used substrate being Arg-NHMec (*131*). The pH optimum for this substrate is 6.8, which reflects the fact that this enzyme is more stable to elevated pH values than either cathepsins B or L.

We have recently found that bovine cathepsin H actually has significant activity against the endopeptidase substrate, Bz-Phe-Val-Arg-NHMec, which is also a good substrate for cathepsin S (see below). Furthermore, Z-Leu-Leu-[^{125}I]Tyr-CHN_2 is a good affinity label of cathepsin H, whereas the excellent inhibitor of cathepsin L, Z-[^{125}I]Tyr-Ala-CHN_2, does not react with bovine cathepsin H (*156*). Diazomethane and fluoromethane analogues of aminopeptidase substrates with rather bulky amino acid side chains in P_1, such as phenylalanine and *O*-benzyl serine, are good inhibitors of rat cathepsin H (*5*), which is somewhat consistent with the reactivity of Z-Leu-Leu-[^{125}I]Tyr-CHN_2. Perhaps more significantly, however, was the observation that inhibitors with a free amino-terminus were rather more specific for cathepsin H, providing a possible mechanism to selectively control the activity of this enzyme (*5*).

The active site of cathepsin H, therefore, clearly differs from cathepsins B and L, having a rather restricted S_2 binding pocket (see Fig. 5). This might explain in part the difficulty experienced in demonstrating any endopeptidase activity of porcine cathepsin H (*142*).

D. Cathepsin S

Purification

Another cysteine proteinase, cathepsin S, has been identified in bovine lymph nodes (*150*). The chromatographic properties of this enzyme are more similar to cathepsin B than cathepsin L. In fact, the enzyme from human or bovine spleen eluted from ion exchange resins at a similar salt concentration to cathepsin B (*78,98*). The separation of cathepsin S from cathepsin B is achieved by chromatofocussing (*78*). The enzyme purified from bovine spleen appears to be a single-chain enzyme of M_r 24,000.

Specificity

The enzyme was characterized by its low activity against Bz-Arg-NH_2. The subsequent identification of cathepsin L by similar criteria led to confusion as to whether cathepsin S was a species variant of cathepsin L. Several features of this enzyme have now been shown to be distinct from those of cathepsin L: (1) the rate constant for the inactivation of cathepsin S by Z-Phe-Phe-CHN_2 was 900 $M^{-1}S^{-1}$, which is 700-fold lower than that for cathepsin L (75,76); (2) the activity against Z-Phe-Arg-NHMec is lower for cathepsin S (*24*); (3) it is more stable than cathepsins L or B at neutral pH (*78*); and (4) the pI of cathepsin S is 6.2 to 6.8, which is higher than that of cathepsin L (*78*). Cathepsin S prefers small neutral amino acids in P_2 and P_3 (*24*), but can accommodate phenylalanine in P_3 rather well.

In spite of these studies on the activity and characterization of cathepsin S, great care still needs to be taken in the characterization of the individual enzymes. We have recently found that by carefully following the purification procedure described

Table 2. Minimum Requirements for Distinguishing between Cathepsins B, H, L, and S

Activity against Z-Arg-Arg-NHMec at pH 6.0, higher activity against Z-Phe-Arg-NHMec, but little activity against Arg-NHMec	cathepsin B
Activity against Arg-NHMec at pH 6.8, low activity against Z-Phe-Arg-NHMec and Z-Arg-Arg-NHMec	cathepsin H
*Require more than 250 mM NaCl to elute from cation exchange column at pH 5.5, high activity against Z-Phe-Arg-NHMec but low activity against Z-Arg-Arg-NHMec and Arg-NHMec at pH 5.5. Some activity against Bz-Phe-Val-Arg-NHMec	cathepsin L
*Require less than 250 mM NaCl to elute from cation exchange column at pH 5.5, high activity against Bz-Phe-Val-Arg-NHMec but low activity against Z-Arg-Arg-NHMec and Arg-NHMec at pH 5.5. Some activity against Z-Phe-Arg-NHMec	cathepsin S

Notes: The following criteria are based on results found for the enzymes of human, rat, and bovine origin. All of the enzymes should be inhibited 99% by 10 μM E-64 in 30 min at 30°C. *Although the criteria of chromatographic properties may prove not to hold true for all species variants of cathepsin S, this is probably the best criteria for distinguishing between these enzymes. There is clearly overlap in activities for many of these enzymes and obviously further criteria as described in the text could be used to be absolutely certain of the identity and composition of any given enzyme preparation.

for cathepsin S from bovine spleen, we can actually obtain pure cathepsin H and only partially purified cathepsin S (*78,156*). The cathepsin H that we purified from bovine spleen shares many characteristics of cathepsin S; such as pI in the range 6.5 to 7.0, low activity against Z-Phe-Arg-NHMec but higher activity against Bz-Phe-Val-Arg-NHMec, and chromatographically similar to cathepsin B. It does, as might be expected, hydrolyze Arg-NHMec, distinguishing it from cathepsin S, but clearly failure to measure all criteria for each enzyme could result in erroneous identification. Table 2 shows the minimum criteria needed to enable each enzyme to be distinguished.

Cathepsin S can actually accommodate iodo-tyrosine in S_2 and even S_1 (but only poorly), and thus in overall ability to accommodate bulky side chains in S_2 and S_1, cathepsin S is very similar to cathepsin L; the major difference is the affinity of binding in these sites (see Fig. 5).

E. Other Cathepsins

The above four enzymes are the best characterized of the lysosomal cysteine proteinases, but there may be others. Cathepsins N and T have been studied by two laboratories (*57,89*), but have not yet been clearly distinguished from cathepsins S and L. A major problem may be mixtures of two enzymes giving an apparent third enzyme with intermediate activity. A single report documented the separation of cathepsins B, L, S, and N from rabbit spleen, but did not provide sufficient material for sequence data to be obtained to show that they were all distinct enzymes (*89*).

Cathepsins J and K were names given to two cysteine proteinase activities of apparent high molecular weight (*85*), but these were not well characterized, and may have been cathepsins B or L associated with other proteins. Indeed, such aggregates have been identified in tissue extracts (*32*).

Cathepsin M has been described as a membrane-bound enzyme that appears to be on the outside of the lysosomal membrane (*44*). This work has not been confirmed by other laboratories as yet, and given the "sticky" nature of the lysosomal cysteine proteinases, it may yet prove to be another form of cathepsin B.

Even though it appears that several of the inadequately studied cryptic cysteine proteinases could be different forms of the better known enzymes, it cannot be ruled out that more cysteine proteinases remain to be discovered. The history of cathepsin S certainly makes this point clear.

V. STRUCTURE OF THE LYSOSOMAL CYSTEINE PROTEINASES

A. Cathepsin B

Mature rat cathepsin B consists of a polypeptide chain of 252 amino acids. The purified single chain enzyme migrates as a protein of M_r 29,000 in SDS/polyacrylamide gel electrophoresis; the increased size compared to that predicted from the sequence may be partially due to glycosylation. Cathepsin B is sometimes cleaved near the N-terminal end of the polypeptide, between amino acids Asn 47 and Val 48 for the rat enzyme, resulting in the active site cysteine being in the light chain of the proteinase (*143*). The cleavage of rat cathepsin B to the two-chain form results in the loss of two amino acids at the cleavage site (*129*). The full length clones are available for human, rat, and mouse cathepsin B (*25*), and these predict that the enzymes have N-terminal extensions that are probably signal and pro-peptides.

B. Cathepsin L

Cathepsin L purified from most species consists of two chains: a heavy chain, M_r 22,000 to 25,000, and a light chain, M_r 5000 to 7000 on SDS reducing gels (*91*). Active-site-labeling with a radioactive analogue of E-64 showed that the active-site cysteine was in the heavy chain of human liver cathepsin L (*94*).

The full cDNA sequences for rat, mouse, and human cathepsin L are known (*54,65,69,119*). The first full cDNA clone for cathepsin L was obtained from the mouse macrophage cell line, J774.2. This clone was 87% identical to the available N-terminal sequences of the heavy and light chains of cathepsin L, identifying it as the mouse equivalent of this enzyme (*96,119*). The full amino acid sequence for human cathepsin L (*126*), chicken cathepsin L (*42*), and rat liver cathepsin L (*65*) have since been published and confirm that the mouse sequence was cathepsin L.

The cDNA sequences indicate that cathepsin L contains a signal peptide of 17 amino acids and a pro-peptide of 96 amino acids. The N-terminal sequence of the secreted (i.e. precursor) form of mouse cathepsin L has been sequenced and determined to start immediately after the putative signal peptide sequence (*149*). Both the pre- and pro-peptides are absent from the mature enzyme, indicating posttranslational modifications of cathepsin L. The cDNA sequence also shows that the two-chain enzyme is derived from a single polypeptide precursor.

C. Cathepsin H

Cathepsin H can also exist as a two-chain enzyme. Rat cathepsin H was purified as a 1:1 ratio of single- to two-chain form and consists of a 220 amino-acid polypeptide chain (*143*). The deduced molecular mass from the protein sequence is 24,000, but on SDS-reducing gels its M_r is 28,000. The increased size is due largely to glycosylation (*143*). The cleavage site of single chain cathepsin H is at Asn 177-Gly 178, which results in the two-chain enzyme. The active site cysteine is in the heavy chain (*143*). Mature human cathepsin H has been sequenced and a partial cDNA cloned (*52,126*). The isolated human enzyme was shown to have an octapeptide, disulfide-linked to the protein. This has been shown from the cDNA to be part of the cleaved pro-sequence (*52*). The analogous octapeptide was not found for the rat enzyme and therefore its role is uncertain (*66,143*).

D. Cathepsin S

To date, cathepsin S has only been purified from bovine spleen and its N-terminal sequence has been published (*96*). It is distinct from the other cysteine proteinases, but closely related to cathepsin L. The cDNA for mouse cathepsin S has also been sequenced but has not yet been published (*122*).

E. Sequence Comparisons

A comparison of the sequences for cathepsins B, L, H, and the available sequence of cathepsin S, is shown in Figure 6. The regions around the active site cysteine and histidine of the proteins show the greatest number of identical residues. Cathepsin L is more similar to cathepsin H (40% identical) and to cathepsin S (71% identical) than it is to cathepsin B (27% identical). The pro-regions of the proteinases are not as similar, indicating that the evolutionary constraints on this part of the protein are either less strong or not the same. Indeed, it is difficult to align these sequences. An alignment of the complete pro-sequence of rat cathepsin H has been made with rat cathepsin L, and does show 25% identity, but this is much less than the 44% identity with the mature portions of these enzymes (*65*). This seems to discount the idea that a general sequence in this region is involved in the recognition of these proteins as lysosomal. Interestingly the pro-regions of mouse and human cathepsin B are more similar to each other than are the mature regions (*25*), and the 28 amino acids deduced from the published cDNA for the pro-region of human cathepsin H are exactly identical to the rat pro-region amino acids, whereas the

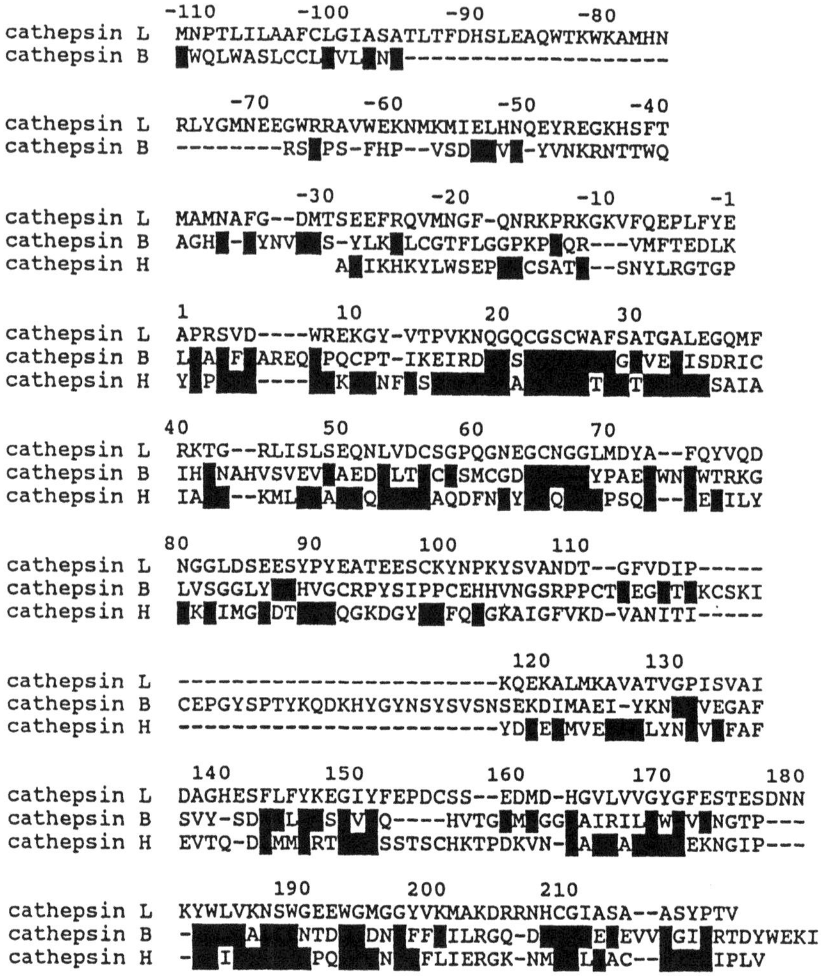

Figure 6. Prepro-sequences of human lysosomal cysteine proteinases. Sequences were obtained from references 25, 52, and 54 and aligned to give best sequence matches. Blanked out residues are those that are identical to aligned residues in cathepsin L and the numbering is according to the sequence of cathepsin L. Important features to note are the lack of similarities in the pro-peptides, but highly conserved sequences around the active-site cysteine (*25*) and histidine (*163*).

pro-regions of human and mouse cathepsin L are less similar to each other than are the mature portions. It could be that the pro-regions of cathepsins B and H serve important, but different, functions for these two enzymes in a variety of mammals, but the lower conservation of sequence does not support a common function for the pro-regions of cathepsin L in mice and man. The role of the pro-regions as activation peptides is discussed later.

VI. THE BIOSYNTHESIS OF LYSOSOMAL PROTEINS

A. The Steps Involved

In the last 10 years many workers have sought to elucidate the steps involved in the biosynthesis of lysosomal proteins. The soluble proteins have been shown to be targeted mainly via the mannose 6-phosphate receptors (reviewed in *82,152*, and Chapter 4 of this volume). The targeting of the lysosomal cysteine proteinases is essentially identical to that of other soluble lysosomal proteins. The biosynthesis of proteinases, however, poses an additional problem to the cell: they must remain inactive during synthesis so as not to degrade the biosynthetic machinery.

The initial stage in the biosynthesis of lysosomal proteins is similar to that of secretory proteins in that it occurs on ribosomes bound to the endoplasmic reticulum. Lysosomal proteins contain a hydrophobic N-terminal region known as the signal peptide, which targets them to the endoplasmic reticulum by a mechanism similar if not identical to the targeting of secretory proteins. The more recent elucidation of cDNA sequences has enabled the demonstration of a putative signal sequence in a great number of lysosomal proteins, including the cysteine proteinases, cathepsin L (*54,70,119*), cathepsin B (*25*), and cathepsin H (*66*).

Soluble lysosomal proteins have been shown to be glycosylated at asparagine residues. This occurs in the lumen of the endoplasmic reticulum and is probably cotranslational. The recognition sequence for this so-called *N*-linked glycosylation has been identified as Asn-X-Ser/Thr, where X can be any amino acid except Pro or Asp. Glycosylation of these proteins proceeds via the addition of a high-mannose sugar unit.

Processing of the carbohydrate moiety occurs soon after its addition. Removal of the terminal glucose residue by glucosidase I occurs within minutes and may precede complete translation of the protein (*7*). The next stage in the processing is the removal of the remaining glucose residues by glucosidase II; this is thought to occur more slowly (*64*). Further processing can occur via the removal of mannose residues by α-mannosidase, yielding octa-mannosyl chains (*19*).

The transfer of proteins from the endoplasmic reticulum to the *cis* face of the Golgi is thought to occur via smooth (non-clathrin-coated) vesicles (*47*).

In the Golgi, *N*-acetyl glucosamine phosphate is transferred to the mannose residues of lysosomal proteins followed by hydrolysis of the *N*-acetyl glucosamine unit to reveal mannose-6-phosphate residues that are recognized by the mannose 6-phosphate receptor (see Chapter 4). This is one of the major targeting mechanisms for mammalian lysosomal proteins. It has been argued that the transferase is the critical step in the marking of proteins for packaging into the lysosome, but the signal that distinguishes between these and nonlysosomal proteins is not yet known.

B. Studies on the Biosynthesis of Cysteine Proteinases

Comparisons of cDNA sequences with the amino acid sequences of mature lysosomal cysteine proteinases has demonstrated the presence of two N-terminal regions which are removed during maturation of these enzymes. The first of these regions corresponds to a typical signal peptide which is removed during import into the endoplasmic reticulum. The second region corresponds to the pro-peptide which is thought to be removed during transport of the proteins between the Golgi and the lysosome. A further processing event must also occur resulting in the generation of the two-chain enzymes, which are seen in purified preparations of cathepsins B, H, and L (*13,73,94,131*).

Cathepsin B

The biosynthesis and processing of cathepsin B has been shown in human fibroblasts: the precursor was shown to be M_r 44,500, and was rapidly converted to a M_r 46,000 and 33,000 forms (*61*). Rat macrophages showed the processing of cathepsin B from a M_r 39,000 form to M_r 29,000 and 25,000 within 21 hr (*81*). Rat liver cathepsin B processing *in vivo* has also been shown (*110*). It was synthesized as a precursor of M_r 39,000 and then processed to M_r 29,000 and 25,000. The two lower M_r forms were seen in the lysosomal fraction, and may indicate that processing occurs in lysosomes *in vivo*. Interestingly, processing is not the same in all cell types. In KNIH 3T3 fibroblasts, cathepsin B is synthesized as a precursor of M_r 39,000 and processed to a stable mature form of M_r 35,000 (*99*), whereas in mouse liver the mature form is a two-chain enzyme of M_r 25,000 + 5000 (*155*). The reason for this is not clear; it may be that the enzyme that processes cathepsin B to the two-chain form is not present in the transformed cells, or that the processing enzyme is not in the same compartment as cathepsin B. Generation of a two-chain form of cathepsin B is not essential, as the sheep liver enzyme is not processed to a two-chain form, and single- and two-chain forms of cathepsin B are catalytically identical (*90*). Cathepsin B also undergoes some C-terminal processing, but again this is most likely as a consequence of being in the hydrolytic environment of the lysosome, rather than being related any essential posttranslational modification.

Cathepsin L

Processing of cathepsin L was first shown for the mouse enzyme in Kirsten virus-transformed NIH 3T3 cells (KNIH 3T3). Using an antibody to the secreted protein in [^{35}S] methionine pulse-chase experiments in cell culture, it was shown that cathepsin L was synthesized as a precursor of M_r 39,000 and processed to a mature form of M_r 20,000 via an intermediate of M_r 29,000 (*55*). The precursor form of cathepsin L was also shown to be secreted from KNIH 3T3 cells. Subsequently, the biosynthesis and processing of cathepsin L has also been demonstrated in rat macrophages and hepatocytes (*81,112*). In both cases, cathepsin L is synthesized as a precursor of M_r 39,000 and processed rapidly to mature forms of M_r 25,000 in the case of hepatocytes, and 23,000 in macrophages. The processing in both cell types was via an intermediate of M_r 30,000. A large proportion of the precursor was secreted: 30% in macrophages and >40% in hepatocytes. Human macrophages also secrete a significant proportion of their newly synthesized precursor cathepsin L (*124*). Biosynthetic studies have shown that the major mature form of cathepsin L in these cells is the single-chain enzyme, with processing to the two-chain form occurring very slowly.

Mouse cathepsin L contains only one octaphospho-mannosyl unit, which is proposed to account for its reduced affinity for the cation-independent mannose 6-phosphate receptor resulting in its secretion (*40*). This may not be the case for all species variants of cathepsin L, although this enzyme does seem to be much less efficiently packaged into lysosomes than cathepsin B (*99*).

Cathepsin H

Rat cathepsin H is synthesized as a pro-enzyme of M_r 41,000 and then processed to the mature form, M_r 28,000, within 60 min in perfused rat liver (*111*). The conversion to the two-chain form was not seen; this is thought to occur more slowly, and again may not occur in all species.

In summary, it appears that the essential steps in the biosynthesis of lysosomal cysteine proteinases are the modification of asparagine residues to give mannose 6-phosphate residues for effective targeting to the lysosome, and removal of the pro-peptide to give the mature enzyme. Any further processing is likely due to the susceptibility of the enzyme to proteolytic cleavage in the lysosome by the high concentration of proteinases in this environment.

VII. THE CONTROL OF PROTEINASES BY ACTIVATION

The assay of lysosomal cysteine proteinases *in vitro* requires the reduction of the active-site by reducing agents such as dithiothreitol or cysteine. From this standpoint, it would therefore seem that oxidation-reduction could control the activity of these enzymes *in vivo*. Using an active-site-directed inhibitor, Z-[^{125}I]Tyr-Ala-CHN_2, that only reacts with active enzyme, it was shown that in cells cathepsins B and L are fully active, and addition of reducing agents could not enhance reactivity with the inhibitor (*99,155*). By contrast, when cells were lysed, addition of a reducing agent was required for these enzymes to react with the inhibitor. Thus it would appear that although the environment in the lysosome is sufficiently oxidizing to maintain the disulfide bonds in these proteinases, it is not sufficient to oxidize the active-site cysteine. Indeed reducing agents are not required for the activity of all cysteine proteinases *in vitro*, as both calpain and pro-cathepsin L can exhibit activity without such additions.

From the biosynthetic studies, it is seen that the lysosomal proteinases are synthesized as pro-enzymes. By analogy to the digestive pro-enzymes trypsinogen and pepsinogen, one might expect the pro-enzymes to be inactive. This appears to be the case. Also, like pepsinogen, they seem to undergo auto-activation at low pH. Auto-activation of mouse precursor cathepsin L occurs at pH 3.0 (*53,97*). A latent form of cathepsin L has been reported in guinea pig sperm. The activation of this at pH 3.0 was inhibited by cysteine proteinase inhibitors such as leupeptin, Z-Phe-Ala-CHN_2, and $HgCl_2$ (*102*). This suggests that an auto-activation mechanism could also be responsible for activation of guinea pig cathepsin L. Evidence for a pepstatin-sensitive enzyme, possibly cathepsin D, being involved in the activation of rat pro-cathepsins L and B *in vitro* has been shown (*113*). Another study, however, showed that the activation of secreted rat pro-cathepsin L is independent of proteinases, but can be enhanced by the addition of cathepsin D (*154*).

Other investigations with whole cells during culture have demonstrated the involvement of several types of proteinases in this activation. 1,10-Phenanthroline, a potent metallo-proteinase inhibitor, prevents processing of precursor rat cathepsins L and B to their single-chain forms (*62*). Since this inhibitor has been shown to interfere with membrane fusion events, it may be mediating its effect through inhibition of lysosome formation (*62*). Another group has shown that rat cathepsin D may be the processing enzyme for rat cathepsin L by using pepstatin to inhibit precursor activation *in vivo* (*114*). One strong argument against this idea is that cathepsin D is also synthesized as a precursor that requires activation and thus it is unclear how the cathepsin D itself is activated. Of course it must be remembered that conditions *in vitro* are not the same as *in vivo* where the concentration of proteinases is much higher, so autocatalytic processing may well proceed faster as newly processed enzymes contribute to a proteolytic cascade.

Thus, it has been demonstrated that there are two possible mechanisms for the processing of cathepsin L: (1) auto-activation, which has been shown to occur *in vitro*; and (2) proteolysis by proteinases other than cathepsin L, evidence of which comes from both *in vivo* and *in vitro* studies. The proteinases involved have been shown to be either aspartic or metallo in nature. The results so far are contradictory and further work is needed to establish what happens *in vivo*.

VIII. CONTROL OF PROTEINASES BY INHIBITION

There are several naturally occurring inhibitors which are thought to control the activity of the lysosomal cysteine proteinases in the cytosol of the cell or extracellularly. α_2-Macroglobulin and the cystatins are the two main types of inhibitors of cysteine proteinases identified to date.

A. α_2-Macroglobulin (α_2M)

This is one of the major serum proteins found in vertebrates and in some invertebrate species (*139*). It has a M_r of 725,000 and has the ability to trap all classes of endopeptidases by a conformational change in the protein after being cleaved by the enzymes in a so-called "bait" region. The interactions of human cathepsins B, H, and L (*92*) with α_2M have all been demonstrated. Once α_2M has bound a proteinase, sites on the inhibitor are exposed that are recognized by receptors on macrophages, and then taken up rapidly. The concentration of this inhibitor is sufficiently high in serum to control the activity of any lysosomal proteinases that get into the circulation.

B. Cystatins

The first demonstration of a heat-stable inhibitor of cathepsin B was from rat liver cytosol (*50*). Since then, several inhibitory activities have been identified from rat tissues (*63,68*). An inhibitor of papain and ficin was also identified in egg white; this was called cystatin (*133*). This name has now been adopted for the group of inhibitors related to this protein by sequence (*15*). The common characteristics of such inhibitors were their low M_r and their stability to extreme conditions of pH and temperature.

Categories

Protein and cDNA sequencing has enabled the complex range of similar proteins to be more clearly characterized and distinguished. Thus all of the cystatins are related by amino acid sequence, suggesting that they may be derived from common a evolutionary ancestral protein (*123*). The type I cystatins are shown to have no disulfide bonds and no signal sequence, suggesting that they are cytoplasmic proteins. Type II cystatins have disulfide bonds and signal peptides, and are thus most probably secreted proteins. The type III proteins contain three cystatin-like units and in fact are the kininogens which are secreted proteins required for release of bradykinin. Sequence data has enabled the identification of type IV cystatins which have two cystatin-like domains, but actually have not been shown to be inhibitory.

Not all of the cystatins fit well into these four categories. The most striking impression gained from this study is that there are many naturally occurring cysteine proteinases found in mammalian tissues, both intracellularly and extracellularly. The individual inhibitors do not seem to have specifically evolved to inhibit individual host proteinases, as they all seem to be good inhibitors of cathepsins L and H, but rather poor inhibitors of cathepsin B (*15*). The concentration of several of the secreted inhibitors have been determined to be in micromolar concentrations in body fluids, and thus as their K_i values for inhibition are nanomolar or lower for all of the lysosomal cysteine proteinases, this should be more than sufficient to control the activity of any enzymes that get out of cells (2). Similar control of activity could be expected by the cytoplasmic inhibitors. The physiological role for these inhibitors is unclear, but the array of inhibitors found in saliva, suggests that they may be effective in the control of parasitic protozoa, many of which utilize cysteine proteinases as essential digestive enzymes (*116*). The high concentrations of cystatins found in tissues would also be a significant obstacle to the extra-lysosomal activities of host cysteine proteinases.

The Mechanism of Inhibition

Chicken cystatin exhibits tight, competitive, reversible binding to papain (*108*). There is no observed cleavage of the inhibitor on interaction with cystatin. The blocking of the active site with small thiol agents does not prevent binding to cystatin, although the affinity of the interaction is reduced (*20*).

The first site of binding proposed for cystatins was deduced from sequence comparisons. There is a highly conserved region in the cystatins (QVVAG at residues 53–57 in chicken cystatin) that was thought to be essential for inhibitory activity (*18*). Inhibitory peptides have been produced which have this sequence (*145*). Although a conserved sequence more likely serves a structural role as it cannot alter the specificity of an inhibitor, the crystal structure of a complex of cystatin B with papain shows that the QVVAG region binds near to the S′ sites of papain (*140*).

The importance of the N-terminal part of the inhibitors was noted when it was found that the truncated form of chicken cystatin has a lower affinity for papain; the K_i is 1000-fold higher (*3*). The reduced inhibition of the human lysosomal cysteine proteinases by removal of the N-terminal portion of human cystatin C is related to the inhibition of these proteinases by analogous peptidyl diazomethane inhibitors, suggesting that the N-terminal peptide of cystatin C fits into the specificity pockets of the enzymes (*4*). The conserved glycine (position 9 in chicken cystatin) appears to be in the position P_1 (*140*). This might explain the inability of the cystatins to inhibit serine proteinases, as the specificity of these enzymes is primarily determined by amino acids in P_1. Although the N-terminus seems to be important to the binding of the secreted cystatins, this does not seem to be the case for the intracellular type I cystatins, as truncation of rice cystatin does not affect papain inhibition (*1*). The N-terminal portion of cystatin B, another type I cystatin, does contact papain and presumably contributes to binding (*140*).

The third binding site was detected in chicken cystatin by fluorescent changes in tryptophan 104 when binding to papain of chicken cystatin (*86*). The crystal structure of this protein demonstrated the formation of an inhibitory wedge-structure which brings together the three inhibitory sites (*21*). Computer-docking experiments demonstrated that this "wedge" could be accommodated by the active site of papain (*21*). The homologous regions of cystatin B were all shown to bind to papain (*140*).

IX. PHYSIOLOGICAL AND PATHOLOGICAL ROLES

A. Lysosomal Protein Turnover

Sub-cellular fractionation (*73*) and immuno-localization (*45*) have demonstrated the lysosomal location of cathepsin L. Cathepsin L is able to degrade a wide range of protein substrates, including components of the extracellular matrix (*74,95*).The determination of the role of lysosomes in protein turnover has been facilitated by the use of lysomotrophic agents (*132*). These compounds passively diffuse into the lysosome, where they become protonated, and hence, trapped. On accumulation of such agents, the lysosomal pH is disrupted and the result is a loss of lysosome function. Therefore the inhibition of protein turnover by such agents is proposed to indicate the amount of degradation due to the lysosomes (*132*). However, since the discovery that at least one lysosomal enzyme is active and stable at neutral pH (cathepsin S), it is not certain that all of the lysosomal enzymes will be inactive. It may be that disruption of the pH gradient may actually prevent membrane fusion, and hence block protein turnover by preventing entry of substrates into the lysosome rather than directly inhibiting the proteinases.

The role of cysteine proteinases in lysosomal turnover of proteins has been demonstrated by the use of synthetic inhibitors. Leupeptin, an inhibitor of both serine and cysteine proteinases, inhibits basal protein turnover by 30 to 40% (*34*). However, use of this inhibitor to control the activity of lysosomal proteolysis is questionable, as we have found that this inhibitor cannot cross membranes (*155*) and hence its only route into lysosomes must be by either pinocytosis or phagocytosis. It is therefore uncertain how effective it would be in inhibiting all lysosomal cysteine proteinase activity. The more specific inhibitors of cysteine proteinases, the diazomethanes, have enabled the conclusive demonstration of cysteine proteinase involvement in lysosomal protein turnover.

Z-Phe-Ala-CHN_2 inhibits macrophage protein turnover by 40% (*134*) and inhibits protein turnover in hepatocytes by 50 to 70% (*60*). The inhibition produced by this compound was not additive to that produced by lysomotrophic agents. Therefore it was concluded that the inhibition was due to inhibition of lysosome function. Total inhibition of basal protein turnover is not seen even with lysomotrophic agents and it seems likely that some protein turnover occurs via another system, possibly occurring in the cytosol. The maximum inhibition produced by lysomotrophic agents is about 70%; inhibition of cysteine proteinases never produces a total reduction in lysosomal protein turnover. The inhibition of lysosomal degradation caused by inhibitors of cysteine proteinases is additive to the inhibition produced by pepstatin, an inhibitor of cathepsin D (*134*). Thus, cysteine proteinases have been shown to contribute to the lysosomal degradation of proteins. The degradation of proteins is also achieved by other proteinases, such as cathepsin D.

B. Developing Embryos

Lysosomal protein turnover is often thought to be only involved in the recycling of amino acids in mammals, but for the embryo this is the sole source of nutrition.

It has been found that cysteine proteinase inhibitors prevent the growth of, and are teratogenic for, developing embryos (*33,49*). Significant levels of pro-cathepsin L have been found in guinea pig sperm (*82*), indicating a possible role for this enzyme in either fusion with the egg or development of the embryo. The role of lysosomal proteinases in the embryo is probably more akin to the role of cysteine proteinases in parasitic protozoa than to the role in mature cells; that is, the enzymes are probably directly involved in providing the nutritional requirements of the embryo.

C. Bone Resorption

Bone resorption is a process involving the removal of both the mineral (calcium phosphate) and the organic (collagen) components of bone. This occurs in an acidic extracellular phago-lysosome-like structure on the ruffled border of the osteoclast (*12*). The pH occurring in this specialized structure is 3 to 4 (*136*). This acidic pH is suitable for the activity and stability of the cysteine proteinases and also for the activation of precursor cathepsin L.

Evidence for the roles of cathepsins B and L comes from the use of the specific inhibitors, Z-Phe-Ala-CHN_2 and E-64. Z-Phe-Ala-CHN_2 inhibited the osteoclastic resorption of dentine by 75% (*36*). An inhibitor of collagenase, a metallo-enzyme with collagenolytic activity, was used in the same experiments and shown to have no effect on resorption. The collagenase inhibitor has been shown to be effective, however, in the inhibition of resorption by mouse calvaria (*35*). The effect of mouse lysosomal enzymes and bone collagenase has been shown to be additive for the degradation of bone collagen at pH 5.5 (*43*).

D. Tumor Invasion

A role for proteinases in the invasion of host tissue and metastasis of tumors has been suggested (*87*). The enzymes responsible for this are largely unknown, but in recent years a role for lysosomal cysteine proteinases has been proposed (*138*). Investigators have suggested that the ability of tumor cells to invade basement membrane is related to their ability to degrade type IV collagen, a major component of basement membrane. Basement membrane, however, contains several other components, such as laminin, fibronectin, and proteoglycan, and therefore the degradation of these components may be important (*138*).

Lysosomal cysteine proteinases are able to degrade a broad range of substrates and the degradation of basement membrane by cathepsin L has been shown (*10*). Therefore, lysosomal proteinases could potentially be important in the process of tumor invasion. Recent demonstration of extracellular forms of cathepsins L and B has made their role in such processes more likely.

Levels of lysosomal cathepsin B have been shown to correlate with the metastatic potential of mouse tumor cell lines (*137*). However, in order for cells to invade tissues, extracellular proteinases would seem more appropriate. Two secreted, high M_r forms of human cathepsin B have been identified. A M_r 41,000-secreted form was identified in ascitic fluid. This form was rendered active by treatment with pepsin, resulting in a form of M_r 33,000 (*105*). A second high M_r, secreted form of cathepsin B has also been identified in the culture medium from breast tumor cells. This form had activity against low-M_r synthetic substrates (*104*). Rabbit V2 carcinoma cells also secrete a high M_r form of cathepsin B (43,000), which can also be activated by pepsin treatment to a M_r 34,000 form (*8*). Clearly, the role of these high M_r forms and their relationship to active forms needs to be defined before the role of such forms in extracellular events can be verified.

There is increasing evidence that cathepsin L exists outside the cell, often as a precursor (*58,81,112*). The secretion of precursor cathepsin L from NIH 3T3 fibroblasts was increased when the cells were transformed by various tumor-promoting agents, such as phorbol ester (4-*o*-tetradecanoyl phorbol-13-acetate; TPA), platelet derived growth factor (PDGF), SV40 virus, and kirsten sarcoma virus (*58,59*). This precursor form of cathepsin L was purified and shown to be active (*97*). It would now appear that the purification procedure for this enzyme resulted in the expression of activity, as the precursor secreted by the cells only exhibits activity and reactivity with inhibitors if treated to low pH (*99*). Furthermore, the expression and secretion of cathepsin L has been correlated to the metastatic potential of murine fibroblasts, indicating a role for this enzyme in tumor invasion (*38*) and the expression of cathepsin L is elevated in human tumors (*30*). A major question that remains to be answered is how can the secreted precursor forms of cysteine proteinases contribute to tissue degradation if they are inactive zymogens? Although progress is being made on the elucidation of the mechanism of activation in the lysosome, it is still unclear how this occurs outside the cell.

Secreted forms of proteinases are important in the demonstration of the potential involvement in extracellular events. Plasma membrane-associated forms of proteinases, however, may also be important, particularly in pericellular events. Indeed, a pericellularly fixed cysteine proteinases is a particularly attractive idea as this could help to keep it isolated from the extracellular inhibitors of the se enzymes. Cathepsin L is known to be a soluble enzyme, although evidence has been presented indicating it to have a membrane association in some metastatic mouse cell lines (*128*). Similarly, there is evidence for a membrane-bound form of cathepsin B which is elevated in metastatic tumor cell lines (*127*). Another group has demonstrated a membrane association of cathepsin D in rabbit macrophages. The interaction has been shown to be independent of mannose 6-phosphate receptors, but the nature of the interaction is unknown (*39*). There is no evidence for any *trans-* membrane domain in the structure of mouse cathepsin L, mouse cathepsin B, or cathepsin D, and therefore this type of association may represent a novel type of membrane interaction.

E. Pulmonary Emphysema

Pulmonary emphysema is characterized by the destruction of lung parenchyma cells in the bronchiolar walls (*67*). The disease results in extensive degradation of the connective framework of the lung, which consists of elastin, collagen, and other matrix components (*109*). Work with experimental animals has suggested that the degradation of elastin plays a central role in the onset of emphysema (*67*).

The major enzyme proposed to be involved in emphysema is neutrophil elastase, largely because of the observation that individuals that have a genetic deficiency in what is probably the physiological inhibitor of this enzyme, α_1-proteinase inhibitor, are particularly prone to developing this disease (*83*). However, not all individuals with this genetic defect develop the disease and not all persons develop the disease have the defective gene. It is therefore possible that other enzymes also play a role in the onset of this disease. Neutrophils have been shown to constitute a minor portion of the leukocytes in the lung; 2% in both smokers and nonsmokers (*26*).

Macrophages constitute the largest proportion of the cells obtained from lung lavage of both nonsmokers (63%) and smokers (95%) (*26,120*). The proteinases synthesized by macrophages and the ability of alveolar macrophages to degrade elastin has been the subject of recent work.

The cysteine proteinases have been shown to degrade a number of protein substrates (*16*). The role of these enzymes in elastin degradation has been addressed; first by looking at the action of purified enzymes against elastin, and second by the use of specific inhibitors to inhibit the degradation of elastin by human alveolar macrophages.

Purified cathepsin L was shown to be 100 times more active than cathepsin B in the hydrolysis of elastin (*95*). Cathepsin L had similar elastinolytic activity to porcine pancreatic elastase at its pH optimum, and is therefore a powerful elastase (*95*). Alveolar macrophages have been shown to synthesize and process cathepsin L (*124*) and cathepsin B (*117*). There is increasing evidence that cysteine proteinases are responsible for the degradation of elastin by alveolar macrophages. The degradation of elastin by human alveolar macrophages was found to be dependent on cell contact and could be inhibited up to 80% by Z-Phe-Phe-CHN_2, an inhibitor of cysteine proteinases (*27,28*). Since cathepsin L is the major elastinolytic cysteine proteinase, the observed inhibition was probably due to inhibition of this enzyme.

The major site of activity of cathepsin L is clearly lysosomal. However, in the cell culture models and the lung, the elastin substrate is obviously extracellular, and hence in order to be taken up into cells, some limited degradation of this insoluble protein must occur outside the cell. The pro-form of cathepsin L is known to be secreted by macrophages (*124*), but as yet a mechanism for activation of this protein outside cells is not known. Furthermore, macrophages are also known to synthesize and secrete cystatin C, a potent inhibitor of cathepsin L (*29*).

Neutrophil elastase can partially inactivate cystatin C by cleavage of the N-terminus of this inhibitor, and thus potentially prevent this protein from blocking the activity of cathepsin L (*4*). Interestingly, cathepsin L is also able to inactivate α_1-proteinase inhibitor (the physiological inactivator of neutrophil elastase) and this may point to a cooperative role for these two proteinases in increasing lung proteolysis (*69*).

Therefore, cathepsin L could have several roles in increasing elastin degradation in the diseased lung: (1) by degrading elastin extracellularly; (2) by inactivating α_1-proteinase inhibitor, leading to increased neutrophil elastase activity; or (3) by degrading endocytosed elastin fragments in the lysosome. The precise location of elastin degradation by cathepsin L is unclear and it is interesting to speculate that it may occur in a pericellular location, similar to that seen in the osteoclast.

It therefore appears that the macrophage is a good source of elastases which are capable of the observed elastin degradation seen in emphysema, and the lysosomal cysteine proteinases appear to play a major role in the elastinolytic activity of these cells.

X. CONCLUSION

In addition to the above functions, cysteine proteinases have also been proposed to play a role in the degenerative processes of arthritis (*148*), muscular dystrophy (*80*), and glomerular nephritis (*11*). From these observations, it is clear that the lysosomal proteinases could play a role in pathological processes at sites other than the lysosome, whereas the major physiological roles are in the lysosome. The precise mechanism by which the secreted proteinases contribute to the degradation of extracellular protein still remains to be established.

NOTE ADDED IN PROOF

Since submission of this manuscript, several important discoveries have been made.

Cathepsin J has been purified from rat liver and identified as cathepsin C (*109a*). This is somewhat surprising as cathepsin C had been characterized as a dipeptidyl peptidase with little endopeptidase-like activity. Nikawa et al., (*109a*) found that cathepsin C accounts for as much as 15% of the total activity of lysosomal extracts against Z-Phe-Arg-NHMec, suggesting that it could be a significant endopeptidase. A more detailed study of the endopeptidocytic activity of this enzyme is warranted.

The crystal structure of human cathepsin B has been determined (*107a*). The complete nucleotide sequence of human cathepsin S has been determined (*135a, 153a*), confirming that it is most closely related to cathepsin L in sequence. The propeptide of cathepsin B has been synthesized, and determined to be a tight binding reversible inhibitor of cathepsin B (*48a*) and a recombinant form of pro-cathepsin B can be processed by mature forms of cathepsins B, D, and L (*126a*). Pro-cathepsin L has been shown to undergo autolysis at pH 6.0 in the presence of charged surfaces, indicating that activation in the lysosome can occur at the pH of early lysosomes and does not require activity of another proteinase (*99a*). This is similar in mechanism to activation of the serine proteinases factor XII and factor XI of the blood coagulation pathway. It remains to be seen whether this mechanism is general for all of the lysosomal proteinases, and the nature of an *in vivo* charged surface is not known.

ACKNOWLEDGMENT

This work was supported in part by a grant from the Council for Tobacco Research, USA.

REFERENCES

1. Abe, K., Emori, Y., Kondo, H., Aria, S., and Suzuki, K. (1988). The NH_2-terminal 21 amino acid residues are not essential for the papain-inhibitory activity of oryzacystatin, a member of the cystatin superfamily. Expression of oryzacystatin cDNA and its truncation fragments in Escherichia Coli. *J. Biol. Chem.* 263: 7655–7659.
2. Abrahamson, M., Barrett, A.J., Salvesen, G., and Grubb, A. (1986). Isolation of six cysteine proteinase inhibitors from human urine. *J. Biol. Chem.* 261: 11282–11289.
3. Abrahamson, M., Ritonja, A., Brown, M.A., Grubb, A., Machleidt, W., and Barrett, A.J. (1987). Identification of the probable inhibitory reactive sites of the cysteine proteinase inhibitors human cystatin C and chicken cystatin. *J. Biol. Chem.* 262: 9688–9694.
4. Abrahamson, M., Mason, R.W., Hansson, H., Buttle, D.J., Grubb, A., and Ohlsson, K. (1991). Human cystatin C: Role of the N-terminal segment in the inhibition of human cysteine proteinases and in its inactivation by leucocyte elastase. *Biochem. J.* 273: 621–626.
5. Angliker, H., Wikstrom, P., Kirschke, H., and Shaw, E. (1989). The inactivation of the cysteinyl exopeptidases cathepsin-h and cathepsin-C by affinity-labelling reagents. *Biochem. J.* 262: 63–68.
6. Aronson, N.N. and Barrett, A.J. (1978). The specificity of cathepsin B. Hydrolysis of glucagon at the C-terminus by a peptidyldipeptidase mechanism. *Biochem. J.* 171: 759–765.
7. Atkinson, P.H. and Lee, J.T. (1984). Cotranslational excision of α-glucose and α-mannose in nascent vesicular stomatitis virus G protein. *J. Cell Biol.* 98: 2245–2249.
8. Baici, A. and Knopfel, M. (1986). Cysteine proteinases produced by cultured rabbit V2 carcinoma cells and rabbit skin fibroblasts. *Int. J. Cancer* 38: 753–761.
9. Baker, E.N. (1980). Structure of actinidin, after refinement at 1.7 Å resolution. *J. Mol. Biol.* 141: 441–444.
10. Baricos, W.H., Zhou, Y., Mason, R.W., and Barrett, A.J. (1988). Human kidney cathepsins B and L: characterization and potential role in degradation of glomerular basement membrane. *Biochem. J.* 252: 301–304.
11. Baricos, W.H. and Shah, S.V. (1989). Role of cathepsin B and L in anti-glomerular basement membrane nephritisin rats. *Renal. Physiol. Biochem.* 12: 400–405.
12. Baron, R., Neff, L., Louvard, D., and Courtoy, P.J. (1985). Cell-mediated extracellular acidification and bone resorption: evidence for a low pH in resorbing lacunae and localization of a 100-kD lysosomal membrane protein at the osteoclast ruffled border. *J. Cell Biol.* 101: 2210–2222.
13. Barrett, A.J. (1973). Human cathepsin B1. Purification and some properties of the enzyme. *Biochem. J.* 131: 809–822.
14. Barrett, A.J. (1980). Fluorimetric assays for cathepsin B and cathepsin H with methylcoumarylamide substrates. *Biochem. J.* 187: 909–912.
15. Barrett, A.J. (1987). The cystatins: a new class of peptidase inhibitors. *Trends in Biochemical Sciences* 12: 193–196.
16. Barrett, A.J. and Kirschke, H. (1981). Cathepsin B, cathepsin H, and cathepsin L. *Meth. Enzymol.* 80: 535–561.
17. Barrett, A.J., and McDonald, J.K. (1980). *Mammalian Proteinases: A Glossary and Bibliography,* (Vol. I.) *Endopeptidases.* Academic Press, London.
18. Barrett, A.J., Rawlings, N.D., Davies, E.M., Machleidt, W., Salvesen, G., and Turk, V. (1986). Cysteine proteinase inhibitors of the cystatin superfamily. In: *Proteinase Inhibitors* (Barrett, A.J. and Salvesen, G., eds.), pp. 515–569. Elsevier Science, Cambridge.
19. Bischoff, J. and Kornfeld, R. (1983). Evidence for an alpha-mannosidase in endoplasmic reticulum of rat liver. *J. Biol. Chem.* 258: 7907–7910.
20. Bjork, I. and Ylinenjarvi, K. (1989). Interaction of chicken cystatin with inactivated papains. *Biochem. J.* 260: 61–68.
21. Bode, W., Engh, R., Musil, D., Thiele, U., Huber, R., Karshikov, A., Brzin, J., Kos, J., and Turk, V. (1988). Crystal structure of cystatin. *EMBO J.* 7: 2593–2599.
22. Bond, J.S. and Barrett, A.J. (1980). Degradation of fructose-1,6-bisphosphate aldolase by cathepsin B. *Biochem. J.* 189: 17–25.

23. Bromme, D., Steinert, A., Fittkau, S., Kirschke, H. (1987). Action of rat liver cathepsin B on bradykinin and on the oxidized insulin A-chain. *FEBS. Lett.* 219, 441–444.
24. Bromme, D., Steinert, A., Friebe, S., Fittkau, S., Wiederanders, B. and Kirschke, H. (1989). The specificity of bovine spleen cathepsin S—a comparison with rat liver cathepsin L and cathepsin B. *Biochem. J.* 264, 475–481.
25. Chan, S.J., San Segundo, B.S., McCormick, M.B. and Steiner, D.F. (1986). Nucleotide and predicted amino acid sequences of cloned human and mouse preprocathepsin B cDNAs. *Proc. Natl. Acad. Sci. USA* 83: 7721–7725.
26. Chang, J.C., Lesser, M., Yoo, O.H., and Orlowski, M. (1986). Increased cathepsin B-like activity in alveolar macrophages and bronchoalveolar lavage fluid from smokers. *Am. Rev. Respir. Dis.* 134: 538–541.
27. Chapman, H.A. and Stone, O.L. (1984). Comparison of live human neutrophil and alveolar macrophage elastinolytic activity *in vitro*. *J. Clin. Invest.* 74: 1693–1700.
28. Chapman, H.A., Stone, O.L., and Vavrin, Z. (1984). Degradation of fibrin and elastin by intact human alveolar macrophages *in vitro*. *J. Clin. Invest.* 73: 806–815.
29. Chapman, H.A., Reilly, J.J., Yee, R., and Grubb, A. (1990). Identification of cystatin C, a cysteine proteinase inhibitor, as a major secretory product of human alveolar macrophages *in vitro*. *Am. R. Resp. D.* 141: 698–705.
30. Chauhan, S.S., Goldstein, L.J., and Gottesman, M.M. (1991). Expression of Cathepsin-L in Human Tumors. *Cancer Res.* 51: 1478–1481.
31. Crawford, C., Mason, R.W., Wikstrom, P. and Shaw, E. (1988). The design of peptidyldiazomethane inhibitors to distinguish between the cysteine proteinases calpain II, cathepsin L and cathepsin B. *Biochem. J.* 253: 751–758.
32. Dahlmann, B., Kuehn, L., Heinrich, P.C., Kirschke, H., and Wiederanders, B. (1989). ATP-activated, high-molecular-mass proteinase-I from rat skeletal muscle is a cysteine proteinase-alpha 1-macroglobulin complex. *Biochim. Biophys. Acta* 991: 253–262.
33. Daston, G.P., Baines, D., Yonker, J.E., and Lehmanmckeeman, L.D. (1991). Effects of lysosomal proteinase inhibition on the development of the rat embryo *in vitro*. *Teratology* 43: 253–261.
34. Dean, R.T. (1979). Macrophage protein turnover. *Biochem. J.* 180: 339–345.
35. Delaisse, J.M., Eeckhout, Y., Sear, C., Galloway, A., McCullagh, K., and Vaes, G. (1985). A new synthetic inhibitor of mammalian tissue collagenase inhibits bone resorption in culture. *Biochem. Biophys. Res. Commun.* 133: 483–490.
36. Delaisse, J.M., Boyde, A., Maconnachie, E., Ali, N.N., Sear, C.H.J., Eeckout, Y., Vaes, G., and Jones, S.J. (1987). The effects of inhibitors of cysteine-proteinases and collagenase on the resorptive activity of isolated osteoclasts. *Bone* 8: 305–313.
37. DeMartino, G.N., Doebber, T.W., and Miller, L.L. (1977). Pepstatin-insensitive proteolytic activity of rat liver lysosomes. *J. Biol. Chem.* 252: 7511–7516.
38. Denhardt, D.T., Greenberg, A.H., Egan, S.E., Hamilton, R.T., and Wright, J.A. (1987). Cysteine proteinase cathepsin L expression correlates closely with the metastatic potential of H-ras-transformed murine fibroblasts. *Oncogene* 2: 55–59.
39. Diment, S., Leech, M.S., and Stahl, P.D. (1988). Cathepsin D is membrane-associated in macrophage endosomes. *J. Biol. Chem.* 263: 6901–6907.
40. Dong, J., Prence, E.M., and Sahagian, G.G. (1990). Basis for low affinity binding of a lysosomal cysteine protease to the cation-independent mannose 6-phosphate receptor. *J. Biol. Chem.* 265: 4210–4217.
41. Drenth, J., Kalk, K.H., and Swen, H.M. (1976). Binding of chloromethyl ketone substrate analogues to crystalline papain. *Biochemistry* 15: 3731–3738.
42. Dufour, E., Obled, A., Valin, C., Bechet, D., Ribadeau-Dumas, B., and Huet, J.C. (1987). Purification and amino acid sequence of chicken liver cathepsin L. *Biochemistry* 26: 5689–5695.
43. Eeckhout, Y. (1990). possible role and mechanism of action of dissolved calcium in the degradation of bone collagen by lysosomal cathepsins and collagenase. *Biochem. J.* 272: 529–532.
44. Erickson-Viitanen, S., Balestreri, E., McDermott, M.J., Horecker, B.L., Melloni, E., Pontremoli, S. (1985). Purification and properties of rabbit liver cathepsin M and cathepsin B. *Arch. Biochem. Biophys.* 243: 46–61.

45. Etherington, D.J., Mason, R.W., Taylor, M.A., and Wardale, R.J. (1984) Production of a monospecific antiserum to cathepsin L: the histochemical location of enzyme in rabbit fibroblasts. *Biosci. Rep.* 4: 121–127.
46. Etherington, D.J., Taylor, M.A.J., and Henderson, B. (1988). Elevation of cathepsin L levels in the synovial linings of rabbits with antigen-induced arthritis. *British J. Exp. Med.* 69: 281–289.
47. Farquhar, M.G. and Palade, G.E. (1981). The Golgi apparatus (complex)-(1954–1981)-from artifact to center stage. *J. Cell Biol.* 91: 77–103.
48. Finkle, B.J. and Smith, E.L. (1958). Crystalline papain: number and reactivity of thiol groups; chromotographic behavior. *J. Biol. Chem.* 230: 669–690.
48a. Fox, T., Demiguel, E., Mort, J.S., and Storer, A.C. (1992). Potent slow-binding inhibition of cathepsin-B by its propeptide. *Biochemistry* 31: 12571–12576.
49. Freeman, S.J. and Lloyd, J.B. (1983). Inhibition of proteolysis in rat yolk sac as a cause of teratogenesis. Effects of leupeptin in vitro and in vivo. *J. Embyrol. Exp. Morphol.* 78: 183–193.
50. Frinkelstadt, J.T. (1957). Intracellular distribution of proteolytic enzymes in rat liver tissue. *Proc. Fed. Am. Soc. Exp. Biol.* 95: 302–304.
51. Fruton, J.S., Irving, G.W., and Bergmann, M. (1941). On the proteolytic enzymes from animal tissues: III. the proteolytic enzymes of beef kidney, beef kidney and swine kidney. Classification of the cathepsins. *J. Biol. Chem.* 141: 763–774.
52. Fuchs, R., Machleidt, W., and Gassen, H.G. (1988). Molecular cloning and sequencing of a cDNA coding for mature human kidney cathepsin H. *Biol. Chem. Hoppe-Seyler* 369: 469–475.
53. Gal, S. and Gottesman, M.M. (1986b). The major excreted protein of transformed fibroblasts is an activable acid-protease. *J. Biol. Chem.* 261: 1760–1765.
54. Gal, S. and Gottesman, M.M. (1988). Isolation and sequence of a cDNA for human pro-(cathepsin L). *Biochem. J.* 253: 303–306.
55. Gal, S., Willingham, M.C., and Gottesman, M.M. (1985). Processing and lysosomal localization of a glycoprotein whose secretion is transformation stimulated. *J. Cell Biol.* 100: 535–544.
56. Glazer, A.N. and Smith, E.L. (1971). Papain and other plant sulphydryl proteolytic enzymes. In: *The Enzyme*, 3rd edition (Boyer, P.D., ed.). Academic Press, New York.
57. Gohda, E. and Pitot, H.C. (1981). A new thiol proteinase from rat liver. *J. Biol. Chem.* 256: 2567–2572.
58. Gottesman, M.M. (1978). Transformation-dependent secretion of a low M_r protein by murine fibroblasts. *Proc. Natl. Acad. Sci. USA* 75: 2767–2771.
59. Gottesman, M.M. and Sobel, M.E. (1980). Tumor promoters and kirsten sarcoma virus increase synthesis of a secreted glycoprotein by regulating levels of translatable mRNA. *Cell* 19: 449–455.
60. Grinde, B. (1983). The thiol proteinase inhibitors, Z-Phe-PheCHN_2, inhibit lysosomal protein degradation in isolated rat hepatocytes. *Biochim. Biophys. Acta* 757: 15–20.
61. Hanewinkel, H., Glossl, J., and Kresse, H. (1987). Biosynthesis of cathepsin B in normal and I-cell fibroblasts. *J. Biol. Chem.* 262: 12351–12355.
62. Hara, K., Kominami, E., and Katunuma, N. (1988). Effect of proteinase inhibitors on intracellular processing of cathepsin B, H and L in rat macrophages. *FEBS Lett.* 231: 229–231.
63. Hirado, M., Iwara, D., Niinobe, M., and Fujii, S. (1981). Purification and properties of thiol protease inhibitor from rat liver cytosol. *Biochim. Biophys. Acta* 669: 21–27.
64. Hubbard, S.C. and Robbins, P.W. (1979). Synthesis and processing of protein-linked oligosaccharides *in vivo*. *J. Biol. Chem.* 254: 4568–4576.
65. Ishidoh, K., Towatari, T., Imajoh, S., Kawasaki, H., Kominami, E., Katunuma, N., and Suzuki, K. (1987). Molecular cloning and sequencing of cDNA for rat cathepsin L. *FEBS Lett.* 223: 69–73.
66. Ishidoh, K., Imajoh, S., Emori, Y., Ohno, S., Kawasaki, H., Minami, Y., Kominami, E., Katunuma, N., and Suzuki, K. (1987). Molecular cloning and sequencing of cDNA for rat cathepsin H. *FEBS. Lett.* 226: 33–37.
67. Janoff, A. (1985). Elastases and emphysema. *Am. Rev. Respir. Dis.* 132: 417–433.
68. Jarvinen, M. and Hopsu-Havu, V.K. (1975). Alpha-N-benzoylarginine-2-naphthylamide hydrolase (cathepsin B1?) from rat skin. II. Purification of the enzyme and demonstration of two inhibitors in the skin. *Acta Chem. Scand.* B29: 772–780.

69. Johnson, D.A., Barrett, A.J., and Mason, R.W. (1986). Cathepsin L inactivates alpha-1-proteinase inhibitor by cleavage in the reactive site region. *J. Biol. Chem.* 261: 14748–14751.
70. Joseph, L.J., Chang, L.C., Stamenkovich, D., and Sukhatme, V.P. (1988) Complete nucleotide and deduced amino acid sequences of human and murine preprocathepsin L. *J. Clin. Invest.* 81: 1621–1629.
71. Kamphius, I.G., Kalk, K.H., Swarle, M.B.A., and Drenth, J. (1985) Structure of papain at 1.65 Å resolution. *J. Mol. Biol.* 179: 233–256.
72. Kirschke, H. and Barrett, A.J. (1987). Chemistry of lysosomal proteases. In: *Lysosomes: Their Role in Protein Breakdown* (Glaumann, H., and Ballard, F.J., eds.) pp. 193–238. Academic Press, London.
73. Kirschke, H., Langner, J., Wiederanders, B., Ansorge, S., and Bohley, P. (1977). Cathepsin L: a new proteinase from rat liver lysosomes. *Eur. J. Biochem.* 74: 293–301.
74. Kirschke, H., Kembhavi, A.A., Bohley, P., and Barrett, A.J. (1982). Action of rat liver cathepsin L on collagen and other substrates. *Biochem. J.* 201: 367–372.
75. Kirschke, H., Locnikar, P., and Turk, V. (1984). Species variations among lysosomal cysteine proteinases. *FEBS Lett.* 174: 123–127.
76. Kirschke, H., Schmidt, I., and Wiederanders, B. (1986). Cathepsin S. *Biochem. J.* 240: 455–459.
77. Kirschke, H., Wikstrom, P. and Shaw, E. (1988). Active centre differences between cathepsin L and B: the S_1 binding region. *FEBS Lett.* 228: 128–130.
78. Kirschke, H., Wiederanders, B., Bromme, D. and Rinne, A. (1989). Cathepsin S from bovine spleen—purification, distribution, intracellular localization and action on proteins. *Biochem. J.* 264: 467–473.
79. Koga, H., Yamada, H., Nishimura, Y., Kato, K., and Imoto, T. (1990). Comparative study on specificities of rat cathepsin-L and papain-amino acid differences at substrate-binding sites are involved in their specificities. *J. Biochem.* 108: 976–982.
80. Kominami, E., Ii, K., and Katunuma, N. (1987). Activation of the intramyofibral autophagic lysosomal system in muscular dystrophy. *Am. J. Physiol.* 127: 461–466.
81. Kominami, E., Tsukahara, T., Hara, K., and Katunuma, N. (1988). Biosyntheses and processing of lysosomal cysteine proteinases in rat macrophages. *FEBS Lett.* 231: 225–228.
82. Kornfeld, S. and Mellman, I. (1989). The biogenesis of lysosomes. *Annu. Rev. Cell Biol.* 5: 483–525.
83. Laurell, C.B. and Eriksson, S. (1963). The electrophoretic alpha 1-globulin pattern of serum in alpha 1-antitrypsin deficiency. *Scand. J. Clin. Invest.* 15: 132–140.
84. Lee, X., Ahmed, F.R., Hirama, T., Huber, C.P., Rose, D.R., To, R., Hasnain, S., Tam, A., and Mort, J.S. (1990). Crystallization of recombinant rat cathepsin B. *J. Biol. Chem.* 265: 5950–5951.
85. Liao, J.C.R. and Lenney, J.F. (1984). Cathepsin J and K: high molecular weight cysteine proteinases from human tissues. *Biochem. Biophys. Res. Commun.* 124: 909–916.
86. Lindahl, P., Alriksson, E., Jornvall, H. and Bjork, I. (1988). Interaction of the cysteine proteinase inhibitor chicken cystatin with papain. *Biochemistry* 27: 5074–5082.
87. Liotta, L.A. (1986). Tumor invasion and metastases—role of the extracellular matrix. *Cancer Res.* 46: 1–7.
88. Lowe, G. and Williams, A. (1965). Direct evidence for an acylated thiol as an intermediate in papain and ficin catalyzed hydrolyzes. *Biochem. J.* 96: 189–193.
89. Maciewicz, R.A. and Etherington, D.J. (1988). A comparison of four cathepsins (B, L, N and S) with collagenolytic activity from rabbit spleen. *Biochem. J.* 256: 433–440.
90. Mason, R.W. (1986). Species variations amongst lysosomal cysteine proteinases *Biomed. Biochim. Acta.* 45: 1433–1440.
91. Mason, R.W. (1986). Species variants of cathepsin L and their immunological identification. *Biochem. J.* 240: 285–288.
92. Mason, R.W. (1989). Interaction of lysosomal cysteine proteinases with α_2-macroglobulin: conclusive evidence for the endopeptidase activities of cathepsins B and H. *Arch. Biochem. Biophys.* 273: 367–374.
93. Mason, R.W., Taylor, M.A., and Etherington, D.J. (1984). The purification and properties of cathepsin L from rabbit liver. *Biochem. J.* 217: 209–217.

94. Mason, R.W., Green, G.D.J., and Barrett, A.J. (1985). Human liver cathepsin L. *Biochem. J.* 226: 233–241.
95. Mason, R.W., Johnson, D.A., Barrett, A.J., and Chapman, H.A. (1986). Elastinolytic activity of human cathepsin L. *Biochem. J.* 233: 925–927.
96. Mason, R.W., Walker, J.E., and Northrop, F.D. (1986). The N-terminal amino acid sequences of the heavy and light chains of human cathepsin L. *Biochem. J.* 240: 373–377.
97. Mason, R.W., Gal, S., and Gottesman, M.M. (1987). The identification of the major excreted protein (MEP) from a transformed mouse fibroblast cell line as a catalytically active precursor form of cathepsin L. *Biochem. J.* 248: 449–454.
98. Mason, R.W., Bartholomew, L.T., and Hardwick, B.S. (1989). The use of benzyloxycarbonyl-[^{125}I]-iodotyrosylalanyldiazomethane as a probe for active cysteine proteinases in human tissues. *Biochem. J.* 263: 945–949.
99. Mason, R.W., Wilcox, D., Wikstrom, P., and Shaw, E.N. (1989). The identification of active forms of cysteine protienases in Kirsten-virus-transformed mouse fibroblasts by use of a specific radiolabelled inhibitor. *Biochem. J.* 257: 125–129.
99a. Mason, R.W., and Massey, S.D. (1992). Surface activation of pro-cathepsin L. *Bichem. Biophys. Res. Commun.* 189: 1659–1666.
100. McDonald, J.K. (1985). An overview of protease specificity and catalytic mechanisms: aspects related to nomenclature and classification. *Histochem. J.* 17: 773–785.
101. McDonald, J.K. and Barrett, A.J. (eds.) (1986). *Mammalian Proteases: A Glossary and Bibliography,* Vol. 2, *Exopeptidases*. Academic Press, London.
102. McDonald, J.K. and Kadkhodayan, S. (1988). Cathepsin L—a latent proteinase in guinea pig sperm. *Biochem. Biophys. Res. Commun.* 151: 827–835.
103. McKay, M.J., Marsh, M.W., Kirschke, H., and Bond, J.S. (1984). Inactivation of fructose-1,6-bisphosphate aldolase by cathepsin L. Stimulation by ATP. *Biochim. Biophys. Acta* 784: 9–15.
104. Mort, J. and Recklies, A.D. (1986). Interrelationship of active and latent secreted human cathepsin B precursors. *Biochem. J.* 233: 57–63.
105. Mort, J.S., Leduc, M.S., and Recklies, A.D. (1983). Characterization of a latent cysteine proteinase from ascitic fluid as a high molecular weight form of cathepsin B. *Biochim. Biophys. Acta* 755: 369–375.
106. Mort, J.S., Recklies, A.D., and Poole, A.R. (1984). Extracellular presence of the lysosomal proteinase cathepsin B in rheumatoid synovium and its activity at neutral pH. *Arth. Rheum.* 27: 509–515.
107. Murata, M., Miyashita, S., Yokoo, C., Tamai, M., Hanada, K., Hatayama, K., Towatari, T., Nikawa, T., and Katunuma, N. (1991). Novel epoxysuccinyl peptides—Selective inhibitors of cathepsin-B, *in vitro. FEBS. Lett.* 280: 307–310.
107a. Musil, D., Zucic, D., Turk, D., Engh, R.A., Mayr, I., Huber, R., Popovic, T., Turk, V., Towatari, T., Katunuma, N., and Bode, W. (1991). The refined 2.15-A X-ray crystal structure of human liver cathepsin-B—the structural basis for its specificity. *EMBO J.* 10: 2321–2330.
108. Nicklin, M.J.H. and Barrett, A.J. (1984). Inhibition of cysteine proteinases and dipeptidyl peptidase 1 by egg white cystatin. *Biochem. J.* 223: 245–253.
109. Niewoehner, D.E. (1988). Cigarette smoking, lung inflammation, and the development of emphysema. *J. Lab. Clin. Med.* 111: 15–27.
109a. Nikawa, T., Towatari, T., and Katunuma, N. (1992). Purification and characterization of cathepsin-J from rat liver. *Eur. J. Biochem.* 204: 381–393.
110. Nishimura, Y. and Kato, K. (1987). Intracellular transport and processing of lysosomal cathepsin B. *Biochem. Biophys. Res. Commun.* 148: 254–259.
111. Nishimura, Y. and Kato, K. (1987). Intracellular transport and processing of lysosomal cathepsin H. *Biochem. Biophys. Res. Commun.* 148: 329–334.
112. Nishimura, Y., Furuno, K., and Kato, K. (1988). Biosynthesis and processing of lysosomal cathepsin L in primary cultures of rat hepatocytes. *Arch. Biochem. Biophys.* 263: 107–116.
113. Nishimura, Y., Kawabata, T., and Kato, K. (1988). Identification of latent pro-cathepsins B and L in microsomal lumen: characterization of enzymatic activation and protelytic processing *in vitro. Arch. Biochem. Biophys.* 261: 64–71.

114. Nishimura, Y., Kawabata, T., Furuno, K., and Kato, K. (1989). Evidence that aspartic proteinase is involved in the proteolytic processing event of procathepsin L in lysosomes. *Arch. Biochem. Biophys.* 271: 400–406.
115. North, M.J. (1982). Comparative biochemistry of the proteinases of eukaryotic microorganisms. *Microbiol. Rev.* 46: 308–340.
116. North, M.J., Mottram, J.C., and Coombs, G.H. (1990). Cysteine Proteinases of Parasitic Protozoa. *Parasitol. Today.* 6: 270–275.
117. Orlowski, M., Orlowski, J., Lesser, M., and Kilburn, K.H. (1981). Proteolytic enzymes in bronchopulmonary lavage fluids: cathepsin B-like activity and prolyl endopeptidase. *J. Lab. Clin. Med.* 97: 467–476.
118. Polgar, L. (1973). On the mode of activation of the catalytically essential sulfhydryl group of papain. *Eur. J. Biochem.* 33: 104–109.
119. Portnoy, D.A., Erickson, A.H., Kochan, J., Ravetch, J.V., and Unkeless, J.C. (1986). Cloning and characterization of a mouse cysteine proteinase. *J. Biol. Chem.* 261: 14697–14703.
120. Pratt, S.A., Finley, T.A., Smith, M.H., and Ladman, A.J. (1969). A comparison of alveolar macrophages and pulmonary surfactant obtained from the lungs of human smokers and non-smokers by endobronchial lavage. *Anat. Rec.* 163: 497–508.
121. Press, E.M., Porter, R.R., and Cebra, J. (1960). The isolation and properties of a proteolytic enzyme, cathepsin D, from bovine spleen. *Biochem. J.* 74: 501–514.
122. Qian, F., Bajkowski, A.S., Steiner, D.F., Chan, S.J., Frankfater, A. (1989). Expression of 5 Cathepsins in Murine Melanomas of Varying Metastatic Potential and Normal Tissues. *Cancer Res.* 49: 4870–4875.
123. Rawlings, N.D. and Barrett, A.J. (1990). Evolution of proteins of the cystatin superfamily. *J. Mol. Evol.* 30: 60–71.
124. Reilly, J.J., Mason, R.W., Chen, P., Joseph, L.J., Sukhatme, V.P., and Yee, R., and Chapman, H.A. (1989). Synthesis and processing of cathepsin L, an elastase, by human alveolar macrophages. *Biochem. J.* 257: 493–498.
125. Rich, D.H., Brown, M.A., and Barrett, A.J. (1986). Purification of cathepsin B by a new form of affinity chromotography. *Biochem. J.* 235: 731–734.
126. Ritonja, A., Popovic, T., Kotnik, M., Machleidt, W., and Turk, V. (1988). Amino acid sequences of the human kidney cathepsins H and L. *FEBS Lett.* 228: 341–345.
126a. Rowan, A.D., Mason, P., Mach, L., and Mort, J.S. (1992). Rat procathepsin-B—proteolytic processing to the mature form *in vitro*. *J. Biol. Chem.* 267: 15993–15999.
127. Rozhin, J., Robinson, D., Stevens, M.A., Lah, T.T., Honn, K.V., Ryan, R.E., and Sloane, B.F. (1987). Properties of a plasma membrane-associated cathepsin B-like cysteine proteinase in metastatic B16 melanoma variants. *Cancer Res.* 47: 6620–6628.
128. Rozhin, J., Wade, R.L., Honn, K.V., and Sloane, B.F. (1989). Membrane-associated cathepsin L: a role in metastasis of melanomas. *Biochem. Biophys. Res. Commun.* 164: 556–561.
129. San Segundo, B., Chan, S.J., and Steiner, D.F. (1985). Identification of cDNA clones encoding a precursor of rat liver cathepsin B. *Proc. Natl. Acad. Sci. USA* 82: 2320–2324.
130. Schechter, I. and Berger, A. (1967). On the active site of proteases. I. Papain. *Biochem. Biophys. Res. Commun.* 27: 157–161.
131. Schwartz, W.N. and Barrett, A.J. (1980). Human cathepsin H. *Biochem. J.* 191: 487–497.
132. Seglen, P.O. (1983). Inhibitors of lysosomal function. *Meth. Enzymol.* 96: 737–764.
133. Sen, L.C., and Whitaker, J.R. (1973). Some properties of a ficin-papain inhibitor from avian egg white. *Arch. Biochem. Biophys.* 158: 623–632.
134. Shaw, E. and Dean, R.T. (1980). The inhibition of macrophage protein turnover by a selective inhibitor of thiol proteinases. *Biochem. J.* 186: 385–390.
135. Shaw, E., Wikstrom, P., and Ruscica, J. (1983). An exploration of the primary specificity site of cathepsin B. *Arch. Biochem. Biophys.* 222: 424–429.
135a. Shi, G.P., Munger, J.S., Meara, J.P., Rich, D.H., and Chapman, H.A. (1992). Molecular cloning and expression of human alveolar macrophage cathepsin-S, an elastinolytic cysteine protease. *J. Biol. Chem.* 267: 7258–7262.

136. Silver, I.A., Murrills, R.J., and Etherington, D.J. (1988). Microelectrode studies on the acid microenvironment beneath adherent macrophages and osteoclasts. *Exp. Cell Res.* 175: 266–276.
137. Sloane, B.F., Dunn, J.R., and Honn, K.V. (1981) Lysosomal cathepsin B: correlation with metastatic potential. *Science* 212: 1151–1153.
138. Sloane, B.F., Rozhin, J., Johnson, K., Taylor, H., Crissman, J.D., and Honn, K.V. (1986). Cathepsin B: association with plasma membrane in metastatic tumors. *Proc. Natl. Acad. Sci. USA* 83: 2483–2487.
139. Starkey, P.M., and Barrett, A.J. (1982) Evolution of α_2-macroglobulin. *Biochem. J.* 205: 91–95.
140. Stubbs, M.T., Laber, B., Bode, W., Huber, R., Jerala, R., Lenarcic, B., Turk, V. (1990). The refined 2.4A X-ray crystal structure of recombinant human stefin-B in complex with the cysteine proteinase papain—A novel type of proteinase inhibitor interaction. *EMBO J.* 9: 1939–1947.
141. Takahashi, T., Yonezawa, S., Dehdarani, A.H., and Tang, J. (1986). Comparative studies of two cathepsin B isozymes from porcine spleen. *J. Biol. Chem.* 261: 9375–9381.
142. Takahashi, T., Dehdarani, A.H., and Tang, J. (1988). Porcine spleen cathepsin H hydrolysis oligopeptides solely by aminopeptidase activity. *J. Biol. Chem.* 263: 10952–10957.
143. Takio, K., Towatari, T., Katunuma, N., Teller, D. and Titani, K. (1983). Homology of amino acid sequences of rat liver cathepsins B and H with that of papain. *Proc. Natl. Acad. Sci. USA* 80: 3666–3670.
144. Tallan, H.H., Jones, M.E., Fruton, J.S. (1952). On the proteolytic enzymes of animal tissues: X. Beef spleen cathepsin C. *J. Biol. Chem.* 194: 793–805.
145. Teno, N., Tsuboi, S., Itoh, N., Okamoto, H., and Okada, Y. (1987). Significant effects of Z-Gln-Val-Val-OME, common sequences of thiol proteinase inhibitors on thiol proteinases. *Biochem. Biophys. Res. Commun.* 143: 749–752.
146. Towatari, T. and Katunuma, N. (1983). Selective cleavage of peptide bonds by cathepsin L and B from rat liver. *J. Biochem.* 93: 1119–1128.
147. Towatari, T., Tanaka, K., Yoshikawa, D., and Katunuma, N., (1978). Purification and properties of a new cathepsin from rat liver. *J. Biochem.* 84: 659–671.
148. Trabandt, A., Aicher, W.K., Gay, R.E., Sukhatme, V.P., Nilsonhamilton, M., Hamilton, R.T., Mcghee, J.R., Fassbender, H.G., and Gay, S. (1990). Expression of the collagenolytic and ras-induced cysteine proteinase cathepsin-L and proliferation-associated oncogenes in synovial cells of MRL/I mice and patients with rheumatoid arthritis. *Matrix.* 10: 349–361.
149. Troen, B.R., Gal, S., Gottesman, M.M. (1987). Sequence and expression of the cDNA for MEP (major excreted protein), a transformation-regulated secreted protein. *Biochem. J.* 246: 731–735.
150. Turnsek, T., Kregar, I., and Lebez, D. (1975). Acid sulphydryl protease from calf lymph nodes. *Biochim. Biophys. Acta* 403: 514–520.
151. Varughese, K.I., Ahmed, F.R., Carey, P.R., Hasnain, S., Huber, C.P., and Storer, A.C. (1989). The crystal structure of a papain-E-64 complex. *Biochemistry* 28: 1330–1332.
152. Von Figura, K. and Hasilik, K. (1986). Lysosomal enzymes and their receptors. *Ann. Rev. Biochem.* 55: 167–194.
153. Watanabe, H., Green, G.D., Shaw, E. (1979). A comparison of the behavior of chymotrypsin and cathepsin B towards peptidyl diazomethyl ketones. *Biochem. Biophys. Res. Commun.* 89: 1354–1360.
153a. Wiederanders, B., Bromme, D., Kirschke, H., von Figura, K., Schmidt, B., and Peters, C. (1992). Phylogenetic conservation of cysteine proteinases—cloning and expression of a cDNA coding for human cathepsin-S. *J. Biol. Chem.* 267: 13708–13713.
154. Wiederanders, B. and Kirschke, H. (1989) The processing of a cathepsin L precursor *in vitro*. *Arch. Biochem. Biophys.* 272: 516–521.
155. Wilcox, D. and Mason, R.W. (1992). Inhibition of cysteine proteinases in lysosomes and whole cells. *Biochem. J.* 285: 495–502.
156. Xin, X.-Q., Gunesekera, B., and Mason, R.W. (1992). The specificity and elastinolytic activities of bovine cathepsin-S and cathepsin-H. *Arch. Biochem. Biophys.* 299: 334–339.

MECHANISM AND REGULATION OF AUTOPHAGIC DEGRADATION OF CELLULAR PROTEINS

William A. Dunn, Jr.

I. INTRODUCTION

Cellular proteins are continually being degraded and resynthesized. Protein degradation occurs by both nonlysosomal and lysosomal mechanisms. In addition to lysosomes, proteinases have been localized to virtually every cellular compartment

Advances in Cell and Molecular Biology of Membranes
Volume 1, pages 117–138

ISBN: 1-55938-362-3

including cytosol, mitochondria, endosomes, plasma membrane, and endoplasmic reticulum (*4,10,27,51*). Cellular proteins may be selectively ubiquitinated by soluble enzymes at a lysine residue internal to a promoting N-terminal amino acid. Poly-ubiquitinated proteins are then rapidly degraded by large ATP-dependent proteasomes or within lysosomes (*12,25,36,46*). This proteolytic system is restricted to the selective degradation of cytosolic proteins. The remainder of the cellular proteins (as much as 70% of the total) are degraded by lysosomal means (*48*). In addition, lysosomal protein degradation has been implicated in the cellular remodeling that occurs during differentiation, metamorphosis, aging, and transformation, and in the removal of abnormal cellular components that may occur in cell toxicity or death (*8,27,73,79,80,81*). Liver growth during regeneration has been attributed to an inhibition of lysosomal protein degradation, rather than accelerated protein synthesis (*54,68*).

Mechanisms of protein entry into lysosomes include: endocytosis, crinophagy, hsc73 (defined below) -mediated transport, and autophagy. Endocytosis is a process for the delivery of exogenous and plasma membrane proteins to lysosomes and is discussed in more detail in other chapters. Crinophagy is the fusion of lysosomes with secretory vesicles resulting in the ultimate degradation of secretory proteins. We have recently shown that as much as 40% of newly synthesized serum proteins are degraded prior to secretion by this mechanism (*38*). Dice and co-workers have identified a "receptor-mediated" route for entry of cytosolic proteins into lysosomes. They propose that a peptide recognition receptor selectively binds to a protein and by, an as yet unknown mechanism, inserts the protein into a lysosome. This soluble receptor belongs to the family of heat shock proteins and has been recently identified as heat-shock cognate protein, hsc73 (*6,9,51*). Autophagy is a well-recognized mechanism for the delivery of cellular components to lysosomes. This process occurs in all mammalian cells and can be classified into two categories: microautophagy and macroautophagy. Microautophagy is the sequestration of small portions of cytoplasm by an invagination of the lysosome membrane. It is this mechanism which has been implicated in the accumulation of vesicles within secondary lysosomes identified as multivesicular bodies (*27*). During macroautophagy, organelles and cytosol are sequestered away from the remaining cytoplasm within a double-membraned compartment and subsequently degraded within the lysosomal vacuolar system. For the purposes of this review, the discussion will be limited to macroautophagy.

II. STRUCTURE AND FUNCTION OF AUTOPHAGIC VACUOLES

The occurrence of autophagic vacuoles has been described in many cell types. Depending upon environmental conditions and metabolic state of the cell, autophagic vacuoles may account for as much as 1.5% of the cellular volume. Based solely on morphology, autophagic vacuoles can be classified into two categories:

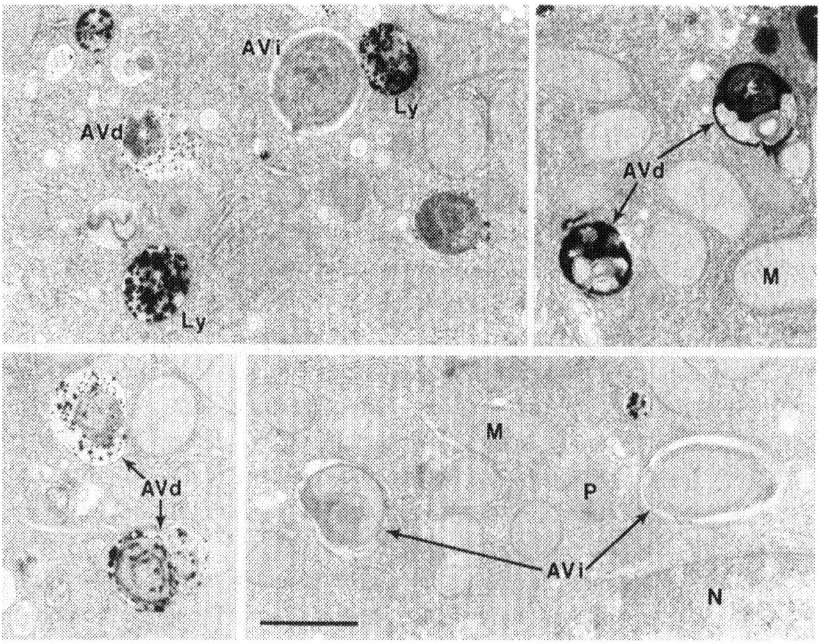

Figure 1. Morphology of autophagic vacuoles. Livers were cyclically perfused with amino acid-depleted medium containing glucagon. At 30 min of perfusion the livers were aldehyde-fixed and processed for acid phosphatase cytochemistry (*14*). At this time, autophagosomes or nascent autophagic vacuoles (AVi) and autolysosomes or degradative autophagic vacuoles (AVd) are routinely observed. N, nucleus; M, mitochondria; P, peroxisome; and Ly, lysosome. Bar, 1.0 μm.

autophagosomes and autolysosomes (Fig. 1). Autophagosomes or nascent autophagic vacuoles are 0.3 to 1.3 μm in diameter and bound by at least two membranes. Vacuolar contents are comprised of undegraded cytoplasmic ground substance and recognizable cellular organelles (e.g., mitochondria and peroxisomes). Autolysosomes or degradative autophagic vacuoles are similar in size to autophagosomes but bound by only one membrane. Their matrix is heterogenous containing electron-lucent regions interdispersed between cellular components in various states of degeneration. In addition, one finds remnants of partially degraded membranes sometimes present as myelinic whorls. Autolysosomes, not autophagosomes, contain acid phosphatase justifying their classification as secondary lysosomes. Upon a stimulus (e.g., amino acid deprivation), autophagosomes appear as early as 5 to 10 min followed by the occurrence of autolysosomes at 10 to 20 min (*14,65*). Within 30 to 45 min, the fractional volumes of these vacuoles reach a steady state whereby the volumes and numbers of both types of vacuoles are equal. At this time, the rate of autophagosome formation equals the rate of loss of autolysosomes. The contents of the autolysosome are hydrolyzed resulting in a release of amino acids, monosaccharides, fatty acids, and mononucleosides coincident with a reduction in vacuole size. This hydrolytic autolysosome eventually becomes a smaller residual or dense body (e.g., secondary lysosome) containing only undegraded byproducts (e.g., lipofuscin). This process has been shown to occur with a half-life of 8.0 min (*65*).

There have been many approaches used for the isolation of lysosomes and autophagic vacuoles. Of the most successful is separation on the basis of differing densities. Autophagic vacuoles migrate at two densities on Percoll gradients. Utilizing morphological criteria, the lighter fraction is enriched for autophagosomes while the heavy fraction contains mostly autolysosomes (*19*). However, the conditions of the gradient fail to resolve autolysosomes from multivesicular and residual bodies. Lysosomal enzyme markers migrate at two densities on Ludox, Nycodenz, and Metrizamide gradients (*12,42,72*). One reason for the density separation of lysosomal enzymes is that autolysosomes migrate at lower densities than residual or dense bodies.

Lysosomal sequestration is responsible for the normal degradation of as much as 70% of cellular proteins and the enhanced degradation of proteins in response to nutrient-deprivation (*34,48*). Enhanced protein degradation that occurs during nutrient deprivation can be efficiently inhibited by propylamine, a lysosomotropic amine, and 3-methyladenine, which suppresses formation of autophagic vacuoles (*34,70*). On the other hand, the role of lysosomes in "basal" degradation of proteins has been the subject of some controversy. It has been reported that lysosomotrophic amines inhibit endogenous protein degradation by only 20 to 40% in cells cultured under optimal growth conditions (*1,34,50*). However, the experimentally-defined "basal" state obtained when culturing cells in hormone and amino acid enriched medium may not reflect *in vivo* conditions. Due to diurnal feeding rhythms, circulating levels of essential amino acids fluctuate resulting in variations in the occurrence of autophagic vacuoles (*52,56,58*). Lysosomal degradation of endogenous proteins is regulated by circulating levels of amino acids (*48*). The sources of essential amino acids *in vivo* are limited to nutritional intake and degradation of endogenous proteins. Under starvation conditions, essential amino acids are obtained solely from the degradation of existing proteins. Starvation enhances the formation of autophagic vacuoles and transport of cellular proteins into lysosomes (*52*). Therefore, autophagy and lysosome protein degradation is carefully regulated such that a constant supply of essential amino acids necessary for protein synthesis is maintained. Autophagy-mediated protein degradation is presumed to be responsible for the removal of cellular organelles that occurs during cell remodeling (e.g., differentiation and transformation) and injury (*8,27,73,79,80,81*). Finally, autophagy has been shown to be inversely related to cell growth (*57*). That is, autophagy is depressed under conditions of rapid cell growth (e.g., liver regeneration) (*54,68*).

III. FORMATION OF AUTOPHAGIC VACUOLES

In the past, researchers have suggested that autophagic vacuoles originated from Golgi cisternae, endoplasmic reticulum, plasma membrane, lysosomes, and phagophores (*27,55,61,69*). These earlier conclusions were based on membrane morphology and enzyme cytochemistry. With the refinement of immunological and lectin probes and the techniques for using these probes at the ultrastructural level, we are now able to define more accurately the membrane source of autophagosomes. We have performed a systematic examination of the presence and absence of organelle-specific antigens within autophagic vacuoles (Table 1). Our data reveals that antigens of rough endoplasmic reticulum and lysosomes are found in autophagic

Table 1. Localization of Cellular Antigens and DAMP to Autophagic Vacuoles

	Cellular Localization[1]								
	PM	*Go*	*SER*	*RER*	*ENDO*	*LYS*	*AVi*	*AVi/d*	*AVd*
Antigens									
Plasma membrane:									
EGF receptor	++++	+			++				
antigen CE9	++++								
antigen HA4	++++								
Golgi apparatus:									
sialyl-transferase		+++							
antigen 53Fc3		++							
mannose 6-phosphate receptor	+	+++			++			+	
Endoplasmic reticulum:									
ribophorin II		++		++++			++		
RER antigens				++++			+++	++	+
KDEL sequence				++++			++		+
Lysosome:									
LGP120						++++		++	++++
LIMP I						++++			++++
LIMP IV						++++			++++
Cathepsin L						++++		+	++++
Cathepsin D						++++		+	++++
DAMP (acidification)						++++		++++	++

Note: [1] PM, plasma membrane; Go, Golgi apparatus; SER, smooth endoplasmic reticulum; RER, rough endoplasmic reticulum; ENDO, endosomes; LYS, multivesicular bodies and secondary lysosomes; AVi, nascent autophagic vacuole or early autophagosome; AVi/d, intermediate autophagic vacuole or late autophagosome; AVd, degradative autophagic vacuole or autolysosome.

vacuoles. We have identified membrane antigens of the rough endoplasmic reticulum to both the inner and outer membranes of the autophagosome (*14*). The cisternal space between the inner and outer membranes of autophagosomes contains proteins recognized by antibodies specific for the C-terminal tetrapeptide sequence (KDEL) of lumenal proteins of the rough endoplasmic reticulum (Fig. 2). In addition, this cisternal space contains glucose 6-phosphatase and inosine-diphosphatase (*3,18*). On the other hand, cytochrome P-450 appears to be absent from the limiting membranes of these vacuoles (*84*). Yamamoto and associates have proposed autophagosomes are formed from "isolation membranes" that contain lysosomal membrane proteins (*83,84*). Less than 50% of the autophagosomes, identified by a limiting double membrane, contain acid hydrolases and lysosomal membrane proteins (i.e., LGP120, LIMP I, LIMP V, and LGP107) (*15,21*). When present, LGP120 is restricted to the outer membrane (*15*). The distribution of sialylated LGP120 justifies the binding of cationized ferritin and wheat germ agglutinin to the inner surface of this membrane (*15,62,83*). The data are consistent with the autophagosome being formed from the rough endoplasmic reticulum and rapidly fusing with a LGP120-containing vesicle resulting in the deposition of this protein at the outer membrane.

The difficulties in evaluating the events of vacuole formation are twofold. First, different promoters of autophagy are being used to study this process. For example, it is not known if vacuole formation is the same under different environmental stresses (e.g., nutrient deprivation vs. vinblastine toxicity). Second, we are using morphologic observations to study a very dynamic cellular event. Indeed, lysosomes rapidly fuse with newly formed vacuoles. At steady state, we estimate that only 25–30% of the total number of autophagic vacuoles are autophagosomes bound by membranes arising solely from the rough endoplasmic reticulum. The membranes of the remaining autophagosomes (20–25% of total AV's) and autolysosomes (50% of total AV's) contain not only portions of the rough endoplasmic reticulum but also lysosome (e.g., LGP120) and endosome (e.g., mannose 6-phosphate receptor) membrane proteins (see below). Therefore, these membranes are hybrids from more than one source and thereby unique. Indeed, the characterization of the limiting membranes of autophagic vacuoles revealed major differences between these membranes and other cellular membranes. For example, freeze fracture replicas of autophagic vacuoles were found to display a unique distribution of intramembrane particles not found elsewhere in the cell (*26,61*). In addition, the content of unsaturated fatty acids is highest in vacuolar membranes (*60*).

Results provided by us and others support the concept that autophagosomes are formed from smooth-regions of the rough endoplasmic reticulum (*3,14,18,21*). The inner and outer membranes of autophagosomes arise from an invagination of the rough endoplasmic reticulum. Rough endoplasmic reticulum membrane proteins are found in both limiting membranes (*14*), while lumenal proteins are restricted to the region between these membranes (Fig. 2). The absence of ribosomes from the limiting membranes of autophagosomes suggest that protein synthesis is not occurring at the site of vacuole formation. This scenario is further substantiated by

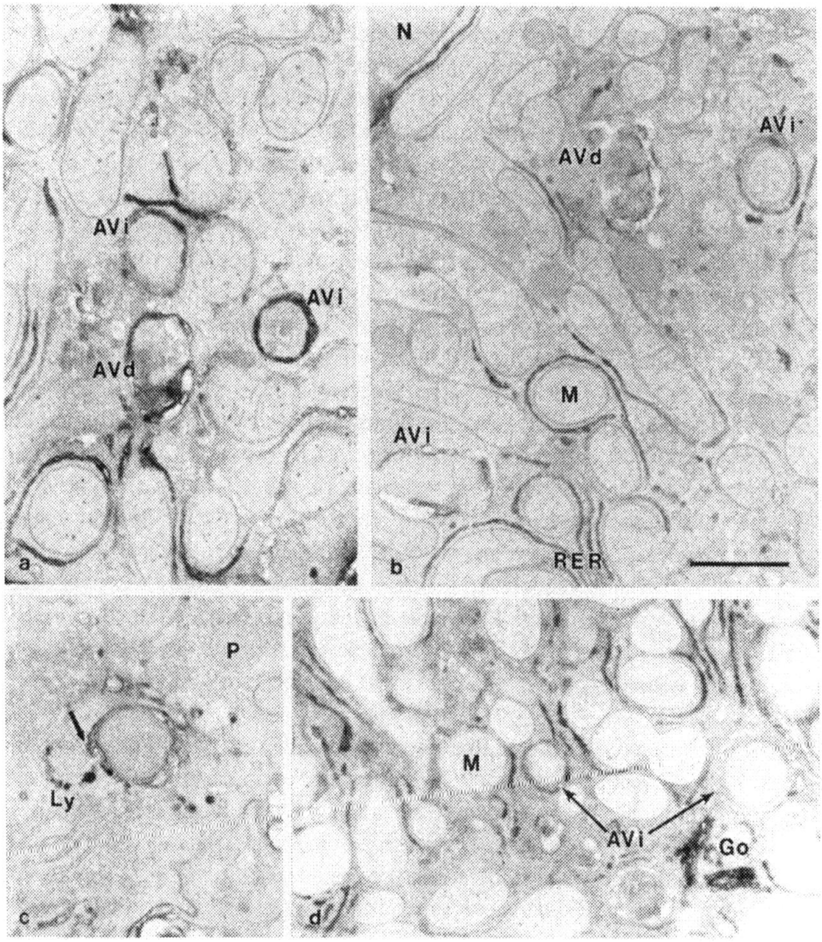

Figure 2. Immunoperoxidase localization of rough endoplasmic reticulum (RER) and secretory proteins to autophagic vacuoles. Livers previously perfused with nutrient-depleted medium were fixed and processed for immunoperoxidase cytochemistry as previously described (*14*). Membrane antigens (14, 25, and 40 kDa) recognized by anti-RER antibodies were found at the limiting membranes of autophagosomes (AVi) and occasionally within autolysosomes (AVd) (**a**,**c**). Lumenal RER proteins recognized by an anti-KDEL antibody (kindly provided by Dr. David Vaux, see ref. *77*) were present within the space between the inner and outer membranes of autophagosomes (**b**). Lysosomes (Ly) can be seen fusing with autophagosomes (arrow) (**c**). Serum albumin was present in some, not all, autophagosomes (**d**). N, nucleus; M, mitochondria; P, peroxisome; Go, Golgi apparatus; and RER, rough endoplasmic reticulum. Bar, 1.0 μm. [Panels **a**, **c**, and **d** are reproduced from the Journal of Cell Biology (Vol. 110, pp (1930 and 1932) by copyright permission of the Rockefeller University Press.]

the absence of serum albumin from 65% of newly formed autophagosomes (Fig. 2). It is possible that the release of membrane-bound ribosomes is required to allow vacuole formation. Indeed, cycloheximide which prevents the loss of ribosomes suppresses autophagy and protein degradation in hepatocytes and cells in culture (*17,66*). However, pactamycin which promotes ribosome loss from membranes also inhibits protein degradation and autophagy in cultured cells (*17,66*, Jasionowski and Dunn, unpublished observations). The inhibitory effects of these drugs on autophagy suggest two possibilities: (1) the onset of autophagy requires the synthesis of a short-lived protein; or (2) autophagy is inhibited by the accumulation of amino acids upon the absence of protein synthesis. Autophagy can occur without protein synthesis. Cycloheximide, pactamycin and puromycin failed to inhibit vacuole formation in isolated kidney tubules (*71*). In addition, cycloheximide inhibition of protein degradation in liver can be reversed by the addition of glucagon, a promoter of autophagy (*82*). As will be discussed in detail below, the onset of autophagy is modulated by a set of "regulatory" amino acids. In the absence of protein synthesis, these amino acids will accumulate ultimately resulting in the inhibition of autophagy. Therefore, it is likely that inhibitors of protein synthesis suppress autophagy by promoting the accumulation of regulatory amino acids.

The effects of temperature on protein degradation and vacuole formation have also been examined. Low temperatures (<20°C) inhibit protein degradation and the formation of autophagic vacuoles in hepatocytes (*22,49*). On the other hand, high temperatures (39.5°C) enhance lysosome-mediated protein degradation in CHO and FM3A cells in culture (*24*). However, enhanced degradation of proteins was not coincident with a significant increase in the occurrence of autophagic vacuoles (*39*).

The cytoskeleton has a major influence in the movements of organelles within the cell. Indeed, the organization of the rough endoplasmic reticulum and lysosomes within the cell is closely associated with microtubules (*45,63,74*). Vinblastine treatment increases the occurrence of autophagic vacuoles in liver and apparently stimulates cellular lipolysis and protein degradation (*40,41*). However, microtubule poisons have been shown to inhibit endogenous protein degradation in isolated hepatocytes (*33,67*) and cultured normal rat kidney cells (*2*). Whether this drug stimulates vacuole formation or inhibits its maturation has been the subject of some debate. Vinblastine does appear to increase the cytoplasmic volume fraction of autophagosomes while slightly decreasing the volume of autolysosomes (*34,70*). For comparison, proteinase inhibitors increase the number and size of autolysosomes by preventing intralysosomal hydrolysis (*34*). Interestingly, the accumulation of autolysosomes in the presence of leupeptin can be prevented by phalloidin treatment (*76*). We have found that cytochalasins B and D depress protein degradation by efficiently inhibiting the formation of autophagosomes (*2*). These results suggest that the disassembly of microfilaments inhibits the formation of autophagic vacuoles. Finally, intermediate filaments have been implicated in targeting proteins into autophagic vacuoles. Cytosolic and plasma membrane proteins whose degradation is inhibited by lysosomal perturbants become associ-

ated with intermediate filaments prior to their sequestration into autophagosomes (*11,16*).

The formation of the autophagosome is a regulated event (see below) and results in the sequestration of cellular components for eventual degradation. In addition, this event requires ATP and can be inhibited by GTP-γ-S (*29,71*). By some yet undefined mechanism, 3-methyladenine inhibits the formation of autophagic vacuoles (*70*). Autophagosomes have been shown to sequester for degradation internalized gap junctions, peroxisomes, mitochondria, endoplasmic reticulum, ribosomes, and cytosolic ground substance (*30,43,44,53,55*). Although the sequestration of cellular components within autophagosomes is thought to be nonselective (*32*), some results suggest that macromolecules are selectively degraded within autolysosomes (*31,35*). For example, the sequestration of ribosomes and degradation of rRNA are reduced in the presence of glucagon, a potent activator of vacuole formation (*35*). In addition, the specific activity of cytosolic proteins in autolysosomes isolated from leupeptin-treated animals differ (*31*). This difference is unrelated to the half-life of the protein. Two possible mechanisms have been proposed for the selective uptake of cytosolic proteins into lysosomes. First, sequestration of cytosolic proteins may be dependent on their ability to associate with intermediate filaments (*11,16*). Second, Dice and coworkers have shown that some cytosolic proteins are selectively transported into lysosomes by a receptor-mediated event (*9,51*). This receptor belongs to the 70 kDa family of heat-shock proteins and recognizes a pentapeptide sequence (KFERQ) found in many proteins including 30% of the cytosolic proteins (*6,9*). Upon binding, the receptor initiates transport of the protein into a pre-lysosome (e.g., endosome or autophagosome) or lysosome (e.g., multivesicular body or autolysosome) compartment ultimately resulting in complete degradation of the protein.

IV. MATURATION OF AUTOPHAGIC VACUOLES

After formation of the autophagosome, this vacuole acquires lysosomal membrane proteins and hydrolases thereby becoming a autolysosome, which, in turn, transforms into a residual body. Coincident with this acquisition is a loss of the inner membrane presumably due to hydrolysis. Many studies have shown profiles of secondary lysosomes fusing with autophagosomes (see Figs. 1 and 2c). Utilizing immunological and morphological criteria, we have defined a third vacuolar compartment with characteristics of both autophagosomes and autolysosomes (*15*). Similar to the autophagosome, this intermediate vacuole is bound by a double membrane containing RER antigens and lacks acid hydrolases (Table 1). However, this vacuole is acidic and LGP120, a lysosomal membrane protein, can be identified at its outer limiting membrane (Fig. 3). We have classified this vacuole as a late autophagosome distinct from nascent or early autophagosomes and autolysosomes (Table 1). In addition, based on its characteristics, we propose that the late autophagosome is an intermediate stage in the maturation of early autophagosomes

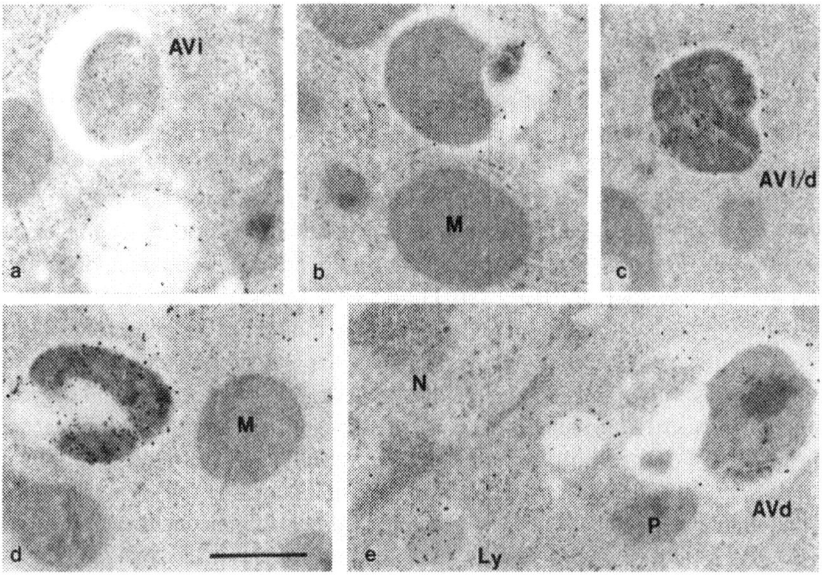

Figure 3. Maturation of autophagosomes into autolysosomes. Livers were perfused with nutrient-depleted medium containing DAMP for 30 min. The liver was then fixed by perfusion, dehydrated, and embedded in LRGold resin (*14*). DAMP (5 nm) and LGP120 (10nm) were localized by double-labeling immunogold cytochemistry (**a–e**). We have defined immunologically three distinct vacuolar compartments: newly-formed or early autophagosomes (AVi), intermediate or late autophagosomes (AVi/d), and autolysosomes (AVd). Early autophagosomes are bound by a double membrane and are neither acidic nor contain lysosomal proteins (**a** and Fig. 1). Late autophagosomes are bound by a double membrane but, unlike early autophagosomes, are acidic and contain LGP120 at its outer limiting membrane (**c**). However, these vacuoles do not contain acid hydrolases (*15*). Autolysosomes are acidic and contain lysosomal membrane proteins and acid hydrolases (**e** and Fig. 1). We also visualized vacuoles which are not yet acidic but contain LGP120 (**b**). In other vacuoles the inner membranes are in the process of being hydrolyzed (**d**). N, nucleus; M, mitochondria; P, peroxisome; and Ly, lysosome. Bar, 0.5 μm.

into autolysosomes (Fig. 4). Newly-formed autophagosomes mature into residual bodies in the following manner: (1) formation of a late autophagosome upon acidification of an early autophagosome which presumably occurs upon the acquisition or activation of a proton pump; (2) delivery of acid hydrolases by either mannose 6-phosphate receptor-mediated transfer from endosomes or fusion with pre-existing lysosomes thereby transforming the late autophagosome into a autolysosome; and (3) hydrolysis of autolysosome contents resulting in its maturation into a residual body (Fig. 4).

The maturation of autophagosomes may differ from that proposed for endosomes (see Chapter 1). Mannose 6-phosphate receptor delivers acid hydrolases to en-

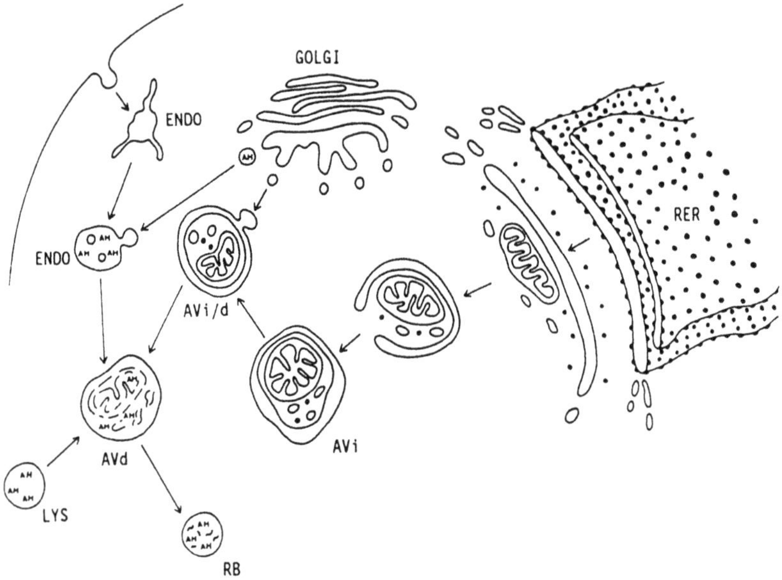

Figure 4. Proposed mechanism of autophagy. Early autophagosomes (AVi) are formed from ribosome-free regions of the rough endoplasmic reticulum (RER). These vacuoles then fuse with a vesicles containing LGP120 and proton pumps (i.e., "prelysosome") possibly derived from the Golgi apparatus to form an acidic late autophagosome (AVi/d). This vacuole then acquires acid hydrolases (AH) from lysosomes (LYS) or endosomes (ENDO) thereby becoming an autolysosome (AVd). The autolysosome is then transformed into a residual body (RB) through the action of hydrolytic enzymes. {Portions of this figure is reproduced from the Journal of Cell Biology (Vol. 110, p. 1944) by copyright permission of the Rockefeller University Press.}

dosomes. These endosomes then acquire lysosomal membrane proteins thereby becoming multivesicular bodies (e.g., secondary lysosomes). The mannose 6-phosphate receptor is not normally visualized in lysosomes and therefore, is presumed to recycle out of the late endosomes prior to acquisition of lysosome membrane proteins. This recycling is dependent upon vesicle acidification (*33*). The mannose 6-phosphate receptor has been identified in autophagic vacuoles characterized by a limiting double-membrane (e.g., early or late autophagosome), but not autolysosomes. The absence of acid hydrolases from newly-formed autophagosomes suggests that these receptor-positive vacuoles are probably late autophagosomes. Its presence implies a role for this receptor in the delivery of newly synthesized acid hydrolases to autophagic vacuoles. However, when mannose 6-phosphate recognition is disrupted by tunicamycin inhibition of glycosylation, the amount of cathepsins D and L present in an isolated fraction of autophagic vacuoles was not affected (*15*). In addition, tunicamycin had little effect on protein

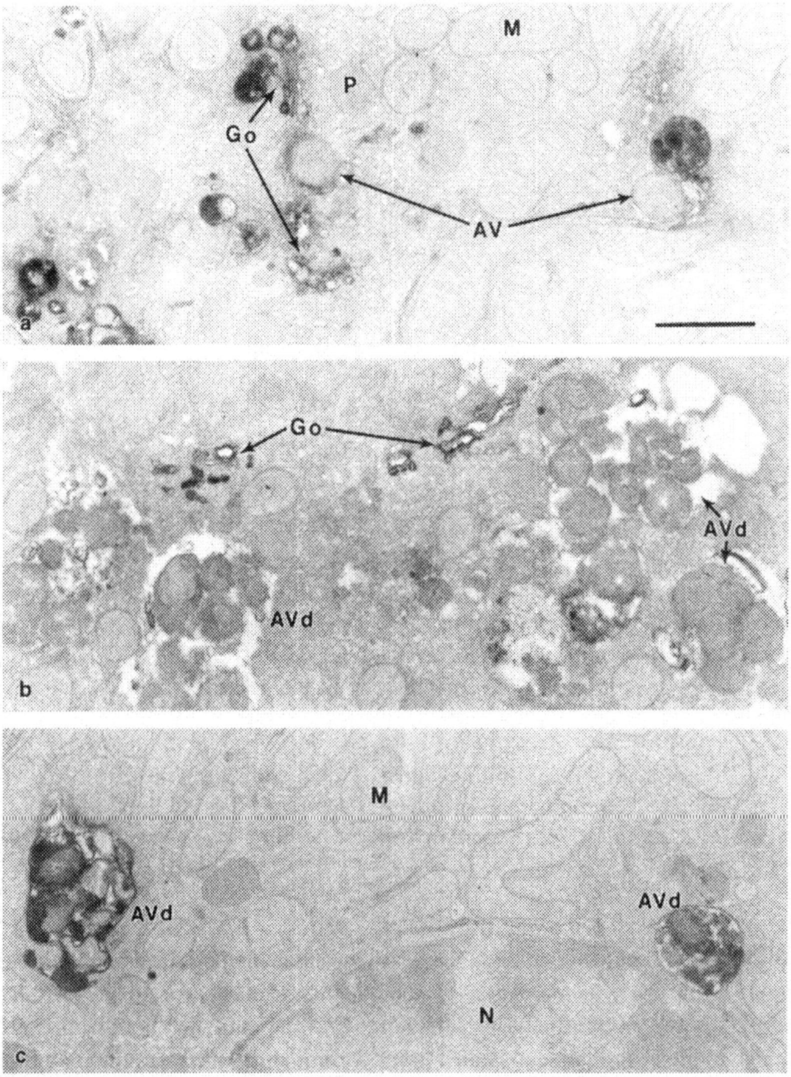

Figure 5. Immunoperoxidase localization of mannose 6-phosphate receptor to autophagic vacuoles. Livers were perfused with nutrient-depleted medium for 30 min (**a**), or with medium containing leupeptin (0.5 mM) for 120 min (**b**), or with medium containing leupeptin (0.5 mM) and primaquine (0.3 mM) for 120 min (**c**). Upon fixation, the mannose 6-phosphate receptor was localized using immunoperoxidase procedures (*14*). This receptor was normally found in Golgi cisternae and small vesicles presumed to be endosomes. Although the mannose 6-phosphate receptor can be found occasionally in a vacuole which resembles an autophagosome (AV), this receptor is not found in autolysosomes (AVd) unless the liver was treated with primaquine and leupeptin. N, nucleus; M, mitochondria; P, peroxisome; and Go, Golgi apparatus. Bar, 1.0 μm.

degradation which had been induced by amino acid deprivation (*37*). Combined, the results suggest that a majority of the acid hydrolases are delivered to autophagic vacuoles by fusion with pre-existing lysosomes. The absence of the mannose 6-phosphate receptor from autolysosomes can be interpreted as a loss of receptor due to degradation or recycling. When protein degradation is inhibited by leupeptin, the receptor did not accumulate within autolysosomes (Fig. 5b). However, when the acidic lumen of the autophagic vacuole is neutralized by primaquine in combination with leupeptin inhibition of proteinases, the mannose 6-phosphate receptor is found in large autolysosomes (Fig. 5c). The data suggest that the mannose 6-phosphate receptor delivers newly formed acid hydrolases to autophagic vacuoles and, in turn, the acidic nature of the intermediate compartment allows the receptor to be reutilized.

The presence of the mannose 6-phosphate receptor in autophagic vacuoles (Table 1 and Fig. 5a) suggests that either endosomes (pre-lysosomes) and/or Golgi-derived vesicles are fusing with these vacuoles. In order to evaluate the relationship between endocytosis and autophagy, we have followed the distribution of internalized ligand in hepatocytes under conditions whereby autophagy is enhanced and proteolysis inhibited by leupeptin treatment. Upon internalization, epidermal growth factor and its receptor are transported from endosomes to multivesicular bodies (e.g., secondary lysosomes) and autolysosomes (*13,14*). Since neither epidermal growth factor nor its receptor are found in autophagosomes, entry of these proteins into autolysosomes does not appear to be from these vacuoles but rather endosomes or multivesicular bodies (*14*). Tooze and associates have provided morphologic evidence that early endosomes can fuse with acid phosphatase-positive autophagic vacuoles (e.g., autolysosomes), but not with newly formed vacuoles (*75*). Lawrence and Brown have also shown that endocytosed colloidal gold particles are not present in autophagosomes but found in autolysosomes (*37*). However, the slow entry of the endocytosed gold into autolysosomes suggests this may be occurring by fusion of lysosomes with autophagosomes. Gordon and Seglen have suggested that internalized proteins can enter the autophagic pathway by fusion of the endosome with an intermediate vacuole termed amphisome (*23*). This acidic compartment does not contain significant hydrolytic activity and thus, is similar, if not identical, to the late autophagosome vacuole we defined biochemically and morphologically (*15,32,70*). Although transfer of endocytic tracers to degradative autolysosomes occurs, this does not appear to be a frequent event. Only 5 to 10% of the autolysosomes contain nonhydrolyzable colloidal gold which had been endocytosed by cultured rat liver cells (*37*). Therefore, these degradative pathways appear to be independent of one another with endocytosed material degraded primarily in multivesicular bodies or secondary lysosomes and cellular proteins sequestered within autophagic vacuoles degraded within autolysosomes. However, the degradation of some cytosolic proteins may occur by entry of these proteins into multivesicular bodies and/or autolysosomes via microautophagy or hsc73-mediated transport (*6,9,12,27,36,51*).

The maturation of the amphisome (late autophagosome) into an autolysosome is apparently suppressed by asparagine and vinblastine (*28,70*). We have reported the accumulation of autophagic vacuoles upon incubating cells with nocodazole (*2*). These vacuoles are acidic, but lack lysosomal hydrolases. These results support our original concept that vacuole acidification and acquisition of hydrolytic enzymes are separate events. Therefore, microtubules are required for the delivery of acid hydrolases to late autophagosomes which presumably occurs by vesicle fusion. However, the acidification of autophagosomes appears to occur independent of microtubules. This suggests that acidification does not require vesicle fusion for delivery of the proton pump or that the fusion event does not require microtubules. A alternative possibility is that acidification occurs upon activation of a proton pump which was originally present in early (non-acidic) autophagosomes.

Upon acquisition of the hydrolytic enzymes, the sequestered contents are digested resulting in the transformation of the autolysosome into a smaller residual body. This hydrolytic event can be inhibited by leupeptin (*34*). In addition, we and others have suggested that protein ubiquitination is somehow required for this transformation. Free ubiquitin and high molecular weight ubiquitin conjugates have

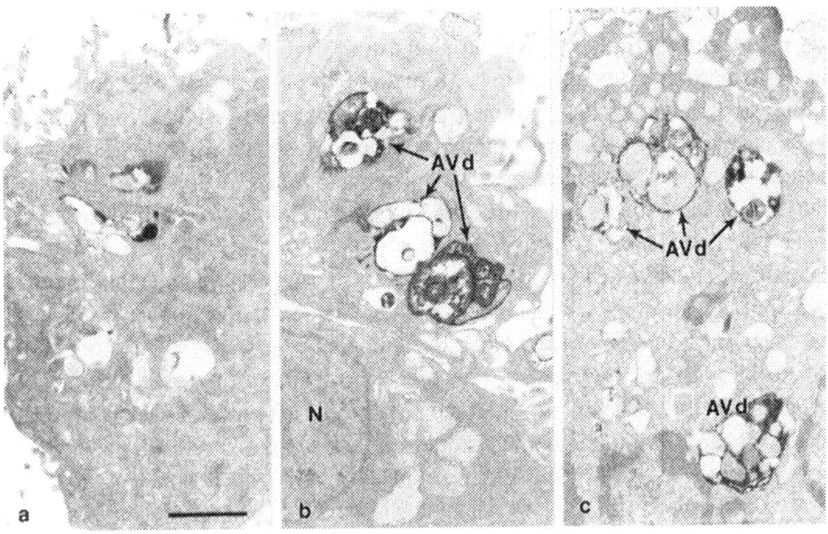

Figure 6. Accumulation of autolysosomes in CHO cells defective in protein ubiquitination. E36 (**a**) and ts20 (**b**) and (**c**) CHO cells were grown to confluency, warmed at 41°C for 1 hr to inactivate existing E1 activity, and then incubated at 39.5°C (nonpermissive temperature) for 6 hr in either complete medium (**a**) and (**b**) or medium depleted of serum and amino acids (**c**). Cells were aldehyde fixed and processed for acid phosphatase cytochemistry (*13*). Under heat-stressed conditions, autolysosomes (AVd) accumulated in cells deficient in protein ubiquitination. N, nucleus. Bar, 1.0 μm.

been identified in lysosomes and autolysosomes (*12,36,64*). In fact, ubiquitin conjugates accumulate in lysosomes upon treatment of fibroblasts with inhibitors of cysteine proteinases (*12*). The enzymes of ubiquitination have not yet been identified within lysosomes. Nevertheless, these conjugates can be synthesized in the cytosol and subsequently transported to lysosomes by such routes as hsc73-mediated transport and/or autophagic sequestration (i.e., micro- or macroautophagy). The presence of these conjugates within lysosomes suggests a possible role for ubiquitination in the intralysosomal degradation of proteins. Mutants defective in ubiquitination are unable to degrade short-lived and abnormal proteins at their non-permissive temperatures (*5*). The data support the concept that ubiquitination is required for the degradation of these proteins. These studies have recently been expanded to show that these temperature-sensitive (ts) mutants are also defective in the degradation of long-lived proteins (*24*). The degradation of these proteins is sensitive to lysosomotropic amines suggesting these proteins are degraded normally by lysosomal mechanisms. In addition, at the non-permissive temperature, large autophagic vacuoles containing acid phosphatase accumulate in the ts20 mutant cells compared to the E36 parent CHO cells (Fig. 6). These vacuoles are acidic and contain cathepsin L suggesting that the formation of the autophagosome and its maturation into an autolysosome is unaltered in the mutant cells (*39*). However, the accumulation of autolysosomes suggests that their transformation into residual bodies was suppressed at non-permissive temperatures. Combined, the data suggest that protein ubiquination is necessary for intralysosomal degradation of long-lived endogenous proteins.

V. REGULATION OF AUTOPHAGY

Protein turnover is necessary for the normal maintenance of the cell. In addition, amino acid availability is regulated to ensure continued protein synthesis. In the absence of an exogenous supply of amino acids, the cell quickly responds by increasing protein degradation and thereby, mobilizing amino acids. Protein degradation and autophagy is modulated by a variety of environmental factors, the most influential being amino acids (*48*). Acute regulation of endogenous protein degradation is mediated by modulation of autophagosome formation. Within 5 to 10 min after removal of amino acids from the medium, formation of an autophagosome is initiated and protein degradation begun (*14,65*). Just as quickly and efficiently, autophagosome formation is inhibited upon the addition of amino acids (*65*). The regulation of protein degradation and autophagy by amino acids have been established using hepatocytes. We have extended these observations to investigate the effects of amino acids on the regulation of protein degradation in primary cell cultures (rat astrocytes, rat neurons, and human endometrial fibroblasts) and cultures of established cell lines (normal rat kidney and rat hepatoma cells) (Fig. 7). In general, amino acid deprivation significantly enhanced protein degradation in all cultures studied. Of the twenty-two amino acids, five essential (leucine,

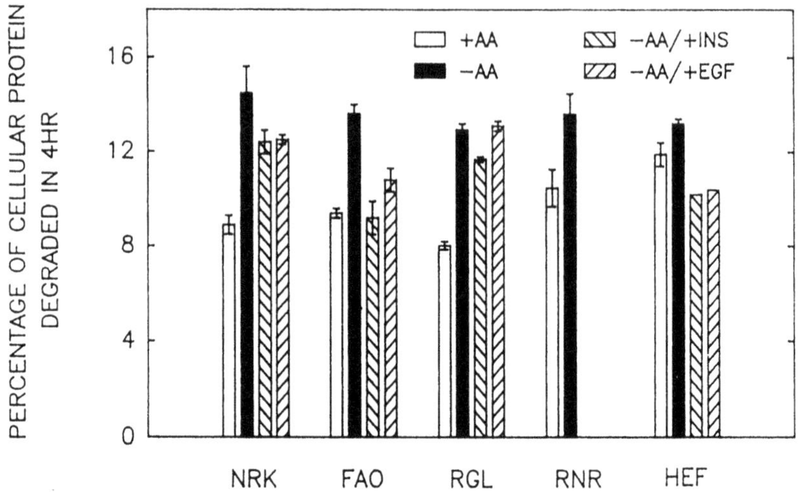

Figure 7. Regulation of protein degradation in cultured cells. Cells were grown to confluency (e.g., stationary growth) and their proteins radiolabeled by incubating with ^{14}C-valine for 16 hours. Cells were washed and then incubated in a Kreb's Henseleit medium containing 15 mM valine chase and either: (1) no additions (–AA); (2) amino acids (+AA); (3) 0.1 μM insulin (–AA/INS); or (4) 60 nM epidermal growth factor (–AA/EGF). The amount of acid-soluble radioactivity produced between 1 and 5 hr of incubation was measured and expressed as a percentage of the total radiolabeled cellular protein at time zero. The values represent the mean ± SEM of 3 to 14 determinations. NRK, normal rat kidney; FAO, rat hepatoma; RGL, rat brain astrocytes; RNR, rat brain neurons; and HEF, human endometrial fibroblasts.

methionine, tryptophan, phenylalanine, and histidine) and four non-essential (tyrosine, glutamine, proline, and alanine) amino acids are the primary regulators with leucine having the greatest influence on the modulation of hepatic autophagy (for review see ref. *48*). Although amino acids inhibit autophagosome formation, the mode and site of this action has yet to be evaluated. Mortimore and associates have proposed a intricate mechanism of regulation including both positive and negative modulation (*48*). In addition, Scornik and coworkers have suggested that the amino acid regulation is imposed through intracellular levels of aminoacyl-tRNA or amino acid metabolites (*67,69*).

Insulin and glucagon have opposite effects on autophagy. Insulin inhibits the formation of autophagosomes, while glucagon stimulates formation via a cAMP second message (*48,53,82*). Isoproterenol stimulates autophagy in the liver but inhibits this process in the heart (*7*). In cultured cells, we have found that insulin can depress protein degradation which had been previously enhanced by amino acid deprivation (Fig. 7). In addition, epidermal growth factor is capable of inhibiting protein degradation. However, the magnitude of the cellular response to insulin and epidermal growth factor inhibition varies depending upon the cell type.

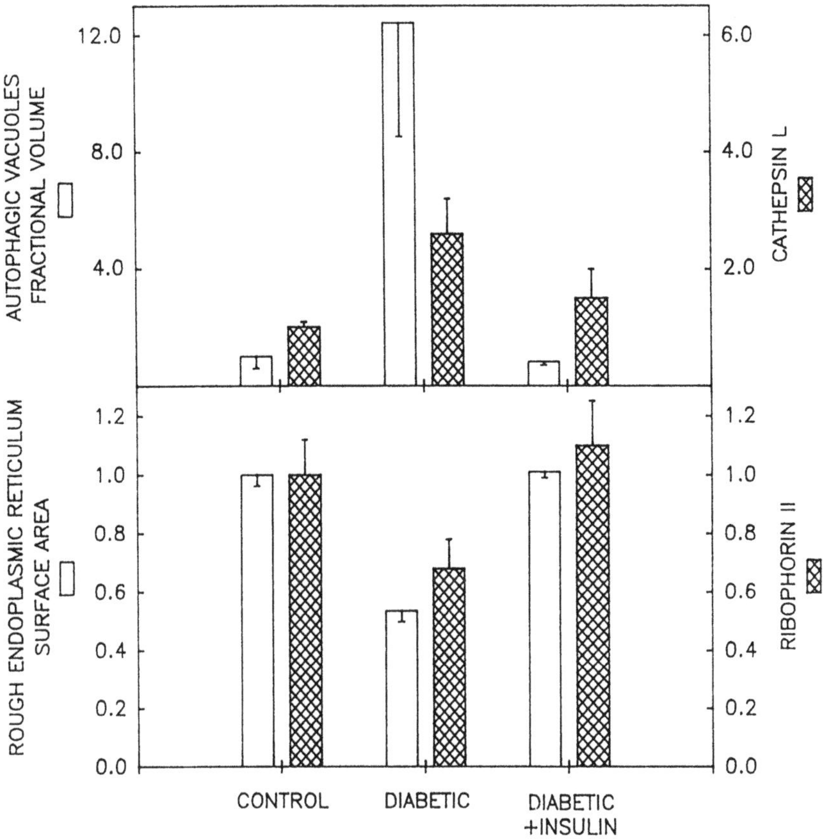

Figure 8. Effects of insulin on autophagy and the rough endoplasmic reticulum. Livers from untreated rats (CONTROL) or rats injected with streptozocin (90–120 mg/kg) (DIABETIC) or with streptozocin followed by daily insulin (2 units) injections (DIABETIC + INSULIN) were either homogenized or aldehyde fixed. The relative amounts of cathepsin L and ribophorin II in liver homogenates were quantified on Western blots. The fractional volume represented by autophagic vacuoles and the surface areas of the rough endoplasmic reticulum were quantified morphometrically (*14,78*). The values represent a mean ± SEM of 4 to 11 determinations and expressed relative to control values.

We also examined the effects of insulin on autophagy *in vivo* (Fig. 8). This was done by destroying the insulin producing B-cells with a single injection of streptozocin. The dramatic increase in circulating glucose to 350 to 450 mg/dl serum suggested that serum insulin levels were depressed. When livers were examined, we found a 12-fold increase in the occurrence of autophagic vacuoles and a fivefold elevation in cellular levels of cathepsin L in livers from rats previously treated with streptozocin (Fig. 8). Coincident with lysosome activation, the surface area of the rough endoplasmic reticulum decreased by over 40%. The depletion of rough endoplasmic reticulum was comparable to losses in membrane proteins (e.g., ribophorin II). Upon replacing insulin with daily injections, serum glucose approached normal levels (100–150 mg/dl). In these rats, autophagic vacuoles were absent and the rough endoplasmic reticulum appeared normal (Fig. 8). In addition, hepatic levels of cathepsin L and ribophorin II approached normal values. The results suggest that insulin modulates autophagy as well as the status of the rough endoplasmic reticulum *in vivo*. Whether this regulation is coordinated independently or that the loss of the rough endoplasmic reticulum is due to autophagy-mediated degradation is unknown.

Little is known about the nature of the signals responsible for the regulation of autophagy. Since these effectors inhibit or promote autophagosome formation, it is likely that the site of recognition or action of an amino acid or a second messenger would be either the rough endoplasmic reticulum or microfilaments, which are required for vacuole formation. However, indirect evidence suggests that leucine may recognize a "regulatory" complex situated at or in close proximity to the plasma membrane and not associated with the site of vacuole formation (*59*). Clearly, further studies are necessary to evaluate this rather complex mechanism developed by the cell to regulate autophagy.

VI. SUMMARY

Based on data reported here and elsewhere, autophagy proceeds in four distinct events (Fig. 4): (1) formation of an early autophagosome from ribosome-free portions of the RER; (2) acquisition or activation of a proton pump and acidification of the autophagosome, thereby becoming a late autophagosome; (3) maturation of the late autophagosome into an autolysosome which occurs by fusion of the late autophagosome with vesicles containing acid hydrolases; and (4) hydrolysis of autolysosome contents resulting in its transformation into a residual body. Step #1 is the site of regulation of autophagy. Vacuole formation is suppressed by amino acids, insulin, and growth factors. In the absence of these modulators, protein degradation and autophagy is enhanced in many cell types. Vacuole formation (Step #1) requires functional microfilaments. Maturation of late autophagosomes into autolysosomes (Step #3) requires intact microtubules. Finally, protein ubiquitination is necessary for the hydrolysis of autolysosome contents (Step #4) to proceed.

ACKNOWLEDGMENTS

I am greatly indebted to my colleagues, Dr. Susan Lenk, Tammy Jasionowski, Daniel Tuttle, Andrew Aplin, Daniel Fisher, and William Blakeney, whose experimental work has contributed to our current understanding of the mechanisms and regulation of protein degradation by the cell. In addition, I acknowledge the collaborative assistance of Drs. Alan Schwartz and Aaron Ceichanover in providing the cells defective in protein ubiquitination. These studies have been supported over the past eight years by an NIH grant DK33326.

REFERENCES

1. Amenta, J.S. and Brocher, S.C. (1981). Mechanisms of protein turnover in cultured cells. *Life Sci.* 28: 1195–1208.
2. Aplin, A., Jasionowski, T., Tuttle, D.L., Lenk, S.E., and Dunn, Jr., W.A. (1992). Cytoskeletal elements are required for the formation and maturation of autophagic vacuoles. *J. Cell. Physiol.* 152: 458–466.
3. Arstila, A.U. and Trump, B.F. (1968). Studies on cellular autophagocytosis. The formation of autophagic vacuoles in the liver after glucagon administration. *Amer. J. Pathol.* 53: 687–733.
4. Bond, J.S. and Butler, P.E. (1987). Intracellular proteases. *Ann. Rev. Biochem.* 56: 333–364.
5. Ciechanover, A., Finley, D., and Varshavsky, A. (1984). Ubiquitin dependence of selective protein degradation demonstrated in the mammalian cell cycle mutant ts85. *Cell* 37: 57–66.
6. Chiang, H.-L., Terlecky, S.R., Plant, C.P., and Dice, J.F. (1989). A role for a 70-kilodalton heat shock protein in lysosomal degradation of intracellular proteins. *Science* 246: 382–385.
7. Dammrich, J. and Pfeifer, U. (1982). Acute effects of isoproterenol on cellular autophagy. *Virchows Arch. B Cell Pathol.* 38: 209–218.
8. deDuve, C. and Wattiaux, R. (1966). Functions of lysosomes. *Ann. Rev. Physiol.* 28: 435–492.
9. Dice, J.F. and Chiang, H.-L. (1988). Peptide signals for protein degradation within lysosomes. *Biochem. Soc. Symp.* 55: 45–55.
10. Diment, S. and Stahl, P. (1985). Macrophage endosomes contain proteases which degrade endocytosed protein ligands. *J. Biol. Chem.* 260: 15311–15317.
11. Doherty, F.J., Wassell, J.A., and Mayer, R.J. (1987). A putative protein-sequestration site involving intermediate filaments for protein degradation by autophagy. Studies with microinjected purified glycolytic enzymes in 3T3-L1 cells. *Biochem. J.* 241: 793–800.
12. Doherty, F.J., Osborn, N.U., Wassell, J.A., Heggie, P.E., Lászl6, L., and Mayer, R.J. (1989). Ubiquitin-protein conjugates accumulate in the lysosomal system of fibroblasts treated with cysteine proteinase inhibitors. *Biochem. J.* 263: 47–55.
13. Dunn, W.A., Connolly, T.P., and Hubbard, A.L. (1986). Receptor-mediated endocytosis of epidermal growth factor by rat hepatocytes: receptor pathway. *J. Cell Biol.* 102: 24–36.
14. Dunn, Jr., W.A. (1990). Studies on the mechanisms of autophagy: formation of the autophagic vacuoles. *J. Cell Biol.* 110: 1923–1933.
15. Dunn, Jr., W.A. (1990). Studies on the mechanisms of autophagy: maturation of the autophagic vacuoles. *J. Cell Biol.* 110: 1935–1945.
16. Earl, R.T., Mangiapane, E.H., Billett, E.E., and Mayer, R.J. (1987). A putative protein-sequestration site involving intermediate filaments for protein degradation by autophagy. Studies with transplanted Sendai-viral envelope proteins in HTC cells. *Biochem. J.* 241: 809–815.
17. Epstein, D., Elias-Bishko, S., and Hershko, A. (1975). Requirement for protein synthesis in the regulation of protein breakdown in cultured hepatoma cells. *Biochemistry* 14: 5199–5204.
18. Ericsson, J.L.E. (1969). Studies on induced cellular autophagy. II. Characterization of the membranes bordering autophagosomes in parenchymal liver cells. *Exper. Cell Res.* 56: 393–405.
19. Fellinger, E., and Réz, G. (1990). Isolation of pancreatic autophagic vacuoles induced by vinblastine or neutral red. *Eur. J. Cell Biol.* 51: 220–228.
20. Forgac, M. (1989). Structure and function of vacuolar class of ATP-driven proton pumps. *Physiol. Rev.* 69: 765–796.
21. Furuno, K., Ishikawa, T., Akasaki, K., Lee, S., Nishimura, Y., Truji, H., Himeno, M., and Kato, K. (1990). Immunocytochemical study of the surrounding envelope of autophagic vacuoles in cultured rat hepatocytes. *Exper. Cell Res.* 189: 261-268.
22. Gordon, P.B., Kovács, A.L., and Seglen, P.O. (1987). Temperature dependence of protein degradation, autophagic sequestration and mitochrondrial sugar uptake in rat hepatocytes. *Biochim. Biophy. Acta* 929: 128–133.
23. Gordon, P.B. and Seglen, P.O. (1988). Prelysosomal convergence of autophagic and endocytic pathways. *Biochem. Biophys. Res. Comm.* 151: 40–47.

24. Gropper, R., Brandy, R.A., Elias, S., Bearer, C.F., Mayer, A., Schwartz, A.L., and Ciechanover, A. (1991). The ubiquitin-activating enzyme, E1, is required for stress-induced lysosomal degradation of cellular proteins. *J. Biol. Chem.* 266: 3602–3610.
25. Hershko, A. (1988). Ubiquitin-mediated protein degradation. *J. Biol. Chem.* 263: 15237–15240.
26. Hirsimäki, Y. and Hirsimäki, P. (1982). Vinblastine-induced autophagic vacuoles in mouse liver and Ehrlich ascites tumor cell as assessed by freeze-fracture electron microscopy. *Eur. J. Cell Biol.* 27: 298–301.
27. Holtzman, E. (1989). *Lysosomes*, pp. 439, Plenum Publishing, New York.
28. Høyvik, H., Gordon, P.B., Berg, T.O., Strømhaug, P.E., and Seglen, P.O. (1991). Inhibition of autophagic-lysosomal delivery and autophagic lactolysis by asparagine. *J. Cell Biol.* 113: 1305–1312.
29. Kadowaki, M., Venerando, R., Miotto, G., Siliprandi, N., and Mortimore, G.E. (1991). *De novo* formation of macroautophagic vacuoles in the permeabilized isolated rat hepatocyte. *FASEB J.* 5: A1178.
30. Knecht, E., Martinez-Ramon, A., and Grisolia, S. (1988). Autophagy of mitochondria in rat liver assessed by immunogold procedures. *J. Histochem. Cytochem.* 36: 1433–1440.
31. Kominami, E., Hashida, S., Khairallah, E.A., and Katunuma., N. (1983). Sequestration of cytoplasmic enzymes in an autophagic vacuole lysosomal system induced by injection of leupeptin. *J. Biol. Chem.* 258: 6093–6100.
32. Kopitz, J., Kisen, G.Ø., Gordon, P.B., Bohley, P., and Seglen, P.O. (1990). Nonselective autophagy of cytosolic enzymes by isolated rat hepatocytes. *J. Cell Biol.* 111: 941–953.
33. Kornfeld, S. and Mellman, I. (1989). The biogenesis of lysosomes. *Annu. Rev. Cell Biol.* 5: 483–525.
34. Kovács, A.L., Reith, A., and Seglen, P.O. (1982). Accumulation of autophagosomes after inhibition of hepatocytic protein degradation by vinblastine, leupeptin or a lysosomotropic amine. *Exper. Cell Res.* 137: 191–201.
35. Lardeux, B.R., and Mortimore, G.E. (1987). Amino acid and hormonal control of macromolecular turnover in perfused rat liver. *J. Biol. Chem.* 262: 14514–14519.
36. László, L., Doherty, F.J., Osborn, N.U., and Mayer, R.J. (1990). Ubiquintinated protein conjugates are specifically enriched in the lysosomal system of fibroblasts. *FEBS Letters* 261: 365–368.
37. Lawrence, B.P. and Brown, W.J. (1992). Autophagic vacuoles rapidly fuse with pre-existing lysosomes in cultured hepatocytes. *J. Cell Science* 102: 515–526.
38. Lenk, S.E., Fisher, D.L., and Dunn, Jr., W.A. (1991). Regulation of protein secretion by crinophagy. *Eur. J. Cell Biol.* 56: 201–209.
39. Lenk, S.E., Dunn, Jr., W.A., Trausch, J.S., Ciechanover, A., and Schwartz, A.L. (1992). Ubiquitin-activating enzyme, E1, is associated with maturation of autophagic vacuoles. *J. Cell Biol.* 118: 301–308.
40. Marzella, L. and Glaumann, H. (1980). Increased degradation in rat liver induced by vinblastine. I. Biochemical characterization. *Lab. Invest.* 42: 8–17.
41. Marzella, L. and Glaumann, H. (1980). Increased degradation in rat liver induced by vinblastine. II. Morphologic characterization. *Lab. Invest.* 42: 18–27.
42. Marzella, L., Ahlberg, J., and Glaumann, H. (1982). Isolation of autophagic vacuoles from rat liver: morphological and biochemical characterization. *J. Cell Biol.* 93: 144–154.
43. Masaki, R., Yamamoto, A., and Tashiro, Y. (1987). Cytochrome P-450 and NADPH-cytochrome P-450 reductase are degraded in the autolysosomes in rat liver. *J. Cell Biol.* 104: 1207–1215.
44. Matsushita, M., Saito, T., Keino, H., Kuruhara, Y., Abe, S., and Ogawa, K. (1982). Autophagy of peroxisomes in hepatic parenchymal cells. *Acta Histochem. Cytochem.* 15: 277–283.
45. Matteoni, R. and Kreis, T.E. (1987). Translocation and clustering of endosomes and lysosomes depends on microtubules. *J. Cell Biol.* 105: 1253–1265.

46. Matthews, W., Tanaka, K., Driscoll, J., Ichihara, A., and Goldberg, A.L. (1989). Involvement of the proteasome in various degradative processes in mammalian cells. *Proc. Natl. Acad. Sci. USA* 86: 2597–2601.
47. Mortimore, G.E., Pösö, A.R., Kadowaki, M., and Wert, Jr., J.J. (1987). Multiphasic control of hepatic protein degradation by regulatory amino acids. *J. Biol. Chem.* 262: 16322–16327.
48. Mortimore, G.E., Pösö, A.R., and Lardeux, B.R. (1989). Mechanism and regulation of protein degradation in liver. *Diabetes/Metabolism Rev.* 5: 49–70.
49. Neff, N.T., DeMartino, G.N., and Goldberg, A.L. (1979). The effect of protease inhibitors and decreased temperature on the degradation of different classes of proteins in cultured hepatocytes. *J. Cell Physiol.* 101: 439–458.
50. Ohkuma, S., Chudzik, J., and Poole, B. (1986). The effects of basic substances and acidic ionophores on the digestion of exogenous and endogenous proteins in mouse peritoneal macrophages. *J. Cell Biol.* 102: 959–966.
51. Olson, T.S. and Dice, J.F. (1989). Regulation of protein degradation rates in eukaryotes. *Curr. Opin. Cell Biol.* 1: 1194–1200.
52. Pfeifer, U. (1973). Cellular autophagy and cell atrophy in the rat liver during long-term starvation. *Virchows Arch. Abt. B ZellPath.* 12: 195–211.
53. Pfeifer, U. (1978). Inhibition by insulin of the formation of autophagic vacuoles in rat liver. *J. Cell Biol.* 78: 152–167.
54. Pfeifer, U. (1979). Inhibited autophagic degradation of cytoplasm during compensatory growth of liver cells after partial hepatectomy. *Virchows Arch. B Cell Path.* 30: 313–333.
55. Pfeifer, U. (1980). Autophagic sequestration of internalized gap junctions in rat liver. *Eur. J. Cell Biol.* 21: 244–246.
56. Pfeifer, U. (1987). Functional morphology of the lysosomal apparatus. In: *Lysosomes, Their Role in Protein Breakdown.* (Glaumann, H. and Ballard, F.J., eds.), pp. 3–59. Academic Press, New York.
57. Pfeifer, U. and Dunker, R. (1985). Inhibition of protein degradation and autophagy in growing as compared to density inhibited fibroblasts. *Progr. Clin. Biol. Res.* 180: 659–661.
58. Pfeifer, U. and Strauss, P. (1981). Autophagic vacuoles in heart muscle and liver. A comparative morphometric study including circadian variations in meal-fed rats. *J. Mole. Cell. Cardiol.* 13: 37–49.
59. Pösö, A.R., Wert, J.J., and Mortimore, G.E. (1982). Multifunctional control by amino acids of deprivation-induced proteolysis in liver: Role of leucine. *J. Biol. Chem.* 257: 12114–12120.
60. Reunanen, H., Punnonen, E.-L., and Hirsimäki, P. (1985). Studies on vinblastine-induced autophagocytosis in mouse liver V. A cytochemical study on the origin of membranes. *Histochemistry* 83: 513–517.
61. Réz, G. and Meldolesi, J. (1980). Freeze-fracture of drug-induced autophagocytosis in mouse exocrine pancreas. *Lab. Invest.* 43: 267–277.
62. Sakai, M., Araki, N., and Ogawa, K. (1989). Lysosomal movements during heterophagy and autophagy: with special reference to nematolysosome and wrapping lysosome. *J. Elect. Micros. Tech.* 12: 101–131.
63. Scheel, J., Matteoni, R., Ludwig, T., Hoflack, B., and Kreis, T.E. (1990). Microtubule depolymerization inhibits transport of cathepsin D from the golgi apparatus to lysosomes. *J. Cell Sci.* 96: 711–720.
64. Schwartz, A.L., Ciechanover, A., Brandt, R.A., and Geuze, H.J. (1988). Immunoelectron microscopic localization of ubiquitin in hepatoma cells. *EMBO J.* 10: 2961–2966.
65. Schworer, C.M., Shiffer, K.A., and Mortimore, G.E. (1981). Quantitative relationship between autophagy and proteolysis during graded amino acid deprivation in perfused rat liver. *J. Biol. Chem.* 256: 7652–7658.
66. Scornik, O.A. (1974). *In vivo* rates of deaggregation of polyribosomes in normal and regenerating liver after the injection of pactamycin. *Biochim. Biophys. Acta* 374: 76–81.

67. Scornik, O.A. (1983). Faster protein degradation in response to decreased steady state levels of amino acylation of $tRNA^{His}$ in Chinese hamster ovary cells. *J. Biol. Chem.* 258: 882–886.
68. Scornik, O.A. and Botbol, V. (1976). Role of changes in protein degration in the growth of regenerating livers. *J. Biol. Chem.* 251: 2891–2897.
69. Scornik, O.A., Ledbetter, M.L.S., and Malter, J.S. (1980). Role of aminoacylation of histidyl-tRNA in the regulation of protein degradation in Chinese hamster ovary cells. *J. Biol. Chem.* 255: 6322–6329.
70. Seglen, P.O. (1987). Regulation of autophagic protein degradation in isolated liver cells. In: *Lysosomes, Their Role in Protein Breakdown.* (Glaumann, H. and Ballard, F.J., eds.), pp, 371–414. Academic Press, New York.
71. Shelburne, J.D., Arstila, A.U., and Trump, B.F. (1973). Studies on cellular autophagocytosis. *Amer. J. Physiol.* 73: 641–670.
72. Surmacz, C.A., Pösö, A.R., and Mortimore, G.E. (1987). Regulation of lysosomal fusion during deprivation-induced autophagy in perfused rat liver. *Biochem. J.* 242: 453–458.
73. Swift, H. and Hruban, Z. (1964). Focal degradation as a biological process. *Fed. Proc.* 23: 1026–1037.
74. Terasaki, M., Chen, L.B., and Fujiwara, K. (1986). Microtubules and the endoplasmic reticulum are highly interdependent structures. *J. Cell Biol.* 103: 1557–1568.
75. Tooze, J., Hollinshead, M., Ludwig, T., Howell, K., Hoflack, B., and Kern, H. (1990). In exocrine pancreas, the basolateral endocytic pathway converges with the autophagic pathway immediately after the early endosome. *J. Cell Biol.* 111: 329–345.
76. Ueno, T., Watanabe, S., Hirose, M., Namihisa, T., and Kominami, E. (1990). Phalloidin-induced accumulation of myosin in rat hepatocytes is caused by suppression of autolysosome formation. *Eur. J. Biochem.* 190: 63–69.
77. Vaux, D., Tooze, J., and Fuller, S. (1990). Identification by anti-idiotype antibodies of an intracellular membrane protein that recognizes a mammalian endoplasmic reticulum retention signal. *Nature* 345: 495–502.
78. Weibel, E.R. (1969). Stereological principals for morphometry in electron microscopic cytology. *Inter. Rev. Cytol.* 26: 235–302.
79. Weissmann, G. (1965). Lysosomes. *N. Engl. J. Med.* 273: 1084–1090
80. Weissmann, G. (1965). Lysosomes (Concluded). *N. Engl. J. Med.* 273: 1143–1149.
81. Wharton, S.A. and Riley, P.A. (1984). Lysosomes and ageing. In: *Lysosomes in Biology and Pathology*, Vol. 7. (Dingle, J.T., Dean, R.T., and Sly, W., eds.), pp. 401–419. Elsevier Science,
82. Woodside, K.H., Ward, W.F., and Mortimore, G.E. (1974). Effects of glucagon on general protein degradation and synthesis in perfused rat liver. *J. Biol. Chem.* 249: 5458–5463.
83. Yamamoto, A., Masaki, R., and Tashiro, Y. (1990). Characterization of the isolation membranes and the limiting membranes of autophagosomes in rat hepatocytes by lectin cytochemistry. *J. Histochem. Cytochem.* 38: 573–580.
84. Yamamoto, A., Masaki, R., Fukui, Y., and Tashiro, Y. (1990). Absence of cytochrome P-450 and presence of autolysosomal membrane antigens on the isolation membranes and autophagosomal membranes in rat hepatocytes. *J. Histochem. Cytochem.* 38: 1571–1581.

CELL-FREE SYSTEMS FOR ENDOCYTOSIS

William A. Braell

I. INTRODUCTION

Endocytosis—the mechanism by which cells internalize and transport macromolecular materials within themselves—is a complex and multistep process. In its simplest terms, it can be considered to consist of the following six stages: (1)

Advances in Cell and Molecular Biology of Membranes
Volume 1, pages 139–168

ISBN: 1-55938-362-3

formation of endocytic vesicles; (2) delivery of those vesicles to an early endocytic compartment, termed early endosomes; (3) sorting of receptors from lysosomally-targeted materials; (4) recycling (or transcytosis) of receptors; (5) transport of materials to late endosomal compartments; and (6) delivery to lysosomes for degradation (*95,109*). As will be described below, each of these steps has been investigated by various strategies involving cell-free or permeabilized cell systems, in hopes of defining and characterizing the cellular participants mediating these events.

The purpose of this chapter is to review the state of the literature on the reconstitution by cell-free systems of events in the endocytic pathway, and the prevailing concepts in the field, as of the time of preparation (early 1992). Given the limitations of space, little attempt has been made to deal comprehensively with questions of general interest to the problem of membrane trafficking overall. This will be left to previous reviews of cell-free trafficking (*31,41,106*), and the remaining chapters of this volume. Instead, specific questions where cell-free systems have provided insight are introduced just sufficiently for clarity, before addressing the cell-free results themselves. For the student new to this field, some attempt has been made to deal with the rationale and philosophy of the cell-free approach, both in general and as it applies to the specific steps of endocytosis.

II. ADVANTAGES AND LIMITATIONS OF CELL-FREE RECONSTITUTION

There are two basic approaches available for defining the participants in a complex, multistep pathway such as exists in endocytosis. The first involves defining the pathway by saturating that pathway with genetic mutants defective in its various aspects, and then determining the genetic interrelationships between those genes. The other approach is to develop cell-free assays that reconstruct the individual steps in the pathway, and then determine by biochemical means the components needed for that reconstitution. Each approach has its strengths and weaknesses. One advantage provided by the cell-free assay approach is its more focused examination of a given process than in genetic or other approaches involving entire cells, reducing the danger of observing events caused by side effects from unrelated activities. These sorts of pleiotropic effects have been observed, for example, in studies of yeast secretory mutants. The *sec*53 mutation affects protein insertion into the endoplasmic reticulum, but its gene product was found in reality to be the enzyme phosphomannomutase (*51*). Apparently, alterations in glycosylation of critical proteins in the endoplasmic reticulum, due to the loss of mannose 1-phosphate formed by this enzyme, affected the insertion process indirectly.

Another advantage is that, because a cell-free assay is generally based upon *function*, it will investigate critical components for that function by definition. This is somewhat in contrast to genetic approaches, where critical components are identified on genetic grounds, but their actual function often remains unidentified.

Indeed, most of the *sec* mutants whose functions are currently known had that functional role defined by observing their sequence homology to proteins that had already been characterized biochemically. These include GTP-binding proteins, sugar metabolism enzymes, NSF, and SNAP proteins, and phospholipid exchange proteins (*5,51,83,87*). Assuming that the assay is an accurate representation of the *in vivo* process, the cell-free system approach obviates the question of identifying the function.

The cell-free system also is amenable to isolation of the individual elements critical for function. The basic scheme for this approach is as follows: an individual element of the cell-free system is inactivated or removed by a treatment. If the missing element can by resupplied to the system, thereby reconstituting the cell-free function, a modified cell-free assay has been created that can be used not only to identify an important element, but can also be theoretically used (in an ideal world) to isolate that component. Once that element is purified, the system can be supplemented with the pure component, enabling further critical elements to be isolated by an iteration of the entire process. With luck and patience, a series of proteins (perhaps the entire lot) necessary for the cell-free event will be isolated.

Naturally there are limitations to the cell-free approach. First, there is no guarantee that a reconstitution system can be created for every process. Even if it is possible to reconstitute a process, the protein and membrane elements comprising it may present special difficulties to their isolation and identification. In any event, the varied properties of those elements will preclude any hope of a cut-and-dried isolation methodology analogous to the systematic genetic identification approach. Finally, proteins important for *in vivo* function may be functionally bypassed by the cell-free formulation, so that these proteins are not uncovered by that assay. For these reasons, there are complementary roles for both the genetic and cell-free approaches in membrane trafficking. It is worth noting that where both approaches have been brought to bear on the same problem, as in the secretory pathway, they have provided a gratifying synergistic effect. For this reason, if none other, it would be of great value to develop a coterie of mutants defective in the endocytic pathway to equal those available in secretion.

III. CONCEPTS IN DEVELOPING CELL-FREE ASSAYS FOR ENDOCYTOSIS

By the term "cell-free", one implies a system in which no remnant cells are present. In fact, this is an extreme interpretation of the term, when applied to the kinds of systems used for trafficking. What is actually required for useful studies is a system where ready access to the cytoplasmic interior of the cell is achieved. This need not imply disruption of total cell architecture. The earliest systems for endocytic vesicle fusion (*14,21,39*) did actually employ cell fractions, or isolated vesicles, for assays, and this is still a valid approach for many systems. However, the removal of the cytoskeletal framework of the cell is problematic for studies of directional events, such as in polarized cell trafficking, and might also remove important elements in the transport of nonpolarized elements as well. This problem has been addressed by the development of techniques for preparing permeabilized cell systems (also called "semi-intact" cells) (*27*). Simons and Virta found that stripping MDCK cells with adherent nitrocellulose could remove portions of the cell membrane, following which soluble proteins and free vesicles, but not those attached to cytoskeleton, were released (*92*). Beckers and co-workers developed a more rapid method for adherent cells, involving scraping of hypotonically swollen cells, that created holes in cells by shearing them at their attachment points with the substratum (*7*). Chemical permeabilization of the cell membrane can be performed with a reagent like streptolysin-O, which creates pores by destabilizing the bilayer (*35*). This latter method is less desirable than physical methods, as the chemical reagent may cause intracellular damage and since it is not trivial to produce pores that can pass macromolecules by this method, but it may be the only available method for nonadherent cell lines.

While it may seem a tautology to state that one first needs a good assay before one can do biochemistry, the problem has special implications for endocytosis and other membrane trafficking events. The actual physiological events occurring in endocytosis are difficult to detect or to re-enact in biochemical terms. Vesicle fusion events presumably occur in milliseconds, on a stochastic basis at the vesicle level (*94*). Vesicle formation involves a topological transformation of the cell membrane from a sheet to an encapsulated volume. These sorts of events are difficult to translate into common biochemical terms compatible with protein purification, so investigators have resorted to assays that operate "by analogy" to gain insight in these cases.

This creation of cell-free systems that operate "by analogy" is the underlying foundation for reconstituting trafficking events. The basic idea is to envision detectable consequences for a specific probe, when it undergoes a reconstituted endocytic process, and to assume that the detection of that consequence is equivalent to the process itself. Thus, a protease-sensitive marker may resist protease action when enveloped in the lipid bilayer of a newly formed endocytic vesicle. The hazard in this analogy is that other, unforeseen, processes may provide the same detectable signal without necessarily involving the actual desired event. For example, a probe that is resistant to protease because it is trapped in aggregated membranes might be scored as having been endocytosed, although no such event has occurred. A second, more subtle point is that even an assay with fidelity to the underlying event must be sufficiently linear, reproducible, and rapid to be useful for biochemical manipulations, or little of value will come from it.

The strategies generally used for detecting membrane trafficking events are based on the concepts that trafficking requires vesicles to be formed, and then to fuse with a target compartment during the transport process. Biochemical detection of such events typically relies on the production of membrane-delimited compartments that are detected either by sequestration of contents, or by mixing of contents of two vesicles. The basic strategy for detecting vesicle formation is shown in Figure 1A.

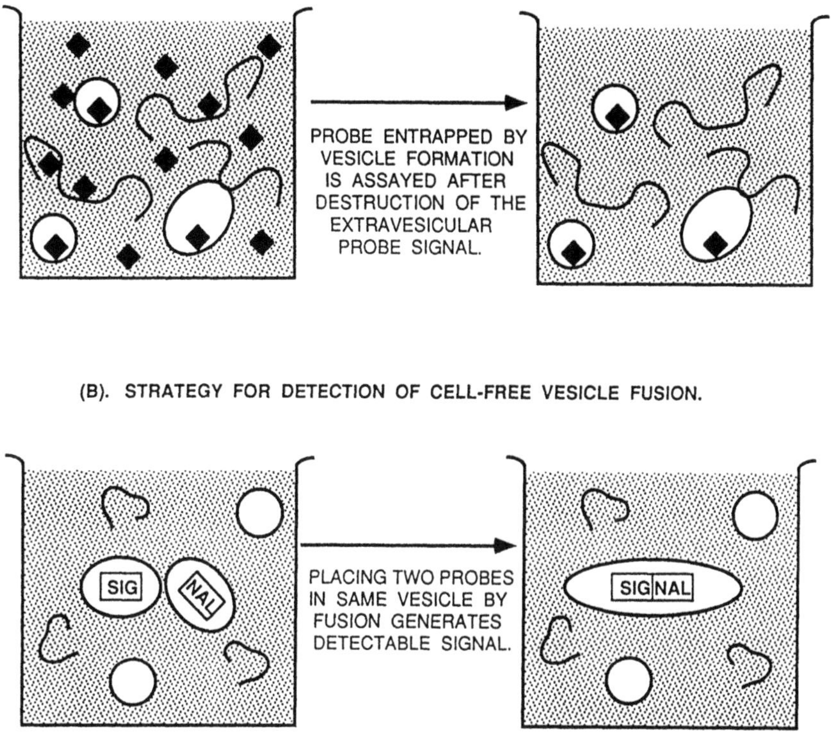

Figure 1. Panel (**A**). Depicted is a general scheme for the detection of cell-free endocytic vesicle formation. Membrane fractions are incubated with a detectable probe, which becomes entrapped in the vesicles as they form. Following this incubation, the fractions are treated to destroy the probe which remains outside vesicles, using a reagent that does not penetrate the vesicles containing endocytosed probe. The undamaged probe residing within those vesicles is then assayed. Panel (**B**). Depicted is a general scheme for the detection of cell-free endocytic vesicle fusion. Membrane vesicle fractions containing two different probes, each previously endocytosed by separated cell populations, are incubated together to promote vesicle fusion. Fusions that place both probes within the same membrane compartment allow the two probes to react with each other, generating the signal that represents endocytic vesicle fusion.

A detectable probe is applied to a membrane and allowed to be internalized into a vesicle. A treatment is then performed with a membrane-impermeant reagent that destroys the capability of any extravesicular probe to provide a detectable signal. Any remaining probe is then measured, and considered equivalent to vesicle formation. Vesicular transport to a new compartment is generally monitored by assays that detect the mixing of the contents of the two interacting vesicular compartments—transport vesicles and target compartments—after they have fused. In such strategies, a signal is not generated unless two probes, one donated by each interacting vesicular compartment, are placed in the same compartment as a consequence of this fusion (Fig. 1B). Although not a theoretical requirement for detecting these fusions, treatments that destroy extravesicular probes are frequently employed here as well. This reduces the chance that the two probes will generate signal extravesicularly (for example, because of vesicle lysis), since only those probes that remain protected within the same membrane compartment will then provide a detectable signal.

IV. FORMATION OF ENDOCYTIC VESICLES

The initiation of endocytosis proceeds with the formation and budding of the endocytic vesicle. This process has been best studied, both *in vivo* and *in vitro*, as it occurs in receptor-mediated endocytosis via clathrin coated pits (*15,74,95,109*). Less well studied are the endocytic events involving nonclathrin-mediated internalizations (*2,42,88*).

Receptor-mediated endocytosis begins with the clustering of receptors in localized regions of the plasma membrane. The clustering appears concomitant with the assembly of latticeworks of clathrin and clathrin-associated proteins on the cytoplasmic face of the membrane (*74,109*). Is clustering in clathrin-coated pits stimulated by ligand association (and therefore involves preferential recognition and clustering of receptors bearing ligands), or is receptor clustering stochastic and independent of receptor occupancy? Currently, the evidence appears to favor the latter for many receptors, especially those with relatively short cytoplasmic tails (*15,74*). Certain receptors with larger cytoplasmic domains, such as tyrosine kinase domains, may internalize preferentially when hormone is bound (*57*). Since hormone-induced tyrosine kinase signaling can affect many processes, however, it is unclear for these receptors if this increase in internalization is due to increases in clustering in coated pits, or in overall endocytic rates. Complicating this issue would be any potential alterations in the rate of recycling of these receptors, and the maintenance of the relative balance between internalization and return to the plasma membrane. What is clear thus far is that the cytoplasmic tails of receptors do appear to interact with the clathrin lattice, apparently via the clathrin-associated proteins (adaptors) (*15,59,74,79*). Thus, clustering in coated pits could be either induced, via alterations of the receptor tails that improve interactions with adaptors, or constitutive, if the assembly-competent conformations are maintained throughout the life of the receptor. Thus, either mode of uptake may fit the knowledge of the molecular make-up of the clathrin-coated pit.

Does clathrin direct coated-pit assembly, or do clustered receptors favor clathrin assembly? This question has been addressed both *in vivo* and *in vitro* (*15,74*). The studies of Anderson and colleagues nicely favor the view that clustered receptors preferentially promote clathrin binding to localized regions of the membrane (*58,65*). Following the stripping of clathrin coats from immobilized membrane, these researchers found that assembly of clathrin back onto the stripped membranes occurred preferentially at those sites which had originally borne clathrin coats. Further work identified high-affinity sites for AP-2 adaptor binding on the stripped plasma membranes, although membrane binding of AP-2 complexes was itself insufficient to allow complete coated-pit assembly (*59*). Recently, studies by Miller and colleagues of cell lines expressing different levels of transferrin receptors indicates that similar rules of assembly are followed *in vivo* (*64*). Cell lines expressing high numbers of transferrin receptors were found to possess increased amounts of flat clathrin lattice on their plasma membrane, although the frequency of invaginated coated entities per cell was essentially unchanged from the normal situation. This data might be explained by the increased formation of receptor-clathrin lattice clusters, due to the increased receptor number, but with a disparately regulated number of sites per cell for invagination of coated pits. Thus, the formation of nascent receptor-clathrin lattices may not necessarily be physically coupled directly to the invagination machinery, but may be separately regulated.

Evidence for the cell-free formation of coated vesicle structures comes from two different styles of assays. First, the studies of the Anderson group have implied that such vesicles are budded *in vitro*, in that coated pits labeled with receptor-bound LDL-gold disappear from the associated plasma membrane when incubated under permissive conditions (*56*). This sort of "disappearance assay" is of course negative evidence, and less convincing than positive evidence of coated vesicle formation. However, loss of receptors from the membrane was coincident with loss of clathrin and AP-2 adaptor complexes as well, as would be expected for a budding process. The receptor loss exhibited a requirement for ATP and cytosol, as well as high levels of calcium (above 150 μM). This high calcium level is 1000-fold higher than the basal calcium concentration within cells, giving rise to concern over the physiological relevance of the receptor loss. However, it remains possible that localized transient concentrations of calcium might reach such levels right at the surface of the membrane where the pits reside, due to ion import from channels, so it cannot be ruled out that this process is relevant to actual coated-pit budding.

More positive evidence for cell-free vesicle budding comes from assays designed to directly detect the formation of intact vesicles. Smythe and co-workers used a system based on permeabilized A431 cells that had receptor-bound ^{125}I-labeled (or peroxidase-conjugated) transferrin (*93*). Vesicles were scored as having been budded if, following incubation, the transferrin's label became inaccessible to exogenously added anti-transferrin antibody. Using the peroxidase tag, morphological evidence for vesicle formation was obtained. A low (10–15 %) efficiency of budding, relative to total bound label, was observed; more troubling was the observation that, although ATP, cytosol and elevated temperatures aided the sequestration of the label, there remained a component of sequestration that did not require one or more of these agents. This concern may have its answer in the relative inefficiency of the antibody-based detection method of internalization.

Schmid and Carter have recently shown that antibody inaccessibility is not the optimal means of detecting vesicle budding, as deeply invaginated coated pits appear to exclude the antibody, and thus are scored as having budded (*89*). This raises problems for interpretation of results with the antibody-based assay, as it confuses pit invagination with budding itself. The Schmid group obtained better results by using an assay for inaccessibility in which transferrin, labeled with disulfide-linked biotin, could be stripped of its biotin by cell surface treatment with a membrane-impermeant reducing agent (glutathione). Use of this biotin-based assay on intact cells proved a more stringent criterion for vesicle budding, given the smaller size (308 mol.wt.) of the glutathione reagent, compared to the size of an antibody molecule (150,000 mol.wt.). The biotin-based assay has also proved amenable to examination of cell-free coated-pit budding (*90*). Differences in ATP requirements, and in sensitivity to anti-clathrin and anti-transferrin receptor tail antibodies, were observed between the *in vitro* processes of coated pit invagination and actual vesicle formation, as assessed by antibody inaccessibility and resistance to reducing agent, respectively. These distinct requirements imply that separable stages of coated-pit invagination and budding can be recognized *in vitro*. Nonetheless, one must recognize that even the disulfide-linked biotin resistance assay cannot *ensure* that a vesicle has actually budded, only that the coated pit containing the biotinylated probe excludes relatively small molecules. This factor may explain a partial overlap of properties, including a fractional component of ATP- and cytosol-independent "budding", which was observed between the antibody exclusion and the biotin resistance methodologies.

There is evidence for a vesicle budding pathway which is mechanistically independent of the clathrin-related pathway (*42,88*). These vesicles should be deficient in the receptor proteins that commonly populate coated pits. The formation of these vesicles, which are slightly smaller than coated structures (in Hep2 cells), can be differentially modulated relative to coated-vesicles by cytochalasin D treatment, or by acidification of the cytosol. To date, cell-free reconstitution of the clathrin-independent uptake, distinct from coated-vesicle formation, has not been reported, presumably because of the greater technical problems associated with setting up an assay that avoids receptors. One possibility, however, would be to use as a marker one of the GPI-linked proteins, which are excluded from coated pits, and thus might be endocytosed primarily in an alternative manner (*74*). The folate receptor, a GPI-linked protein, has been identified in a nonclathrin surface structure (termed plasmalemmal vesicles, or "caveolae") by the Anderson group (*2*), and may represent a structure capable of one type of nonclathrin-mediated endocytosis (potocytosis).

V. FUSION OF VESICLES WITH EARLY ENDOSOMES

The most extensive work in cell-free endocytic systems has examined the fusion of early endocytic vesicles and endosomes with each other. There have been several assays developed to measure this type of vesicle fusion, all based on the following concept of detecting fusion. Preexisting endocytic vesicles are isolated from two populations of cells; the vesicles are preloaded with different endocytic probes prior to the isolation. These two types of vesicles are then coincubated under conditions promoting fusion. If fusion should occur, then the mixing of the two probes within the product compartment generates a detectable signal. Typically, a membrane-impermeant scavenging reagent is added outside the vesicles, to prevent any mixing of the endocytosed probes due to vesicle breakage. Depending on the exact details of the probes used, different advantages are accrued.

The first endosome–endosome fusion system, developed by Davey et al., detected the release of labeled sialic acid from endocytosed virus, when it encountered endocytosed neuraminidase inside the target vesicles (*21*). An inhibitor of neuraminidase added outside the vesicles improved the signal-to-noise ratio of this assay. This assay demonstrated that endosome-endosome fusion did occur, but was hampered by the need for a finite timeframe following fusion, during which the enzyme catalyzed sialic acid release. This complicated the interpretation of the results: different reagents or conditions might affect the enzyme activity inside the vesicles, rather than the rate of vesicle fusion. An assay that exploited the ability to observe the reacting vesicles morphologically was developed by Gruenberg and Howell, using endocytic vesicles bearing VSV glycoprotein, which could be immunoisolated on beads coated with antibody to the glycoprotein cytoplasmic tail (*39,40*). Here, fusion was detected by the ability to radioiodinate this glycoprotein with lactoperoxidase that had been endocytosed into a separate set of vesicles. Again, the signal generated required enzyme function for a finite time after fusion, within vesicles whose internal milieu was ill-defined. Braell developed an ELISA assay based on the avidin-biotin binding reaction: endocytosed avidin-β-galactosidase reporter enzyme would associate with endocytosed biotin-IgG upon vesicle fusion (*14*). The amount of avidin-β-galactosidase: biotin-IgG complex which resulted, detected in a second stage ELISA procedure, represented the level of vesicle fusion. Biotin–insulin was added outside the vesicles to scavenge any avidin released by vesicle breakage. The assay thus relied on a diffusion-controlled association between avidin and biotin, which occurred rapidly and essentially irreversibly after vesicle fusion, and was relatively insensitive to variations in the vesicle's internal conditions. As a result, a close to real-time measurement of vesicle fusion could be achieved. This two-stage detection strategy, using a reaction involving protein–protein association within the fused vesicles, followed by measurement thereafter of the complexes formed, has been adapted in the formulation of several other fusion assays (*10,23,38,60,75,112*).

For all these systems, a similar set of properties characterize the early endosome fusion process. The probe enters the reactive vesicle compartment rapidly upon internalization, and exits that compartment minutes later. Fusion requires incubation above 15 to 20 °C, ATP, and membrane and cytosolic proteins, including some inhibited by N-ethylmaleimide (NEM). Also, the nonhydrolyzable analogue GTPγS had inhibitory effects on cell-free fusion at μM levels.

Are clathrin coated vesicles *bona fide* precursors in the pathway to endosomes? Woodman and Warren were able to demonstrate the ability of coated vesicles to participate in fusion with endosome structures in a cell-free system, thus providing evidence in favor of this point (*113*). This fusion required cytosolic proteins sensitive to NEM, as well as ATP, in congruence with the properties of other cell-free systems measuring early endocytic vesicle fusion. Significantly, fusion appeared to be specific for the early endocytic compartments: poor reactivity was observed with vesicles derived from cell surface. This fact argues in favor of the relevancy of this fusion process to the *in vivo* situation. Thus, this study provided a link between the cell-free budding processes and those processes relating to the endosomal compartment.

Once a clathrin-coated vesicle has budded, the clathrin coat is a problem for fusion of that vesicle with its intended target membrane. Theoretical studies imply that any membrane fusion process first requires a close apposition of the membranes of the two fusing bodies, for which a proteinaceous coat would present a barrier. These considerations have also been shown to apply specifically to the clathrin coat itself. The studies of Alstiel and Branton showed that, even under forcing conditions for vesicle fusion (nonphysiologically high Ca^{2+} levels), coated vesicles were unable to fuse with lysosomes unless the coat was first removed (*1*). In the case of the coated vesicle work by Woodman and Warren, clathrin was released by the action of "uncoating ATPase" on those coated vesicles (*113*). As the cytosol contains this protein, clathrin would presumably have been released by its function, and thereby allow fusion to proceed.

This would seem to imply a role for uncoating ATPase in the endocytic process, although this is a point that remains to be proven directly for the *in vivo* situation. The uncoating ATPase, now known to be a member of the 70-kDa heat-shock "chaperonin" family, is able to promote the ATP-dependent removal of the clathrin coat from coated structures (*82,84*). However, this enzyme only removes clathrin, not the adaptor complexes, from the vesicles (*74*). The role of this protein in clathrin metabolism might also involve maintaining a dynamic equilibrium between soluble and lattice-associated clathrin triskelions. Approximately half the clathrin in a typical healthy cell is not in coated structures, but exists in soluble form (*34*). As treatments that cause the formation of aberrant clathrin structures, drawing clathrin out of the soluble pool, also inhibit the production of normal clathrin lattices (*44,45*), the maintenance of a substantial soluble pool of recycled clathrin for forming fresh coated pits may be an alternative role for this protein (*15,74,89*). It is worth mentioning that Anderson and colleagues observed an ATP-dependent release of clathrin, without adding uncoating ATPase, from the clathrin lattice structures that they could create *in vitro* (*65*). Apyrase was able to stop the release, presumably by hydrolysis of the ATP to ADP and then AMP. These observations, however, are made problematic by the studies that indicate that *in vitro* coated vesicle budding, which resembles clathrin release in the Anderson group's system, has a component that is independent of exogenously added cytosolic proteins (*56,90,93*).

A recent discovery is that, in some cell types, the early endosomes can form a tubular network, or "endosomal reticulum", in which boluses of fluid appear to migrate, analogous to the functioning of a peristaltic pump (*47*). This has led to some concern in the field regarding exactly what the cell-free fusion of early endosomes represents. This is, however, only a conceptual problem if one envisions early endosomal trafficking as exclusively involving the trafficking of small, spherical vesicles. Indeed, given the resultant surface-to-volume constraints that would occur when many small vesicles are fused together, it is difficult to imagine that the fusion product would maintain a vesicular shape. The relative kinetics of entry and exit of vesicles into the endosomal system, and the relative sizes of those vesicles, will determine the dynamic size of that system. Thus, that system might be vesicular or tubular, depending on whether budding rates are comparable to, or slower than, fusions that add membrane to the system. Since cell-free fusion of early endocytic vesicles is often very efficient (*107*), membrane donation may well be quite rapid *in vivo*, and tubules might well result, unless membrane removal is equally rapid.

A corollary to this point is that the relative contribution of tubular and vesicular elements in the early endosomes can vary depending on cell type or activity. Tooze and Hollinshead have found that tubular endosomal elements predominate in epithelial-like cells, but are infrequent in fibroblastic cells (*98*). Moreover, depending on cell activation, the degree of tubulation may vary. Here, an analogy can be made to the formation of "tubular lysosomes", which are induced upon macrophage activation (*97*). These structures, which may actually be formed from late endosomal structures (*78*), can be observed to elongate in a microtubule-dependent fashion (*46*). Loss of early endosome tubules should also occur, if membrane entry into a dynamic system declined. In the *Drosophila shibire* mutant, budding of new endocytic vesicles is blocked at the nonpermissive temperature (*52,55*). A progressive loss of endosomal tubules is observed upon instituting the temperature block, and the tubules return at permissive temperature (*52*). This phenomenon is better explained by the dynamic maintenance of the tubular endosomal system via vesicle entry and exit, than by a static system. The crux of the issue is that, without knowledge of the balance between membrane entry and exit into the early endosomal system, the specific morphology of that endosome system cannot indicate the role of vesicle fusion in its creation.

VI. RECONSTITUTION OF ENDOSOMAL SORTING AND MATURATION

Once receptors and their associated proteins have entered the early endosomal compartment, the ligands must be separated from the receptors prior to receptor recycling, if these ligands are to progress to the lysosome (*109*). This is often accomplished by acidification of the endocytic compartment, which releases the ligand from its receptor. For transferrin, an exception, acidification causes the release of its bound iron, while the apoprotein recycles to the cell surface along with its receptor (*20*). This property has made the transferrin receptor a convenient probe for addressing details of receptor recycling, because the receptor can be followed by following its bound transferrin. Ligands released from their receptors, now in the fluid phase, are transported via late endosomes to the lysosomes (*54*). The receptors recycle to the surface in smaller, tubulovesicular processes; the smaller volume of these entities allows the recycling of equivalent areas of membrane to that originally endocytosed, while much of the endocytosed fluid volume remains within the cell. This constitutes a segregation of receptors and fluid phase, concomitant to the sorting process, which should be detectable. Note that segregation need not require actual transfer into separate vesicle systems, only a regional separation into vesicular vs. tubulovesicular regions.

Wessling-Resnick and Braell were able to observe the cell-free segregation of transferrin receptors from fluid-phase material destined to reach lysosomes, using the endosomal vesicle fusion system developed in K562 cells (*107*). These cells were shown *in vivo* to mature the fluid-phase probe, avidin-β-galactosidase, to lysosomes, while receptor-bound biotin–transferrin, the other probe, was recycled properly. In the corresponding cell-free system, these two probes could meet with high efficiency (>50 %) in early endosomes, as judged by both the fusion assay itself, and by their comigration on Percoll density gradients. Following this meeting, that portion of the avidin-β-galactosidase which had not been entrained by the receptor-bound biotin–transferrin was observed, upon continued incubation, to migrate with increased density on the Percoll gradients. This resulted in a physical separation of that denser portion of the fluid-phase probe from the receptor-bound biotin–transferrin probe (and its bound avidin-β-galactosidase), which remained in the light vesicle region of the gradient, consistent with its presence in early endosomal and recycling vesicles. This constituted a direct demonstration of a cell-free segregation of the recycling receptor probe from the lysosomally-directed fluid-phase probe. That a separation of the two markers was observed, without any density shift of the transferrin peak, ruled out an artefactual density shift due to nonspecific aggregation of vesicles. The conditions optimal for this separation mirrored those for vesicle fusion, consistent with entry into the endosomal compartment being a prerequisite for the segregation process. It must be remembered, however, that since late endosomes and lysosomes were not excluded from these incubations, that the segregation might have involved transport of AvβGal into a preexisting late endocytic compartment.

One possible argument that actual *in vitro* segregation had not been observed in this work was that an alternative scenario could be imagined, wherein that portion of avidin-β-galactosidase which did not react with biotin–transferrin had actually entered a parallel, but mechanistically distinct, pathway that permitted cell-free densification of the avidin-β-galactosidase-populated vesicles, but that the transferrin receptors had avoided this pathway. This question has been addressed by an experiment assuring that avidin-β-galactosidase *must* have entered by the same route as the transferrin receptors.

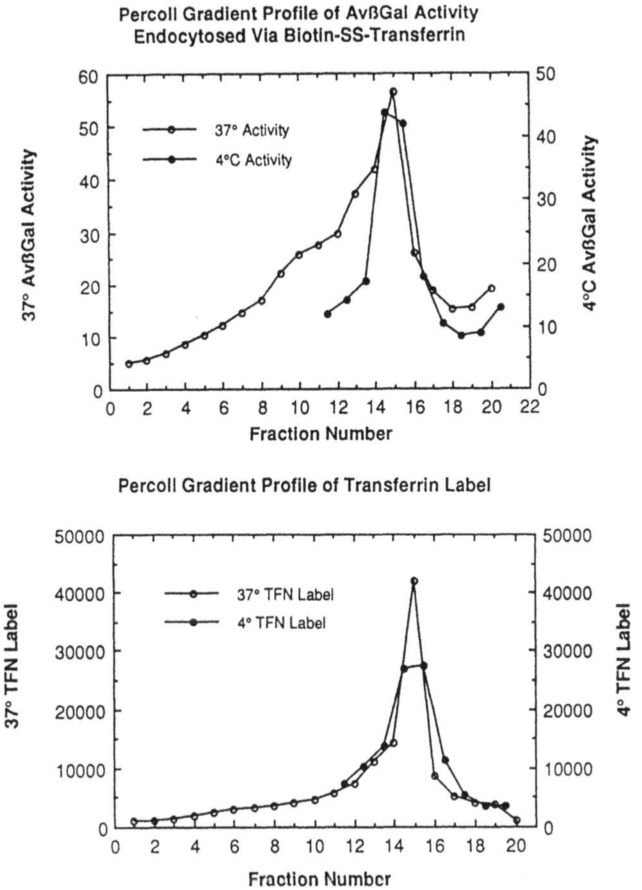

Figure 2. K562 cells were permitted to endocytose avidin-β-galactosidase (AvβGal) conjugates that were associated, via the avidin, with biotin groups disulfide-linked to transferrin (biotin-SS-transferrin). Greater than 95% of the internalized β-galactosidase activity entered these cells in "piggyback" form via receptor-mediated endocytosis of the biotin-SS-transferrin, which was trace-labeled with ^{125}I. Isolated endosomes from these cells were treated with membrane-permeant reducing agent (2-mercaptoethylamine) to separate the AvβGal from the transferrin by disulfide bond reduction *in situ*. The treated endosomes were then incubated at 37 °C, under conditions that permit cell-free maturation to occur, or mock-incubated at 4 °C, and Percoll density gradient analysis was performed. Upon incubation at 37 °C, over 50% of the AvβGal label shifted to denser regions of the gradient (leftward), when compared to its position after mock incubation (upper panel). On the same gradients, transferrin label did not shift position during this 37 °C incubation from the mock incubation position (*lower panel*). Higher percentage shifts of AvβGal could not be obtained, as the *in situ* reduction of this biotin disulfide conjugate is slow (half-time: 5–10 min at 37 °C), and the segregation process occurs at a comparable rate. Nevertheless, segregation of most of the AvβGal away from the transferrin which had ferried this enzyme into the endosomes was observed to occur *in vitro*. Therefore, this segregation process can indeed occur for two probes originating from a single endosomal compartment.

Figure 2 shows the segregation of materials derived from within the same endosomal vesicles. Transferrin, modified with a cleavable disulfide-linked biotin, was bound to cell surface receptors at 0 °C. Avidin-β-galactosidase was then bound to this cleavable biotin moiety at 0 °C, and then permitted to "piggyback" into endocytic vesicles when uptake of the receptor-bound transferrin was performed at 37 °C. Controls demonstrated that over 95% of the internalized β-galactosidase activity entered by receptor-mediated uptake, thereby ensuring that the vast bulk of the avidin-β-galactosidase entered cells via the identical pathway as the transferrin probe. Once inside the cells, endosomal vesicles were isolated, and the segregation assay was performed at 37 °C, in the presence of a membrane-permeant reducing agent (2-mercaptoethylamine) that split the disulfide linkage *in situ*, thereby converting the avidin-β-galactosidase probe back to a fluid-phase marker. Although at 37 °C the rate of splitting of the disulfide linkages within the intact vesicles was comparable to the rate of segregation, so that segregation was completed before 100% of cleavage had occurred, nevertheless about 50% of the piggybacked avidin-β-galactosidase was released in time for it to be segregated on Percoll density gradients away from the transferrin probe (Fig. 2). Hence, the cell-free segregation of this avidin-β-galactosidase away from the receptor-bound transferrin *must* have occurred from the same compartment. This would argue that cell-free segregation of recycling receptors from lysosomally-directed materials was observed in this system. The exact form of this segregation, whether involving placement of the two probes into distinct vesicles, or whether regionalized separation occurred within a fragile single compartment, that fragmented upon centrifugation, has not yet been determined.

Roederer and colleagues were able to observe an ATP-dependent densification of isolated, ^{125}I-EGF-labeled endosomes, as assessed by Percoll density gradients (*80*). By 160 minutes of incubation, this increase in density was approaching that of lysosomes. Densification seemed dependent on acidification of the endosomes, and occurred in the absence of added preexisting lysosomes. This was taken as evidence of maturation of endosomes directly into lysosomes, as the morphology of the resultant incubated vesicles revealed an electron-dense interior, reminiscent of the dense lysosomal "matrix" observed by Buckmaster and co-workers (*16*). Since these criteria are not necessarily proof of lysosomal identity, however, the evidence is also consistent with the idea that these observed bodies represent matured and condensed entities on their way to, but not yet arriving at lysosomes.

VII. RECEPTOR RECYCLING PROCESSES

Once receptors have been sorted from their released ligands, cells typically rescue these receptors from degradation in the lysosomes, by recycling them to the cell surface. It is not uncommon for receptors like the LDL receptor to be recycled hundreds of times by a cell, before finally being degraded. Recycling is thought to occur at least partly via the *trans*-Golgi network (TGN) (*37*). This fact has allowed the partial reconstitution of at least one step in that recycling process. The TGN is known to possess a sialyl transferase activity that adds terminal sialic acid to glycoproteins (*81*). Goda and Pfeffer used this fact to recreate the recycling process to the TGN, by using the sialyl transferase activity to mark the arrival of their probe, the cation-independent mannose 6-phosphate receptor (CI-MPR), in the TGN (*30*). Sialyl transferase is inactive in the CHO 1021 mutant cell line; thus, all glycoproteins, including CI-MPR, lack sialic acid in these cells. Radiolabeled CI-MPR was prepared in 1021 cells, and the cells were permeabilized, prior to adding TGN vesicles from wild-type cells. Upon entering the wild-type TGN vesicles, the asialo-CI-MPR was sialylated by the resident sialyl transferase. Thus, labeled CI-MPR that was bound to slug lectin resin (which binds sialylated glycoproteins) was scored as having reached the TGN compartment. The donor vesicles from 1021 cells were derived from endosomes, since CI-MPR radioiodinated *in situ* with endocytosed lactoperoxidase was sialylated by the *in vitro* system, directly demonstrating endosome-to-TGN transport. The cell-free system monitoring endosome-to-TGN transport proved to require ATP, and was inhibited by GTPγS. There was also a requirement for cytosolic proteins, including at least one protein sensitive to N-ethylmaleimide (ETF-1), but which is distinct from the NSF/SEC18p protein purified by Rothman and co-workers (*32*).

The final event in recycling is the fusion of the recycling vesicle with the cell surface. Such a process is difficult to define biochemically in a cell-free context, for several reasons. First, the fusion necessarily will result in the eversion of the fusing vesicle, releasing its contents into the medium. This release, even if reconstituted, is difficult to distinguish from simple breakage of the original vesicle, adding to the difficulty of detecting specific release. Second, the target membrane for the fusion event is the plasma membrane, not another vesicle, and its integrity perforce must be breached if access to the intracellular milieu is to be achieved. Consequently, certain properties of the plasma membrane (ion gradients, pH gradients, asymmetric distribution of diffusible components) cannot be maintained, and may affect any reconstituted activity. Finally, even if isolated plasma membrane vesicles are used in the assay, these are not representative of the normal state of that membrane, and may even be oriented in reverse from that orientation desired for a cell-free assay (cytoplasmic face outward). These considerations make any simplistic formulation of an efficient vesicle recycling assay difficult.

An elegant approach to dealing with many, if not all, of these concerns was developed by Podbilewicz and Mellman (*76*). These researchers examined the recycling of transferrin receptors in the polarized MDCK cell line. By selectively permeabilizing the apical surfaces of these cells, the interior of the cell was made accessible, yet the basolateral face of the cell monolayer, where the transferrin receptors undergo endocytosis and recycling, was left intact afterwards. Thus, the issue of accessing the cell interior without breaching (basolateral) membrane integrity was addressed. Also, permeabilization left the MDCK cell tight junctions still intact. This allowed some measure of the specificity of fusion, since any breakage of the recycling vesicles would release the vesicle contents into the apical medium. Hence, release only into the medium on the basolateral side was likely to represent true vesicle fusion with the membrane.

To actually perform the recycling assay, receptor-bound transferrin was internalized by intact cells under conditions of ATP depletion, which preferentially inhibited recycling. The apical surface was then breached, and recycling to the basolateral surface judged by the ATP and cytosol-dependent release of the internalized transferrin. Little horseradish peroxidase, which recycles poorly in MDCK cells *in vivo*, was released from endocytic vesicles under these *in vitro* conditions, indicative of there being little artefactual vesicle lysis. Interestingly, although NEM treatment of the total system did inhibit recycling, the sensitive components were not in the cytosolic fraction of that system, as treated cytosol was still functional for transport. Thus, cytosolic factors resistant to NEM were implicated in the recycling process. One concern regarding the data was the ATP-independence observed for transferrin internalization by MDCK cells. The apparent uptake may thus have included a component of transferrin present in deeply invaginated coated pits that had not budded (*89*). This might have yielded an artefactual signal resembling recycling, if those coated pits had then later reopened, and allowed their entrapped label to exit.

VIII. TRANSCYTOSIS

In cells such as the MDCK and Caco-2 epithelial cell lines, and in hepatocytes, delivery of membrane to the plasma membrane assumes an additional complexity, in that delivery is directed either to the apical domain, or the basolateral domain, of the cell (*48,66*). This delivery complication not only involves secretory membrane transport, but also has an endocytic component. Some receptors, such as the polymeric IgA receptor, are able to be endocytosed at one surface (here, the basolateral surface), and delivered across the cell to the apical surface (transcytosis), rather than being recycled to the same surface. The sorting process that distinguishes between transcytotic delivery, and normal recycling, appears to operate at the level of the early endosomes or TGN. The latter compartment is implicated in polarized delivery during secretion as well, with distinct vesicles targeted to the two surfaces budding from the TGN. Permeabilized cell and cell-free systems addressing the problem of polarized delivery have been constructed, although most address the polarized exocytic transport from TGN to surface, rather than by endocytically derived markers (*8,35,86*). Although brief mentions of cell-free reconstitution of transcytotic transport have been made in the literature (*66*), this is a problem just beginning to be addressed experimentally.

IX. LATE ENDOSOMAL PROCESSES

One of the more difficult endocytic events to observe *in vitro* has been the transfer of material from early endosomes to late endosomes; the reasons for this are currently unclear. Lateral transfer among late endosomes has apparently been reconstituted in a meaningful way, using polarized epithelial (MDCK) cells. The MDCK cell line can be made to form tight monolayers, which allows segregation of the apical and basolateral surfaces of the cells (*66*). It has been observed *in vivo* that ligands endocytosed from the apical and basolateral surfaces, respectively, can eventually meet in a common compartment. This compartment appears to correspond to the late endosomes, and the fusion events *in vivo* that permit the ligands to mix occur between late endosomes (*11*).

Bomsel and colleagues have provided evidence for cell-free reconstruction of this lateral mixing process in a permeabilized MDCK system (*10*). Avidin, endocytosed from the apical surface, and biotinylated horseradish peroxidase, endocytosed from the basolateral surface, were the probes for this assay. Mixing of the contents of either early or late endosomes was scored by peroxidase activity immunoprecipitated with anti-avidin serum. Early apical MDCK endosomes were found to readily fuse among themselves *in vitro*, without need of cytoskeletal elements. Similarly, early basolateral–basolateral endosomal fusion occurred *in vitro*. However, apical and basolateral early endosomes did not fuse with each other. Apical–basolateral mixing due to *in vitro* fusion *did* occur after the probes had been chased into late endosomes. This late endosomal fusion required the presence of intact, polymerized microtubules: 20 μM of the microtubule stabilizing drug taxol, or 5 μM taxol in the presence of brain microtubules, provided the assembled microtubules needed to promote the fusion. An inhibition by GTPγS was also observed, although it was not established whether this effect was due to GTP-binding proteins other than tubulin itself.

Given the need for microtubules in the system, Bomsel et al. examined the role of microtubule motor proteins in fusion (*10*). Both kinesin and cytoplasmic dynein were found to promote the cell-free fusion of the late endosomes, as removal of either motor protein reduced fusion levels almost 50%. Inhibition of kinesin with specific antisera, or destruction of the dynein with vanadate plus UV light (*91*), abrogated that portion of the late endosomal fusion. A concern about the role of microtubule motors in this system stems from the equal *in vitro* efficacy of the two different motors, thought to direct transport in opposite directions within the cell. Nonspecific entities, such as fluorescent beads, can be transported by loosely adsorbed microtubule motors in cell-free microtubule transport assays (*91,100*). It remains to be proven that the stimulation of fusion provided by these motors was due to specific interactions, and not an effect of simply increasing the frequency of encounters between late endosomes in a dilute cell-free milieu. Hence, any conclusions regarding the relative contributions of these two motor proteins in late endosome transport and fusion, based on the cell-free work, remain tentative.

X. LYSOSOMAL FUSION EVENTS

The final destination for endocytosed ligands is the secondary lysosome, wherein degradation by lysosomal hydrolases occurs. Paradoxically, this event was probably the first stage in endocytosis to be reconstituted, but has remained one of the less-well studied. The pioneering reconstitution study, by Oates and Touster, involved the cell-free fusion of lysosomes with phagosomes, the vesicles formed by macrophage ingestion of external bodies, from *Acanthamoeba* (*72*). They were able to record a dependence on ATP and cytosolic components for these fusions, in accord with work dating much later. Others soon followed suit using mouse macrophages (*29*).

Recently, Mayorga and colleagues have detected fusion of phagosomes with endosomal elements in intact J774 macrophage-like cells, and in a cell-free system based on those cells (*60*). Fusion was detected by association of anti-DNP antibody-coated, fixed *Staph. aureus cells* (in the phagosomes) with endosomal DNP-β-glucuronidase. Kinetics studies using 5-nm gold in lysosomes, and 20-nm gold in endosomes, implied that endosomes, rather than lysosomes, were the fusogenic bodies. As short labeling times, favoring early endosomes, placed gold label with phagosomes, the conclusion was made that early endosomes fused with phagosomes; no fusion with classic lysosomes was detectable. Interestingly, GTPγS stimulated, rather than inhibited, the fusion at low cytosol levels: inhibition was seen only at high cytosol levels. Moreover, similar paradoxical effects of GTPγS have been observed by this group for early endosomal fusion itself (*61*). Their data fits a model where GTPγS promotes the binding of a dilute cytosolic factor needed for fusion to the membranes, but at high concentrations, this factor can "jam the works", never cycling off the membranes. Alternatively, two antagonistic GTP-binding factors are present, that are differentially sensitive to GTPγS. The endosomes fusing with phagosomes had a tubular component to them. As recent morphological studies of macrophage phagosome–endosome fusions seem to imply that the so-called "tubular lysosomes" that fuse with phagosomes in activated macrophages might in fact derive from elements classically considered late endosomal elements (*78,97*), it may well be that fusion with these entities, rather than early endosomes, accounts for the GTPγS-stimulatable fusion with phagosomes in the J774 system. These considerations may well indicate that more than one form of vesicle–vesicle fusion is being detected in this system: a constitutive endocytic fusion process, and one relating to the inducible phagocytic degradative function of macrophage type cells.

Fusion of lysosomes with late endosomes, exclusive of phagosomes, has been reconstituted by Mullock and colleagues (*68*). These workers took advantage of the available subcellular fractionation techniques for rat liver organelles, to provide them with separated pools of the participant organelles (late endosomes and lysosomes) for their studies. Radiolabeled asialofetuin, endocytosed *in vivo* into liver cells by the ASGP receptor, was chased sufficiently long into the cells to permit its advance into late endosomes. These endosomes, apparently corresponding to the receptor-negative endosomes of Mueller and Hubbard (*67*), were the only active endosomes for fusion with lysosomes: early endosomes (labeled with asialofetuin or pIgA in shorter uptake times) did not transfer their contents to lysosomes. The endosomes, after isolation, were incubated with purified lysosomes, ATP, and cytosol at 37 °C. The radiolabeled ligand, after the incubation, was observed to comigrate with the lysosomes on Nycodenz gradients, and was digested into TCA-soluble fragments upon continued incubation at 37 °C with ATP, consistent with lysosomal degradation. The transport to lysosomes was dependent on incubation with ATP, and at elevated temperature. Cytosolic proteins appeared to be required, as trypsin treatment or boiling inhibited the stimulation by cytosol, but the required proteins were resistant to treatment with N-ethylmaleimide, in contrast to the situation earlier in the endocytic pathway. One question not resolved in this study, due to the nature of detection of the transfer, was which factors were required for actual transfer to lysosomes, and which facilitated the subsequent breakdown of ligand. For example, it was not made clear whether the requirement for ATP was not due simply to a need for acidification of the lysosomes, via ATP-driven proton pumping, in order to effect asialofetuin hydrolysis by the acid proteases of the lysosome.

What can be gleaned from the cell-free assays observing densification of lysosomally directed entities? These assays do not definitively address the controversial question of whether endosomes and lysosomes are gradually remodeled in a maturation process during endocytosis, or whether they exist as pre-existing compartments through which materials are transported (*36,69*). An intriguing possibility consistent with either view of the endocytic pathway, however, is that the cell-free densification may be a reconstitution of a localized maturation process of packaging lysosomally directed materials for further transport. In this regard, it is interesting to note that reversible condensation of the matrix of secretory chromaffin granules can be accomplished *in vitro* (*28*). Lysosomal contents are able to condense into a dense matrix (*16*) that may not only aid them in resisting autolysis, but may provide a means of retrieval of lysosomal components from the degradative site, by packaging them back into lysosomes (*69*). This would be analogous to recycling of cell surface receptors for repeat usage. An analogous process might be generally used for the concentration of lysosomally-directed materials themselves.

XI. CELL PROTEINS INVOLVED IN ENDOCYTOSIS

The premier characteristic of studying cellular events in cell-free or permeabilized cell situations is that the internal cellular components participating in that process become accessible to the investigator. Thus, the agents responsible for catalyzing these processes become feasible objects of study. This will naturally include those proteins which are governing endocytic events, and indeed several proteins have been identified which appear to be good candidates for either mechanistic, or regulatory, roles in endocytosis.

The proteins involved in endocytosis which have been best characterized are those involved in clathrin-coated vesicle production. These include clathrin, the proteins of the adaptor complexes, "uncoating ATPase", and of course the receptors needed for receptor-mediated uptake (*15,74,84*). Ironically, little ground-breaking evidence for their role in endocytosis has derived from cell-free reconstitution work of the type described here, since their probable involvement was already highlighted either by *in vivo* studies (*25*), or by their biochemical isolation directly from coated vesicles themselves (*15,74*). Rather, the cell-free systems have served a confirmatory role, as described above, in showing that these proteins are indeed required for vesicle production in the cell-free milieu (*26,56,90*).

More notable results have come from the assays dependent on vesicle fusion events. The Rothman group first identified a protein required for Golgi vesicle transport in the secretory pathway, termed "NSF" (NEM-Sensitive Fusion protein), as a target for the alkylating agent NEM (*9*). This protein, once sequenced, was found to be homologous to the yeast SEC18 gene product, originally identified on genetic grounds as being crucial for ER-to-Golgi transport (*110*). Subsequently, it was shown that NSF/SEC18p was also involved in ER-to-Golgi transport in mammalian systems as well (*6*). This protein, thought to be involved in the vesicular fusion process itself, has been postulated to be a general requirement for vesicle trafficking events requiring fusion (*83*). In line with this, NEM sensitivity has been a consistent feature of the early endosome fusion assays. At least one cell-free endocytic vesicle fusion system has been found to specifically involve the NSF protein. Diaz and colleagues found that vesicle fusion in a system based on the J774 (macrophage-like) cell line, in which fusion was detected by an anti-DNP antibody: DNP-β-glucuronidase ELISA assay, could be inhibited with antibody directed against NSF (*24*). Significantly, supplementing this system, following pretreatment with NEM, with purified NSF would restore vesicle fusion. Thus, in this fusion system, NSF appears to be the only relevant cytosolic protein that is inactivated by N-ethylmaleimide.

Other NEM-sensitive proteins than the NSF protein have also been implicated in endocytic trafficking. Wessling-Resnick found that, although NEM treatment inhibited endocytic vesicle fusion in a K562 cell-based system, the fusion activity could be restored using cytosol proteins treated under conditions that inactivate NSF for Golgi transport activity (*108*). Thus, either NSF itself is not a crucial protein in this system, or its function in this system differs from that observed in the Golgi vesicle transport system. Goda and Pfeffer have determined that the transport of mannose 6-phosphate receptors from endocytic vesicles to the *trans*-Golgi network is sensitive to NEM, but immunodepletion of NSF from cytosol did not remove the activity (*32*). Partial purification of the protein which restores that endosome-to TGN transport activity (ETF-1) defines a protein of about 50- to 100-kDa total molecular weight by gel filtration; moreover, this protein functions prior to a step inhibited by GTPγS. As the NSF protein is a tetramer of 76-kDa molecular weight subunits, and acts distal to the GTPγS-sensitive step in Golgi transport (*83*), this seems to indicate a requirement for a distinct protein from NSF in recycling. It is worth noting here that N-ethylmaleimide, while fairly specific for sulfhydryl alkylation, would be expected to modify the cysteine sidechains of many cytosolic proteins. Thus, it is by no means a diagnostic reagent; indeed, it is fortunate for the field that so few relevant proteins turn out to be inactivated by this reagent, or its usefulness would be slight!

Membrane-bound proteins have been implicated in promoting the fusion of early endocytic vesicles. Evidence for such proteins stemmed originally from the observation that endocytic vesicle preparations pretreated with trypsin were inactivated for subsequent fusion, despite addition of fresh cytosolic proteins to the fusion reactions (*23,112*). This suggested that trypsin-sensitive proteins on the vesicle surfaces were required for that fusion. Recently, Colombo and colleagues have been able to demonstrate the presence of a *trans*-acting factor on endosomes derived from J774 cells, which is able to restore fusion competence to trypsinized endosomes from that cell line (*19*). The restorative function had the properties of a loosely bound peripheral membrane protein, as it could restore fusion activity when added either from a high salt extract of endosomal membranes, or directly from endosomal membranes themselves. This factor appeared to be distinct from NSF protein, a required protein for endocytic vesicle fusion in this system (*24*) that is known to be loosely bound to membranes, as it was resistant to NEM, and migrated on gel filtration at about 200 kDa molecular weight. The large size would seem to rule out small GTP-binding proteins as the source of the restorative activity, although it must be remembered that such proteins in soluble form might be present with other proteins as a complex. To date, there is a lack of evidence for any integral membrane proteins, which probably would not act in *trans*, that catalyze endocytic vesicle fusion *in vitro*.

XII. REGULATION OF ENDOCYTIC TRAFFICKING

It has been known for decades that, during mitosis, the nuclear envelope breaks down, facilitating the bipolar movement of the chromosomes into the regions of the cell destined to become the daughter cells. This observation was extended by Warren, who found that, at mitosis, a general halting of membrane trafficking occurred (*103,104*). In addition to the nuclear envelope breakdown, vesiculation of the endoplasmic reticulum and partial fragmentation of the Golgi ensued. This cessation was found to be uncoupled from membrane biosynthesis directly, and occurred both for the exocytic and the endocytic pathways. For the endocytic pathway, the trafficking halt at mitosis extends to both vesicle formation and vesicle fusion.

As the $p34^{cdc2}$ kinase is integral to the coordinate regulation of the various subcellular processes occurring during mitosis (*71*), its role in regulating endocytosis has been investigated. Tuomikoski and colleagues discovered that the action of this kinase was able to prevent vesicle fusion in his cell-free endosome fusion system (*99*). This was evidence, if not direct, for a role for that kinase in regulating endocytic transport *in vivo*. Pypaert and co-workers found that invagination of coated pits in permeabilized cells was inhibited by mitotic cytosol, and that this effect could be partially reproduced using added $p34^{cdc2}$ kinase (*77*). Further evidence that phosphorylation accounted for the inhibition was given by the ability of the phosphatase inhibitor okadaic acid to decrease the reversal of the inhibition. Together, these data argue for a role for the $p34^{cdc2}$ kinase in regulating the cessation of endocytic trafficking during mitosis. However, it is likely that other protein kinases also contribute to the regulation of endocytic vesicle transport. Woodman et al. found that the phosphatase inhibitors, okadaic acid and microcystin-LR, induced a progressive inhibition of cell-free endosomal vesicle fusion in their A431 cell-based system (*111*). This inhibition could be reversed by addition of purified protein phosphatase 2A, but not protein phosphatase 1. Interestingly, the effect of the phosphatase inhibitors on $p34^{cdc2}$ kinase activity was not sufficient to explain this inhibition, and using cytosol depleted of $p34^{cdc2}$ kinase in the cell-free assay did not prevent the inhibition by the phosphatase inhibitors. This would imply that at least one other protein kinase participates in the inhibition of endocytic vesicle fusion.

There appears to be a role for small GTP-binding proteins in the regulation of endocytic transport. The first inkling that such proteins might be involved in trafficking events came from the secretory field. Salminen and Novick showed that the SEC4 gene product, needed for secretory vesicle transport in yeast, was related to the *ras* family of small GTP-binding proteins (*87*). Since that time, a family of related proteins, termed rab proteins, has been implicated in membrane trafficking in general, and this function is conserved across the eukaryotic kingdom (*4,12*). The genes for more than 20 such proteins have been cloned at this time, and more may yet be found.

The rab proteins have the capability of specific binding to subcellular membrane domains, which is dictated by their isoprenylated carboxyl termini (*4,53*). When chimeric rab gene constructs are expressed in cells, the resultant chimeric rab proteins become localized to the membrane organelle appropriate for the rab protein donating the C-terminal portion of the chimera (*18*). Interestingly, overexpression of these chimeric rabs in cells did not cause derangement of their organelle morphology, at least at the level of immunofluorescence. Assuming that the chimeras were not deactivated, this suggests that either the catalytic domains of different rabs perform similar functions for their respective organelle targets, or else the rabs are not crucial for organelle function. By immunolocalization, at least three rab proteins are found affiliated with endocytic membranes, implying a role for these proteins in endocytosis. These are rab 4, rab 5, and rab 7. The rab 4 protein was found by van der Sluijs and colleagues to be associated with the early endosome population involved in the recycling of the transferrin receptor (*102*). Chavrier and co-workers found that the rab 5 protein was also found associated with an early endosome population, while the rab 7 protein was associated with late endosomes

(*17*). Interestingly, the rab 4 protein has a phosphorylation site for the p34^{cdc2} kinase, and is hyperphosphorylated during mitosis (*3*). Since recycling is arrested at mitosis (*85,105*), this implies a possible route by which that kinase could regulate transferrin recycling.

Because they have distinct localizations on endocytic subcompartments, the immunolocalization data for rabs 4, 5, and 7 has been cited as evidence for pre-existing compartmentation of the endocytic system, under the assumption that they bind via hypothetical residential proteins (*36*). This conclusion is dangerous, however, if the theory that these rab proteins aid in targeting of membranes during trafficking is valid (*4,12*). The rabs probably cycle onto and off the cytoplasmic surfaces of the endosomes while performing their function (*4*). Thus, the rabs might have a transient role in an endocytic maturation process, because the rab targeting activity could in essence dictate, and therefore define, the maturation event itself.

The first evidence that GTP-binding proteins might be involved in endocytic trafficking came from cell-free systems, in which the nonhydrolyzable analogue GTPγS was shown to affect vesicle fusion (*61,62*). This effect of GTPγS has turned out to be general for cell-free trafficking systems at all stages, implying a role for these proteins in many trafficking steps. With regard to cell-free endosome fusion, Wessling-Resnick showed that the role of this nucleotide analogue in the inhibition of endosomal vesicle fusion was to cause a cytosolic component, presumably a protein, to bind to the endosome membrane and prevent vesicle fusion (*108*). Upon subsequent incubation with excess untreated cytosol and GTP, the membrane-bound inhibitory factor was removed. This evidence was consistent with the cyclic binding and subsequent release of a cytosolic GTP-binding protein, as envisioned for the role of such proteins for membrane trafficking in general (*4,13*).

The one direct test of the question of the role of these proteins in regulation of endocytosis to date has been performed by Gorvel et al. (*33*). These researchers overexpressed normal and mutated forms of rab 5, and used these proteins in their cell-free endosome fusion system to observe the requirement for these proteins. They found that mutated versions of rab 5, unable to bind GTP properly and thus presumably nonfunctional, did indeed inhibit the vesicle fusion process, as did antibodies against rab 5. Cytosol containing overexpressed wild-type rab 5 stimulated fusion, while cytosols with overexpressed mutant rab 5 lacking the C-terminal membrane binding domain, or antibodies against rabs 2 and 7 (not associated with early endosomes) had little or no effects on fusion. Finally, the antibody inhibition by anti-rab 5 serum could be overcome by adding excess wild type, but not mutant, rab 5 protein, suggesting that the antibody inhibited rab 5 function specifically. A significant point about these studies is that removal, or inhibition of function, of rab 5 caused inhibition of endosomal vesicle fusion. Since the early biochemical evidence for GTP-binding protein involvement in these events showed that nonhydrolyzable analogues, which should hyperactivate these sorts of proteins, also inhibited vesicle fusion, the conclusion is that rab 5 would seem to need to be able to cycle efficiently between the inactive and activated states, in order to promote endocytic vesicle fusion effectively. Locking the protein in either one of these two states results in a loss of fusion competence.

In addition to the rab proteins, other GTP-binding proteins have been implicated in trafficking events. Clearly, those processes involving microtubules (*10,11,49*) will require tubulin, which is a GTP-binding protein. The ARF protein, a small GTP-binding protein which is myristoylated rather than isoprenylated, appears to be involved in transport from endoplasmic reticulum to Golgi, as it is incorporated into budding ER vesicles (*83*). A related protein in yeast, Sar1p, has been linked genetically to secretory transport from the endoplasmic reticulum to the Golgi (*22,70*). As there are multiple ARF-related genes, there may be a role for ARF-like proteins in endocytosis as well as secretion. Finally, dynamin, a large (100,000 mol. wt.) protein, has been associated with the defect in the *drosophila* temperature sensitive mutation *shibire*, in which endocytic vesicles fail to bud from nerve synapses at high temperature (*101*). This protein is homologous to a novel family of GTP- binding proteins, the Mx proteins (*73*). Dynamin, however, may not have a regulatory role, considering that it has a high GTP turnover rate in association with microtubules (*63,73*), more akin to microtubule motor proteins than to *ras*-like proteins.

Do the larger heterotrimeric GTP-binding proteins, of the type more commonly associated with receptor-mediated signal transduction, participate in the regulation of endocytosis? This possibility must be kept in mind. Firstly, it has been found that effects of both G_s and G_i type proteins on endocytic or transcytotic trafficking have been observed (*43,66,96*). Second, in cell-free systems, effects have been observed that are difficult to relegate to small GTP-binding proteins. It has been shown that small GTP-binding proteins, while activated by nonhydrolyzable GTP analogues, are insensitive to activation by GDP plus AlF_4^- ion treatment, a treatment known to activate heterotrimeric G proteins (*50*). Wessling-Resnick, among others, has nevertheless observed inhibition of endosomal vesicle fusion in a cell-free system by GDP plus AlF_4^-, but not by either agent alone (*108*). This would argue for a non-rab directed regulatory aspect to the fusion of these vesicles, suggestive of a role for the heterotrimeric G proteins.

XII. FUTURE PROSPECTS AND ISSUES

Cell-free endocytic assays are now reaching a mature stage of development, in that, with a few exceptions, the fact of successful reconstitution as an end in itself is no longer the goal. These assays are now being refined to a point where they will be useful biochemical tools for unearthing fresh insight into endocytic events. The most exciting stage of work with these systems—defining the important proteins involved in catalyzing endocytosis, and discerning their mechanisms of action—is only just now beginning to be addressed. It will be important to determine how many proteins are implicated in vesicle fusion, and their relation to the paradigm NSF protein. Equally vital is the question of exactly how intracellular membrane fusion is evoked by such proteins: it must certainly differ from that of fusogenic viral proteins, which involves irreversible denaturation of the fusogen. Another question begging to be answered is what are the interrelations between the various GTP-binding proteins, rabs or others, and exactly what functions they are providing to trafficking vesicles—targeting, or otherwise. Finally, for the sake of more rapid progress, it is to be fervently wished that the sort of genetic information which the *sec* mutants have provided for secretion could be provided for endocytic processes as well. The benefits of having two productive avenues of approach for determining the molecular mechanisms underlying endocytosis would doubtless enrich the entire field.

ACKNOWLEDGMENTS

The author would like to thank Drs. Bruce Schnapp and Marianne Wessling-Resnick for helpful comments. The work from the author's laboratory has been supported under grant GM 34635 from the National Institutes of Health.

REFERENCES

1. Altstiel, L. and Branton, D. (1983). Fusion of coated vesicles with lysosomes: measurement with a fluorescent assay. *Cell* 32: 921–929.
2. Anderson, R.G.W., Kamen, B.A., Rothberg, K.G., and Lacey, S.W. (1992). Potocytosis: sequestration and transport of small molecules by caveolae. *Science* 255: 410–411.
3. Bailly, E., McCaffrey, M., Touchot, N., Zahraoui, A., Goud, B., and Bornens, M. (1991). Phosphorylation of two small GTP-binding proteins of the rab family by $p34^{cdc2}$. *Nature* 350: 715–718.
4. Balch, W.E. (1990). Small GTP-binding proteins in vesicular transport. *Trends Biochem. Sci.* 15: 473–476.
5. Bankaitis, V.A., Aitken, J.F., Cleves, A.E., and Dowhan, W.F. (1990). An essential role for a phospholipid transfer protein in yeast Golgi function. *Nature* 347: 561–562.
6. Beckers, C.J.M., Block, M.R., Glick, B.S., Rothman, J.E., and Balch, W.E. (1989). Vesicular transport between the endoplamic reticulum and the Golgi stack requires the NEM-sensitive fusion protein. *Nature* 339: 397–398.
7. Beckers, C.J.M., Keller, D.S., and Balch, W.E. (1987). Semi-intact cells permeable to macromolecules: use in reconstitution of protein transport from the ER to the Golgi complex. *Cell* 50: 523–534.
8. Bennet, M.K., Wandinger-Ness, A., and Simons, K. (1989). Release of putative exocytic transport vesicles from perforated MDCK cells. *EMBO J.* 7: 4075–4085.
9. Block, M.R., Glick, B.S., Wilcox, C.A., Wieland, F.T., and Rothman, J.E. (1988). Purification of an NEM-sensitive protein NSF catalyzing vesicular transport. *Proc. Natl. Acad. Sci. USA* 85: 7852–7856.
10. Bomsel, M., Parton, R., Kuznetsov, S.A., Schroer, T.A., and Gruenberg, J. (1990). Microtubule- and motor-dependent fusion *in vitro* between apical and basolateral endocytic vesicles from MDCK cells. *Cell* 62: 719–731.
11. Bomsel, M., Prydz, K., Parton, R.G., Gruenberg, J., and Simons, K. (1989). Endocytosis in filter-grown Madin-Darby canine kidney cells. *J. Cell Biol.* 109: 3243–3258.
12. Bourne, H.R. (1988). Do GTPases direct membrane traffic in secretion? *Cell* 53: 669–671.

13. Bourne, H.R., Sanders, D.A., and McCormick, F. (1990). The GTPase superfamily: a conserved switch for diverse cell functions. *Nature* 348: 125–132.
14. Braell, W.A. (1987). Fusion between endocytic vesicles in a cell-free system. *Proc. Natl. Acad. Sci. USA* 84: 1137–1141.
15. Brodsky, F.M. (1988). Living with clathrin: its role in intracellular membrane traffic. *Science* 242: 1396–1402.
16. Buckmaster, M.J., Ferris, A.L., and Storrie, B. (1988). Effects of pH, detergent, and salt on aggregation of Chinese hamster ovary cell lysosomal enzymes. *Biochem. J.* 249: 921–923.
17. Chavrier, P., Parton, R.G., Hauri, H.P., Simons, K., and Zerial, M. (1990). Localization of low molecular weight GTP binding proteins to exocytic and endocytic compartments. *Cell* 62: 317–329.
18. Chavrier, P., Gorvel, J.-P., Stelzer, E., Simons, K., Gruenberg, J., and Zerial, M. (1991). Hypervariable C-terminal domain of rab proteins acts as a targeting signal. *Nature* 353: 769–772.
19. Colombo, M.I., Gonzalo, S., Weidman, P., and Stahl, P. (1991). Characterization of trypsin-sensitive factor(s) required for endosome-endosome fusion. *J. Biol. Chem.* 266: 23438–23445.
20. Dautry-Varsat, A., Ciechanover, A., and Lodish, H.F. (1983). pH and the recycling of transferrin during receptor-mediated endocytosis. *Proc. Natl. Acad. Sci. USA* 80: 2258–2262.
21. Davey, J., Hurtley, S.M., and Warren, G. (1985). Reconstitution of an endocytic fusion event in a cell-free system. *Cell* 43: 643–652.
22. d'Enfert, C., Wuestehube, L.J., Lila, T., and Schekman, R. (1991). Sec12p-dependent membrane binding of the small GTP-binding protein Sar1p promotes the formation of transport vesicles from the ER. *J. Cell Biol.* 114: 663–670.
23. Diaz, R., Mayorga, L., and Stahl, P. (1988). *In vitro* fusion of endosomes following receptor-mediated endocytosis. *J. Biol. Chem.* 263: 6093–6100.
24. Diaz, R., Mayorga, L.S., Weidman, P.J., Rothman, J.E., and Stahl, P.D. (1989). Vesicle fusion following receptor-mediated endocytosis requires a protein active in Golgi transport. *Nature* 339: 398–400.
25. Doxsey, S.J., Brodsky, F.M., Blank, G.S., and Helenius, A. (1987). Inhibition of endocytosis by anti-clathrin antibodies. *Cell* 50: 453–463.
26. Draper, R.K., Goda, Y., Brodsky, F.M., and Pfeffer, S.R. (1990). Antibodies to clathrin inhibit endocytosis but not recycling to the *trans*-Golgi network *in vitro*. *Science* 248: 1539–1541.
27. Featherstone, C. (1988). Perforated cell systems to study membrane transport. *Trends Biochem. Sci.* 13: 284–286.
28. Fernandez, J.M., Villalon, M., and Verdugo, P. (1991). Reversible condensation of mast cell secretory products *in vitro*. *Biophys. J.* 59: 1022–1027.
29. Geisow, M., D'Arcy Hart, P., and Young, M.R. (1982). Extracellular fusion of macrophage phagosomes with lysosomes. *Cell Biol. Int. Rep.* 6: 361–367.
30. Goda, Y. and Pfeffer, S.R. (1988). Selective recycling of the mannose 6-phosphate/IGF-II receptor to the trans Golgi network *in vitro*. *Cell* 55: 309–320.
31. Goda, Y. and Pfeffer, S.R. (1989). Cell-free systems to study vesicular transport along the secretory and endocytic pathways. *FASEB J.* 3: 2488–2494.
32. Goda, Y. and Pfeffer, S.R. (1991). Identification of a novel, NEM-sensitive cytosolic factor required for vesicular transport from endosomes to the *trans*-Golgi network *in vitro*. *J. Cell Biol.* 112: 823–832.
33. Gorvel, J.-P., Chavrier, P., Zerial, M., and Gruenberg, J. (1991). rab5 controls early endosome fusion *in vitro*. *Cell* 64: 915–925.
34. Goud, B., Huet, C., and Louvard, D. (1985). Assembled and unassembled pools of clathrin: a quantitative study using an enzyme immunoassay. *J. Cell Biol.* 100: 521–527.
35. Gravotta, D., Adesnik, M., and Sabatini, D.D. (1990). Transport of influenza HA from the *trans*-Golgi network to the apical surface of MDCK cells permeabilized in their basolateral plasma membranes: energy dependence and involvement of GTP-binding proteins. *J. Cell Biol.* 111: 2893–2908.

36. Griffiths, G. and Gruenberg, J. (1991). The argument for pre-existing early and late endosomes. *Trends Cell Biol.* 1: 5–9.
37. Griffiths, G. and Simons, K. (1986). The *trans* Golgi network: sorting at the exit site of the Golgi complex. *Science* 234: 438–443.
38. Gruenberg, J., Griffiths, G., and Howell, K.E. (1989). Characterization of the early endosome and putative endocytic carrier vesicles *in vivo* and with an assay of vesicle function *in vitro*. *J. Cell Biol.* 108: 1301–1316.
39. Gruenberg, J. and Howell, K.E. (1986). Reconstitution of vesicle fusions occurring in endocytosis with a cell-free system. *EMBO J.* 5: 3091–3101.
40. Gruenberg, J. and Howell, K.E. (1987). An internalized transmembrane protein resides in a fusion-competent endosome for less than 5 minutes. *Proc. Natl. Acad. Sci. USA* 84: 5758–5762.
41. Gruenberg, J. and Howell, K.E. (1989). Membrane traffic in endocytosis: insights from cell-free assays. *Annu. Rev. Cell Biol.* 5: 453–481.
42. Hansen, S.H., Sandvig, K., and van Deurs, B. (1991). The preendosomal compartment comprises distinct coated and uncoated vesicle populations. *J. Cell Biol.* 113: 731–741.
43. Haraguchi, K. and Rodbell, M. (1991). Carbachol-activated muscarinic (M1 and M3) receptors transfected into Chinese hamster ovary cells inhibit trafficking of endosomes. *Proc. Natl. Acad. Sci. USA* 88: 5964–5968.
44. Heuser, J. (1989). Effects of cytoplasmic acidification on clathrin lattice morphology. *J. Cell Biol.* 108: 401–411.
45. Heuser, J.E. and Anderson, R.G.W. (1989). Hypertonic media inhibit receptor-mediated endocytosis by blocking clathrin-coated pit formation. *J. Cell Biol.* 108: 389–400.
46. Hollenbeck, P.J. and Swanson, J.A. (1990). Radial extension of macrophage tubular lysosomes is supported by kinesin. *Nature* 346: 864–866.
47. Hopkins, C.R., Gibson, A., Shipman, M., and Miller, K. (1990). Movement of internalized ligand-receptor complexes along a continuous endosomal reticulum. *Science* 346: 335–339.
48. Hubbard, A.L. and Stieger, B. (1989). Biogenesis of endogenous plasma membrane proteins in epithelial cells. *Annu. Rev. Physiol.* 51: 755–770.
49. Hunziker, W., Male, P., and Mellman, I. (1990). Differential microtubule requirements for transcytosis in MDCK cells. *EMBO J.* 9: 3515–3525.
50. Kahn, R.A. (1991). Fluoride is not an activator of the smaller (20–25 kDa) GTP-binding proteins. *J. Biol. Chem.* 266: 15595–15597.
51. Kepes, F. and Schekman, R. (1988). The yeast SEC53 gene encodes phosphomannomutase. *J. Biol. Chem.* 263: 9155–9161.
52. Kessell, I., Holst, B.D., and Roth, T.F. (1989). Membrane intermediates in endocytosis are labile, as shown in a temperature-sensitive mutant. *Proc. Natl. Acad. Sci. USA* 86: 4968–4972.
53. Kinsella, B.T. and Maltese, W.A. (1991). rab GTP-binding proteins implicated in vesicular transport are isoprenylated at cysteines within a novel carboxyl-terminal motif. *J. Biol. Chem.* 266: 8540–8544.
54. Kornfeld, S. and Mellman, I. (1989). The biogenesis of lysosomes. *Annu. Rev. Cell Biol.* 5: 483–525.
55. Kosaka, T. and Ikeda, K. (1983). Reversible blockage of membrane retrieval and endocytosis in the garland cell of the temperature-sensitive mutant of *Drosophila, shibire*. *J. Cell Biol.* 97: 499–507.
56. Lin, H.C., Moore, M.S., Sanan, D.A., and Anderson, R.G.W. (1991). Reconstitution of clathrin-coated pit budding from plasma membranes. *J. Cell Biol.* 114: 881–891.
57. Lund, K.A., Opresko, L.K., Starbuck, C., Walsh, B.J., and Wiley, H.S. (1990). Quantitative analysis of the endocytic system involved in hormone-induced receptor internalization. *J. Biol. Chem.* 265: 15713–15723.
58. Mahaffey, D.T., Moore, M.S., Brodsky, F.M., and Anderson, R.G.W. (1989). Coat proteins isolated from clathrin coated vesicles can assemble into coated pits. *J. Cell Biol.* 108: 1615–1624.

59. Mahaffey, D.T., Peeler, J.S., Brodsky, F.M., and Anderson, R.G.W. (1990). Clathrin-coated pits contain an integral membrane proteins that binds the AP-2 subunit with high affinity. *J. Biol. Chem.* 265: 16514–16520.
60. Mayorga, L.S., Bertini, F., and Stahl, P.D. (1991). Fusion of newly formed phagosomes with endosomes in intact cells and in a cell-free system. *J. Biol. Chem.* 266: 6511–6517.
61. Mayorga, L.S., Diaz, R., Colombo, M.I., and Stahl, P.D. (1989). GTPγS stimulation of endosome fusion suggests a role for a GTP-binding protein in the priming of vesicles before fusion. *Cell Regulation* 1: 113–124.
62. Mayorga, L.S., Diaz, R., and Stahl, P.D. (1989). Regulatory role for GTP-binding proteins in endocytosis. *Science* 244: 1475–1477.
63. McIntosh, J.R. and Pfarr, C.M. (1991). Mitotic motors. *J. Cell Biol.* 115: 577–585.
64. Miller, K., Shipman, M., Trowbridge, I.S., and Hopkins, C.R. (1991). Transferrin receptors promote the formation of clathrin lattices. *Cell* 65: 621–632.
65. Moore, M.S., Mahaffey, D.T., Brodsky, F.M., and Anderson, R.G.W. (1987). Assembly of clathrin-coated pits onto purified plasma membrane. *Science* 236: 558–563.
66. Mostov, K., Apodaca, G., Aroeti, B., and Okamoto, C. (1992). Plasma membrane protein sorting in polarized epithelial cells. *J. Cell Biol.* 116: 577–583.
67. Mueller, S.C. and Hubbard, A.L. (1986). Receptor-mediated endocytosis of asialoglycoproteins by rat hepatocytes: receptor-positive and receptor-negative endosomes. *J. Cell Biol.* 102: 932–942.
68. Mullock, B.M., Branch, W.J., van Schaik, M., Gilbert, L.K., and Luzio, J.P. (1989). Reconstitution of an endosome–lysosome interaction in a cell-free system. *J. Cell Biol.* 108: 2093–2100.
69. Murphy, R.F. (1991). Maturation models for endosome and lysosome biogenesis. *Trends Cell Biol.* 1: 77–82.
70. Nakano, A. and Muramatsu, M. (1989). A novel GTP-binding protein, Sar1p, is involved in transport from the endoplasmic reticulum to the Golgi apparatus. *J. Cell Biol.* 109: 2677–2691.
71. Nurse, P. (1990). Universal control mechanism regulating onset of M-phase. *Nature* 344: 503–508.
72. Oates, P.J. and Touster, O. (1980). *In vitro* fusion of *Acanthamoeba* phagolysosomes. III. Evidence that cyclic nucleotides and vacuole subpopulations respectively control the rate and the extent of vacuole fusion in *Acanthamoeba* homogenates. *J. Cell Biol.* 85: 804–810.
73. Obar, R.A., Collins, C.A., Hammarback, J.A., Shpetner, H.S., and Vallee, R.B. (1990). Molecular cloning of the microtubule-associated mechanochemical enzyme dynamin reveals homology with a new family of GTP-binding proteins. *Nature* 347: 256–261.
74. Pearse, B.M.F. and Robinson, M.S. (1990). Clathrin, Adaptors, and Sorting. *Annu. Rev. Cell Biol.* 6: 151–171.
75. Pitt, A. and Schwartz, A.L. (1991). Reconstitution of human hepatoma endosome-endosome fusion *in vitro*: potential roles for an endoprotease and a phosphoprotein phosphatase. *Eur. J. Cell Biol.* 55: 328–335.
76. Podbilewicz, B. and Mellman, I. (1990). ATP and cytosol requirements for transferrin recycling in intact and disrupted MDCK cells. *EMBO J.* 9: 3477–3487.
77. Pypaert, M., Mundy, D., Souter, E., Labbe, J.-C., and Warren, G. (1991). Mitotic cytosol inhibits invagination of coated pits in broken mitotic cells. *J. Cell Biol.* 114: 1159–1166.
78. Rabinowitz, S., Horstmann, H., Gordon, S., and Griffiths, G. (1992). Immunocytochemical characterization of the endocytic and phagolysosomal compartments in peritoneal macrophages. *J. Cell Biol.* 116: 95–112.
79. Robinson, M.S. (1987). 100-kDa coated vesicle proteins: molecular heterogeneity and intracellular distribution studied with monoclonal antibodies. *J. Cell Biol.* 104: 887–895.
80. Roederer, M., Barry, J.R., Wilson, R.B., and Murphy, R.F. (1990). Endosomes can undergo an ATP-dependent density increase in the absence of dense lysosomes. *Eur. J. Cell Biol.* 51: 229–234.

81. Roth, J., Taatjes, D.J., Lucocq, J.M., Weinstein, J., and Paulson, J.C. (1985). Demonstration of an extensive *trans*-tubular network continuous with the Golgi apparatus that may function in glycosylation. *Cell* 43: 287–295.
82. Rothman, J. (1989). Polypeptide binding proteins: catalysts of protein folding and related processes in cells. *Cell* 59: 591–601.
83. Rothman, J.E. and Orci, L. (1992). Molecular dissection of the secretory pathway. *Nature* 355: 409–415.
84. Rothman, J.E. and Schmid, S.L. (1986). Enzymatic recycling of clathrin from coated vesicles. *Cell* 46: 5–9.
85. Sager, P.R., Brown, P.A., and Berlin, R.D. (1984). Analysis of transferrin recycling in mitotic and interphase HeLa cells by quantitative fluorescence microscopy. *Cell* 39: 275–282.
86. Salamero, J., Sztul, E.S., and Howell, K.E. (1990). Exocytic transport vesicles generated *in vitro* from the *trans*-Golgi network carry secretory and plasma membrane proteins. *Proc. Natl. Acad. Sci. USA* 87: 7717–7721.
87. Salminen, A. and Novick, P.J. (1987). A ras-like protein is required for a post-Golgi event in yeast secretion. *Cell* 49: 527–538.
88. Sandvig, K. and van Deurs, B. (1990). Selective modulation of the endocytic uptake of ricin and fluid phase markers without alteration in transferrin endocytosis. *J. Biol. Chem.* 265: 6382–6388.
89. Schmid, S. and Carter, L.L. (1990). ATP is required for receptor-mediated endocytosis in intact cells. *J. Cell Biol.* 111: 2307–2318.
90. Schmid, S.L. and Smythe, E. (1991). Stage-specific assays for coated pit formation and coated vesicle budding *in vitro*. *J. Cell Biol.* 114: 869–880.
91. Schroer, T.A. and Sheetz, M.P. (1991). Functions of microtubule-based motors. *Annu. Rev. Physiol.* 53: 629–652.
92. Simons, K. and Virta, H. (1987). Perforated MDCK cells support intracellular transport. *EMBO J.* 6: 2241–2247.
93. Smythe, E., Pypaert, M., Lucocq, J., and Warren, G. (1989). Formation of coated vesicles from coated pits in broken A431 cells. *J. Cell Biol.* 108: 843–853.
94. Stegmann, T., Doms, R.W., and Helenius, A. (1989). Protein-mediated membrane fusion. *Annu. Rev. Biophys. Biophys. Chem.* 18: 187–211.
95. Steinman, R.M., Mellman, I.S., Muller, W.A., and Cohn, Z.A. (1983). Endocytosis and the recycling of plasma membrane. *J. Cell Biol.* 96: 1–27.
96. Stow, J.L., de Almeida, J.B., Narula, N., Holtzman, E.J., Ercolani, L., and Ausiello, D.A. (1991). A heterotrimeric G protein, $G\alpha_i$-3, on Golgi membranes regulates the secretion of a heparan sulfate proteoglycan in LLC-PK_1 epithelial cells. *J. Cell Biol.* 114: 1113–1124.
97. Swanson, J.A. and Silverstein, S.C. (1988). Pinocytic flow through macrophages. In: *Processing and Presentation of Antigen*, pp. 15–27. (Pernis, B., Silverstein, S., and Vogel, H., Eds.). Academic Press, London.
98. Tooze, J. and Hollinshead, M. (1991). Tubular early endosomal networks in AtT20 and other cells. *J. Cell Biol.* 115: 635-653.
99. Tuomikoski, T., Felix, M.-A., Doree, M., and Gruenberg, J. (1989). Inhibition of endocytic vesicle fusion *in vitro* by the cell-cycle control protein kinase cdc2. *Nature* 342: 942–945.
100. Vale, R.D. (1987). Intracellular transport using microtubule-based motors. *Annu. Rev. Cell Biol.* 3: 347–378.
101. van der Bliek, A.M. and Meyerowitz, E.M. (1991). Dynamin-like protein encoded by the *Drosophila shibire* gene associated with vesicular traffic. *Nature* 351: 411–414.
102. van der Sluijs, P., Hull, M., Zahraoui, A., Tavitian, A., Goud, B., and Mellman, I. (1991). The small GTP-binding protein rab 4 is associated with early endosomes. *Proc. Natl. Acad. Sci. USA* 88: 6313–6317.
103. Warren, G. (1985). Membrane traffic and organelle division. *Trends Biochem. Sci.* 10: 439–443.
104. Warren, G. (1989). Mitosis and membranes. *Nature* 342: 857–858.

105. Warren, G., Davoust, J., and Cockcroft, A. (1984). Recycling of transferrin receptors in A431 cells is inhibited during mitosis. *EMBO J.* 3: 2217–2225.
106. Warren, G., Woodman, P., Pypaert, M., and Smythe, E. (1988). Cell-free assays and the mechanism of receptor-mediated endocytosis. *Trends Biochem. Sci.* 13: 462–465.
107. Wessling-Resnick, M. and Braell, W.A. (1990). The sorting and segregation mechanism of the endocytic pathway is functional in a cell-free system. *J. Biol. Chem.* 265: 690–699.
108. Wessling-Resnick, M. and Braell, W.A. (1990). Characterization of the mechanism of endocytic vesicle fusion *in vitro*. *J. Biol. Chem.* 265: 16751–16759.
109. Wileman, T., Harding, C., and Stahl, P. (1985). Receptor-mediated endocytosis. *Biochem. J.* 232: 1–14.
110. Wilson, D.W., Wilcox, C.A., Flynn, G.C., Chen, E., Kuang, W.-J. (1989). A fusion protein required for vesicle-mediated transport in both mammalian cells and yeast. *Nature* 339: 355–359.
111. Woodman, P.G., Mundy, D.I., Cohen, P., and Warren, G. (1992). Cell-free fusion of endocytic vesicles is regulated by phosphorylation. *J. Cell Biol.* 116: 331-338.
112. Woodman, P.G. and Warren, G. (1988). Fusion between vesicles from the pathway of receptor-mediated endocytosis in a cell-free system. *Eur. J. Biochem.* 173: 101–108.
113. Woodman, P.G. and Warren, G. (1991). Isolation of functional, coated, endocytic vesicles. *J. Cell Biol.* 112: 1133–1141.

GENETIC ANALYSIS OF MEMBRANE TRAFFIC IN MAMMALIAN CELLS

Penelope A. Colbaugh and Rockford K. Draper

Advances in Cell and Molecular Biology of Membranes
Volume 1, pages 169–198

ISBN: 1-55938-362-3

I. INTRODUCTION

There are more than 15 distinct membranes in mammalian cells, most of which are stations on either the secretory or endocytic pathways. The processes by which material is transported among membranes, and how membrane traffic is controlled, are complex and poorly understood. One approach to gain information about complicated processes is to isolate and analyze mutants with defects in discrete steps of the processes. A variety of mammalian cell mutants have been isolated that express lesions affecting endocytosis or secretion, or both, and it is the purpose of this article to catalog and review work on such mutants. Except for the low density lipoprotein (LDL) receptor mutants, we have not included mutants believed to be defective solely in receptor function.

II. PATHWAYS OF VESICULAR TRANSPORT

There have been recent reviews on the endocytic and secretory pathways of vesicular transport (*4,9,17,44,53,84,99*) and only a brief summary will be presented here.

A. Secretion

Membrane proteins and proteins destined for secretion translocate across the endoplasmic reticulum (ER) membrane during translation, with proteolytic cleavage of signal peptides, addition of N-linked carbohydrate side chains and formation of disulfide bonds occurring before translation is complete. The polypeptides fold into transport-competent tertiary and quaternary structures and are exported to a compartment intermediate between the ER and the *cis* Golgi (*104,105,108–110*). Proteins are transported from this intermediate compartment sequentially to the *cis*, *medial* and *trans* Golgi cisternae. Trimming and modification of N-linked and O-linked oligosaccharide side chains by enzymes resident in the different compartments occurs during transport (*8*). The classical model for the mechanism of transport between compartments is vesicular transport where material is carried by vesicles that bud from donor membranes and fuse with a target membrane (*79,100*). In addition, tubular transport, where a membrane tubule originating from the donor organelle extends to and fuses with a recipient membrane, may occur in some situations (*21,80*).

Adjacent to the *trans* Golgi is the *trans* Golgi network (TGN) (*39*), a mildly acidic organelle (*2*) in which proteins destined for constitutive secretion, regulated secretion, or lysosomes are sorted and sent to their appropriate compartments. The mechanisms by which proteins are sorted and packaged into vesicles for constitutive or regulated secretion are poorly understood (*9*). Most soluble mammalian lysosomal enzymes carry a mannose 6-phosphate recognition marker that is added after the enzymes leave the ER (*53*). This targeting signal is recognized by mannose 6-phosphate receptors in the TGN. Mannose 6-phosphate receptors occupied by lysosomal enzymes concentrate in clathrin-coated regions of the TGN and are packaged into vesicles that carry receptors and enzymes to a late endosomal compartment (also called a prelysosome) (*38*). Lysosomal enzymes dissociate from the mannose 6-phosphate receptor in response to the more acidic interior of the late endosome. Receptors then recycle to the TGN where they engage in another round of transport. How mature lysosomes derive from the prelysosome compartment is not clear, but maturation of the compartment to form a lysosome, fusion of the prelysosome with preexisting lysosomes, or vesicular transport between prelysosomes and mature lysosomes might be involved. Other than the mannose 6-phosphate marker, a sorting sequence has been identified on the cytoplasmic tail of an integral membrane protein of lysosomes that participates in delivery to lysosomes (*85*). At least some lysosomal integral membrane proteins lacking the mannose 6-phosphate marker are probably first secreted to the plasma membrane and reach lysosomes via the endocytic pathway (*85*).

B. Endocytosis

Macromolecules and fluid from the extracellular space are brought into the cell by endocytosis, which begins by invagination of the plasma membrane to form vesicles. Two types of vesicles have been observed to form: those that are clathrin coated and those that are not (*17*). The role of vesicles that do not have clathrin is poorly understood, but ligands internalized by uncoated or coated vesicles are reported to follow a common intracellular pathway (*123*). Receptors internalized by clathrin-coated vesicles include those that mediate delivery of growth factors, nutrients such as iron and cholesterol, and other vital macromolecules, into the cell. For clathrin-coated vesicles, it has been suggested that the cytoplasmic tails of cell surface protein receptors interact with clathrin adaptor molecules (*34,83*). Clathrin molecules assemble with these adaptors into polyhedral coat structures to form coated vesicles (*67*). After coated vesicles form, the clathrin is shed almost immediately due to the action of an uncoating ATPase (*7,16*). Small uncoated vesicles then fuse with, or form by fusion, a larger endosome.

Endosomes are acidic organelles (*65*) and many receptors release their ligands in response to the acidic interior of the endosome. These receptors then recycle back to the plasma membrane while the ligands within the fluid phase continue on to lysosomes. Other receptors, such as the epidermal growth factor receptor, transport their ligands deeper into the endocytic apparatus and eventually reach lysosomes. The mechanism by which constituents are sorted within early endosomes are unknown. Also unknown is how material transfers from early endosomes to late endosomes and eventually lysosomes, although evidence for models for both maturation of early endosomes to form later compartments and vesicular transport between endosomes has been presented (*30,40,95,119*).

The vacuolar proton-translocating ATPase has been characterized and consists of eight to 13 subunits (*1,3,117*). Proton transport by the enzyme complex is electrogenic and is limited by the membrane potential and the luminal pH. Properties of the endosome independent of the proton pump have been demonstrated to regulate proton pumping and luminal pH. For example, endosome membranes are permeable to protons (*28*), which leads to dissipation of the pH gradient and stimulation of the proton pump. The lipid composition of the membrane apparently modulates the activity of the proton pump (*130*), and both the plasma membrane Na^+/K^+ ATPase (*11,29*) and a calmodulin-activated Ca^{2+} pump (*71*) are proposed to have a role in regulating the acidification of endosomes by generating an interior positive membrane potential that opposes further proton transport. In addition, vesicle acidification is facilitated by a chloride transporter in vesicles that dissipates the interior-positive membrane potential generated during proton pumping by transport of Cl^- ions in parallel to H^+ transport into the vesicle lumen (*129*).

Lysosomes are the terminal compartment of the endocytic pathway and have the most acidic interior of any component of the vacuolar apparatus (*65*). Lysosomes are dynamic structures and exchange contents with each other by fusion and fission (*27*).

III. ESSENTIALS OF SOMATIC CELL GENETICS

A thorough treatment of somatic cell genetics and the practice of isolating mutants from mammalian cells in culture is beyond the scope of this article. The reader is referred to previous reviews for background information and technical details (*37,94,115,120*). We will consider some basic aspects of mutant isolation and analysis.

A. Choosing a Cell Type

The Chinese hamster ovary (CHO) cell line has been a popular choice for genetic analysis of mammalian processes (*87*). CHO cells are convenient for laboratory studies because they have a short generation time, a high plating efficiency, and can be grown attached to culture dishes or in suspension (*36*). CHO cells also have fewer chromosomes than most other cell lines (modal chromosome number of 20 to 21), have maintained a stable karyotype for many years, are easily mutagenized, and produce spontaneous mutants with relatively high frequency (*114*). Many auxotrophic CHO mutants have been isolated, so simple linkage experiments are possible (*87*). In addition, CHO cells are suitable for replica plating techniques (*26,69,93*).

Reported frequencies of spontaneous mutations at a single locus in CHO cells vary widely, and probably reflect both the selection technique and the gene of interest. The frequencies at which recessive mutants are isolated at a single locus depend on whether the gene is functionally haploid or diploid and the stringency of the selection technique. As a guide to a reasonable recovery of mutants, spontaneous mutants at the hypoxanthine-guanine phosphoribosyl transferase (HGPRT) locus of CHO cells, which is X-linked and therefore present in a single copy, are recovered with a frequency of approximately 10^{-5} (*13*), and use of mutagens increases the recovery of mutants to about 10^{-4}. Isolation of mutants having lesions in diploid genes requires the use of mutagens, but is still in a practical range with proper selection schemes; for example, CHO cell mutants with lesions in the diploid ldlA locus (the LDL receptor gene) were isolated by Kingsley and Krieger at a frequency of about 10^{-6} after mutagenesis with ethylmethane sulfonate (*50*).

Cell lines can differ greatly in the types of mutants recovered in response to a given selection procedure. This becomes important if mutants in the same complementation group are isolated over and over again from a particular cell line. For example, endosomal acidification-defective CHO cell mutants of the End1 and End2 complementation groups have been found by numerous researchers, even though very different types of selection procedures were used (see Table 1). The

Table 1. CHO Cell Mutants in the End1–End5 Complementation Groups

Group	*Mutant*	*References*
End1	RPE.44	(*19,66,92,102*)
	RPE.28	(*19,66*)
	RPE.51	(*19*)
	P1R1.P50	(*19*)
	DTG1-5-4	(*91,102,131–133*)
	B3853	(*94,96,97,107*)
	G.8.1	(*14,62,112*)
	31.1	(*14*)
End2	RPE.1A	(*92,102*)
	DTF1-5-1	(*52,91,93,102,131–133*)
	I223	(*96*)
	M311	(*94,96,97,102,107*)
	MPL 3-4	(*102*)
	MPL 2-5	(*102*)
	25.2	(*14*)
End3	G.7.1	(*14,22,62,81,118,121,125*)
End4	DS28-6	(*69,137*)
	V.24.1	(*14,15,46a,86,126*)
	HRP-1	(*125,126a*)
End5	42.2	(*14*)

reason these mutants are isolated at such a high frequency is unclear, but perhaps the genetic loci involved are functionally haploid. If new mutants defective in endosomal acidification were desired, it would be advantageous to use cells other than CHO cells.

B. Temperature-Sensitive Mutants

Selection procedures can be designed to recover either constitutive or conditional mutants. There are, however, important advantages to isolating temperature-sensitive conditional-lethal mutants; for example, cells carrying lethal mutations can be recovered. In addition, temperature-sensitive mutants facilitate complementation analysis and the isolation of revertants. Cloning genes affected in mutants should also be facilitated based on strategies that complement the lethal phenotype with transfected DNA containing normal gene copies.

C. Complementation Analysis

Complementation analysis with somatic cell mutants is usually performed by fusing two cell types to form a hybrid cell containing the combined genetic information of both parents. Hybrids are then tested to determine whether they have the mutant or the wild-type phenotype. An absolute prerequisite for complementation tests is that the phenotypes of the two mutants be recessive. Whether a trait is dominant or recessive is tested by fusing a mutant with wild-type cells and assessing whether the hybrid cells have the mutant or wild-type phenotype. Techniques for fusing cells are not 100% efficient and it is usually (although not always, see *49*) necessary to eliminate unfused parental cells from the population. For constitutive mutants, this is done by placing genetic markers in the cells to be fused so that under selective conditions only hybrid cells will thrive. Details of different methods for selecting hybrids are discussed in reviews on somatic cell genetics cited at the beginning of this section. Obtaining hybrids is considerably facilitated if the mutants being tested have recessive conditional-lethal traits because hybrids can be selected simply by incubating fused cells at the restrictive temperature. For example, if testing whether a trait is dominant or recessive, the temperature-sensitive mutant is fused to a cell type that carries an unrelated recessive temperature-sensitive marker mutation (for an example, see *14*). If there are no survivors after the hybrid cells have been formed and incubated at the restrictive temperature, the mutant probably contains a dominant lesion that is expressed in the hybrid cells. However, if the hybrid cells grow the lesion is recessive. Complementation analysis can proceed by fusing two recessive mutants and selecting hybrids at the restrictive temperature.

IV. METHODS OF MUTANT ISOLATION

Diverse schemes to isolate mutants defective in endocytosis or secretion have been devised. Most mutants have been isolated using either protein toxins that require membrane traffic to express a lethal activity or cytotoxic drugs that interfere with some aspect of endocytosis or secretion. Another approach has been to rapidly screen cells to identify those with a mutant phenotype.

A. Protein Toxins

Protein toxins that require receptor-mediated endocytosis to kill cells have been popular for selecting mutants defective in the endocytic or secretory pathways. The toxins include the bacterial proteins diphtheria toxin and *Pseudomonas aeruginosa* exotoxin A, and the plant toxins modeccin, ricin, and abrin. Several features are common to the toxins: They all bind to cell surface receptors, are internalized into endocytic compartments, and enter the cytosol from a compartment along the endocytic pathway. They all contain a binding subunit (or domain) responsible for binding to cell surface receptors, and an enzymatic subunit (or domain) that catalytically inhibits protein synthesis. The toxins differ, however, in their receptors, enzymatic activities, requirements for acidification or processing after internalization, and compartment from which they escape to the cytosol.

Modeccin, ricin, and abrin are plant toxins that are similar in structure and mechanism of action (*73,74*). They all bind galactose residues, but modeccin preferentially binds O-linked oligosaccharides whereas ricin and abrin bind galactose and N-acetyl galactosamine residues present on a variety of glycoproteins and glycolipids. The three toxins modify a specific ribonucleotide of the 28S rRNA (*25*). This modification inactivates the 60S ribosomal subunit by preventing binding of the ribosome to elongation factor 2.

Diphtheria toxin and *Pseudomonas aeruginosa* exotoxin A both inhibit protein synthesis in sensitive cells by catalyzing the transfer of the ADP-ribose moiety of NAD^+ to elongation factor 2 (*23,128*). This modification inactivates elongation factor 2 and thus blocks protein synthesis. Diphtheria toxin and exotoxin A bind unique but poorly characterized cell surface receptors (*24*). Diphtheria toxin appears to penetrate to the cytosol through the membrane of an early endosome compartment in contrast to ricin, abrin, and modeccin, which apparently intoxicate from later compartments of the endocytic pathway (*63,103*). Evidence has been presented that ricin penetrates to the cytosol from elements of both the Golgi region and the lysosomal pathway (*18,116,136*).

It is not known how the enzymatic subunits of protein toxins translocate across membranes of the endocytic apparatus, but the acidic environment of endosomal compartments is important for some toxins (*70,74*). The cytotoxic activities of diphtheria toxin, modeccin, and exotoxin A are greatly diminished in cells in which the pH of internal vesicles has been elevated. It is believed that these toxins undergo a conformational change at low pH that allows the toxins to insert into the vesicle membranes (*74*). Translocation of the enzymatic subunits to the cytosol apparently does not immediately follow, since a short lag period is observed between the pH-dependent step and inhibition of protein synthesis (*45,63*). This lag suggests that a processing step occurs between insertion of the toxin into the membrane and translocation of the enzymatic activity into the cytosol. Ricin and abrin, in contrast to the other toxins, do not require acidification to intoxicate cells and are even somewhat more cytotoxic when the pH in intracellular vesicles is elevated (*74*).

The initial members of the End1 through End5 complementation groups, which will be discussed in a later section, were obtained using protein toxins. When cells are exposed to a single toxin, mutants with lesions ranging from defects in the initial interaction with receptors to defects in the intracellular targets for the toxins could be expected. To avoid mutants in receptors and in intracellular targets, cells have been simultaneously exposed to two toxins that have different receptors and different intracellular targets (for example, see reference *62*).

Protein toxins have also been used to recover mutants defective in the secretory pathway. Nakano and co-workers (*68,69*) isolated two temperature-sensitive CHO cell mutants unable to replace diphtheria toxin receptors and cell adhesion proteins after pretreatment with trypsin. Cells were first mildly trypsinized, incubated in culture dishes for 7 hr at 39.5 °C, then nonadherant cells were incubated with diphtheria toxin.

B. Cytotoxic Drugs

Two basic strategies have been used to isolate mutants in membrane traffic that are resistant to cytotoxic drugs. The first is to choose a drug that has a direct adverse influence on some aspect of endocytosis or secretion and isolate mutants resistant to the drug. For example, monensin and chloroquine elevate the pH in intracellular vesicles, although by different mechanisms, and have been used to directly isolate mutants with perturbed regulation of vesicular pH (*10,75,77,78*).

The second strategy uses cytotoxic drugs that ordinarily do not require the endocytic or secretory pathways to be cytotoxic, but conditions are manipulated so that endocytosis or secretion is necessary to express cytotoxicity. For example, methotrexate is highly cytotoxic, but does not require endocytosis or secretion to express cytotoxicity. Ryser et al. (*102*) coupled methotrexate to poly(L-lysine) so that endocytosis and proteolytic degradation were required to release methotrexate, and isolated CHO cell mutants that were defective in degrading the methotrexate-poly(L-lysine). Kreiger et al. (*56*) prepared LDL containing the toxic sterol 25-hydroxycholesteryl oleate and selected mutants that were defective in LDL uptake. In a more complicated scheme, Krieger and co-workers used two cytotoxic drugs to isolate mutants with defects in the LDL receptor (*50,57*). Cells were first pre-incubated in a lipoprotein-deficient medium containing mevalonate, LDL, and compactin. In this medium, *de novo* synthesis of cholesterol was blocked by the drug compactin, so that the only source of cholesterol was from the receptor-mediated endocytosis of LDL. Cells able to endocytose LDL, and thus acquire cholesterol, were killed upon addition of amphotericin B, which forms toxic complexes with cholesterol. Interestingly, all the mutants isolated by this scheme were in the gene for the LDL receptor.

C. Screening Methods

Fluorescence-activated cell sorting can be used to screen large numbers of cells for a particular trait provided the trait can be coupled in some manner to fluorescence intensity. Cain et al. (*12*) shifted mutagenized cells to 41 °C, incubated the cells with fluorescent transferrin, chased with unlabeled transferrin, and sorted out those cells that retained a high fluorescence signal. This approach resulted in the isolation of a mutant that fails to recycle transferrin at high temperature. Hearing et al. (*41,42*) used fluorescence-activated cell sorting to recover mutants defective in secretion. CHO cells that stably expressed the influenza virus hemagglutinin protein on their surface were prepared by transfecting the hemagglutinin gene into CHO cells. The hemagglutinin was then cleared from the cell surface and degraded in lysosomes by addition of anti-hemagglutinin. Cells unable to synthesize and replace the hemagglutinin were obtained by sorting for cells that did not bind fluorescent anti-hemagglutinin. Not all screening approaches have used fluorescence-activated cell sorting. For example, Kaiden and Krag (*46*) directly measured radioactive proteins secreted into the culture medium to identify mutants that were defective in secretion.

V. MUTANTS OF THE END1-END5 COMPLEMENTATION GROUPS

Five complementation groups (End1-End5) of temperature-sensitive CHO cell mutants with defects in endocytosis or secretion, or both, have been defined using protein toxins as selective agents (see Table 1). The prominent phenotype of End1, End2, and End3 mutants is defective endosome acidification. End4 mutants are defective in the transport of secretory material from the ER to the *cis* Golgi, and have impaired lysosome biogenesis and reduced fluid-phase endocytosis. The End5 group is represented by a single mutant and is not well characterized.

A. Groups End1–End3

Mutants in the End1, End2, and End3 complementation groups are all defective in endosome acidification and exhibit similar, but not identical, phenotypes. They are cross-resistant to protein toxins requiring endosomal acidification (diphtheria toxin, *Pseudomonas* toxin and modeccin) and are hypersensitive to abrin and ricin. In addition, they are resistant to infection by RNA viruses that require endocytosis into acidic organelles.

End1

There are at least eight independent isolates of mutants belonging to the End1 complementation group (Table 1). End1 mutants exhibit multiple phenotypic changes affecting the vacuolar apparatus. Prominent among these is loss of endosomal acidification activity, and other traits are likely to result from the effect of elevated pH in the lumen of intracellular compartments. It is not known whether the primary lesion of End1 cells is in vesicle acidification or whether the lack of acidification is secondary to some other cellular alteration.

The acidification activity of End1 mutants has been investigated both *in vivo* and *in vitro*. Yamashiro and Maxfield (*132,133*) measured the pH of early endosomes *in vivo* in an End1 mutant (DTG 1-5-4) by a null point method that took advantage of the pH-dependent fluorescence of fluorescein isothiocyanate (FITC). The average pH of endosomes labeled with FITC-transferrin was significantly higher in the mutant than in parental cells (pH 6.7 compared to pH 6.2 after 5 min of FITC-transferrin internalization). Measurements of the acidification of later endocytic compartments using internalized FITC-dextran indicated the probe localized in progressively more acidic environments with time after uptake. Three morphologically distinct sets of endosomes were distinguished: (1) small vesicular and tubular early endosomes labeled by a short pulse of fluorescent probe; (2) small vesicles and tubules near the Golgi apparatus labeled by FITC-transferrin; and (3) large, more acidic, endosomes labeled by FITC-α_2-macroglobulin and FITC-dextran en route to lysosomes after segregation from transferrin. In DTG 1-5-4 cells, the acidification of early small vesicles and tubules was most impaired. Measured pH values of large endosomes were heterogeneous, and recycling compartments and lysosomes appeared unaffected.

Schmid et al. (*107*) monitored acidification of endocytic vesicles in the End1 mutant B3853 *in vivo* by determining the kinetics of acquisition of the pH-dependent conformational changes of the E1 spike glycoprotein from wild-type and fus-1 mutant Semliki Forest virus. The conversion of E1 protein to a trypsin-resistant form, which requires a pH of $\leq$ 6.2 or $\leq$ 5.3 for wild-type and fus-1 viruses, respectively (*47*), was delayed in the mutant at nonpermissive temperature. Half-times for conversion of wild-type virus spike glycoprotein were 8 minutes in the mutant compared to 5 minutes in wild-type cells, and conversion using the mutant virus occurred with a half-time of 23 minutes in the End1 mutant compared to 12 minutes in wild-type cells.

Endosome acidification is completely lost in cell-free vesicle preparations from End1 cells. Merion et al. (*66*) determined that endosomes from two End1 mutants (RPE.28 and RPE.44) were devoid of ATP-dependent acidification activity and that acidification of lysosomes was much less impaired. Robbins et al. (*91*), as well as Roff et al. (*96*), found that endosomes from mutant DTG 1-5-4 lacked acidification activity although dense and light lysosomes acidified normally. Acidification of FITC-transferrin containing endosomes from mutant B3853 was negligible when prepared from cells grown at the nonpermissive temperature (*96,107*).

How does the elevated pH of early endosomes in End1 mutants affect endocytic function? Internalization via clathrin-coated pits and recycling of receptors is apparently unaffected since trafficking of both the transferrin receptor and the α_2-macroglobulin receptor are normal in End1 cells (*112,131*). The pH dependent extraction of Fe^{+3} from transferrin is impaired (*96,97,112*), however, and since supplemental iron in the growth medium permits End1 mutants to grow at the restrictive temperature, this appears to be responsible for the conditionally-lethal defect of End1 mutants (*112*). Uptake by means of the mannose 6-phosphate receptor is markedly decreased (to 4% the normal level); however, this reflects an unexplained loss in binding activity of the mannose 6-phosphate receptor that is also observed in NH_4Cl treated cells (*92,97*). ^{35}S-secretions internalized by the mannose 6-phosphate receptor in B3853 cells (96) localized almost totally in early endosomes, consistent with inability to dissociate from their receptors in less acidic endosomes, and degradation of α_2-macroglobulin was observed to be slower in mutant DTG 1-5-4 than in parental cells (*131*). Transport of soluble macromolecules from endosomes to lysosomes is partially inhibited in End1 mutants; Robbins et al. (*91*) noted in mutant DTG 1-5-4 that only 50% of FITC-dextran chased for 1 to 2 hours after endocytosis sedimented in Percoll gradients with lysosomes while the remainder sedimented with more buoyant compartments.

Functions of the *trans* Golgi cisternae and the TGN are also altered in End1 mutants. Targeting of lysosomal hydrolases to the prelysosomal compartment is aborted, with 90 to 95% of newly synthesized hydrolases being secreted (*91,107*). Galactosylation of Sindbis virus E1 and E2 proteins (a function ascribed to the *trans* Golgi) in DTG 1-5-4 cells is reduced (*91*), although galactosylation of VSV G protein and major cellular glycoproteins apparently occurs normally. Sialylation of some secreted proteins is incomplete in mutant B3853 (*97*). In Sindbis virus infected DTG 1-5-4 cells, viral proteins accumulate in cisternae near the Golgi apparatus (probably the TGN), the posttranslational proteolytic cleavage of viral precursor protein pE2 to E2 is inhibited, and virion release is only 5% of normal (*91*).

In cells with multiple phenotypic changes it is difficult to identify which are primary and which are secondary to the genetic lesion. Roff et al. (*96,97*) studied in temperature-sensitive mutants the onset of phenotypic changes and subsequent recovery to temporally order the appearance of altered traits. After mutant B3853 was shifted to a nonpermissive temperature the first aberrant trait observed was block of transport of endocytosed acid hydrolases from early endosomes to lysosomes. Other phenotypic changes occurred quickly and synchronously, except for the block in release of Sindbis virus from infected cells, which was long delayed over other changes in function. Also, loss of cell surface-associated mannose 6-phosphate binding activity and loss of glycosylation in the Golgi were delayed events. Acidification of endosomes was quickly restored, as were all endocytic activities, when B3853 cells were returned to permissive temperature after manifesting the mutant phenotype. Recovery of mannose 6-phosphate transport, however, was a late event requiring biosynthesis of the total pool of cellular mannose 6-phosphate receptors.

The cause of the acidification defect in the End1 mutants has been extensively investigated but is not yet known. Robbins et al. (*91*) determined that the affinity of the proton pump of mutant DTG 1-5-4 for ATP was identical to that of wild-type cells. Yamashiro and Maxfield (*133*) measured the ability of intact DTG 1-5-4 cells to reestablish endosomal proton gradients after treatment with the ionophore monensin and concluded that proton pumping in DTG 1-5-4 cells was similar to that of wild-type cells. Schmid et al. (*107*) determined that the proton efflux from endosomes of B3853 cells was not significantly different from that of parental cells. To examine whether the acidification defect in End1 mutants results from altered permeability of vacuoles to anions or cations, Schmid et al. (*107*) added valinomycin (a K^+ ionophore) in the presence of K^+-containing buffer to endosomes from mutant B3853 to shunt any existing membrane potential. The proton pumping activity of endosomes of B3853 was not restored by valinomycin, which suggests that End1 acidification activity is not inhibited by a membrane potential.

A functional Na^+/K^+-ATPase activity inhibits acidification of CHO cell endosomes by electrogenic Na^+ transport (*11,29*). This inhibition of acidification is partially reversed by adding vanadate or ouabain, inhibitors of the Na^+/K^+-ATPase. Vanadate equally reversed the inhibition of acidification of endosomes from parent cells and B3853 cells, suggesting End1 acidification activity is not excessively inhibited by a Na^+/K^+-ATPase (*107*).

Chloride transport into vacuoles is required to generate a proton gradient in that Cl^- helps dissipate the membrane potential caused by electrogenic proton transport. If chloride is replaced in an assay medium with an impermeant anion, proton pumping should be inhibited. Replacement of Cl^- with gluconate partially inhibited the acidification of endosomes of mutant B3853 grown at permissive temperature (*107*), suggesting that the lesion of B3853 does not stem from altered regulation of the endosome membrane potential by Cl^-.

End2

Mutants of the End2 complementation group phenotypically resemble End1 mutants. There are, however, several major differences. First, End2 mutants are not as severely affected with respect to acidification as End1 mutants, although each group contains individual members showing a spectrum of severity (*14,96*). Second, an investigation into the recovery of temperature-sensitive End1 and End2 mutants after return to permissive temperature indicated that the two handle the mannose 6-phosphate receptor differently (*97*). Third, transport of endocytosed markers from endosomes to dense lysosomes is not impaired in End2 mutants, but is inhibited in mutants of the End1 group (*91*).

As with mutants of the End1 group, the acidification of endocytic compartments in End2 mutants has been investigated both in intact cells and in isolated vesicles. Yamashiro and Maxfield (*133*) assessed the pH of vesicles of mutant DTF 1-5-1 *in vivo*. Cells internalized either FITC-transferrin to label early endosomes and endosomes of the receptor recycling pathway, or FITC-labeled dextran or α_2-

macroglobulin, which follow the degradative path of endocytosis. Measurements of fluorescence by individual vesicles containing these probes indicated that endosomes of End2 mutant DTF 1-5-1 did not acidify as rapidly or to the same extent as endosomes in wild-type cells. In comparison with the End1 mutant DTG 1-5-4, acidification of vesicles of the End2 mutant was intermediate to that of End1 and wild-type cells. Analysis of endosome populations distinguishable by fluorescence microscopy (*132*) indicated the average pH of early small endosomes as well as large endosomes was elevated. Interestingly, two populations of large endosomes were distinguished. Acidification was normal in one population representing about half of the large endosomes, and the other population of endosomes was barely acidic. The *para*-Golgi recycling endosomes and lysosomes had pH values similar to those of parental cells.

Schmid et al. (*107*) monitored the pH-dependent conversion of wild-type and fus-1 Semliki Forest virus E1 protein to trypsin resistance in the temperature-sensitive End2 mutant M311. The kinetics of conversion of wild-type E1 protein in mutant M311 were almost identical to those of the End1 mutant B3853; conversion of wild-type E1 took twice as long in M311 cells as in parental cells. Fus-1 E1 protein conversion was insignificant in the mutants, indicating that late endosomes fail to acidify to pH 5.3.

In cell-free assays of acidification activity of endosomes from several different End2 mutants (MPL 3-4, DTF 1-5-1, M311 and I223), endosome acidification was partially impaired whereas lysosome acidification was unaltered (*91,96,102,107*). Measurements of acidification in vesicles of a postnuclear supernatant fraction of the End2 mutant cell 25.2 by ATP-stimulated accumulation of acridine orange indicated acidification activity was reduced in this cell line about 50% relative to parental activity (*14*).

To determine the cause of the End2 acidification defect, Schmid et al. (*107*) examined acidification of early (FITC-transferrin containing) endosomes from M311 cells. The H^+ efflux from acidified endosomes was normal, but conditions that increased the membrane potential completely inhibited ATP-stimulated proton pumping. Replacing permeant Cl^- in the assay buffer with gluconate inhibited M311 acidification, and acidification was completely inhibited in Na^+ and K^+ containing buffer (which favors Na^+/K^+-ATPase activity and partially inhibits proton pumping in parental endosomes). This later inhibition could be reversed by addition of vanadate, an inhibitor of the Na^+/K^+-ATPase. Schmid et al. (*107*) concluded that the proton pump of End2 cells was hypersensitive to a membrane potential.

Characterizing the endocytic functions of End2 cells, Klausner et al. (*52*) found that binding, internalization, and recycling of transferrin was normal, but that iron extraction from transferrin was reduced. Yamashiro and Maxfield (*133*) reported that iron extraction from transferrin in DTF 1-5-1 cells was 46% that of parental cells, and a similar reduction was seen by Roff et al. (*96*) using the End2 mutants M311 and I223. Accumulation and degradation of α_2-macroglobulin were found to be markedly decreased in a number of End2 mutants (DTF 1-5-1, M311, I223,

MPL 3-4, and MPL 2-5) (*96,131*), although recycling of α_2-macroglobulin receptors was unaffected in DTF 1-5-1 (*131*).

Robbins and coworkers (*93,96,97*) compared the activity of cell surface mannose 6-phosphate receptors in internalizing extracellular hydrolases in End1 and End2 cells. Both complementation groups had reduced uptake of ligand and this resulted from failure of ligand to dissociate from the receptor in early endosomes and from reduced binding activity of mannose 6-phosphate receptors at the cell surface. Temperature-sensitive End2 mutants recovered mannose 6-phosphate receptor binding activity and ligand uptake upon return to permissive temperature, in contrast to temperature-sensitive End1 mutants in which restoration of binding and uptake in End1 mutants required resynthesis of mannose 6-phosphate receptors. In both End1 and End2 mutants, the transport of lysosomal enzymes to lysosomes during biosynthesis was restored before endocytic transport from the cell surface.

The uptake of ^{3}H-inulin, which enters cells via fluid phase endocytosis, and ^{3}H-methotrexate-poly(lysine), which binds nonspecifically to the plasma membrane prior to endocytosis, was normal in two different End2 mutants (MPL 3-4 and MPL 2-5) isolated for resistance to methotrexate-poly(lysine) (*102*). Degradation of ^{3}H-methotrexate-poly(lysine) by both mutants was reduced to about half of parental levels.

In contrast to End1 mutants, endosome to lysosome transfer is apparently not affected in End2 cells. Robbins et al. (*91*) compared the localization of FITC-dextran in Percoll gradients after the label was chased into DTF 1-5-1 cells for one hour. The distribution of dextran in the gradient was almost identical to that of parental cells, with most of the dextran being in lysosomal fractions. Also, Schmid et al. (*107*) determined that the time of arrival of internalized Semliki Forest virus in lysosomes, as estimated by the appearance of degraded viral proteins in the media, was the same as in parental cells.

Some Golgi functions are altered in End2 cells. In mutant M311, sialylation of secreted proteins was particularly affected (*96*). In DTF 1-5-1 cells, release of Sindbis virions was reduced to 20% of parental levels. Studies on the recovery of function in the temperature-sensitive M311 indicated that the block in release of Sindbis virus from infected cells was clearly secondary to other altered functions (*97*).

End3

The End3 complementation group is defined by a single temperature-sensitive mutant, G.7.1, isolated by Marnell et al. (*62*). Selected for resistance to both modeccin and diphtheria toxin, mutant G.7.1 has a phenotype characteristic of cells treated with agents that abolish pH gradients within the cell. The primary lesion in this mutant appears to be in the vacuolar proton-translocating ATPase.

Marnell et al. (*62*) observed that the ATP-stimulated acidification activity of vesicles in a postnuclear supernatant from G.7.1 cells grown and harvested at the permissive temperature was heat-labile. This demonstrated that the vesicles themselves contained a thermolabile constituent. Stone et al. (*118*) compared clathrin-coated vesicle fractions from G.7.1 and wild-type cells for properties of the vacuolar proton-translocating ATPase. The vanadate- and oligomycin-insensitive, N-ethylmaleimide-sensitive ATPase activity, $^{32}P_i$-ATP exchange activity, and proton pumping were all thermolabile, consistent with a mutation that impaired the vacuolar proton pump. Phosphatidylserine, an activator of the proton pump, did not protect against heat inactivation.

Other phenotypes of G.7.1 cells appear to be secondary to the acidification defect. Although transferrin binding and internalization by G.7.1 cells is normal (*121*) the rate of transferrin-mediated iron accumulation is half that of parental cells. The transport of horseradish peroxidase from early endosomes to lysosomes was measured by Wang (*125*). As was seen with cells of the End1 complementation group (*91*), and in cells treated with amines to elevate endosomal pH (*125*), transport from endosomes to lysosomes was inhibited. Park et al. (*81*) characterized the synthesis, processing and targeting of the lysosomal enzymes α-glucosidase and cathepsin D in G.7.1 cells. At the nonpermissive temperature the mutant secreted both enzymes in the precursor form, a result expected if the enzymes were not efficiently delivered to the prelysosomal compartment due to the acidification defect. Mature hydrolases synthesized prior to shifting cells to the nonpermissive temperature had a greatly reduced half-life (3–5 fold) relative to that of wild-type cells at the same temperature, which might be explained if preexisting lysosomes retained their acidic interior and were not accessible to delivery of new material from biosynthetic or endocytic paths. Electron dense bodies appearing to contain undigested material were also noted in G.7.1 cells by electron microscopy (*81*).

End1 and End2 mutants exhibit impaired Golgi function. Such loss of Golgi activity has not been observed in G.7.1 cells. Sialylation of proteins secreted by G.7.1 cells appeared normal, as judged by binding to wheat germ agglutinin (*81*). In addition, normal glycosylation of α-glucosidase and cathepsin D during biosynthesis, normal acquisition of *endo* H resistance by the mannose 6-phosphate receptor during biosynthesis, and phosphorylation of α-glucosidase (a function of the intermediate compartment or the *cis* Golgi) were observed.

In contrast to mutants of the End1 and End2 complementation groups, infection of G.7.1 cells by Sindbis virus was found to be unaltered (22). Also, in contrast to mutants of the End1 and End2 groups, G.7.1 cell surface mannose 6-phosphate receptors exhibited the same binding of mannose 6-phosphate ligand as those of wild-type cells (*81*). Accumulation of mannose 6-phosphate bearing ligands was greatly reduced, however, and no degradation of endocytosed ligand was observed.

B. End4

The End4 complementation group was initially defined by CHO cell mutant V.24.1, which was selected using diphtheria toxin in a scheme to isolate cells defective in endocytosis (*14*). Two other mutants, DS28-6 and HRP-1, have been recently added to the End4 group by complementation analysis (*126a*). Mutant

DS28-6 was isolated as a secretion-deficient mutant by Nakano et al. (*69*), and mutant HRP-1 was identified by Wang (*125,126a*) among variants with defective fluid internalization. The lesion in all three cell lines results in a temperature-sensitive, conditionally-lethal phenotype. The primary lesion in End4 mutants appears to be in the secretory path.

Secretion of both soluble and membrane proteins is impaired in End4 mutants (*69,86,126*). Using the biosynthesis of influenza virus hemagglutinin as a model system, Kao and Draper (*46a*) investigated where in the secretory pathway proteins accumulated in V.24.1 cells. By immunofluorescence microscopy the hemagglutinin colocalized with a 58-kDa membrane protein normally present in the intermediate compartment between the ER and the *cis* Golgi, suggesting that material could enter, but not leave, the intermediate compartment. They also found that the N-linked oligosaccharides of influenza hemagglutinin were resistant to endoglycosidase D digestion but sensitive to endoglycosidase H, suggesting that transport to the *cis* Golgi did not occur. Folding and trimerization of the influenza hemagglutinin occurred with kinetics indistinguishable from parental cells, and the morphology of the ER of V.24.1 cells appeared normal.

An intriguing property of mutant V.24.1 is that the Golgi complex vanishes within two hours at the nonpermissive temperature (*46a,137*). The disappearance is reminiscent of the effects of brefeldin A on the Golgi complex of mammalian cells (*20,60,61,80*). It is not clear at present what happens to the Golgi complex in End4 mutants.

Endosome function remains relatively intact in End4 mutants. ATP-stimulated acidification activity of endosome fractions from V.24.1 cells was found to be normal, as was transferrin endocytosis and iron accumulation from transferrin (*15*). End4 mutants are only mildly resistant to some protein toxins that require endocytosis to intoxicate cells, and the infection of End4 mutants by different RNA viruses that enter cells by endocytosis is not impaired as assessed by synthesis of viral proteins (*69,86,126*). Fluid-phase endocytosis is the only parameter of endocytosis known to be affected in End4 cells. Both V.24.1 and HRP-1 cells accumulated horseradish peroxidase by fluid-phase endocytosis at a rate 50 to 70% less than parental cells (*125,126*). The reason for the reduced uptake of fluid-phase markers is not clear.

Lysosomes are much more affected by the defect of End4 mutants than endosomes. Transfer of endocytosed material to dense lysosomes of V.24.1 cells was very inefficient (*15*), and ATP-stimulated acidification activity in lysosomal fractions of V.24.1 cells was almost completely lost after cells were placed at the nonpermissive temperature.

C. End5

Mutant 42.2, selected for resistance to diphtheria toxin, is the sole known member of the End5 complementation group (*14*). Mutant 42.2 is not well characterized, but it is known that endosome and lysosome acidification appear to be normal (*14*).

VI. LDL MUTANTS

Krieger and co-workers have selected CHO cells with recessive defects affecting the function of the LDL receptor using either LDL reconstituted with toxic 25-hydroxycholesteryl oleate (*56*) or amphotericin B in cells treated with compactin (*57*). Genetic analysis of these mutants (*50*) defined four complementation groups: ldlA, ldlB, ldlC, and ldlD. Mutants of the ldlA complementation group carry lesions in the LDL receptor structural gene, whereas the other three groups of mutants have defects that pleiomorphically affect glycosylation of glycoproteins and glycolipids. In addition, Malmstrom and Kreiger (*61a*) recently reported members of the ldlE through ldlI complementation groups, isolated either by radiation suicide or by fluorescence-activated cell sorting. At the time of this writing, the properties of the new mutants have not been described in detail.

The mature LDL receptor contains two N-linked and approximately 18 O-linked oligosaccharide side chains. Most of the O-linked chains are clustered in an extracellular domain near the membrane-spanning domain of the receptor. The receptor is synthesized as a 125-kDa precursor that undergoes extensive processing to reach a 155-kDa mature form. The mature form is expressed on the cell surface and is relatively stable. Among the many (at least 59 have been reported) ldlA mutants, three classes have been distinguished: Class 1 mutants have no detectable LDL receptor, mutants of Class 2 synthesize a normal receptor precursor but processing is abnormal and the receptor quickly degrades, and Class 3 mutants have LDL receptors that are structurally indistinguishable from wild-type but do not bind LDL with high affinity. Heterozygotes of these classes and revertants of ldlA indicate that the LDL receptor gene is diploid (*54,111*).

Mutants of the ldlB, ldlC and ldlD classes express only 5 to 10% of wild-type receptor activity. Mutants of these three complementation groups were found to synthesize normal amounts of the 125-kDa precursor form of the LDL receptor, but the receptors were not processed to the 155-kDa mature form. Unprocessed LDL receptors are transported to the cell surface and are rapidly degraded. LdlB, ldlC and ldlD mutants were found to carry general defects affecting synthesis of N-linked, O-linked and lipid-linked oligosaccharides (*49*). Comparison of LDL receptor activity in the ldl mutants with 17 CHO glycosylation (Lec) mutants indicated the ldlB, ldlC, and ldlD defects are most likely unique; few of the Lec mutants had reduced LDL receptor activity, and none to the extent of the ldl mutants.

The reason for the defective oligosaccharide processing of the LDL receptor has been identified in only the ldlD complementation group. Kingsley et al. (*48*) established that ldlD cells have a deficiency in the activity of UDP-galactose and UDP-*N*-acetylgalactosamine 4-epimerase, which is required in the synthesis of UDP-galactose and UDP-*N*-acetylgalactosamine from glucose. Whether this deficiency is due to a defect in the structural gene of UDP-Gal/UDP-GalNAc 4-epimerase enzyme, or in some cell component required for synthesis or regulation of this enzyme is unknown. The ldlD defect is fully correctable in that addition of galactose and *N*-actylgalactosamine to the growth medium allows ldlD cells to synthesize these UDP-sugars via salvage pathways. Addition of galactose alone corrects N-linked but not O-linked processing defects, and does not correct lost LDL receptor activity. Addition of *N*-acetylgalactosamine alone partially corrects O-linked but not N-linked processing and substantially increases the stability and activity of the LDL receptor, suggesting the O-linked carbohydrate chains of the LDL receptor are critically important for its stability. Time of addition experiments in which N-acetylgalactosamine was added at varying intervals after pulse-labeling the cells indicated O-glycosylation of the LDL receptor occurs posttranslationally. O-glycosylation of the receptors occurred when GalNAc was added as much as 15 minutes after a pulse.

Using ldlD cells, Kozarsky et al. (*55*) established that O-linked deficient receptors on the cell surface are cleaved near the membrane-spanning domain with release of the NH_2-terminal extracellular domain into the medium. Reddy et al. (*89*) demonstrated that the stable cell surface expression of another O-glycosylated membrane protein, decay-accelerating factor, in ldlD cells also depended on restored O-glycosylation, suggesting that O-linked oligosaccharides of cell-surface proteins play an important role in the cell-surface expression and stability of membrane proteins.

LdlB and ldlC mutants exhibit almost identical pleiotropic glycosylation defects. The basis of the defects are unknown, although the ldlB phenotype was reversed by transfection with human liver DNA (*51*). A spontaneously arising suppressor of ldlC has been characterized in which the LDL receptor activity is restored by increased receptor expression. This is not due to amplification of the gene for LDL but rather from an increase in the rate of receptor synthesis. Analysis of hybrids indicated that suppression was a codominant trait that acted in *cis* to the LDL receptor gene, since only the ldlC form of receptor was overproduced (*90*).

VII. OTHER MUTANTS

Table 2 catalogs mutants that have not been integrated into either the End or ldl complementation groups.

A. Mutants Resistant to Protein Toxins

A mutant CHO cell line (DMP^r-2) with a phenotype reminiscent of End1-3 mutants was selected by Ghosh and Wu (*32*). Isolated using both modeccin and diphtheria toxin, the mutant cells were rescued by supplemental iron and showed a pattern of toxin sensitivity similar to mutants with defective endosomal acidification. In addition to exhibiting resistance to modeccin and diphtheria toxin, DMP^r-2 was shown to be cross-resistant to *Pseudomonas* toxin and hypersensitive to ricin.

Wu and co-workers (*88,127*) selected and characterized a spontaneous mutant (termed CHO*pro* 4-10) resistant to ricin and *Pseudomonas* toxin but normally

Table 2. Mutants Not Yet Assigned to Complementation Groups

Parent	*Mutant*	*Type*	*References*
CHO	DMP^r-2	Modeccin-, diphtheria toxin-resistant	(*32*)
CHO	CHO*pro* 4-10	Ricin-resistant	(*31,88,127*)
Namal-wa	$J5r^R6$	Ricin-resistant	(*35*)
L cells	LEFIC	Modeccin-, ricin-, diphtheria toxin-resistant	(*58*)
BHK	Mod^r5 Mod^r6	Modeccin-resistant	(*33*)
HeLa	H20e12 H22b11 Hb11 ET-28	Resistant to conjugates of epidermal growth factor and ricin or *Pseudomonas exotoxin*	(*5,6*)
Swiss 3T3	CHL60-64	Chloroquine-resistant	(*10*)
V79	CQ-22	Chloroquine-resistant	(*75*)
CHO	Mon^r-31	Monensin-resistant	(*76,78,106,113,134,135*)
Balb/3T3	MO-5	Monensin-resistant	(*77*)
LM	F40	Resistant to polyethylene glycol-induced fusion	(*59,101*)
L5178Y	Pgf^R6M2	Resistant to polyethylene glycol-induced fusion	(*82*)
CHO	TfT1.11	Defective in transferrin recycling	(*12*)
$LMTK^-$	AF-192	Defective in transferrin recycling	(*72*)
CHO	4B and 4J	Secretion-defective	(*41,42*)
CHO	LEH6	Secretion-defective	(*46*)
CHO	AS15-1	Secretion-defective (defective oligosaccharide lipid synthesis)	(*68*)
L cells	gro29	Secretion-defective	(*124*)

sensitive to diphtheria toxin. Electron microscopy combined with autoradiography using ^{125}I-ricin suggested that internalization was impaired. Although binding of ricin to the cell surface was normal, only about half as much ricin was internalized, and of that internalized, a larger proportion remained proximal to the surface membrane than in wild-type cells. Complementation analysis indicated that the lesion was a dominant trait, and analysis of hybrid strains with differing chromosome loss due to segregation indicated the loss of toxin resistance associated with increased acylation of two proteins of apparent molecular weights near 30 kDa.

Goldmacher et al. (*35*) described a toxin-resistant mutant (J5r^{R}6) isolated from Namal-wa cells, a human B-cell line. The mutant cells were cross-resistant to ricin, diphtheria toxin and immunotoxins directed against the CALLA antigen on the surface of the cells. The basis for this multiple resistance is not clear, but toxin binding to the cell surface appeared to be normal and the cytoplasmic targets for the toxins were sensitive to the actions of the toxins, suggesting that toxin penetration into the cells was defective.

Laurie and Robbins (*58*) used protein toxins to isolate a mutant (LEFIC) of mouse L cells that was cross-resistant to modeccin, *Pseudomonas* exotoxin A and ricin. The cells did not appear to be defective in endosomal acidification. Interestingly, movement of proteins along the secretory pathway was delayed.

Gleeson and Hughes (*33*) described two modeccin-resistant mutants (Mod^{r}5 and Mod^{r}6) isolated from BHK cells. Both mutants bound normal levels of modeccin but did not internalize the toxin. Both mutants were also fully sensitive to the action of ricin.

Herschman and co-workers (*5,6*) isolated mutants (H20e12 and ET-28) using either ricin A chain or *Pseudomonas aeruginosa* exotoxin A coupled to epidermal growth factor. The reason for toxin resistance is not known, but the number of epidermal growth factor receptors on the surface of the mutant cells was normal. The authors suggested that H20e12 and ET-28 cells were altered in distinct endosomal functions.

B. Chloroquine- and Monensin-Resistant Mutants

Chloroquine is a weak base that accumulates in acidic compartments, elevating the pH of the compartments and disrupting pH-dependent endocytic and lysosomal function. Wild-type cells, when incubated with weak bases, secrete lysosomal enzymes, exhibit resistance to toxins that require acidification for translocation to the cytosol, exhibit reduced accumulation by endocytosis, and have reduced degradation in lysosomes resulting in accumulation of residual bodies. Two mutant cell lines have been isolated for resistance to chloroquine. Ono et al. (*75*) described mutant CQ-22 isolated from Chinese hamster lung V79 fibroblasts, and Cain and Murphy (*10*) described a mutant termed CHL60-64 isolated from mouse 3T3 cells. Lysosomal enzyme levels were approximately normal in both chloroquine-resistant mutants, and there did not appear to be any defect in the trafficking of lysosomal enzymes. The defect of CQ-22 cells did not affect the binding or internalization of ricin, and unlike parental cells, chloroquine only slightly affected ricin internalization. Under serum starvation, which induces autophagic degradation of cellular proteins, turnover of endogenous proteins was reduced in CQ-22 cells.

Conceivably, resistance to chloroquine might result from either altered permeability of vesicle membranes, or from changes in proton pumping or regulation of vesicle acidification. Cain and Murphy extensively characterized the acidification of the endocytic apparatus of CHL60-64. Proton pumping, as measured by ATP-stimulated acridine orange accumulation in vesicles, was found to be decreased in both endosomes and lysosomes that were separated by density gradient centrifugation. *In vivo*, generation of proton gradients was found to be much slower in mutant cells recovering from amine treatment. Steady state proton gradients, however, were unaltered. In measuring the acidification of endocytosed probes, little difference was found in the acidification of transferrin localized in endosomes of the early recycling endocytic path. The acidification of FITC-dextran, which is transported to lysosomes, was markedly slowed; maximal acidification of dextran required 8 hours in CHL60-64 cells in comparison to about 1 hour in parental cells. Cain and Murphy proposed that a regulatory element of proton pumping in early endosomes of CHL60-64 cells mislocalized into the late endocytic path, and demonstrated that the regulator had pharmacological properties of the Na^+/K^+-ATPase (*11*).

Monensin is a carboxylic ionophore that collapses proton gradients within cells. Ono et al. (*78*) isolated a monensin-resistant mutant from CHO cells (Mon^r-31) that has been extensively studied. Mon^r-31 cells have reduced internalization of ricin, insulin, and LDL (*78*) and defective down-regulation of insulin and LDL receptors by insulin and LDL, respectively (*106,122*). Sterol metabolism in Mon^r-31 cells is resistant to the regulatory effect of LDL, although regulation of sterol metabolism by sterols is normal. Analysis of the LDL receptor by Yoshimura et al. (*135*) suggested that receptor function was affected as a result of altered O-linked oligosaccharides in the N-terminal ligand binding domain. When mutant cells were transfected with human LDL receptor cDNA, altered human LDL receptors also were synthesized. As determined by analysis of vesicular stomatitis virus G-protein acquisition of Endo H sensitivity, palmitylation, and sialylation, the synthesis of N-linked oligosaccharides was apparently normal. Genetic analysis indicated the defect in Mon^r-31 cells was recessive. Complementation analysis of Mon^r-31 with a phenotypically similar, compactin-resistant mutant V79 cell line (MF-2, which was isolated by Masuda et al. *64*) placed the two mutants in the same complementation group. This group was called *int*, for internalization defective, and by complementation was shown to be distinct from the ldlA, ldlB, ldlC, and ldlD groups (*113*). Yoshida et al. (*134*) investigated the sensitivity of Mon^r-31 cells to protein toxins in the presence of weak bases, ionophores and drugs that affect Golgi function, and concluded the phenotype of Mon^r-31 could result from elevated pH in the Golgi apparatus. The mutant cells were normally sensitive to diphtheria toxin and modeccin, consistent with a normal endosomal pH.

C. Polyethylene Glycol-Resistant Mutants

Roos and Davidson (*98*) isolated a mutant of mouse L cells, termed F40, that was resistant to the fusogenic effects of polyethylene glycol. Ryser and co-workers (*59,101*) subsequently reported that F40 cells were defective in fluid-phase endocytosis, but not in receptor-mediated endocytosis of ligands known to enter cells via clathrin-coated vesicles. The mutant cells did not recycle endocytosed fluid any faster than the parental cells, indicating that the reduced fluid accumulation could not be attributed to an enhanced rate of recycling. The authors suggested that F40 cells were defective in a clathrin-independent pathway of endocytosis that contributed to fluid-phase uptake. Pauw et al. (*82*) isolated a mutant with a similar phenotype from the mouse lymphocytic cell line L5178Y.

D. Mutants Defective in Recycling

A mutant with temperature-sensitive sequestration of cell surface receptors was isolated by Cain et al. (*12*) using fluorescence-activated cell sorting. Both transferrin receptors and α_2-macroglobulin receptors decreased from the cell surface after the cells were placed at 41°C. The internalization of a fluid marker, FITC-dextran was unaltered in the mutant. Accumulation of the dextran was reduced, however, and this resulted from an increased efflux after uptake.

Another mutant, AF192, with aberrant receptor recycling was isolated from LMTK$^-$ cells by O'Keefe and Draper (*72*). Obtained after exposing cells to a conjugate of transferrin and diphtheria toxin, the cells are cross-resistant to modeccin, ricin, abrin and *Pseudomonas aeruginosa* exotoxin A. AF192 cells recycled 25% of internalized transferrin more slowly than parent cells. ATP-stimulated accumulation of acridine orange into vesicles of a postnuclear supernatant of mutant cells did not differ from that of a parental preparation, suggesting that vacuolar acidification was not impaired. Transferrin binding was normal, and the total number of transferrin receptors (internal and cell surface) was the same in parent and mutant cells.

E. Secretion-Defective Mutants

Hearing et al. (*41,42*) used fluorescence-activated cell sorting to select temperature-sensitive mutants of CHO cells impaired in the transport of the influenza hemagglutinin to the cell surface. Two mutants were characterized and both appeared to be defective in N-linked glycosylation of glycoproteins.

Nakano et al. (*68,69*) selected temperature-sensitive, secretion-defective mutants of CHO cells using a combination of diphtheria toxin resistance and replica plating. One mutant, DS28-6 (*69*), is a member of the End4 complementation group already discussed in detail. The other mutant (AS15-1) is defective in oligosaccharide-lipid synthesis (*68*).

Henry et al. (*43*) used an approach similar to that of Nakano et al. (*69*) to enrich for temperature-sensitive secretion-defective mutants using the rat hepatoma cell line FAO. Cells were incubated at the restrictive temperature to elicit temperature-sensitive lesions, treated with pronase to remove surface proteins, and then exposed to concanavalin A at the permissive temperature to kill cells that were capable of replacing surface glycoproteins. However, none of the temperature-sensitive variants expressed stable phenotypes and are not listed in Table 2.

Kaiden and Krag (*46*) directly assayed the supernatants of CHO cells growing in microtiter plates for the temperature-sensitive appearance of radioactive secreted proteins. They found that 1% of 1250 clones screened had significantly reduced secretion at the restrictive temperature. One mutant, LEH6, was further characterized and reported to secrete 60 to 70% less total protein than parental cells, although the secretion of some proteins was more affected than others. The lesion in LEH6 cells did not impair protein glycosylation.

Tufaro et al. (*124*) described a mutant of mouse L cells, termed gro29, selected for the ability to survive infection with Herpes virus. The growth of Herpes virus and vesicular stomatitus virus was defective in gro29 cells. The authors concluded that the secretory export of newly made glycoproteins was defective, due most likely to a lesion associated with the Golgi apparatus.

VIII. CONCLUSION

Genetic analysis of membrane traffic is not as well developed in mammalian cells as in yeast where there are more powerful genetic tools. Nevertheless, many mammalian mutants are available and the current emphasis in most laboratories is to devise new schemes for mutant isolation and to better characterize the existing mutants. With the exception of the gene for the LDL receptor, none of the genes affected in mutants described in this review have been cloned. However, this is a very active area of research and it is likely that reports of successful cloning will be forthcoming. Analysis of the cloned genes, coupled with physiological and biochemical characterization of mutants, should provide novel information about mechanisms of membrane traffic in mammalian cells.

REFERENCES

1. Al-Awqati, Q. (1986). Proton-translocating ATPases. *Annu. Rev. Cell Biol.* 2: 179–199.
2. Anderson, R.G.W. and Pathak, R.K. (1985). Vesicles and cisternae in the *trans* Golgi apparatus of human fibroblasts are acidic compartments. *Cell* 40: 635–643.
3. Arai, H., Terres, G., Pink, S., and Forgac, M. (1988). Topography and subunit stoichiometry of the coated vesicle proton pump. *J. Biol. Chem.* 263: 8796–8802.
4. Balch, W.E. (1989). Biochemistry of interorganelle transport. *J. Biol. Chem.* 264: 16965–16968.
5. Banker, D. and Herschman, H.R. (1989). HeLa cell mutants resistant to epidermal growth factor ricin A-chain conjugate. *J. Cell Physiol.* 139: 42–50.
6. Banker, D.E., Pastan, I., Gottesman, M.M., and Herschman, H.R. (1989). An epidermal growth factor-ricin A chain (EGF-RTA)-resistant mutant and an epidermal growth factor-*Pseudomonas* endotoxin (EGF-PE)-resistant mutant have distinct phenotypes. *J. Cell Physiol.* 139: 51–57.
7. Braell, W.A., Schlossman, D.M., Schmid, S.L., and Rothman, J.E. (1984). Dissociation of clathrin coats coupled to the hydrolysis of ATP: role of an uncoating ATPase. *J. Cell Biol.* 99: 734–741.
8. Brandli, W. (1991). Mammalian glycosylation mutants as tools for the analysis and reconstitution of protein transport. *Biochem. J.* 276: 1–12.
9. Burgess, T.L. and Kelly, R.B. (1987). Constitutive and regulated secretion of proteins. *Annu. Rev. Cell Biol.* 3: 243–293.
10. Cain, C.C. and Murphy, R.F. (1988). A chloroquine-resistant Swiss 3T3 cell line with a defect in late endocytic function. *J. Cell Biol.* 106: 269–277.

11. Cain, C.C., Sipe, D.M., and Murphy, R.F. (1989). Regulation of endocytic pH by the Na,K-ATPase in living cells. *Proc. Natl. Acad. Sci. USA* 86: 544–548.
12. Cain, C.C., Wilson, R.B., and Murphy, R.F. (1991). Isolation by fluorescence-activated cell sorting of Chinese hamster ovary cell lines with pleiotropic temperature-conditional defects in receptor recycling. *J. Biol. Chem.* 266: 11746–11752.
13. Chinault, A.C. and Caskey, C.T. (1984). The hypoxanthine phosphoribosyltransferase gene: a model for the study of mutation in mammalian cells. *Prog. Nucleic Acid Res. and Mol. Biol.* 31: 295–313.
14. Colbaugh, P.A., Kao, C.-Y., Shia, S.-P., Stookey, M., and Draper, R.K. (1988). Three new complementation groups of temperature-sensitive Chinese hamster ovary cell mutants defective in the endocytic pathway. *Somatic Cell Mol. Genet.* 14: 499–507.
15. Colbaugh, P.A., Stookey, M., and Draper, R.K. (1989). Impaired lysosomes in a temperature-sensitive mutant of Chinese hamster ovary cells. *J. Cell Biol.* 108: 2211–2219.
16. DeLuca-Flaherty, C., McKay, D.B., Parham, P., and Hill, B.L. (1990). Uncoating protein (hsc70) binds a conformationally labile domain of clathrin light chanin LC_a to stimulate ATP hydrolysis. *Cell* 62: 875–887.
17. van Deurs, B., Petersen, O.W., Olsnes, S., and Sandvig, K. (1989). The ways of endocytosis. *Internat. Rev. Cytol.* 117: 131–177.
18. van Deurs, B., Sandvig, K., Petersen, O.W., Olsnes, S., Simons, K., and Griffiths, G. (1988). Estimation of the amount of internalized ricin that reaches the *trans*-Golgi network. *J. Cell Biol.* 106: 253–267.
19. Didsbury, J.R., Moehring, J.M., and Moehring, T.J. (1983). Binding and uptake of diphtheria toxin by toxin-resistant Chinese hamster ovary and mouse cells. *Mol. Cell. Biol.* 3: 1283–1294.
20. Doms, R.W., Russ, G., and Yewdell, J.W. (1989). Brefeldin A redistributes resident and itinerant Golgi proteins to the endoplasmic reticulum. *J. Cell Biol.* 109: 61–72.
21. Duden, R., Allan, V., and Kreis, T. (1991). Involvement of β-COP in membrane traffic through the Golgi complex. *Trends in Cell Biology* 1: 14–19.
22. Edwards, J. and Brown, D.T. (1991). Sindbis virus infection of a Chinese hamster ovary cell mutant defective in the acidification of endosomes. *Virology* 182: 28–33.
23. Eidels, L., and Draper, R.K. (1988). Diphtheria toxin. In: *Handbook of Natural Toxins* (Hardegree, M.C. and Tu, A.T., Eds.), pp. 217–247. Marcel Dekker, New York.
24. Eidels, L., Proia, R.L., and Hart, D.A. (1983). Membrane receptors for bacterial toxins. *Microbiol. Rev.* 47: 596–620.
25. Endo, Y., Mitsui, K., Motizuki, M., and Tsurugi, K. (1987). The mechanism of action of ricin and related toxic lectins on eukaryotic ribosomes: the site and the characteristics of the modification in 28 S ribosomal RNA caused by the toxins. *J. Biol. Chem.* 262: 5908–5912.
26. Esko, J.D. and Raetz, C.R.H. (1978). Replica plating and in situ enzymatic assay of animal cell colonies established on filter paper. *Proc. Natl. Acad. Sci. USA* 75: 1190–1193.
27. Ferris, A.L., Brown, J.C., Park, R.D., and Storrie, B. (1987). Chinese hamster ovary cell lysosomes rapidly exchange contents. *J. Cell Biol.* 105: 2703–2712.
28. Fuchs, R., Male, P., and Mellman, I. (1989). Acidification and ion permeabilities of highly purified rat liver endosomes. *J. Biol. Chem.* 264: 2212–2220.
29. Fuchs, R., Schmid, S., and Mellman, I. (1989). A possible role for Na^+,K^+-ATPase in regulating ATP-dependent endosome acidification. *Proc. Natl. Acad. Sci. USA* 86: 539–543.
30. Geuze, H.J., Slot, J.W., Strous, G.J.A.M., Peppard, J., von Figura, K., Hasilik, A., and Schwartz, A.L. (1984). Intracellular receptor sorting during endocytosis: comparative immunoelectron microscopy of multiple receptors in rat liver. *Cell* 37: 195–204.
31. Ghosh, P.C., Wellner, R.B., and Wu, H.C. (1984). Osmotically induced microinjection of ricin bypasses a ricin internalization defect in a Chinese hamster ovary cell mutant cell line. *Mol. Cell. Biol.* 4: 1320–1325.
32. Ghosh, P.C. and Wu, H.C. (1988). Enhancement of cytoxicity of modeccin by nigericin in modeccin-resistant mutant cell lines. *Exp. Cell Res.* 174: 397–410.

33. Gleeson, P.A. and Hughes, R.C. (1985). Binding and uptake of the toxic lectin modeccin by baby hamster kidney (BHK) cells. Isolation of mutants defective in the internalization of modeccin. *J. Cell Sci.* 76: 283–301.
34. Glickman, J.N., Conibear, E., and Pearse, B.M.F. (1989). Specificity of binding of clathrin adaptors to signals on the mannose-6-phosphate/insulin-like growth factor II receptor. *EMBO J.* 8: 1041–1047.
35. Goldmacher, V.S., Anderson, J., Schulz, M.-L., Blattler, W.A., and Lambert, J.M. (1987). Somatic cell mutants resistant to ricin, diphtheria toxin and immunotoxins. *J. Biol. Chem.* 262: 3205–3209.
36. Gottesman, M.M. (1985). Growth properties of Chinese hamster ovary (CHO) cells. In: *Molecular Cell Genetics* (Gottesman, M.M., Ed.), pp. 139–154. John Wiley and Sons, New York.
37. Gottesman, M.M. (Ed.) (1985). *Molecular Cell Genetics*, 1st ed. John Wiley and Sons, New York.
38. Griffiths, G., Hoflack, B., Simons, K., Mellman, I., and Kornfeld, S. (1988). The mannose 6-phosphate receptor and the biogenesis of lysosomes. *Cell* 52: 329–341.
39. Griffiths, G. and Simons, K. (1986). The *trans* Golgi network: sorting at the exit site of the Golgi complex. *Science* 234: 438–443.
40. Gruenberg, J. and Howell, K.E. (1989). Membrane traffic in endocytosis: Insights from cell-free assays. *Annu. Rev. Cell Biol.* 5: 453–481.
41. Hearing, J., Gething, M.J., and Sambrook, J. (1989). Addition of truncated oligosaccharides to influenza virus hemagglutinin results in its temperature-conditional cell-surface expression. *J. Cell Biol.* 108: 355–365.
42. Hearing, J., Hunter, E., Rodgers, L., Gething, M.J., and Sambrook, J. (1989). Isolation of Chinese hamster ovary cell lines temperature conditional for the cell surface expression of integral membrane glycoproteins. *J. Cell Biol.* 108: 339–353.
43. Henry, I.I.I., K.W., Brown, J.L., and McCracken, A.A. (1990). Selective enrichment for temperature-sensitive secretion mutants of mammalian cells using plant lectin, concanavalin A. *Somatic Cell Mol. Genet.* 16: 297–304.
44. Hubbard, A.L. (1989). Endocytosis. *Cur. Opin. Cell Biol.* 1: 675–683.
45. Hudson, T.H. and Neville, D.M. (1987). Temporal separation of protein toxin translocation from processing events. *J. Biol. Chem.* 262: 16484–16494.
46. Kaiden, A. and Krag, S.S. (1991). Isolation of a mutant of Chinese hamster ovary cells with defective secretion of a subset of secretory proteins. *Somatic Cell Mol. Genet.* 17: 15–33.
46a. Kao, C.-Y. and Draper, R.K. (1992). Retention of secretory proteins in an intermediate compartment and disappearance of the Golgi complex in an End4 mutant of Chinese hamster ovary cells. *J. Cell Biol.* 117: 701–715.
47. Kielian, M. C., Keranen, S., Kaariainen, L., and Helenius, A. (1984). Membrane fusion mutants of Semliki Forest virus. *J. Cell Biol.* 98: 139–145.
48. Kingsley, D.M., Kozarsky, K.F., Hobbie, L., and Krieger, M. (1986). Reversible defects in O-linked glycosylation and LDL receptor expression in a UDP-Gal/UDP-GalNAc 4-epimerase deficient mutant. *Cell* 44: 749–759.
49. Kingsley, D.M., Kozarsky, K.F., Segal, M., and Krieger, M. (1986). Three types of LDL receptor-deficient mutants have pleitropic defects in the synthesis of N-linked, O-linked and lipid-linked carbohydrate chains. *J. Cell Biol.* 102: 1576–1585.
50. Kingsley, D.M. and Krieger, M. (1984). Receptor-mediated endocytosis of low density lipoprotein: somatic cell mutants define multiple genes required for expression of surface-receptor activity. *Proc. Natl. Acad. Sci. USA* 81: 5454–5458.
51. Kingsley, D.M., Sege, R.D., Kozarsky, K.F., and Krieger, M. (1986). DNA-mediated transfer of a human gene required for LDL receptor expression and for multiple Golgi processing pathways. *Mol. Cell. Biol.* 6: 2734–2737.

52. Klausner, R.D., van Renswoude, J., Kempf, C., Rao, K., Bateman, J.L., and Robbins, A.R. (1984). Failure to release iron from transferrin in a Chinese hamster ovary cell mutant pleiotropically defective in endocytosis. *J. Cell Biol.* 98: 1098–1101.
53. Kornfeld, S. and Mellman, I. (1989). The biogenesis of lysosomes. *Annu. Rev. Cell Biol.* 5: 483–525.
54. Kozarsky, K.F., Brush,, H.A., and M. Krieger (1986). Unusual forms of low density lipoprotein receptors in hamster cell mutants with defects in the receptor structural gene. *J. Cell Biol.* 102: 1567–1575.
55. Kozarsky, K.F., Kingsley, D., and Krieger, M. (1988). Use of a mutant cell line to study the kinetics and function of O-linked glycosylation of low density lipoprotein receptors. *Proc. Natl. Acad. Sci. USA* 85: 4335–4339.
56. Krieger, M., Brown, M.S., and Goldstein, J.L. (1981). Isolation of Chinese hamster cell mutants defective in the receptor-mediated endocytosis of low-density lipoprotein. *J. Mol. Biol.* 150: 167–184.
57. Krieger, M., Martin, J., Segal, M., and Kingsley, D. (1983). Amphotericin B selection of mutant CHO cells with defects in the receptor-mediated endocytosis of low-density lipoprotein and cholesterol synthesis. *Proc. Natl. Acad. Sci. USA* 80: 5607–5611.
58. Laurie, S.M. and Robbins, A.R. (1991). A toxin-resistant mouse L-cell mutant defective in protein transport along the secretory pathway. *J. Cell Physiol.* 147: 215–223.
59. Li, W., Ryser, H.J.-P., and Shen, W.-C. (1986). Altered endocytosis in a mutant of LM fibroblasts defective in cell-cell fusion. *J. Cell Physiol.* 126: 161–166.
60. Lippincott-Schwartz, J., Donaldson, J.G., Schweizer, A., Berger, E.G., Hauri, H.P., Yuan, L.C., and Klausner, R.D. (1990). Microtubule-dependent retrograde transport of proteins into the ER in the presence of brefeldin A suggests an ER recycling pathway. *Cell* 60: 821–836.
61. Lippincott-Schwartz, J., Yuan, L.C., Bonifacino, J.S., and Klausner, R.D. (1989). Rapid redistribution of Golgi proteins into the ER in cells treated with brefeldin A: Evidence for membrane cycling from the Golgi to ER. *Cell* 56: 801–813.
61a. Malmstrom, K. and Krieger, M. (1991). Use of radiation suicide to isolate constitutive and temperature-sensitive conditional Chinese hamster ovary cell mutants with defects in the endocytosis of low density lipoprotein. *J. Biol. Chem.* 266: 24025–24030.
62. Marnell, M.H., Mathis, L.S., Stookey, M., Shia, S.-H., Stone, D.K., and Draper, R.K. (1984). A Chinese hamster ovary cell mutant with a heat-sensitive, conditional-lethal defect in vacuolar function. *J. Cell Biol.* 99: 1907–1916.
63. Marnell, M.H., Shia, S.-P., Stookey, M., and Draper, R.K. (1984). Evidence for penetration of diphtheria toxin to the cytosol through a prelysosomal membrane. *Infect. Immun.* 44: 145–150.
64. Masuda, A., Akiyama, S.-I., and Kuwano, M. (1982). Chinese hamster cell mutant resistant to ML236B (compactin) is defective in endocytosis of low density lipoprotein. *Mol. Cell. Biol.* 2: 1354–1361.
65. Mellman, I., Fuchs, R., and Helenius, A. (1986). Acidification of the endocytic and exocytic pathways. *Annu. Rev. Biochem.* 55: 663–700.
66. Merion, M., Schlesinger, P., Brooks, R.M., Moehring, J.M., Moehring, T.J., and Sly, W. (1983). Defective acidification of endosomes in Chinese hamster ovary cell mutants "cross resistant" to toxins and viruses. *Proc. Natl. Acad. Sci. USA* 80: 5315–5319.
67. Morris, S.A., Ahle, S., and Ungewickell, E. (1989). Clathrin-coated visicles. *Cur. Opin. Cell Biol.* 1: 684–690.
68. Nakano, A. and Akamatsu, Y. (1985). A novel type of Chinese hamster ovary cell mutant defective in oligosaccharide-lipid synthesis. *Biochem. Biophys. Acta* 845: 507–510.
69. Nakano, A., Nishijima, M., Maeda, M., and Akamatsu, Y. (1985). A temperature-sensitive Chinese hamster ovary cell mutant pleiotropically defective in protein export. *Biochem. Biophys. Acta* 845: 324–332.
70. Neville, D.M. and Hudson, T.H. (1986). Transmembrane transport of diphtheria toxin, related toxins, and colicins. *Annu. Rev. Biochem.* 55: 195-224.

71. Nunez, M.T., Gaete, V., and Escobar, A. (1990). Endocytic vesicles contain a calmodulin-activated Ca^{2+} pump that mediates the inhibition of acidification by calcium. *Biochem. Biophys. Acta* 1028: 21–24.
72. O'Keefe, D.O. and Draper, R.K. (1988). Two pathways of transferrin recyling evident in a variant of mouse LMTK cells. *Somatic Cell Mol. Genet.* 14: 473-487.
73. Olsnes, S. and Pihl, A. (1982). Toxic lectins and related proteins. In: *Molecular Action of Toxins and Viruses* (Cohen, P. and van Heyningen, S., Eds.), pp. 51–105. Elsevier/North Holland, Amsterdam.
74. Olsnes, S. and Sandvig, K. (1985). Entry of polypeptide toxins into animal cells. In: *Endocytosis* (Pastan, I. and Willingham, M.C., Eds.), pp. 195–234. Plenum Publishing, New York.
75. Ono, M., Ando, M., Shimada, T., Furuno, K., Kato, K., and Kuwano, M. (1983). Isolation of chloroquine-resistant Chinese hamster V79 cell variants that are also resistant to ammonium chloride. *J. Biochem. (Tokyo)* 94: 1493–1503.
76. Ono, M., Mannen, K., Shimada, T., Kuwano, M., and Mifune, K. (1985). Effect of monensin on the synthesis, maturation and secretion of vesicular stomatitis virus proteins in a monensin-resistant Chinese hamster ovary cell line. *Cell Struct. Funct.* 10: 279–294.
77. Ono, M., Mifune, K., Yoshimura, A., Ohnishi, S., and Kuwano, M. (1985). Monensin-resistant mouse Balb/3T3 cell mutant with aberrant penetration of vesicular stomatitis virus. *J. Cell Biol.* 101: 60–65.
78. Ono, M., Yamato, H., Ando, M., and Kuwano, M. (1984). Chinese hamster ovary cell variants resistant to monensin, an ionophoric antibiotic. I. Isolation and altered endocytosis of ricin. *J. Cell Physiol.* 119: 198–203.
79. Orci, L., Malhotra, V., Amherdt, T., Serafini, T., and Rothman, J.E. (1989). Dissection of a single round of vesicular transport: sequential intermediates for intercisternal movement through the Golgi stack. *Cell* 56: 357–368.
80. Orci, L., Tagaya, M., Amherdt, M., Perrelet, A., Donaldson, J.G., Lippincott-Schwartz, J., Klausner, R.D., and Rothman, J.E. (1991). Brefeldin A, a drug that blocks secretion, prevents the assembly of non-clathrin coated buds on Golgi cisternae. *Cell* 64: 1183–1195.
81. Park, J.E., Draper, R.K., and Brown, W.J. (1991). Biosynthesis of lysosomal enzymes in cells of the End3 complementation group conditionally defective in endosomal acidification. *Somatic Cell Mol. Genet.* 17: 137–150.
82. Pauw, P.G., Giagi, K.G., and Stadler, J.K. (1986). Reduced pinosome formation a cell-fusion-impaired mutant of mouse lymphocytic cell line L5178Y. *J. Cell Physiol.* 126: 243–248.
83. Pearse, B.M.F. (1988). Receptors compete for adaptors found in plasma membrane coated pits. *EMBO J.* 7: 3331–3336.
84. Pelham, H.R.B. (1989). Control of protein exit from the endoplasmic reticulum. *Annu. Rev. Cell Biol.* 5: 1–23.
85. Peters, C., Braun, M., Weber, B., Wendland, M., Schmidt, B., Pohlman, R., Waheed, A., and von Figura, K. (1990). Targeting of a lysososmal membrane protein: a tyrosine-containing endocytosis signal in the cytoplasmic tail of lysosomal acid phosphatase is necessary and sufficent for targeting to lysosomes. *EMBO J.* 9: 3497–3506.
86. Presley, J.F., Draper, R.K., and Brown, D.T. (1991). Defective transport of Sindbis virus glycoproteins in End4 mutant of Chinese hamster ovary cells. *J. Virol.* 65: 1332–1339.
87. Puck, T.T. (1985). Development of the Chinese hamster ovary (CHO) cell for use in somatic cell genetics. In: *Molecular Cell Genetics* (Gottesman, M.M., Ed.), pp. 37–64. John Wiley and Sons, New York.
88. Ray, B. and Wu, H. (1982). Chinese hamster ovary cell mutants defective in the internalization of ricin. *Mol. Cell. Biol.* 2: 535–544.
89. Reddy, P., Caras, I., and Krieger, M. (1989). Effects of O-linked glycosylation on the cell surface expression and stability of decay-accelerating factor, a glycophospholipid-anchored membrane protein. *J. Biol. Chem.* 264: 17329–17336.

90. Reddy, P. and Krieger, M. (1989). Isolation and characterization of an extragenic suppressor of the low-density lipoprotein receptor-deficient phenotype of a Chinese hamster ovary cell mutant. *Mol. Cell. Biol.* 9: 4799–4806.
91. Robbins, A., Oliver, C., Bateman, J.L., Krag, S., Galloway, C.J., and Mellman, I. (1984). A single mutation in Chinese hamster ovary cells impairs both Golgi and endosomal functions. *J. Cell Biol.* 99: 1296–1308.
92. Robbins, A.R. (1988). Endocytosis and compartmentalization of lysosomal enzymes in normal and mutant mammalian cells: mannose 6-phosphate-dependent pathways. In: *Protein Transport and Organelle Biogenesis* (Das, R. and Robbins, P., eds.), pp. 464–509. Academic Press, New York.
93. Robbins, A.R., Peng, S.S., and Marshall, J.L. (1983). Mutant Chinese hamster cells pleiotropically defective in receptor-mediated endocytosis. *J. Cell Biol.* 96: 1064–1071.
94. Robbins, A.R. and Roff, C.F. (1987). Isolation of mutant Chinese hamster ovary cells defective in endocytosis. *Methods in Enzymology* 138: 458–470.
95. Roederer, M., Bowser, R., and Murphy, R.F. (1987). Kinetics and temperature dependence of exposure of endocytosed material to proteolytic enzymes and low pH: evidence for a maturation model for the formation of lysosomes. *J. Cell Physiol.* 131: 200–209.
96. Roff, C.F., Fuchs, R., Mellman, I., and Robbins, A.R. (1986). Chinese hamster ovary cell mutants with temperature-sensitive defects in endocytosis. I. Loss of function on shifting to the nonpermissive temperature. *J. Cell Biol.* 103: 2283–2297.
97. Roff, C.F., Hall, C.W., and Robbins, A.R. (1990). Recovery of function in Chinese hamster ovary cell mutants with temperature-sensitive defects in vacuolar acidification. *J. Cell Biol.* 110: 1023–1032.
98. Roos, D.S. and Davidson, R.L. (1980). Isolation of mouse cell lines resistant to the fusion-inducing effect of polyethylene glycol. *Somatic Cell Mol. Genet.* 6: 381–391.
99. Rose, J.K. and Doms, R.W. (1988). Regulation of protein export from the endoplasmic reticulum. *Annu. Rev. Cell Biol.* 4: 257–288.
100. Rothman, J.E. and Orci, L. (1990). Movement of proteins through the Golgi stack: a molecular dissection of vesicular transport. *FASEB J.* 4: 1460–1468.
101. Ryser, H.J.-P., Li, W., Mandel, R., and Shen, W.-C. (1988). Stable variant of LM fibroblast defective in fluid-phase but competent in receptor-mediated endocytosis. *J. Cell Physiol.* 137: 490–496.
102. Ryser, H.J.-P., Mandel, R., Hacobian, A., and Shen, W.-C. (1988). Methotrexate-poly(lysine) as a selective agent for mutants of Chinese hamster ovary cells defective in endocytosis. *J. Cell Physiol.* 135: 277–284.
103. Sandvig, K., Sundan, A., and Olsnes, S. (1984). Evidence that diphtheria toxin and modeccin enter the cytosol from different vesicular compartments. *J. Cell Biol.* 98: 963–970.
104. Saraste, J. and Kuismanen, E. (1984). Pre- and post-Golgi vacuoles operate in the transport of Semliki Forest virus membrane glycoproteins to the cell surface. *Cell* 38: 535–549.
105. Saraste, J. and Svensson, K. (1991). Distribution of the intermediate elements operating in ER to Golgi transport. *J. Cell Sci.* in press.
106. Sato, Y., Ono, M., Takaki, R., and Kuwano, M. (1984). Chinese hamster ovary cell variants resistant to monensin, an ionophoric antibiotic. II. Growth requirements for insulin and altered insulin receptor activity. *J. Cell Physiol.* 119: 204–210.
107. Schmid, S., Fuchs, R., Kielian, M., Helenius, A., and Mellman, I. (1989). Acidification of endosome subpopulations in wild-type Chinese hamster ovary cells and temperature-sensitive acidification-defective mutants. *J. Cell Biol.* 108: 1291–1300.
108. Schweizer, A., Fransen, J., Bachi, T., Ginsel, L., and Hauri, H. (1988). Identification, by a monoclonal antibody, of a 53 kD protein associated with a tubulo-vesicular compartment at the cis-side of the golgi apparatus. *J. Cell Biol.* 107: 1643–1653.

109. Schweizer, A., Fransen, J.A.M., Matter, K., Kreis, T.E., and Ginsel, L. (1990). Identification of an intermediate compartment involved in protein transport from endoplasmic reticulum to Golgi apparatus. *Eur. J. Cell Biol.* 53: 185–196.
110. Schweizer, A., Matter, K., Ketcham, C.M., and Hauri, H.-P. (1991). The isolated ER-Golgi intermediate compartment exhibits properties that are different from ER and cis Golgi. *J. Cell Biol.* 113: 45–54.
111. Sege, R.D., Kozarsky, K.F., and Krieger, M. (1986). Characterization of a family of gamma-ray induced CHO mutants demonstrates that the ldlA locus is diploid and encodes the LDL receptor. *Mol. Cell. Biol.* 6: 3268–3277.
112. Shia, S.-P. (1986). Characterization of a modeccin-resistant mutant [PhD Dissertation]. University of Texas at Dallas.
113. Shite, S., Seguchi, T., Yoshida, T., Kohno, K., Ono, M., and Kuwano, M. (1988). A new class mutation of low density lipoprotein receptor with altered carbohydrate chains. *J. Biol. Chem.* 263: 19286–19289.
114. Siciliano, M.J., Stallings, R.L., and Adair, G.M. (1985). The genetic map of the Chinese hamster and the genetic consequences of chromosomal rearrangements in CHO cells. In: *Molecular Cell Genetics* (Gottesman, M.M., Ed.), pp. 95–135. John Wiley and Sons, New York.
115. Siminovitch, L. (1976). On the nature of heriditable variation in cultured somatic cells. *Cell* 7: 1–11.
116. Simmons, B.M., Stahl, P.D., and Russell, J.H. (1986). Mannose receptor-mediated uptake of ricin toxin and ricin A chain by macrophages. *J. Biol. Chem.* 261: 7912–7920.
117. Stone, D.K., Crider, B.P., and Xie, X.-S. (1990). Structural properties of vacuolar proton pumps. *Kidney Internat.* 38: 649–653.
118. Stone, D.K., Marnell, M., Yang, Y., and Draper, R.K. (1987). Thermolabile proton translocating ATPase and pump activities in a clathrin-coated vesicle fraction from an acidification defective Chinese hamster cell line. *J. Biol. Chem.* 262: 9883–9886.
119. Stoorvogel, W., Strous, G.J., Gueze, H.J., Oorschot, V., and Schwartz, A.L. (1991). Late endosomes derive from early endosomes by maturation. *Cell* 65: 417–427.
120. Thompson, L.H. and Baker, R.M. (1973). Isolation of mutants in cultured mammalian cells. *Methods in Cell Biol.* 6: 209–281.
121. Timchak, L.M., Kruse, F., Marnell, M.H., and Draper, R.K. (1986). A thermosensitive lesion in a Chinese hamster cell mutant causing differential effects on the acidification of endosomes and lysosomes. *J. Biol. Chem.* 261: 14154–14159.
122. Tomita, K., Yoshida, T., Yoshimura, A., Ono, M., and Kuwano, M. (1987). Chinese hamster ovary cell mutant with defective down-regulation of low density lipoprotein receptors. *J. Biol. Chem.* 262: 1398–1404.
123. Tran, D., Carpentier, J.-L., Sawano, F., Gordon, P., and Orci, L. (1987). Ligands internalized through coated or noncoated invaginations follow a common intracellular pathway. *Proc. Natl. Acad. Sci. USA* 84: 7957–7961.
124. Tufaro, F., Snider, M.D., and McKnight, S.L. (1987). Identification and characterization of a mouse cell mutant defective in the intracellular transport of glycoproteins. *J. Cell Biol.* 105: 647–657.
125. Wang, R.-H. (1990). Fluid-phase Endocytosis in Wild-type and Mutant Chinese Hamster Ovary Cells [PhD Dissertation]. University of Texas at Dallas.
126. Wang, R.-H., Colbaugh, P.A., Kao, C.-Y., Rutledge, E.A., and Draper, R.K. (1990). Impaired secretion and fluid-phase endocytosis in the End4 mutant of Chinese hamster ovary cells. *J. Biol. Chem.* 265: 20179–20187.
126a. Wang, R.-H., Colbaugh, P.A., Kuo, P., Bau, M.-Y., Poppe, L.M., and Draper, R.K. (1992). Novel method for isolating mammalian cells defective in fluid-phase endocytosis. *Somatic Cell Mol. Genet.* 118: 543–551.

127. Wellner, R.B., Ray, B., Ghosh, P.C., and Wu, H.C. (1984). Genetic and biochemical analysis of mutation(s) affecting ricin internalization in Chinese hamster ovary cells. *J. Biol. Chem.* 259: 12788–12793.
128. Wick, M.J., Hamood, A.N., and Iglewski, B.H. (1990). Analysis of the structure-function relationship of *Pseudomonas aeruginosa* exotoxin A. *Mol. Microbiol.* 4: 527–535.
129. Xie, X.-S., Crider, B.P., and Stone, D.K. (1989). Isolation and reconstitution of the chloride transporter of clathrin-coated vesicles. *J. Biol. Chem.* 264: 18870–18873.
130. Xie, X.-S. and Stone, D.K. (1986). Isolation and reconstitution of the clathrin-coated vesicle proton translocating complex. *J. Biol. Chem.* 261: 2492–2495.
131. Yamashiro, D.J., Borden, L.A., and Maxfield, F.R. (1989). Kinetics of alpha$_2$-macroglobulin endocytosis and degradation in mutant and wild-type Chinese hamster ovary cells. *J. Cell Physiol.* 139: 377–382.
132. Yamashiro, D.J. and Maxfield, F.R. (1987). Acidification of morphologically distinct endosomes in mutant and wild-type Chinese hamster ovary cells. *J. Cell Biol.* 105: 2723–2733.
133. Yamashiro, D.J. and Maxfield, F.R. (1987). Kinetics of endosome acidification in mutant and wild-type Chinese hamster ovary cells. *J. Cell Biol.* 105: 2713–2721.
134. Yoshida, T., Chen, C., Zhang, M., and Wu, H.C. (1990). Increased cytotoxicity of ricin in a putative Golgi-defective mutant of Chinese hamster ovary cell. *Exp. Cell Res.* 190: 11–16.
135. Yoshimura, A., Yoshida, T., Seguchi, T., Waki, M., Ono, M., and Kuwano, M. (1987). Low binding capacity and altered O-linked glycosylation of low density lipoprotein receptor in a monensin-resistant mutant of Chinese hamster ovary cells. *J. Biol. Chem.* 262: 13299–13308.
136. Youle, R.J. and Colombatti, M. (1987). Hybridoma cells containing intracellular antiricin antibodies show ricin meets secretory antibody before entering the cytosol. *J. Biol. Chem.* 262: 4676–4682.
137. Zuber, C., Roth, J., Misteli, T., Nakano, A., and Moreman, K. (1991). DS28-6, a temperature-sensitive mutant of Chinese hamster ovary cells, expresses key phenotypic changes associated with brefeldin A treatment. *Proc. Natl. Acad. Sci. USA* 88: 9818–9822.

PLASMA MEMBRANE LIPID TRANSPORT IN CULTURED CELLS:

STUDIES USING LIPID ANALOGS AND MODEL SYSTEMS

Michael Koval

Advances in Cell and Molecular Biology of Membranes
Volume 1, pages 199–224

ISBN: 1-55938-362-3

ABSTRACT

This chapter describes studies on plasma membrane lipid transport and metabolism in cultured mammalian cells, with particular emphasis on the use of exogenously added lipid probes to examine these processes. Plasma membrane lipids that are restricted to the extracytoplasmic (outer) leaflet are internalized by vesicle-mediated endocytic transport pathways. Most endocytosed membrane lipid is recycled to the plasma membrane and it appears that vesicular transport of lipid and protein occur by similar, but not identical, mechanisms. In contrast, some lipids are capable of transbilayer movement across the cell surface, as a result of either protein-mediated movement or metabolism to a lipid species capable of spontaneous transbilayer movement. Lipids translocated across the bilayer have access to additional nonvesicular-facilitated transport pathways from the cytoplasmic (inner) leaflet of the plasma membrane to other intracellular membranes. Finally, possible roles for lipids in mediating membrane fusion events are discussed in light of current work-using systems for cell-free reconstitution of vesicular transport. In particular, some arguments are presented suggesting that membrane fusion may be coupled to lipid metabolism.

I. INTRODUCTION

Lipids are an essential and major part of all cell membranes. In addition to providing the basis for membrane structure, there is increasing evidence of lipid involvement in protein function, signal transduction, membrane fusion, and other cellular processes. The importance of lipids is underscored by the fact that each organelle has a unique lipid composition that is required for proper membrane function (*21,28,125,132*).

Phosphoglycerolipids are the most prevalent lipids found in animal cell membranes (*132*). Major phosphoglycerolipids include phosphatidylcholine (PC), phosphatidylethanolamine (PE), phosphatidylserine (PS), and phosphatidylinositol (PI). Each phosphoglycerolipid contains a glycerol backbone linked to two fatty acyl chains, which show considerable variability depending upon the lipid and tissue type examined (*132*). Also, these lipids are distributed asymmetrically across the membrane bilayer, with PE and PS enriched in the cytoplasmic (inner) leaflet and PC enriched in the extracytoplasmic (outer) leaflet (*40*). Another major lipid found in animal cell membranes are sphingolipids, such as sphingomyelin (SM), glycosphingolipids (e.g., gangliosides and cerebrosides), and ceramide (*43,114*). In contrast to the glycerolipids, sphingolipids contain a long-chain sphingosine backbone and a single fatty acyl chain. Sphingomyelin and glycosphingolipids are largely found in the extracytoplasmic leaflet of the plasma membrane (*40*). Cholesterol makes up the majority of the remaining membrane lipid with some membranes, such as the plasma membrane, having cholesterol content of 50 mol% or higher

(*22,88,125*). Cholesterol esters, triacylglycerols, unesterified fatty acids, and other types of lipids may be present in membranes at low levels.

Membrane lipid composition is maintained by a combination of two basic mechanisms. Lipids may be added to or removed from a membrane directly by transport, as discussed below. Alternatively, membrane lipids may be altered by local metabolism. As an example of local metabolism, bis(monoacylglycero)phosphate is synthesized in the lysosomal compartment and is absent from other intracellular membranes (*13,131*). Lipids are also subject to remodeling by enzymatic alteration of either the polar head group or the fatty acyl chains (*28*). Head group exchange and acylation/deacylation reactions appear to occur throughout the cell, however, this area needs to be examined in greater detail. Also, the mechanisms that coordinate lipid metabolism and transport in order to maintain membrane integrity remain largely unknown.

This chapter describes studies on plasma membrane lipid transport and metabolism in cultured mammalian cells, with particular emphasis on the use of exogenously added lipid probes to examine these processes. There is a major focus on vesicular transport and sorting of plasma membrane lipid throughout endocytic compartments. Also, mechanisms involved in the transbilayer movement of lipid between the two leaflets of the plasma membrane will be discussed, as well as nonvesicular-facilitated movement of lipid from the inner leaflet of the plasma membrane to intracellular compartments. Finally, some possible roles for lipids in mediating membrane fusion are examined, in the context of lipid metabolism. A discussion of lipid transport to the cell surface from the major sites of lipid biosynthesis, the endoplasmic reticulum and Golgi apparatus, is beyond the scope of this chapter. In addition, lipid function involving organelles outside of primary vesicle-mediated transport pathways, such as peroxisomes and mitochondria, will not be discussed. For further information, the reader is referred to the following reviews (*6,21,22,28,43,53,80,88,104,114,125,129,137*).

II. USE OF LIPID ANALOGS IN STUDIES OF TRANSPORT AND METABOLISM

In contrast to proteins, relatively few studies have focused on lipid transport. Since it is difficult to monitor endogenous lipids, much of the work on lipid transport in cultured cells has relied on the use of labeled lipid analogs. One approach, developed by Pagano and co-workers (*75,80*), involves the use of lipid molecules where the naturally occurring acyl chain is replaced by a short-chain fluorescent fatty acid, *N*-[7-nitrobenz-2-oxa-1,3-diazol-4-yl] aminohexanoic acid (C_6-NBD-fatty acid).

Most lipids containing C_6-NBD-fatty acid are able to rapidly transfer between membranes, a property that has proven useful in studies of lipid transport (*68,75,80*). To label cultured cells at the plasma membrane with C_6-NBD-lipid, the cells are first cooled to 2 to 7 °C to inhibit vesicular transport and then incubated with liposomes containing a given short chain, fluorescent lipid (Fig. 1). C_6-NBD-

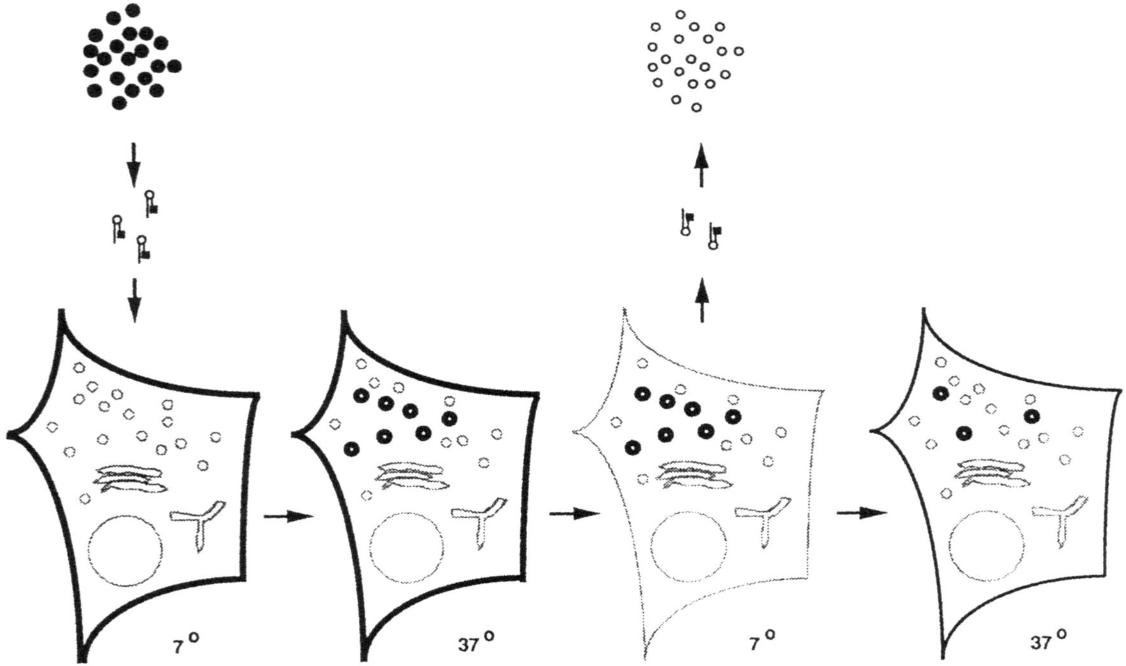

Figure 1. Schematic diagram of a pulse-chase experiment with C_6-NBD-lipids. In the first panel, cells are incubated at 7 °C with liposomes containing C_6-NBD-lipid (*filled circles, top*). This results in specific labeling of the plasma membrane as indicated by the heavy line. The liposomes are then washed away, and the cell is incubated for 37 °C for a given amount of time during which endocytosis occurs and intracellular membranes are labeled (second panel). The third panel depicts the back exchange process, where cells are incubated with unlabeled liposomes (*open circles, top*) at 7 °C to remove C_6-NBD-lipid from the plasma membrane. Further incubation of cells containing intracellular fluorescent lipid at 37 °C (fourth panel) allows visualization of C_6-NBD-lipid recycling to the cell surface.

lipid monomers transfer from the liposomes and integrate into the plasma membrane bilayer (*113*). Alternatively, C_6-NBD-lipid can be delivered to cells as a bovine serum albumin/lipid complex (*77*). Once the cells are labeled, the liposomes are then washed away and the cells are warmed to 37 °C to allow vesicular transport and lipid metabolism to resume. At the desired time, the cells are cooled again and any C_6-NBD-lipid present at the plasma membrane is removed by treatment of the cells at 2 to 7° C with medium containing unlabeled liposomes, a process called back exchange. By removing fluorescent lipid present at the plasma membrane, intracellular compartments labeled with C_6-NBD-lipid become more prominent when viewed by fluorescence microscopy (see Fig. 2).

Furthermore, the back exchange process provides a useful tool for measuring the amount of C_6-NBD-lipid at the cell surface, since the amount of fluorescent lipid removed by back exchange can be readily measured by fluorometry. This procedure should be used with some caution since some short-chain lipids, such as C_6-NBD-ceramide, can be extracted from intracellular membranes by back exchange (*78*). Metabolism of C_6-NBD-lipid to other lipid species can be determined by thin layer chromatographic (TLC) analysis of lipids extracted from labeled cells either before or after back exchange.

Long-chain lipids have also been incorporated into cells (*26,48,96*). Considerable care must be taken that the long-chain lipids are integrated into the membrane, since long-chain lipids are more subject to aggregation and nonspecific adsorption to the cell surface than are short-chain lipids. Also, note that once long-chain lipids are incorporated into a bilayer they are resistant to back exchange (*48*), and thus cannot be used for pulse-chase studies in the manner that short-chain lipids are used. As an example, baby hamster kidney (BHK) cells were labeled with an aqueous dispersion of *N*-(lissamine rhodamine B sulfonyl)-PE (N-Rh-PE) which contains the fluorophore in the polar head group of the molecule (*48*). Proper integration of N-Rh-PE into the plasma membrane bilayer was confirmed by fluorescence recovery after photobleaching (FRAP). Both C_6-NBD-PC and N-Rh-PE incorporated into the plasma membrane of BHK cells showed similar diffusion coefficients (0.27 vs. 0.23 $\mu m^2/s$) and mobile fractions (69% vs 70%).

In principle, one can also incorporate lipids into the plasma membrane by fusion with artificial liposomes. In practice, this is difficult to achieve (*64,73*). One approach has been to use influenza virus-infected Madin-Darby Canine Kidney (MDCK) cells (*45,122, 123*). Since these cells contain the hemagglutinin (HA) fusion protein on their cell surface, they are able to bind liposomes containing ganglioside G_{D1a}. Once the liposomes are bound, fusion is induced by treatment of the cells with an acidic buffer. This technique has been used to incorporate both fluorescent (*122,124*) and radiolabeled (*45*) lipids into the plasma membrane of MDCK cells. To date, this technique has not been used to study lipid transport between the plasma membrane and intracellular compartments.

III. VESICLE-MEDIATED MEMBRANE LIPID TRANSPORT

A. Endocytosis

C_6-NBD-PC (*48,102,105*), C_6-NBD-SM (51, 52, 57), and C_6-NBD-GlcCer (*27, 47*), which are generally restricted to the outer leaflet of the plasma membrane, have been used to study lipid endocytosis in different types of cultured cells. Typically, cells labeled with these C_6-NBD-lipids at 2 to 7 ° C and then subsequently incubated at 37 ° C accumulate intracellular vesicles containing endocytosed fluorescent lipid (see Fig. 3, pathway a). Labeling of intracellular vesicles is temperature-dependent and inhibited by treatment of cells with sodium azide and 2-deoxyglucose (*48,51,102,105*), confirming that these C_6-NBD-lipids are suitable probes for the endocytosis of plasma membrane lipid.

The intracellular distribution of endocytosed C_6-NBD-lipid depends upon the cell type examined (e.g., Fig. 2). In contrast, different C_6-NBD-lipid species that are restricted to the extracytoplasmic leaflet show similar patterns of labeling upon endocytosis in the same cell type. Most cell types examined show intracellular vesicles distributed throughout the cell periphery (*48,52,105*). In each of these cases, the intracellular compartment labeled by endocytosed C_6-NBD-lipid can be readily distinguished from the Golgi apparatus, by both spatial and morphological criteria. In contrast, the dominant feature labeled by endocytosed fluorescent lipid in Chinese hamster fibroblasts is a collection of intracellular vesicles adjacent to the nucleus that is often in the region of the Golgi apparatus (Fig. 2). This labeling pattern has been observed with both Chinese hamster lung (V-79) and ovary (CHO-K1) fibroblasts labeled with C_6-NBD-PC (*102,105*), C_6-NBD-SM (*51,57*) and C_6-NBD-PI (*121*). Note also that CHO cells labeled with rhodamine-conjugated transferrin show a similar pattern of intracellular fluorescence, suggesting that lipid and protein are being endocytosed by a similar mechanism in these cells (see below) (*135*).

Using CHO-K1 cells labeled with C_6-NBD-SM, these vesicles were found to be organized around the centriole (*51*). Consistent with centriole function as a microtubule organizing center, treatment of CHO-K1 with nocodazole caused vesicles containing endocytosed C_6-NBD-SM to be distributed throughout the periphery of the cell.

In some instances, this perinuclear accumulation of intracellular vesicles has been localized to the region of the Golgi apparatus. There is some confusion in the literature as to whether this reflects *bona fide* transport of endocytosed lipid to the Golgi apparatus (*47,48,94,102,105,129*). However, this does not appear to be the case for the following reasons. Many of the colocalization studies involved using fluorescent lectins, such as rhodamine-conjugated wheat germ agglutinin (Rh-WGA), as markers for the Golgi apparatus (*105,128*). However, since lectins bind to molecules on the basis of carbohydrate residues, these are relatively nonspecific reagents. Note that prior to permeabilization and treatment with Rh-WGA, the cell surface must be blocked with concanavilin A to avoid labeling the plasma mem-

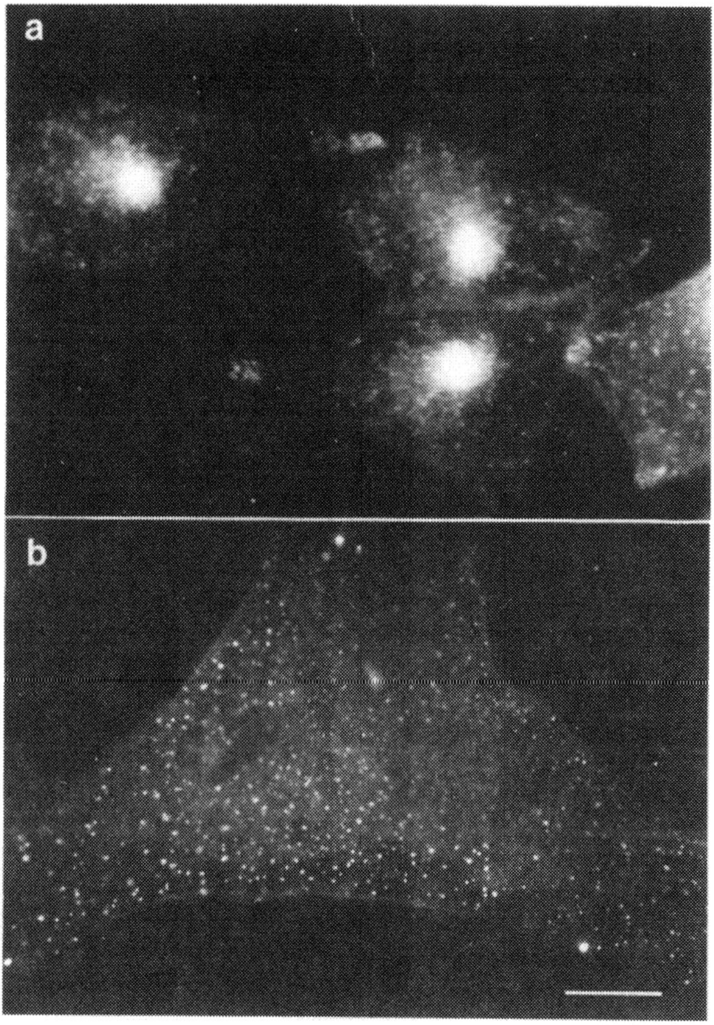

Figure 2. Cell specific differences in the appearance of endocytosed C_6-NBD-SM. Either CHO (**a**) or human skin (**b**) fibroblasts were labeled with C_6-NBD-SM for 30 min at 7 °C, followed by incubation at 37 °C for either 30 min (**a**) or 10 min (**b**), conditions that should label equivalent portions of the endocytic pathway, based on the kinetics of lipid recycling in these cells (*51,52*). The cells were treated with back exchange medium at 7 °C and then photographed using optics suitable for NBD fluorescence. These cell types represent two extremes in the distribution of endocytic vesicles throughout the cell (see text). Bar, 10 μM.

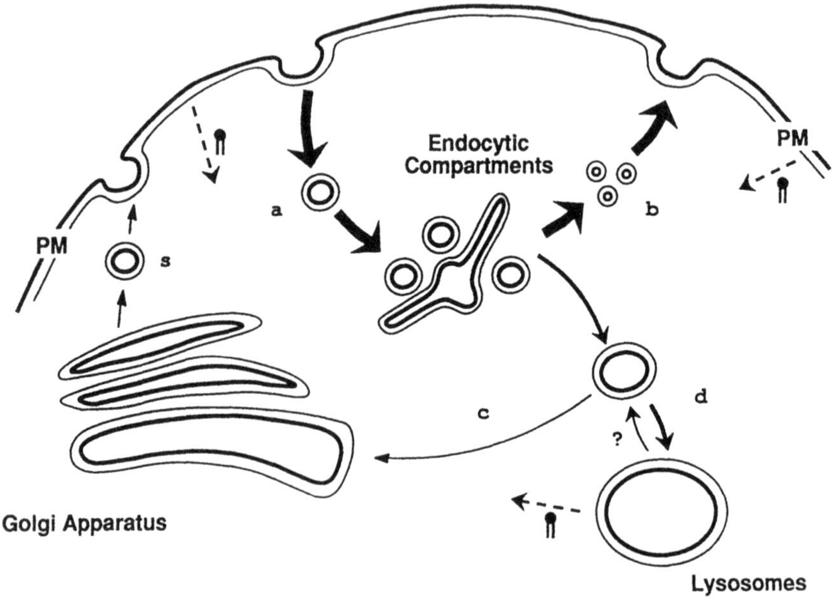

Figure 3. Plasma membrane lipid transport pathways. Shown is a simplified version of a plasma membrane lipid transport pathways, including endocytosis (**a**), direct recycling to the plasma membrane (**b**), recycling through the Golgi apparatus (**c**), transport along the degradative pathway (**d**) and secretory transport (**s**). Membrane bilayers are drawn as double lines with the extracytoplasmic leaflet indicated by the bold line and the cytoplasmic leaflet indicated by the narrow line. Also shown are alternate, nonvesicular lipid transport pathways (*dashed lines*). The relative width of each arrow in the diagram is meant to represent the relative amount of plasma membrane lipid transported along each pathway.

brane (*128*). Despite blocking of the cell surface, Rh-WGA is likely to label significant portions of the endocytic system in addition to the Golgi apparatus, since proteins that bind WGA (*34*) and other lectins (*72*) are endocytosed.

CHO-K1 cells labeled with C_6-NBD-SM show some hydrolysis of fluorescent SM to C_6-NBD-ceramide (*51*). This C_6-NBD-ceramide undergoes rapid transbilayer movement (Fig. 4a) and is subsequently transported to the Golgi apparatus by monomer transfer, a mechanism that is distinct from vesicular transport pathways (Fig. 3, dashed lines). Thus, the amount of fluorescent lipid accumulated at the Golgi apparatus in C_6-NBD-SM labeled cells is related to the amount of SM hydrolysis. Note that Niemann-Pick type A cells, which are largely deficient in SM hydrolysis, show little fluorescent lipid accumulation at the Golgi apparatus after labeling with C_6-NBD-SM (*52*). With extended incubation time at 37 ° C, significant amounts of fluorescent ceramide are produced by CHO-K1 cells, and labeling of the Golgi apparatus is clearly distinguishable from intracellular vesicles in the perinuclear region (*51*).

Finally, there is a mutant CHO cell line, CHO-15B, that endocytoses C_6-NBD-PC; however, the labeled intracellular vesicles do not accumulate in the perinuclear region of the cell and are clearly distinguishable from the Golgi apparatus (*105*). These points taken together suggest that the proximity of endocytosed C_6-NBD-lipid to the Golgi apparatus is not an indication of ***extensive*** overlap between the secretory and endocytic transport pathways for lipids. This does not preclude the possibility of small amounts of vesicular lipid traffic between the plasma membrane and the Golgi apparatus (see recycling, below).

Recently, fluorescent ceramide analogues have been localized to the *trans* Golgi apparatus using excitation of the acyl-chain fluorophore to oxidize diaminobenzidine followed by electron microscopy (EM) (*79,82*). Using this approach, in conjunction with labeling of cells with endocytosed C_6-NBD-lipid, may allow a more definitive morphological examination of lipid transport during endocytosis.

B. Recycling of Endocytosed Plasma Membrane Lipid

In CHO-K1 cells labeled with C_6-NBD-SM, most of the endocytosed fluorescent SM is recycled back to the plasma membrane (Fig. 3, pathway b) and is capable of multiple rounds of endocytosis and recycling (*51*). Disruption of microtubules by treatment of the cells with nocodazole did not inhibit C_6-NBD-SM endocytosis or recycling, despite affecting the intracellular distribution of endocytic vesicles containing fluorescent SM in CHO-K1 cells (see above). This is consistent with a role for microtubules in mediating vesicular transport of lipid and protein along the degradative pathway, rather than along early elements of the endocytic pathway (*11,36*).

Plasma membrane recycling has also been demonstrated using human skin fibroblasts labeled with C_6-NBD-SM (*52*), and BHK cells labeled with C_6-NBD-GlcCer (*47*). Both CHO-K1 cells and BHK cells simultaneously labeled with C_6-NBD-lipid and rhodamine-conjugated transferrin showed extensive, but not exact, colocalization of both probes. Also, the rate of C_6-NBD-SM recycling in CHO-K1 cells ($t_{1/2}$ ~ 15–20 min) and human skin fibroblasts ($t_{1/2}$ <5 min), are comparable to rates of protein receptor recycling measured in these two cell types (*14,135*). These points taken together suggest that lipid and protein recycle between the plasma membrane and intracellular compartments by similar vesicular transport pathways.

As described above, there appears to be little overlap between transport of endocytosed C_6-NBD-lipid and traffic through the Golgi apparatus. Consistent with this notion, endocytosis and recycling of C_6-NBD- SM is not inhibited by treatment of CHO-K1 cells with monensin, an agent that disrupts transport from the *cis* to *medial* cisterna of the Golgi apparatus in animal cells (*51,58,115*).

Studies on glycosphingolipid metabolism provide some evidence that small amounts of lipid transport occurs between the endocytic system and the Golgi apparatus (Fig. 3, pathway c). This was examined using normal human skin fibroblasts that were continuously incubated at 37 °C in medium containing a long-chain ganglioside, G_{M2}, which was tritiated in the sphingoid base portion of the molecule (*44,110*). After 4 days at 37 °C, the cells were harvested and treated with trypsin to remove any G_{M2} micelles that were nonspecifically adsorbed to the cell surface (*96,110*). During the 4 day incubation, 3.5% of the cell-associated ^{3}H-lipid was converted to G_{D1a}, that was formed by elongation of the G_{M2} head group most probably as a result of transport to the Golgi apparatus (*110*).

Nearly 40% of the exogenously added G_{M2} was hydrolyzed in lysosomes by normal cells (*110*). In order to confirm that G_{D1a} was formed from undegraded G_{M2}, fibroblasts isolated from patients with two types of G_{M2} gangliosidosis, Tay-Sachs disease and Sandhoff disease, were examined (*44,110*). In each case, less than 2% of the G_{M2} was hydrolyzed; however, the amount of G_{D1a} formed by gangliosidosis cells was comparable to the amount formed by unaffected cells. Higher order gangliosides are largely retained in lysosomes in the absence of extensive hydrolysis (*97,120*); thus, elongation of G_{M2} is likely to reflect the recycling of lipids from the plasma membrane through the Golgi apparatus. However, back flow of non-metabolized G_{M2} from lysosomes cannot be completely ruled out. In either case, this is probably a relatively minor lipid transport pathway, since the rate of G_{M3} conversion to G_{D1a} is 2 to 3 orders of magnitude slower than the rate of C_6-NBD-SM recycling in the same cell type (*52*). It seems likely that lipid transport from the cell surface might accompany transport of receptors, such as the mannose 6-phosphate receptor, to the Golgi apparatus (*50,109*; also see the chapter in this volume by Hoflack and Lobel).

C. Sorting of Lipid Between the Recycling and Degradative Pathways

As indicated above, exogenously added glycosphingolipids added to cultured cells show significant transport along the degradative pathway to lysosomes (Fig. 3, pathway d). Cells labeled with ^{3}H-G_{M3} that were homogenized and fractionated on Percoll gradients were found to accumulate tritiated ganglioside in a heavy density compartment, characteristic of lysosomes, within 30 to 60 min of incubation at 37 °C (*110*). Transport of N-Rh-PE from the plasma membrane to lysosomes has been demonstrated in BHK cells (*48*). Little, if any, recycling of N-Rh-PE was observed. Head group-labeled fluorescent PE was found to form clusters in the plasma membrane, which may prohibit lipid recycling to the plasma membrane, or may cause efficient targeting of N-Rh-PE to the lysosomes. Note that C_6-NBD-PC, which does not aggregate in the membrane, was localized to intracellular compartments that did not contain N-Rh-PE (*48*).

When C_6-NBD-SM transport was examined in Niemann-Pick type A human skin fibroblasts, which lack lysosomal SMase (A-SMase), the fluorescent lipid was found to label lysosomes as well as recycle between the plasma membrane and intracellular compartments (*52*). In contrast, normal human skin fibroblasts and CHO-K1 cells do not retain C_6-NBD-SM in lysosomes (*51,52*). The lack of lysosomal labeling in normal cells is the result of rapid A-SMase hydrolysis of C_6-NBD-SM to C_6-NBD-ceramide, which is subsequently transported to the Golgi apparatus by monomer transfer (Fig. 3, dashed lines). Since the rate-limiting step of C_6-NBD-SM hydrolysis by A-SMase is delivery to the lysosomes, the metabolism of C_6-NBD-SM by normal cells compared to that of Niemann-Pick cells can be used to measure the rate of lipid transport along the degradative pathway. Using this approach it was found that plasma membrane recycling of C_6-NBD-SM is 18- to 19-fold faster than the rate of fluorescent SM delivery to the lysosomes in human skin fibroblasts (*52*). Thus, most endocytosed C_6-NBD-SM is recycled to the plasma membrane with small amounts of fluorescent SM diverted to the degradative pathway. These observations are consistent with models for the sorting of endocytic traffic where the majority of endosome membrane is returned to the plasma membrane, while the lumenal contents are retained in intracellular vesicles that ultimately fuse with the lysosomal compartment (*24,56,91*; also see the chapter in this volume by Murphy).

IV. NONVESICULAR MOVEMENT OF PLASMA MEMBRANE LIPIDS

A. Transbilayer Movement and Plasma Membrane Lipid Metabolism

As described above, lipids are often restricted to a single membrane leaflet. However, this depends upon both the lipid species and membrane examined. For example, C_6-NBD-PC, which is normally unable to spontaneously transfer across bilayers, shows rapid transbilayer movement across the plasma membrane of a cell line derived from simian virus 40 (SV40) infected human lung fibroblasts (WI-38). Uninfected WI-38 cells retain C_6-NBD-PC at the extracytoplasmic leaflet of the plasma membrane, as do most other cells examined (*105*). The mechanism for C_6-NBD-PC "flip-flop" in SV40-infected cells is not known; however, there is evidence suggesting that plasma membrane of transformed cells may contain neutral lipid domains that could alter its physical properties (*67*). Whether plasma membrane transbilayer movement of C_6-NBD-PC occurs in other instances has not been determined.

In contrast to lipids containing a highly polar head group, cholesterol (*22,88,137*), ceramide (*78*), and DAG (*1,29,76*) are capable of spontaneous transbilayer movement in most, if not all, membrane bilayers. Furthermore, lipid metabolism that results in the removal of the polar head group can generate lipid species capable of spontaneous transbilayer movement (Fig. 4a). For instance, hydrolysis of C_6-NBD-SM to C_6-NBD-ceramide can occur at both the cell surface and lysosomes (see above, *51,52*). Chinese hamster (V-79) fibroblasts hydrolyze C_6-

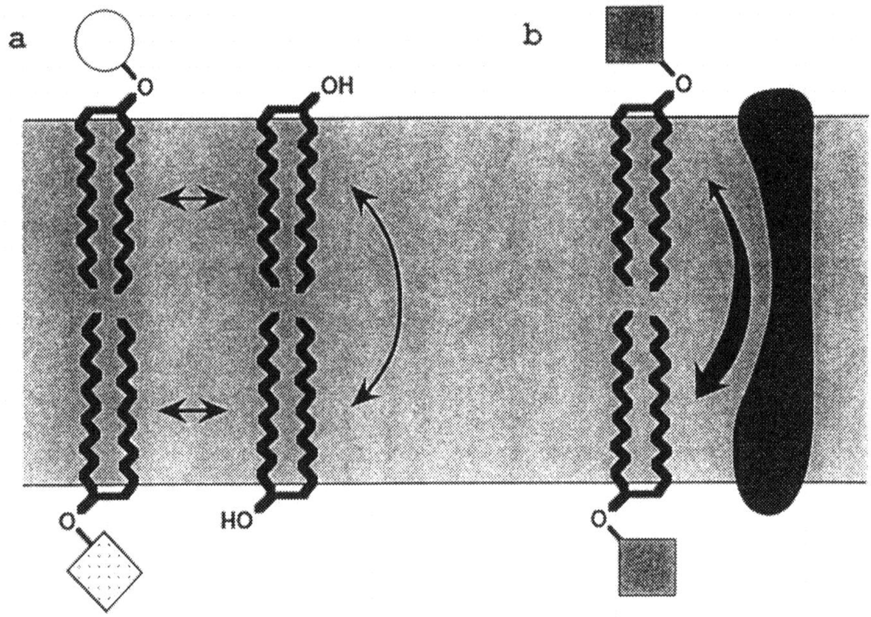

Figure 4. Lipid transbilayer movement at the cell surface. (**a**) Head group hydrolysis may result in the formation of either DAG or ceramide, which can then undergo rapid, spontaneous transbilayer movement. Either of these molecules can be further metabolized by addition of a head group that may subsequently interfere with transbilayer movement. (**b**) Aminophospholipids undergo protein-mediated transbilayer movement at the plasma membrane that leaves the lipid molecule intact. Note that protein facilitated lipid movement occurs preferentially from the extracytoplasmic leaflet to the cytoplasmic leaflet of the plasma membrane. In contrast, spontaneous transbilayer movement is likely to result in equal redistribution of lipid across the bilayer.

NBD-phosphatidic acid (C_6-NBD-PA) to C_6-NBD-DAG at the extracytoplasmic leaflet of the plasma membrane (*76*). Another example of lipid hydrolysis at the outer leaflet of the plasma membrane is a PI-specific phospholipase C, which was identified using Swiss 3T3 cells labeled with C_6-NBD-PI (116). C_6-NBD-DAG was further metabolized to other fluorescent lipids (e.g., C_6-NBD-PA, C_6-NBD-PC) following movement to the cytoplasmic leaflet (*74,116*), while further metabolism of C_6-NBD-ceramide occurs at the Golgi apparatus (*58*).

Transbilayer movement across the plasma membrane can also occur by protein-mediated transport of polar lipids. In particular, a specific plasma membrane aminophospholipid translocase activity has been identified using fluorescent and spin-labeled analogs of PE and PS (reviewed by *80,137*). Consistent with the localization of most PE and PS in the cytoplasmic leaflet of membranes, transbilayer movement of PE and PS occurs preferentially from the outer to the inner leaflet of the plasma membrane; however, transport in the opposite direction also occurs (Fig. 4b). Aminophospholipid translocase activity requires ATP and is inhibited by treatment with *N*-ethylmaleimide (NEM) (*18,46,63,138*). The translocase has not been unequivocally identified; however, several protein candidates have been found (*18,108,136*).

Although transbilayer movement of plasma membrane lipids has been observed, the role of these processes in membrane function remains to be determined. It is possible that directed transbilayer movement of lipids may be required to maintain the proper asymmetric distribution of lipids; that is, move lipid from the "wrong" leaflet to the "correct" leaflet (*95*). Another possibility, suggested by Zachowski and Devaux (*137*), is that transbilayer movement may be used to create regions of membrane that have unequal proportions of lipid in each leaflet. This, in turn, would enable membrane bending with a high radius of curvature and could potentially participate in processes such as vesicle formation. This seems plausible, since dramatic shape changes can be induced in red blood cells by manipulating the amount of lipid in either membrane leaflet (*19*). In addition, treatment of red blood cells with either exogenously added phospholipase C (1) or sphingomyelinase (2) can result in shape changes and membrane endovesiculation. Whether these phenomena reflect the involvement of lipid transbilayer movement in membrane vesicle formation during endocytosis remains to be determined. Also, the extent of lipid transbilayer movement in other endocytic compartments is not known at present.

B. Lipid Transport from the Cytoplasmic Leaflet

In each instance, cells labeled with fluorescent lipids capable of transbilayer movement show labeling of intracellular membranes (such as the endoplasmic reticulum and Golgi apparatus), suggesting that the lipid has been transported from the cytoplasmic leaflet of the plasma membrane to intracellular compartments (*63,76,79,103,116*). Labeled cells incubated at 37 ° C tend to show labeling of endocytic compartments in addition to other intracellular membranes (*103,105*). However, transport from the inner leaflet also occurs at temperatures below 10 °C, where vesicular transport is inhibited (*63,76*). In some cases, redistribution of short-chain lipids from the cytoplasmic leaflet may be due to spontaneous intermembrane transfer of short-chain lipids, such as C_6-NBD-PS and C_6-NBD-PE (*63,103*). In this light, it is worth noting that cells microinjected with C_6-NBD-PC show a similar distribution of fluorescent lipid (*75*).

In contrast to other C_6-NBD-lipids, C_6-NBD-DAG does not rapidly transfer between membranes (*69*). However, C_6-NBD-DAG is normally metabolized by cells into other C_6-NBD-lipids that can spontaneously transfer between membranes (*74*). To avoid this complication, Pagano and Longmuir (*76*) synthesized a C_6-NBD-PA analog containing the polar head group in the β position of the glycerol backbone, instead of the normal α position. V-79 cells hydrolyze C_6-NBD-β-PA to

C_6-NBD-β-DAG; however, the resulting fluorescent β-DAG could not be further metabolized to other fluorescent lipids. Cells treated with C_6-NBD-β-PA show C_6-NBD-β-DAG labeling of intracellular membranes at temperatures where endocytosis is inhibited, suggesting a nonvesicular-facilitated mechanism for lipid movement from the inner leaflet of the plasma membrane (*76*).

Further evidence in support of this type of facilitated movement has been provided by studies with a long-chain fluorescent PS analog, C_{12}-NBD-PS (*46*). Transfer of C_{12}-NBD-lipids between membranes is 150- to 200-fold slower than transfer of the corresponding C_6-NBD-lipids (*69*). Nonetheless, C_{12}-NBD-PS treated human skin fibroblasts and CHO-K1 cells show labeling of intracellular membranes, which is dependent upon the transbilayer movement of C_{12}-NBD-PS and not due to vesicle movement (*46*).

These studies suggest that there exist nonvesicular, facilitated lipid transport pathways between the cytoplasmic leaflet of the plasma membrane and intracellular membranes. Possible mechanisms that could account for this type of transport include protein-mediated monomer lipid transfer (*37,133,140*), or lateral lipid diffusion along contacts that may exist between the plasma membrane and other organelles (*7*). However, the molecular mechanism responsible for redistribution of lipids from the cytoplasmic leaflet of the plasma membrane has not been determined.

V. ROLE OF LIPIDS IN MEMBRANE FUSION

A somewhat "classical" view of traffic along the endocytic system is that molecular transport is accompanied by movement of discrete vesicles between a series of distinct intracellular compartments (*33,36,111*). Recent studies by Hopkins et al. (*39*) suggest that the endocytic compartment may be more like a tubular network rather than a vesiculated compartment. It seems likely that the endocytic system is a combination of both tubular and vesicular membrane elements, which are maintained by membrane fission and fusion (*30,31,117*). Membrane fusion must require lipid involvement at some level, since lipids form the majority of the membrane bilayer. At present, the role of lipids in membrane fusion is unclear; however, some inferences may be made from model systems that examine vesicular transport processes *in vitro*.

Cell-free reconstitution of both secretory and endocytic membrane transport has been used to identify a number of factors required for fusion of intracellular vesicles (*3,36*; also see the chapter in this volume by Braell). It is of particular interest that endomembrane fusion appears to occur by a common mechanism, and thus insights from one reconstitution system are often applicable to other systems. For example, an NEM-sensitive fusion protein (NSF) was originally isolated using a system for reconstitution of transport through the Golgi apparatus (*8*), and was subsequently found to support *in vitro* fusion of endosomes (*23*) and transport between the endoplasmic reticulum and Golgi apparatus (*4*). To date, reconstituted biological systems have not been used to assign definitive roles to membrane lipids in the fusion reaction.

Using a reconstituted fusion system, Rothman and co-workers have found that fatty acyl-Coenzyme A (acyl-CoA) is required for both budding and fusion of Golgi transport vesicles (*84,85*). The range of effective acyl-CoA chain lengths is broad, with palmitoyl-, heptadecanoyl- and steroyl-CoA showing highest activity (*32*). Since a nonhydrolyzable analog of acyl-CoA inhibits *in vitro* transport through Golgi stacks, acyl-CoA is likely to be required for acylation, rather than have direct involvement in altering membrane bilayer conformation (*84*). Proteins represent a likely substrate, particularly since acylated proteins, such as *rab* (*16,35*), ADP-ribosylation factor (ARF; *55,112*) and heterotrimeric G-proteins (*17*), appear to be involved in vesicular transport. However, lipid acylation has not been ruled out. This is particularly relevant in the context of a recent finding that NSF may have phospholipase A_2 activity, resulting in the formation of lyso-PC and unesterified fatty acids (*42*). Although it has been suggested that lyso-PC can act as a fusogenic agent (*61,62*), an extensive role for this lipid in biological fusion has not been found (*15*). Also, concentrations of lyso-PC as low as 1 μM inhibit NSF-dependent fusion *in vitro* (*32*). There is somewhat stronger evidence that *cis*-unsaturated fatty acids may promote membrane fusion (*38*); however, the concentration of either lyso-PC or fatty acid required for membrane fusion is significantly higher than amounts typically found in biological membranes (*15*). Further work will be required to determine whether phospholipase A_2 activity is involved in the membrane fusion reaction.

Most of what is known about lipid behavior during fusion has been determined through the use of model membrane systems. A detailed discussion of fusion between artificial liposomes is beyond the scope of this chapter (for reviews, see references *9,25,15,83,99*). For the purposes of this discussion, membrane fusion occurs when two bilayers are brought into close apposition and destabilized to form a nonbilayer intermediate structure, which is then converted into a single continuous bilayer. The actual steps and intermediates required for biological membrane fusion remain to be determined. A likely candidate for the nonbilayer intermediates are inverted micelles, which have been observed by EM as a "lipidic particles" found in both model and biological membranes (*127*). Lipids that easily undergo the liquid crystalline ($L_{\alpha'}$ bilayer) to inverted hexagonal (H_{II}) phase transition, such as PE, promote formation of inverted micelles as determined by EM and physical studies (*98*). The enrichment of PE in the cytoplasmic leaflet of most organelles is consistent with a possible role for phase transitions involving PE in membrane fusion (*40*).

The presence of DAG in membranes can alter the ability of PE to form inverted micelles (*100*). DAG concentrations as low as 2 mol% can decrease the temperature required for the L_α/H_{II} phase transition from 50 to 60 °C to the physiologically relevant range 34 to 44 °C for a number of different PE-containing model membrane systems (*100*). This was found to correlate with a similar decrease in the temperature required for rapid fusion of monomethylated-dioleoylPE liposomes in the presence of DAG. Consistent with these findings, phospholipase C can act as a catalytic agent to induce fusion of large unilammellar liposomes (*70*). Fusion only occurred between liposomes composed of PC, PE and cholesterol, although the molar ratio of each lipid could vary somewhat. Liposomes composed of 50 mol% PC, 25 mol% PE, and 25 mol% cholesterol were found to fuse at maximal rates when 6% of the total phospholipid was hydrolyzed to form DAG.

One function of DAG is to act as a second messenger to activate protein kinase C following hormone stimulation (for reviews, see *5,87,71*). There is increasing evidence that coupling between receptor binding and phospholipase C activity involves GTP-binding proteins of the *ras*-family (*130*). In particular, cells transformed with oncogenic forms of *ras* often show increased DAG production, resulting from the breakdown of either higher order phosphatidylinositides (*130,134*) or PC (*60,86*).

Another group of GTP-binding proteins related to *ras* are the *rab* proteins (*118,139*). One of these, *rab5*, has been found to be required for endosome fusion in a cell-free system (*35*). It seems likely that other related *rab* proteins are involved in other intracellular fusion reactions (*12,16*). Given that *rab* proteins are in the *ras* superfamily, and that DAG may be involved in membrane fusion, it is tempting to speculate that *rab* proteins may regulate phospholipase C activity as a part of the fusion reaction. In this case, binding between endosomes might activate *rab* in a manner corresponding to agonist activation of *ras*. Instead of acting as a second messenger, resultant DAG might then help destabilize the membrane as a means to initialize fusion. It is also conceivable that *rab* may activate other types of phospholipases (e.g., A_2). To date, however, the precise involvement of *rab* proteins in membrane fusion is not known. Also, potential regulation of lipid turnover during membrane fusion is likely to involve other regulatory GTPases as well, such as ARF (*55*) and G_α (*17*).

The nature of membrane fusion requires that lipids must be involved at some level. It seems plausible that protein-lipid interactions will be elucidated by reconstituting biological membrane fusion events *in vitro*. One possible role for proteins in this capacity is to metabolize lipids in order to change the local physical characteristics of membranes about to fuse. However, if only small amounts of lipid turnover are required for membrane fusion, this type of metabolism may be difficult to detect. Also, it is possible that proteins may have a more direct role in perturbing the membrane bilayer as a means of mediating fusion, as has been suggested for the viral fusion proteins (*9*).

VI. SUMMARY AND FUTURE DIRECTIONS

Labeled lipid analogs have been developed as reliable probes for lipid behavior in cells. Based on the work outlined in this chapter, lipids on the two leaflets of the plasma membrane are subject to distinct fates. Lipids on the extracytoplasmic leaflet appear to be internalized by vesicle-mediated transport, while lipids on the cytoplasmic leaflet are subject to movement by both vesicle-dependent and independent pathways. It seems likely that lipid transport is affected by other factors besides localization in the membrane bilayer, such as protein-lipid and lipid-lipid interactions. For example, C_6-NBD-ceramide and C_{12}-NBD-PS show much greater affinity for Golgi apparatus membranes than lipids such as C_6-NBD-DAG and C_6-NBD-PA (*46,81*).

By and large, endocytosed lipids in the outer leaflet appear to be transported along similar intracellular pathways as the recycling protein transferrin. Endocytosed lipid is not exclusively restricted to recycling, since small amounts of lipid are diverted to the degradative pathway and Golgi apparatus. The basic mechanism involved in sorting of lipid between these pathways remains largely unknown; however, lipid recycling appears to reflect the targeting of most endocytosed membrane-associated material back to the cell surface. Since different lipid species have been shown to label similar intracellular compartments and/or recycle, apparent differences in the transport of endocytosed lipid may be due, in part, to superficial differences in the organization of the endocytic system in various cell types rather than to fundamental differences in lipid transport. Perhaps recycling of plasma membrane lipid between the cell surface and intracellular compartments is, to some extent, a "default" pathway. In this model, endocytosed lipid follows the "bulk flow" of membrane throughout the plasma membrane recycling pathway, with small amounts of lipid diverted to accompany vesicular transport along the degradative pathway (*24,56,91*; also see the chapter in this volume by Murphy).

As is the case with most receptor-mediated endocytosis, transferrin is internalized largely by clathrin-coated pits (*33*). On the other hand, plasma membrane lipid is likely to be internalized by both clathrin-dependent and independent endocytosis (*119*). The extent of lipid internalized by each of these processes is not known at present. This could potentially be determined by using C_6-NBD-lipids in combination with selective inhibition of either clathrin-dependent or independent endocytosis (see *20,93*). Nonetheless, it seems likely that since these pathways intersect at an early point (*119*), lipids internalized by both mechanisms will be transported through similar compartments.

The cytoplasmic tail of protein receptors plays an important role in the sequestration of these molecules into coated pits (*33,41,54,59,65,66*; see also the chapter in this volume by Roth). In contrast, the receptors for Shiga toxin receptors are glycolipids that are also preferentially internalized by clathrin-coated pits, despite the fact that these glycolipids do not span the bilayer (*92,94*). The mechanism involved in this phenomena is unclear, although one possibility is that Shiga toxin binding causes cross-linking of the glycolipid receptors (*92*). Thus, regions of the plasma membrane containing high levels of protein receptors (e.g., coated pits) could more effectively "trap" these clusters of glycolipids. A prediction of this hypothesis is that other lipids that form clusters, such as N-Rh-PE (*48*), should also be preferentially localized to coated pits.

Little is known about the transport of membrane cholesterol from the cell surface (*88,129*). In human skin fibroblasts incubated with cholesterol containing liposomes, increases in plasma membrane cholesterol were detected within 6 hr at 37 °C, however, down regulation of cholesterol biosynthesis (which requires transport to the endoplasmic reticulum) was not observed until 24 hr of incubation (*89*). This suggests that cholesterol is largely retained at the cell surface and may be recycling along with other plasma membrane lipids. Sphingomyelin (SM) may play a role in retaining cholesterol in the plasma membrane recycling pool, since treatment of cells with exogenously added SMase causes rapid redistribution of cholesterol from the cell surface to the endoplasmic reticulum (*106, 107*). Although development of fluorescent cholesterol analogs that are suitable for membrane lipid transport studies is likely to be difficult, this may help further elucidate cholesterol transport pathways from the cell surface. Another possible approach may be the use of EM autoradiography to monitor the redistribution of radiolabeled sterols exogenously added to the plasma membrane.

Most of the work described in this chapter has examined lipid transport in nonpolarized cells in culture. In polarized cells, which contain discrete plasma membrane domains, transcytosis between the apical and basolateral domains is another form of vesicular transport (*90*). Transcytosis has been examined indirectly using cultured colon carcinoma (Caco2) cells labeled with C_6-NBD-ceramide, to produce C_6-NBD-SM and C_6-NBD-GlcCer which were subsequently transported to both the apical and basolateral surfaces (*126*). Using a complex back exchange regimen, these newly synthesized C_6-NBD-sphingolipids appeared to be transcytosed between the two plasma membrane domains with a half time of hours. However, to date, lipid transcytosis has not been demonstrated by examination of polarized cells directly labeled at a given plasma membrane domain.

Another aspect of polarized cells is that the apical and basolateral domains of the plasma membrane differ in lipid composition (*101*). In particular, the apical domain has a higher glycosphingolipid content than the basolateral surface. A major mechanism involved in the generation of a polarized distribution of lipid involves the sorting of newly synthesized lipid in the Golgi apparatus, followed by vectorial delivery to the appropriate plasma membrane domain (*101, 126*). Since each domain maintains a distinct plasma membrane recycling pathway (*10*), lipid composition is likely to be maintained by "default" plasma membrane lipid transport. It remains to be directly shown that multiple distinct plasma membrane lipid recycling pathways exist in polarized cells, although there is some indirect evidence in support of this notion (*126*).

Transcytosis and other specialized transport pathways may also play a role in preserving the lipid composition of plasma membrane domains. In support of possible specialized lipid transport pathways, it was found using undifferentiated HT29 human colon carcinoma cells that C_6-NBD-GlcCer can be specifically transported to the Golgi apparatus without prior hydrolysis to C_6-NBD-Cer. In contrast, C_6-NBD-SM, C_6-NBD-galactosylceramide, and C_6-NBD-lactosylceramide are predominantly transported to endocytic compartments containing co-internalized rhodamine-conjugated transferrin, that is the "default" plasma membrane recycling pathway (*49*). The mechanism of C_6-NBD-GlcCer transport to the Golgi apparatus in these cells was not elucidated and may involve unique endocytic pathways; however, nonvesicular transport of C_6-NBD-GlcCer has not been ruled out. Consistent with these observations, transport of intact GlcCer from the lysosomes to the Golgi apparatus has been demonstrated by examining the metabolism of exogenously added, radiolabeled lipids in rat liver (*120*). Interestingly, using a differentiated cell line clonally isolated from the original HT29 cells, C_6-NBD-GlcCer was no longer transported to the Golgi apparatus (*49*). Rather, in the differentiated cell line, endocytosed C_6-NBD-GlcCer labeled the same intracellular compartments as endocytosed C_6-NBD-SM. This raises the possibility that unique lipid transport pathways may be found in specialized cells and may play a role in cell differentiation. This broad area is just beginning to be addressed.

The studies mentioned in this chapter have provided a framework of possible plasma membrane lipid transport pathways in animal cells. The role of lipids in the control of membrane transport is somewhat less clear. The next step towards understanding how cells coordinate membrane traffic will require examination of lipid-protein and lipid-lipid interactions. In particular, these phenomena will need to be integrated with possible roles for the metabolic alteration of lipids as a means to "tailor" the biophysical properties of membranes.

ABBREVIATIONS

acyl-CoA:	fatty acyl coenzyme A
C_6-NBD-:	*N*-[7-nitrobenz-2-oxa-1,3-diazol-4-yl])aminohexanoyl-
DAG:	diacylglycerol
GlcCer:	glucosylceramide
NEM:	*N*-ethylmaleimide
N-Rh-PE:	*N*-(lissamine rhodamine B sulfonyl)-PE
NSF:	NEM-sensitive fusion protein
PA:	phosphatidic acid
PC:	phosphatidylcholine
PE:	phosphatidylethanolamine
PI:	phosphatidylinositol
PS:	phosphatidylserine
SM:	sphingomyelin
WGA:	wheat germ agglutinin

ACKNOWLEDGMENTS

I would like to thank Richard Pagano, Philip Stahl, and Donna Wilcox for their critical reading of the manuscript. I particularly thank James Lenhard for helpful discussions on the role of *rab* proteins in intracellular transport, and Tony Ting for insightful comments on the organization of this chapter. Michael Koval is an Amgen fellow of the Life Sciences Research Foundation.

REFERENCES

1. Allan D., Thomas, P., and Michell, R.H. (1978). Rapid transbilayer diffusion of 1,2-diacylglycerol and its relevance to control of membrane curvature. *Nature* 276: 289–290.
2. Allan, D. and Walklin, C.M. (1988). Endovesiculation of human erythrocytes exposed to sphingomyelinase C: a possible explanation for the enzyme-resistant pool of sphingomyelin. *Biochim. Biophys. Acta* 938: 403–410.
3. Balch, W.E. (1989). Biochemistry of interorganelle transport. *J. Biol. Chem.* 264: 16965–16968.
4. Beckers, C.J.M., Block, M.R., Glick, B.S., Rothman, J.E., and Balch, W.E. (1989). Vesicular transport between the endoplasmic reticulum and the Golgi stack requires the NEM-sensitive fusion protein. *Nature* 339: 397–398.
5. Berridge, M.J. and Levine, R.R. (1989). Inositol phosphates and cell signalling. *Nature* 341: 197–205.
6. Bishop, W.R. and Bell, R.M. (1988). Assembly of phospholipids into cellular membranes: biosynthesis, transmembrane movement and intracellular interaction. *Ann. Rev. Cell. Biol.* 4: 579–610.
7. Blanchette-Mackie, E.J. and Scow, R.O. (1982). Continuity of intracellular channels with extracellular space in adipose tissue and liver: demonstrated with tannic acid and lanthanum. *Anat Rec.* 203: 205–219.
8. Block, M.R., Glick, B.S., Wilcox, C.A., Wieland, F.T., and Rothman, J.E. (1988). Purification of an N-ethylmaleimide-sensitive protein catalyzing vesicular transport. *Proc. Natl. Acad. Sci. USA* 85: 7852–7856.
9. Blumenthal, R. (1985). Membrane fusion. Curr. *Topics Membr. Transport* 29: 203–254.
10. Bomsel, M., Prydz, K., Parton, R.G., Gruenberg, J., and Simons, K. (1989). Endocytosis in filter-grown Madin-Darby canine kidney cells. *J. Cell Biol.* 109: 3243–3258.
11. Bomsel, M., Parton, M., Kuznetsov, S.A., Schroer, T.A., and Gruenberg, J. (1990). Microtubule- and motor-dependent fusion *in vitro* between apical and basolateral endocytic vesicles from MDCK cells. *Cell* 62: 719–731.
12. Bourne, H.R. (1988). Do GTPases direct membrane traffic in secretion? Cell 53: 669–671.
13. Brotherus, J. and Renkonen, O. (1977). Subcellular distributions of lipids in cultured BHK cells: evidence for the enrichment of lysobisphosphatidic acid and neutral lipids in lysosomes. *J. Lipid Res.* 18: 191–202.
14. Brown, M.S. and J.L. Goldstein (1979). Receptor-mediated endocytosis: Insights from the lipoprotein receptor system. *Proc. Natl. Acad. Sci. USA* 76: 3330–3337.
15. Burger, K.N.J. and Verkleij, A.J. (1990). Membrane fusion. *Experentia* 46: 631–643.
16. Chevrier, P., Parton, R.G., Hauri, H.P., Simons, K., and Zerial, M. (1990). Localization of low molecular weight GTP binding proteins to exocytic and endocytic compartments. *Cell* 62: 317–329.
17. Colombo, M.I., Mayorga, L.S., Casey, P.J., and Stahl, P.D. (1992). Evidence of a role for heterotrimeric GTP-binding proteins in endosome fusion. *Science* 255: 1695–1697.
18. Connor, J. and Schroit, A.J. (1989). Transbilayer movement of phosphatidylserine in the human erythrocyte: evidence that the aminophospholipid transporter is a ubiquitous membrane protein. *Biochemistry* 28: 9680–9685
19. Daleke, D.L. and Heustis, W.H. (1989). Erythrocyte morphology reflects the transbilayer distribution of incorporated phospholipids. *J. Cell Biol.* 108: 1375–1385.
20. Davoust, J., Gruenberg, J., and Howell, K.E. (1987). Two threshold values of low pH block endocytosis at different stages. *EMBO J.* 6: 3601–3609.
21. Dawidowicz, E.A. (1987). Dynamics of membrane lipid metabolism and turnover. *Ann. Rev. Biochem.* 56: 43–61.
22. Dawidowicz, E.A. (1987). Lipid exchange: Transmembrane movement, spontaneous movement and protein-mediated transfer of lipids and cholesterol. *Current Topics in Membrane Transport* 29: 175–202.

23. Diaz, R., Mayorga, L.S., Weidman, P.J., Rothman, J.E., and Stahl, P.D. (1989). Vesicle fusion following receptor-mediated endocytosis requires a protein active in Golgi transport. *Nature* 339: 398–400.
24. Dunn, K.W., McGraw, T.E., and Maxfield, F.R. (1989). Iterative fractionation of recycling receptors from lysosomally destined ligands in an early sorting endosome. *J. Cell Biol.* 109: 3303–3314.
25. Duzgunes, N. (1985). Membrane fusion. In: *Subcellular Biochemistry*, Vol. 11, pp. 195–286 (Roodlyn, D.B., Ed.). Plenum Press, New York.
26. Edidin, M. (1981). Molecular motions and membrane organization and function. In: *Membrane Structure*, pp. 37–82 (Finean, J.B. and Michel, R.H., Eds.). Elsevier Biomedical Press, Amsterdam.
27. Eskelinen, S., Kok, J.W., Sormunen, R., and Hoekstra, D. (1991). Coated endosomal vesicles: sorting and recycling compartment for transferrin in BHK cells. *Eur. J. Cell Biol.* 56: 210–222.
28. Esko, J.D. and C.R.H. Raetz (1983). Synthesis of phospholipids in animal cells. In: *The Enzymes*, Vol. XVI (Boymer, P.D., Ed.). Academic Press, New York.
29. Ganong B.R. and Bell, R.M. (1984). Transmembrane movement of phosphatidylglycerol and diacylglycerol sulfhydryl analogues. *Biochemistry* 23: 4977–4983.
30. Geuze, H.J., Slot, J. W., Strous, G. J. A. M., Lodish, H. F., and Schwartz, A.L. (1983). Intracellular site of asialoglycoprotein receptor-ligand uncoupling: Double-label immunoelectron microscopy during receptor-mediated endocytosis. *Cell* 32: 277–287.
31. Geuze, H.J., Slot, J.W., and Schwartz, A.L. (1987). Membranes of sorting organelles display lateral heterogeneity in receptor distribution. *J. Cell Biol.* 104: 1715–1723.
32. Glick, B.S. and Rothman, J.E. (1987). Possible role for fatty acyl-coenzyme A in intracellular protein transport. *Nature* 326: 309–312.
33. Goldstein, J.L., Brown, M.S., Anderson, R.G.W., Russell, D.W., and Schneider, W.J. (1985). Receptor-mediated endocytosis: concepts emerging from the LDL-receptor system. *Ann. Rev. Cell Biol.* 1: 1–39.
34. Gonatas, N.K., Stieber, A., Hickey, W.F., Herbert, S.H., and Gonatas, J.O. (1984). Endosomes and Golgi vesicles in adsorptive and fluid phase endocytosis. *J. Cell Biol.* 99: 1379–1390.
35. Gorvel, J.-P., Chavrier, P., Zerial, M., Gruenberg, J. (1991). *rab5* controls early endosome fusion *in vitro*. *Cell* 64: 915–925.
36. Gruenberg, J. and Howell, K.E. (1989). Membrane traffic in endocytosis: insights from cell-free systems. *Ann. Rev. Cell Biol.* 5: 453–481.
37. Helmkamp, G.M., Jr. (1986). Phospholipid transfer proteins: mechanism of action. *J. Bioenerg. Biomembr.* 18: 71–91.
38. Hope, M.J. and Cullis, P.R. (1981). The role of nonbilayer lipid structures in the fusion of human erythrocytes induced by lipid fusogens. *Biochim. Biophys. Acta* 640: 82–90.
39. Hopkins, C.R., Gibson, A., Shipman, M., and Miller, K. (1990). Movement of internalized ligand-receptor complexes along a continuous endosomal reticulum. *Nature* 346: 335–339.
40. Houslay, M.D. and Stanley, K.K. (1982). *Dynamics of Biological Membranes*. John Wiley & Sons, New York.
41. Jing, S., Spencer, T., Miller, K., Hopkins, C., and Trowbridge, I.S. (1990). Role of the human transferrin receptor cytoplasmic domain in endocytosis: localization of a specific signal sequence for internalization. *J. Cell Biol.* 110: 283–294.
42. Kasinathan, C., Yamaki, K., Grzelinska, E., Palecz, D., Slomiany, B., and Slomiany, A. (1991). Fusion of transport vesicles is modulated by NSF's phospholipase A_2 activity. *FASEB J.* 5: A1190 (abstract).
43. Kishimoto, Y. (1983). Sphingolipid formation. In: *The Enzymes*, Vol. XVI, pp. 358–409 (Boyer, P.D., Ed.). Academic Press, New York.
44. Klein, D., Leinekugel, P., Pohlentz, G., Schwarzmann, G., and Sandhoff, K. (1988). Metabolism and intracellular transport of gangliosides in cultured fibroblasts. In: *New Trends in Ganglioside*

Research: Neurochemical and Neuroregenerative Aspects, pp. 247–258 (Ledeen, R.W., Hogan, E.L., Tettamanti, G., Yates, A.J., and Yu, R.K., Eds.). Liviana Press, Padova.

45. Knoll, G., Burger, K.N.J., Bron, R., van Meer, G., and Verkleij, A.J. (1988). Fusion of liposomes with the plasma membrane of epithelial cells: fate of incorporated lipids followed by freeze fracture and autoradiography of plastic sections. *J. Cell Biol.* 107: 2511–2521.
46. Kobayashi, T. and Arakawa, Y. (1991). Transport of exogenous fluorescent phosphatidylserine analogus to the Golgi apparatus in cultured fibroblasts. *J. Cell Biol.* 113: 235–244.
47. Kok, J.W., Eskelinen, S., Hoekstra, K., and Hoekstra, D. (1989). Salvage of glucosylceramide by recycling after internalization along the pathway of receptor-mediated endocytosis. *Proc. Natl. Acad. Sci. USA* 86: 9896–9900.
48. Kok, J.W., ter Beest, M., Scherphof, G., and Hoekstra, D. (1990). A nonexchangeable fluorescent phospholipid analog as a membrane traffic marker of the endocytic pathway. Eur. *J. Cell Biol.* 53: 173–184.
49. Kok, J.W., Babia, T., and Hoekstra, D. (1991). Sorting of sphingolipids in the endocytic pathway of HT29 cells. *J. Cell Biol.* 114: 231–239.
50. Kornfeld, S. and Mellman, I. (1989). The biogenesis of lysosomes. *Ann. Rev. Cell Biol.* 5: 483–525.
51. Koval, M. and Pagano, R.E. (1989). Lipid recycling between the plasma membrane and intracellular compartments: transport and metabolism of fluorescent sphingomyelin analogues in cultured fibroblasts. *J. Cell Biol.* 108: 2169–2181.
52. Koval, M. and Pagano, R.E. (1990). Sorting of an internalized plasma membrane between recycling and degradative pathways in normal and Niemann-Pick, type A fibroblasts. *J. Cell Biol.* 111: 429–442.
53. Koval, M. and Pagano, R.E. (1991). Intracellular transport and metabolism of sphingomyelin. *Biochim. Biophys. Acta* 1082: 113–125.
54. Ktistakis, N.T., Thomas, D., and Roth, M.G. (1990). Characteristics of the tyrosine recognition signal for internalization of transmembrane surface glycoproteins. *J. Cell Biol.* 111: 1393–1407.
55. Lenhard, J.M., Kahn, R.A., and Stahl, P.D. (1992). Evidence for ADP-ribosylation factor (ARF) as a regulator of in vitro endosome-endosome fusion. *J. Biol. Chem.* (In press).
56. Linderman, J.L. and Lauffenberger, D.A. (1988). Analysis of intracellular receptor/ligand sorting in endosomes. *J. Theor. Biol.* 132: 203–245.
57. Lipsky, N.G. and Pagano, R.E. (1984). Fluorescent sphingomyelin labels the plasma membrane of cultured fibroblasts. *Ann. N. Y. Acad. Sci.* 435: 306–308.
58. Lipsky, N.G. and Pagano, R.E. (1985). Intracellular translocation of fluorescent sphingolipids in cultured fibroblasts: Endogenously synthesized sphingomyelin and glucosylcerebroside analogues pass through the Golgi apparatus en route to the plasma membrane. *J. Cell Biol.* 100: 27–34.
59. Lobel, P., Fujimoto, K., Ye, R.D., Griffiths, G., and Kornfeld, S. (1989). Mutations in the cytoplasmic domain of the 275-kDa mannose 6-phosphate receptor differentially alter lysosomal enzyme sorting and endocytosis. *Cell* 57: 787–796.
60. Lopez-Barahona, M., Kaplan, P.L., Cornet, M.E., Diaz-Meco, M.T., Larrodera, P., Diaz-Laviada, I., Municio, A.M., and Moscat, J. (1990). Kinetic evidence of a rapid activation of phosphatidylcholine hydrolysis by Ki-ras oncogene. *J. Biol. Chem.* 265: 9022–9026.
61. Lucy, J.A. (1970). The fusion of biological membranes. *Nature* 227: 814–817.
62. Lucy, J.A. and Ahkong, Q.F. (1987). Osmotic forces and the fusion of biomembranes. In: *Molecular Mechanisms of Membrane Fusion*, pp. 163–180 (Ohki, S., Doyle, D., Flanagan, T.D., Hui, S.W., and Mayhew, E., Eds.). Plenum Press, New York.
63. Martin, O.C. and Pagano, R.E. (1987). Transbilayer movement of fluorescent analogs of phosphatidylserine and phosphatidylethanolamine at the plasma membrane of cultured cells. *J. Biol. Chem.* 262: 5890–5898.

64. Margolis, L.B. (1988). Cell interactions with solid and fluid liposomes *in vitro:* lessons for 'liposomologists' and cell biologists. In: *Liposomes as Drug Carriers, Recent Trends and Progress*, pp. 75–92 (Gregoriadis, G., Ed.). John Wiley & Sons, Chichester.
65. Miettinen, H.M., Rose, J.K., and Mellman, I. (1989). Fc receptor isoforms exhibit distinct abilities for coated pit localization as a result of cytoplasmic domain heterogeneity. *Cell* 58: 317–327.
66. Mostov, K.E., de Bruyn Kops, A., and Deitcher, D.L. (1986). Deletion of the cytoplasmic domain of the polymericimmunoglobulin receptor prevents basolateral localization and endocytosis. *Cell* 47: 359–364.
67. Mountford, C.E. and Wright, L.C. (1988). Organization of lipids in the plasma membrane of malignant and stimulated cells: a new model. *Trends Biochem. Sci.* 13: 172–177.
68. Nichols, J.W. and Pagano, R.E. (1981). Kinetics of soluble lipid monomer diffusion between vesicles. *Biochemistry* 20: 2783–2789.
69. Nichols, J.W. and Pagano, R.E. (1982). Use of resonance energy transfer to study the kinetics of amphiphile transfer between vesicles. *Biochemistry* 21: 1720–1726.
70. Nieva, J.-L., Goni, F.M., and Alonso, A. (1989). Liposome fusion catalytically induced by phospholipase C. *Biochemistry* 28: 7364–7367.
71. Nishizuka, Y. (1988). Molecular heterogeneity of protein kinase C and its implications for cellular regulation. *Nature* 334: 661–665.
72. Olsnes, S. and Sandvig, K. (1985). Entry of polypeptide toxins into animal cells. In: *Endocytosis*, pp. 195–234 (Pastan, I. and Willingham, M.C., Eds.). Plenum Press, New York.
73. Pagano, R.E. and Weinstein, J.N. (1978). Interactions of liposomes with mammalian cells. *Annu. Rev. Biophys. Bioeng.* 7: 435–468.
74. Pagano, R.E., Longmuir, K.J., and Martin, O.C. (1983). Intracellular transport and metabolism of a fluorescent phosphatidic acid analogue in cultured fibroblasts. *J. Biol. Chem.* 258: 2034–2040.
75. Pagano, R.E. and Sleight, R.G. (1985). Defining lipid transport pathways in animal cells. *Science* (Wash. DC). 229: 1051–1057.
76. Pagano, R.E. and Longmuir, K.J. (1985). Phosphorylation, transbilayer movement, and facilitated intracellular transport of diacylglycerol are involved in the uptake of a fluorescent analog of phosphatidic acid by cultured fibroblasts. *J. Biol. Chem.* 260: 1909–1916.
77. Pagano, R.E. and Martin, O.C. (1988). A series of fluorescent N-(acyl)-sphingosines: synthesis, physical properties, and studies in cultured cells. *Biochemistry* 27: 4439–4445.
78. Pagano, R.E. (1989). A fluorescent derivative of ceramide: Physical properties and use in studying the Golgi apparatus of animal cells. *Meth. Cell Biol.* 29(Pt A): 75–85.
79. Pagano, R.E., Sepanski, M.A., and Martin, O.C. (1989). Molecular trapping of a fluorescent ceramide analogue at the Golgi apparatus of fixed cells: Interaction with endogenous lipids provides a *trans*-Golgi marker for both light and electron microscopy. *J. Cell Biol.* 109: 2067–2079.
80. Pagano, R.E. (1990). Lipid traffic in eukaryotic cells: mechanisms for intracellular transport and organellar enrichment of lipids. *Curr. Opin. Cell Biol.* 2: 652–663.
81. Pagano, R.E. (1990). The Golgi apparatus: insights from lipid biochemistry. *Biochem. Soc. Trans.* 18: 361–366.
82. Pagano, R.E., Martin, O.C., Kang, H.C., and Haugland, R.P. (1991). A novel ceramide analogue for studying membrane traffic in animal cells: accumulation at the Golgi apparatus results in altered spectral properties of the sphingolipid precursor. *J. Cell Biol.* 113: 1267–1279.
83. Papahadjopoulos, D., Nir, S., and Duzgunes, N. (1990). Molecular mechanisms of calcium-induced membrane fusion. *J. Bioenerg. Biomembr.* 22: 157–179.
84. Pfanner, N., Orci, L., Glick, B.S., Amherdt, M., Arden, S.R., Malhotra, V., and Rothman, J.E. (1989). Fatty acyl-coenzyme A is required for budding of transporty vesicles from Golgi cisternae. *Cell* 59: 95–102.

85. Pfanner, N., Glick, B.S., Alden, S.R., and Rothman, J.E. (1990). Fatty acylation promotes fusion of transport vesicles with Golgi cisternae. *J. Cell Biol.* 110: 955–961.
86. Price, B.D., Morris, J.D.H., Marshall, C.J., and Hall, A. (1989). Stimulation of phosphatidylcholine hydrolysis, diacylglycerol release, and arachidonic acid production by oncogenic *ras* is a consequence of protein kinase C activation. *J. Biol. Chem.* 264: 16638–16643.
87. Rana, R.S. and Hokin, L.H. (1990). Role of phosphoinositides in transmembrane signalling. *Physiol. Rev.* 70: 115–168.
88. Reinhart, M.P. (1990). Intracellular sterol trafficking. *Experentia* 46: 599–610.
89. Robertson, D.L. and Poznansky, M.J. (1985). The effect of nonreceptor-mediated uptake of cholesterol on intracellular cholesterol metabolism in human skin fibroblasts. *Biochem. J.* 232: 553–557.
90. Rodriguez-Boulan, E. and Nelson, W.J. (1989). Morphogenesis of the polarized epithelial cell phenotype. *Science* 245: 718–725.
91. Rome, H.L. (1985). Curling receptors. *Trends Biochem. Sci.* 10: 151.
92. Sandvig, K., Olsnes, S., Brown, J.E., Petersen, O.W., and van, B. Deurs (1989). Endocytosis from coated pits of Shiga toxin: a glycolipid-binding protein from Shigella dysenteriae 1. *J. Cell Biol.* 108: 1331–1343.
93. Sandvig, K. and van Deurs, B. (1990). Selective modulation of the endocytic uptake of ricin and fluid phase markers without alteration in transferrin endocytosis. *J. Biol. Chem.* 265: 6382–6388.
94. Sandvig, K., Prydz, K., Ryd, M., and van, B. Deurs (1991). Endocytosis and intracellular transport of the glycolipid-binding ligand Shiga toxin in polarized MDCK cells. *J. Cell Biol.* 113: 553–562.
95. Schroit, A.J., Madsen, J.W., and Tanaka, Y. (1985). *In vivo* recognition and clearance of red blood cells containing phosphatidylserine in their plasma membranes. *J. Biol. Chem.* 260: 5131–5138.
96. Schwarzmann, G., Hoffmann-Bleihauer, P., Schubert, J., Sandhoff, K., and Marsh, D. (1983). Incorporation of ganglioside analogues into fibroblast membranes. A spin-label study. *Biochemistry* 22: 5041–5048.
97. Schwarzmann, G., Hinrichs, Sonderfeld, S., Marsh, D., and K. Sandhoff (1986). Metabolism of exogenous gangliosides in cultured fibroblasts and cerebellar cells. In: *Enzymes of Lipid Metabolism II*, pp. 553–562. (Freysz, L., Dreyfus, H., Massarelli, R., and Gatt, S. Eds.). Plenum Press, New York.
98. Seddon, J.M. (1990). Structure of the inverted hexagonal (H_{II}) phase, and non-lamellar phase transitions of lipids. *Biochim. Biophys. Acta* 1031: 1–69.
99. Siegel, D.P., Ellens, H., and Bentz, J. (1987). Membrane fusion via intermediates in L_{α}/H_{II} phase transitions. In: *Molecular Mechanisms of Membrane Fusion*, pp. 53–72 (Ohki, S., Doyle, D., Flanagan, T.D., Hui, S.W., and Mayhew, E., Eds.). Plenum Press, New York.
100. Siegel, D.P., Banschbach, J., Alford, D., Ellens, H., Lis, L.J., Quinn, P.J., Yeagle, P.L., and Bentz, J. (1989). Physiological levels of diacylglycerols in phospholipid membranes induce membrane fusion and stabilize inverted phases. *Biochemistry* 28: 3703–3709.
101. Simons, K. and van Meer, G. (1988). Lipid sorting in epithelial cells. *Biochemistry* 27: 6197–6202.
102. Sleight, R.G. and Pagano, R.E. (1984). Transport of a fluorescent phosphatidylcholine analog from the plasma membrane to the Golgi apparatus. *J. Cell Biol.* 99: 742–751.
103. Sleight, R.G. and Pagano, R.E. (1985). Transbilayer movement of a fluorescent phosphatidylethanolamine analog across the plasma membranes of cultured mammalian cells. *J. Biol. Chem.* 260: 1146–1154.
104. Sleight, R.G. (1987). Intracellular lipid transport in eukaryotes. *Ann. Rev. Physiol.* 49: 193–208.
105. Sleight, R.G. and Abanto, M.N. (1989). Differences in intracellular transport of a fluorescent phosphatidylcholine analog in established cell lines. *J. Cell Sci.* 63: 363–374.
106. Slotte, J.P. and Bierman, E.L. (1988). Depletion of plasma-membrane sphingomyelin rapidly alters the distribution of cholesterol between plasma membranes and intracellular cholesterol pools in cultured fibroblasts. *Biochem. J.* 250: 653–658.

107. Slotte, J.P., Hedstrom, G., Rannstrom, S., and Ekman, S. (1989). Effects of sphingomyelin degradation on cell cholesterol oxidizability and steady-state distribution between the cell surface and the cell interior. *Biochim. Biophys. Acta* 985: 90–96.
108. Smith, R.E. and Daleke, D.L. (1990). Phosphatidylserine transport in Rh_{null} erythrocytes. *Blood* 76: 1021–1027.
109. Snyder, M.D. (1989). Remodeling of glycoprotein oligosaccharide after endocytosis: a measure of transport into compartments of the Golgi apparatus. *Meth. Cell Biol.* 32: 339–346.
110. Sonderfeld, S., Conzelman, E., Schwarzmann, G., Burg, J., Hinrichs, U., and Sandhoff, K. (1985). Incorporation and metabolism of ganglioside G_{M2} in skin fibroblasts from normal and G_{M2} gangliosidosis subjects. *Eur. J. Biochem.* 149: 247–255.
112. Stearns, T., Willingham, M.C., Botstein, D., and Kahn, R.A. (1990). ADP-ribosylation factor is functionally and physically associated with the Golgi complex. *Proc. Natl. Acad. Sci. USA* 87: 1238–1242.
111. Steinman, R.M., Mellman, I.S., Muller, W.A., and Cohn, Z.A. (1983). Endocytosis and the recycling of plasma membrane. *J. Cell Biol.* 96: 1–27.
113. Struck, D.K. and Pagano, R.E. (1980). Insertion of fluorescent phospholipids into the plasma membrane of a mammalian cell. *J. Biol. Chem.* 255: 5404–5410.
114. Sweeley, C.C. (1985). Sphingolipids. In: *Biochemistry of Lipids and Membranes*, pp. 361–403. (Vance, D.E. and Vance, J.E., Eds.). Benjamin Cummings Publishing, Menlo Park, CA.
115. Tartakoff, A.M.(1983). Perturbation of vesicular traffic with the carboxylic ionophore monensin. *Cell* 32: 1026–1028.
116. Ting, A.E. and Pagano, R.E. (1990). Detection of a phosphatidylinositol-specific phospholipase C at the surface of Swiss 3T3 cells and its potential role in the regulation of cell growth. *J. Biol. Chem.* 265: 5337–5340.
117. Tooze, J. and Hollinshead, M. (1991). Tubular early endosomal networks in AtT20 and other cells. *J. Cell Biol.* 115: 635–653.
118. Touchot, N., Chardin, P., and Tavitian, A. (1987). Four additional members of the *ras* gene superfamily isolated by an oligonucleotide strategy: molecular cloning of YPT-related cDNAs from a rat brain library. *Proc. Natl. Acad. Sci. USA* 84: 8210–8214.
119. Tran, D., Carpentier, J.L., Sawano, F., Gordon, P., and Orci, L. (1987). Ligands internalized through coated or noncoated invaginations follow a common intracellular pathway. *Proc. Natl. Acad. Sci. USA* 84: 7957–7961.
120. Trinchera, M., Ghidoni, R., Sonnino, S., and Tettamanti, G. (1990). Recycling of glucosylceramide and sphingosine for the biosynthesis of gangliosides and sphingomyelin in rat-liver. *Biochem. J.* 270: 451–457.
121. Uster, P.S. and Pagano, R.E. (1986). Synthesis and properties of fluorescent analogs of cytidine diphosphate-diacylglycerol and phosphatidylinositol. In: *Enzymes of Lipid Metabolism II*, p. 493–500. (Freyz, L., Dreyfus, H., Massarelli, R., and Gatt, S., Eds.). Plenum Press, New York.
122. van Meer, G. and Simons, K. (1983). An efficient method for introducing defined lipids into the plasma membrane of cultured cells. *J. Cell Biol.* 97: 1365–1374.
123. van Meer, G., Davoust, J., and Simons, K. (1985). Parameters affecting low pH-mediated fusion of liposomes with the plasma membrane of cells infected with influenza virus. *Biochemistry* 24: 3593–3602.
124. van Meer, G., Gumbiner, B., and Simons, K. (1986). The tight junction does not allow lipid molecules to diffuse from one epithelial cell to the next. *Nature* 322: 639–641.
125. van Meer, G. (1989). Lipid traffic in animal cells. *Ann. Rev. Cell Biol.* 5: 247–276.
126. van't Hoff, W. and van Meer, G. (1990). Generation of lipid polarity in intestinal epithelial (Caco-2) cells: sphingolipid synthesis in the Golgi complex and sorting before vesicular traffic to the plasma membrane. *J. Cell Biol.* 111: 977–986.
127. Verkleij, A.J., van Ectheld, C.J.A., Gerritsen, W.J., Cullis, P.R., and de Kruijff, B. (1980). The lipidic particle as an intermediate structure in membrane fusion processes and bilayer to hexagonal HII transitions. *Biochim. Biophys. Acta* 600: 620–624.

128. Virtanen, I., Ekblom, P., and Lauria, P. (1980). Subcellular compartmentalization of saccharide moieties in cultured normal and malignant cells. *J. Cell Biol.* 85: 429–434.
129. Voelker, D.R. (1990). Lipid transport pathways in mammalian cells. *Experentia* 46: 569–578.
130. Wakelam, M.J.O., Davies, S.A., Houslay, M.D., McKay, I., Marshall, C.J., and Hall, A. (1986). Normal $p21^{N\text{-}ras}$ couples bombesin and other growth factor receptors to inositol phosphate production. *Nature* 323: 173–176.
131. Wherrett, J.R. and Huterer, S. (1972). Enrichment of bis-(monoacylglycerol) phosphate in lysosomes from rat liver. *J. Biol. Chem.* 247: 4114–4120.
132. White, D.A. (1973). The phospholipid composition of mammalian tissues. In: *Form and Function of Phospholipids*, pp. 441–483. (Ansell, G.B., Hawthorne, J.N., and Dawson, R.M.C., Eds.). Elsevier Scientific Publishing, Amsterdam.
133. Wirtz, K.W.A. and Gadella, Jr., T.W.J. (1990). Properties and modes of action of specific and non-specific phospholipid transfer proteins. *Experientia* 46: 551–658.
134. Wolfman, A. and Macara, I.C. (1987). Elevated levels of diacylglycerol and decreased phorbol ester sensitivity in *ras*-transformed fibroblasts. *Nature* 325: 359–361.
135. Yamashiro, D.J., Tycko, B., Fluss, S.R., and Maxfield, F.R. (1984). Segregation of transferrin to a mildly acidic (pH 6.5) *para*-Golgi compartment in the recycling pathway. *Cell* 37: 789–800.
136. Zachowski, A., Henry, J.P., and Devaux (1989). Control of transmembrane lipid asymmetry in chromaffin granules by an ATP-dependant protein. *Nature* 340: 75–76.
137. Zachowski, A. and Devaux, P.F. (1990). Transmembrane movements of lipids. *Experientia* 46: 644–655.
138. Zachowski, A. and Gaudry-Talarmain, Y.M. (1990). Phospholipid transverse diffusion in synaptosomes: evidence for the involvement of the aminophospholipid translocase. *J. Neuro.* 55: 1352–1356.
139. Zahraoui, A., Touchot, N., Chardin, P., and Tavitian, A. (1989). The human rab genes encode a family of GTP-binding proteins related to yeast YPT1 and SEC4 products involved in secretion. *J. Biol. Chem.* 264: 12394–12401.
140. Zilversmit, D.B. (1984). Lipid transfer proteins. *J. Lipid Res.* 25: 1563–1569.

ENDOSOMES, LYSOSOMES, AND *TRANS*-GOLGI-RELATED SYSTEMS IN CONVENTIONAL NEURONS AND THE FROG RETINA:

SHARDS AND SUPPOSITIONS

Eric Holtzman, Eliene Augenbraun, Robert St. Jules and Maria Santa-Hernandez

Advances in Cell and Molecular Biology of Membranes
Volume 1, pages 225–267

ISBN: 1-55938-362-3

I. INTRODUCTION

Lysosomes, Golgi apparatus, endocytic bodies, and related structures have long been known to be very prominent in neurons and other cells of the nervous system (*69,86,100,158,169*). Membrane recycling is extensive and readily evident in certain neuronal regions, especially the presynaptic terminals responsible for release of neurotransmitters (*38,89,92,126,163,181,232,242*). Some central concepts in current general viewpoints about such cycling, and about lysosome genesis and behavior, originated in work on nervous tissue. Examples include the involvement of distinctive *trans*-Golgi structures (*158,159*), now known as the *trans*-Golgi network or reticulum, and the participation of clathrin-coated vesicles in hydrolase transport (*98*). However, despite such promising beginnings, many key details of lysosome genesis and behavior, and of the endocytic cycles in nervous tissue, remain obscure or in dispute. For example, the daily phagocytosis of the membranes of photoreceptors by the pigment epithelium in the vertebrate retina has yet to be definitively analyzed as regards the responsible receptors and other controlling devices (*14,19,21,204*). Even less is known of the comparable breakdown of photoreceptor membranes in invertebrates, which involves extensive autophagic use of endocytically derived multivesicular bodies (*50,90*). And, although lysosomes or endosomes may be involved in striking pathological phenomena such as the genesis of the amyloid deposits of Alzheimer's disesase (*71b*), the specific details as related to nervous tissues are unclear (cf. *90*).

Our aim here, therefore, is not at providing a comprehensive overview of how lysosomes, endosomes, and *trans*-Golgi structures function in the nervous system. The focus, rather, is on putting the available fragments into the conceptual frameworks emerging from the studies on other cell types, and identifying key areas of present uncertainty. Lack of space, not of interest, forces us to ignore the non-neuronal cells of the nervous system, such as the glia of the central system, or the retinal pigment epithelium. For similar reasons, in considering the origins and fates of neuronal membranes we will deal principally with proteins, even though recent work suggests that the behavior of membrane lipids is significantly different (*69a,114,115,139,148,180,247,255a*).

II. ENDOCYTOSIS BY NEURONS

Neuronal geometry is specialized, usually involving a "cell body" where the nucleus and most of the protein synthetic apparatus are gathered, plus differentiated regions including axonal and dendritic processes, synaptic terminals, the photoreceptive outer segments of retinal photoreceptors (see Fig. 4 below), and the like. Most regions of the neuron—the outer segment may be exceptional—are capable of appreciable endocytosis, though the extent of such activity can vary substantially with physiology and developmental state (*2,97,161*).

Endocytic phenomena comparable, at least superficially, to types seen elsewhere, can be recognized in neurons. For example, the mobile "growth cones" present at the tips of neuronal processes during their extension seem capable both of receptor mediated endocytosis (*104*; for other coated vesicle-mediated uptake see *34*) and of taking in droplets of the surrounding medium by mechanisms akin to "macropinocytic" uptake by ruffled surfaces of mammalian macrophages (*27,173*). Endocytic pits and vesicles bearing coats with the morphology of clathrin coats, are most readily seen in neuronal cell bodies and presynaptic terminals, but clearly are not confined to these locales; other regions, including axons and dendrites also can form such vesicles (*79,86,87,96,169,186*). Similarly, multivesicular bodies accessible to endocytosed tracers, and other presumed intermediates in endocytic cycles, seem to arise in axonal and dendritic terminals, as well as in cell bodies (*86,87,94,124,169,186,223*).

The endocytic intermediates seen along neuronal processes mostly are in transit: structures can move rapidly for long distances in the neuronal cytoplasm (*70,124,223,255*). Considerable attention has been paid to the extensive *retrograde* transport (from the terminals of the axonal and dendritic processes, to the cell bodies) because such movement, of microscopically visible endocytic tracers, has proved a convenient tool for determining where the cell bodies corresponding to a given set of terminals are located (*49,124,255*). Endocytic bodies can also move in the other ("anterograde") direction but this has been studied less intensively.

A. Endocytosis of Ligands

Most studies of endocytosis by neurons have focused on the membrane cycling attendant upon synaptic neurotransmission, and have used tracers, such as horseradish peroxidase (HRP) or metallic colloids, which have no biological relevance to the nervous system. Consequent gaps persist in our appreciation of the importance of the endocytosis that is not linked to synaptic activity, for the daily lives, or life histories of the cells. Some of this endocytosis likely is needed for "nutrition" and physiology in much the same way as in other cells. For example, especially at early stages in their development, neurons take up LDL and transferrin from the extracellular space (*2,103,104,161*). Neurons also can endocytose specific growth factors (*10,225,250*) such as Nerve Growth Factor (NGF). Retinal photoreceptors endocytose "binding proteins" that transport vitamin A derivatives to the photoreceptors for use in the biogenesis of rhodopsin (*85*).

Biologically active "foreign" agents such as cholera toxin, tetanus toxin, and, probably, some of the neurotropic viruses also are endocytosed by neurons, often after binding to specific sets of cell surface molecules (*17,49,144,255*). There have also been reports of large scale transfers of macromolecules from adjacent non-neuronal cells to the large processes of certain invertebrate neurons, but the role of endocytosis in such transfers is still not established (*62,237*).

Receptors corresponding to several of the ligands prominent in other contributions to this book, including LDL, transferrin, and IGF-II, have been identified, at least tentatively, on neurons (*12,136,156,161,250*). Though information about these receptors is still quite fragmentary, no doubt many of them participate in endocytic events much like those elsewhere. Different receptors are concentrated along particular domains of the neuronal surface in patterns assumed to reflect the functional emphases of the corresponding cellular regions. For example, the concentration of transferrin receptors along cell bodies (*31*) presumably relates to the centering of much of the neuron's ordinary metabolism in the cell body.

NGF's interactions with its receptor at axon endings can eventuate in important changes in the cell bodies, far from the endings. Endocytosed NGF is known to reach the cell body by retrograde transport and such movement of the endocytosed growth factor within the cell is widely presumed to be important for the biological effects (*225*). As with other cell types and other hormone-like ligands, consensus has yet to be reached as to the mechanisms of such effects. As far as is known from studies with the usual microscopically detectable conjugates, most of the endocytosed NGF remains within endocytic or degradative bodies. Suspicion has grown, however, that the NGF receptor system includes membrane-associated protein kinases (*178*). Thus, one could imagine that retrograde transport aids the passage to critical cell regions of such kinases, turned "on" by NGF, and carried, exposed, on the cytosolic side of NGF-containing endocytic bodies. Comparable second-messenger production, as well as postendocytic escapes of ligands (*83*) or proteolytic fragments thereof (252), or of viruses, into the cytosol or nucleus are plausible explanations for how diverse factors endocytosed at the tips of growing neurons can affect genetic expression in the cells (*145,250*).

B. Transcytosis and Related Phenomena

Neurons are capable of transcytosis and even of transferring endocytosed material, via synaptic contacts, from one neuron to another (*49,144,255*). Retrograde transport followed by cell-to-cell transfer often is invoked in framing hypotheses about long-range signaling in the nervous system, and is of likely importance as well for moving agents like tetanus toxin from peripheral points to zones of impact (*29,203*). Too little is known as yet of the specifics to compare such phenomena in nervous tissue with transcytosis and related processes in other cell types.

Neurons sometimes have been thought capable of endocytosing adjacent portions of cytoplasm (*185,237*). Such uptake is one possible explanation for the occurrence of configurations in which coated vesicles at the neuronal surface contain projections from neighboring cellular regions (e.g., Fig. 4.8 in *90*; for an interesting variant see *28*). But it is not known whether such configurations generally reflect significant communion, such as the transfer of signals (See *8a,121*), or rather are "accidents" of cell adhesion or membrane reorganization (*113*).

III. RECYCLING OF SYNAPTIC VESICLES

Though there still are a few holdouts (cf. *163*), most neurobiologists are convinced that chemical neurotransmission is based principally on the exocytic release of transmitters from the synaptic vesicles that abound in presynaptic terminals (*32,101,116,260*). Relevant to this conclusion, we found that neurotransmission is accompanied by dramatic increases in the acquisition of extracellularly administered HRP by the synaptic vesicle population and interpreted this as reflecting the retrieval of synaptic vesicle membrane via endocyticlike recycling (*97*). The subsequent demonstrations of the probable reuse of the recycled vesicles for transmission (*33,78,79,223*; see especially the recent studies on living cells in *15*), and observations of recycling in a diversity of neurons, led to the surmise that recycling permits the repeated reuse of the machinery associated with the synaptic vesicle membrane to accumulate and release the small molecules that serve as neurotransmitters. (Different neurons utilize acetylcholine, catecholamines, amino acids, or other small transmitters.) The machinery includes: proteins that help move the transmitters or their precursors into the vesicles; enzymes of transmitter metabolism; agents responsible for transport of nucleotides or of ions such as protons (see below); and other molecules, probably including ones that interact with cytoskeletal elements and with the plasma membrane (*43,92,170,232,261*).

For vertebrate photoreceptors (*55,56,195,196*; see also *184,197*) and certain other neurons (*205*), evaluation of the extent of endocytic uptake of visible tracers into the synaptic vesicles under varying physiological conditions proved useful for monitoring rates of synaptic activity in preparations difficult to approach by more traditional physiological or biochemical means.

Since the cycling of synaptic vesicle membrane is dramatically increased when neurotransmission is evoked, such cycling falls largely into the category of "regulated" endocytic processes, rather than being "constitutive" as seems to be true of some other endocytic cycles. The cycling differs from ligand driven endocytosis in that, in the "resting" state, the pertinent membrane is inside the cell, rather than residing at the cell surface. It differs from the events responsible for reuse of membranes involved in other types of secretion (*58,71,76*) in being more of a local process that takes place close to the sites of exocytosis; most notably, recycling synaptic vesicle membranes in nerve terminals need not travel back to the Golgi apparatus to be reincorporated into secretory structures. Recycled synaptic vesicles can even reenter the exocytic- endocytic cycle without being refilled with transmitter (*32,38*).

A. Models and Mechanisms

Two major models, each with several variants (Fig. 1), have been proposed for recycling. The two are not mutually exclusive; both types of processes could coexist, operating simultaneously or to different extents in different types of neurons or under varying circumstances.

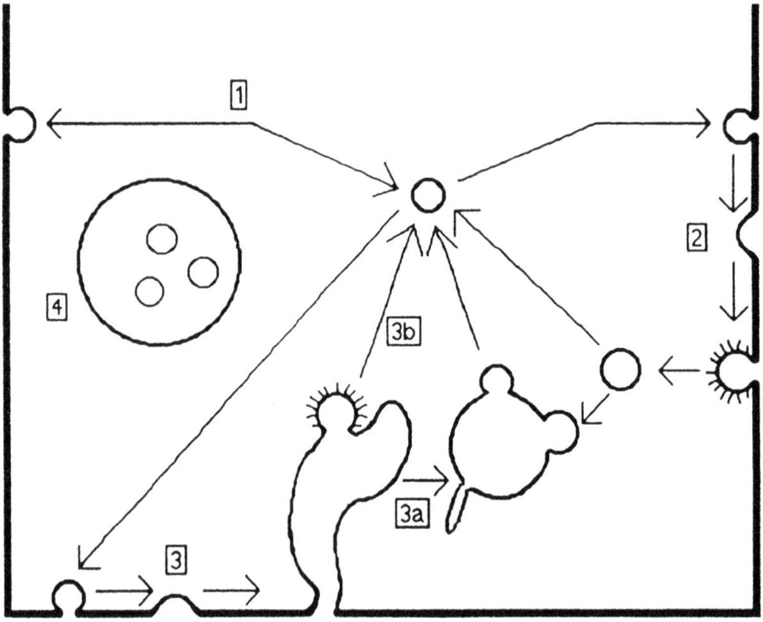

Figure 1. Possible recycling mechanisms for synaptic vesicles: (**1**) Retrieval of membrane after very brief opening of the vesicle at the cell surface. (**2**) Incorporation of the vesicle into the cell surface with later retrieval via clathrin-coated vesicles. (**3**) Retrieval of vesicle membrane via larger compartments: Alternative possibilities illustrated include the formation of vesicles from infoldings of the cell surface (**3b**) and the formation of intracellular intermediate compartments either directly from the cell surface (**3a**) or by fusion of endocytic vesicles. 3b also schematizes the possibility that the budding of vesicles from the larger compartments involves clathrin. Synaptic vesicle membrane may also be degraded in multivesicular bodies (**4**).

"Rapid" Retrieval

The simpler model proposes that an exocytosing synaptic vesicle remains fused to the plasma membrane for a brief time, perhaps sometimes as brief as a fraction of a second (*38*). During this time, the vesicle retains its integrity, as a vesicle or a discrete patch of membrane incorporated in the plasma membrane, but it does establish enough of an opening to the extracellular space that the vesicle's contents can emerge. The opening might sometimes be very small, amounting only to a minute pore (*4,105*). However, for at least a proportion of the vesicles—there is dispute over the numbers (*38,146,211*)—the opening must be sufficiently large at some stage in its history that macromolecular tracers like HRP, dextrans, or antibodies against lumenal domains of synaptic membrane proteins (*142*) can enter. Recycling is envisaged simply as the separation of such a vesicle from the surface, without the involvement of a clathrin coat, or of intermediate recycling compartments comparable to the endosomes of receptor mediated endocytosis. The process might be reminiscent of the rapid cycling of caveolae proposed to explain folate uptake (*188*); the coating on caveolar vesicles is not clathrin (*188A*).

This conception accounts for observations that, even during periods of moderately rapid and extensive release of transmitter, synaptic vesicle membrane antigens need not be extensively transferred to the cell surface, and the axon surface need not change dramatically in area (*243*). It readily explains the apparent "fidelity" of retrieval; the fact that recycled vesicles carry the machinery essential for function but seem not to acquire components from the plasma membrane. The data supporting this last point (e.g., *20,126*) are still only fragmentary. Given the potential complexities of the membrane domains involved, and technical matters such as uncertainties about the extent of association of endocytic tracers with membranes and possibilities of cytochemical diffusion artifacts, it is not yet clear which markers would permit a decision as to when synaptic vesicle components and plasma membrane molecules part company.

The "rapid" retrieval model has elements in common with prevailing models for recycling by gland cells, though these models too are only sketchily supported by definitive evidence. Most importantly, when gland cells exocytose their secretions, the secretion granule's membrane—by mechanisms yet to be delineated—is thought to remain localized as a patch in the cell surface (*71,167*). The patch can be internalized rapidly, or many minutes later. Differences between synapses and gland cells include the involvement of the Golgi apparatus in recycling by gland cells, and the fact that, for the most part, when supposed "fluid-phase" tracers like HRP are taken up by gland cells, the tracers do not reach nascent secretory structures or Golgi sacs. Instead, the tracers accumulate in lysosome-related structures. One of several plausible interpretations (*58,76,90*) is that during recycling in gland cells, the structures carrying membranes retrieved from the cell surface pay a visit to the lysosomes or prelysosomes, where "unwanted" materials acquired from outside the cell can be dumped before the membranes move on to sites of reuse. The proposed synaptic vesicle cycle we are considering here lacks such an "option".

To evaluate the potential implications of these similarities and differences, more must be learned about the molecules that normally are available to enter recycling synaptic vesicles or secretory granules from the extracellular milieu. More must also be learned about the subtle differences between nerve terminals and related gland cells as regards controls of the exocytic-endocytic cycle (e.g. *36*).

Retrieval via Coated Vesicles and Intermediate Structures

The second major model for synaptic vesicle recycling has important features in common with models of receptor mediated endocytosis (*77–79,181,232*). The exocytosing vesicle is supposed to collapse completely into the plasma membrane. Seconds to minutes later, the membrane (or its laterally dispersed components) is retrieved by endocytic-like processes in which clathrin-coated vesicles play central parts. Retrieval can take place at regions of the cell surface distant from the sites of exocytosis, the membrane components having diffused within the cell surface to those regions. If such diffusion disperses components of the vesicle membrane, the dispersion is thought to be counteracted by a process of regathering akin to the clustering of receptors in the coated pits of receptor-mediated endocytosis.

Sometimes retrieval occurs from transient infoldings of the plasma membrane (*40,66*). Of more particular importance for the theme of this book is the proposition that retrieval can also involve the endocytic formation of intracellular sacs or vacuoles—sometimes much larger than synaptic vesicles, sometimes only modestly so—which serve as intermediate structures from which the synaptic vesicles eventually form by budding or a comparable "maturation". These intermediates are often referred to as "cisternae" (*77–79*), though this term has also been used for other structures involved in recycling, including the plasma membrane folds just mentioned. For the nerve terminals of electric fish, a system being intensively investigated from these viewpoints, intermediate stages of retrieval may involve a period of many minutes during which the vesicles, still not filled with transmitter, are smaller than the mature vesicles and exhibit distinctive biophysical characteristics (*256,262*).

There is little doubt that synaptic vesicles can recycle via coated pits and intermediate structures, and that there are situations in which much of the vesicle membrane retrieved from the surface enters such pathways (e.g. *79,132*). This type of recycling is, however, most readily observed under abnormal circumstances: after stimulation of neurotransmission at very rapid rates (*79*); under circumstances where retrieval is slowed temporarily by low temperature (*79,197*); in the *shibire* mutant of *Drosophila*, a temperature-sensitive mutant that affects several types of endocytosis (*120*); in synaptosomes (*60*); or in photoreceptors exposed to abnormal ionic environments (*132*). Does this mean that the pathway is a special one that comes into play only under circumstances where exocytosis and endocytosis are severely imbalanced? Or is it a question of abnormal circumstances exaggerating a normal mechanism, merely making it more evident? From present evidence, the agnostic conclusion, simply that this sort of recycling *can* coexist with that described above, is probably wiser than a vehement defense of positions. Perhaps it will turn out that there is a continuum of potential combinations of "rapid" and "delayed" retrieval; both may be normal mechanisms that usually occur simultaneously, with the emphasis under given conditions depending upon the relations between the rates at which vesicles collapse fully into the plasma membrane, and those at which retrieval mechanisms can generate new vesicles from larger surfaces (plasma membrane or intracellular intermediate compartments).

Key areas of distinction between the two models lie in the involvement of clathrin and of lateral regrouping of membrane molecules. Proponents of rapid retrieval will be heartened by a recent report that native complexes of synaptic vesicle membrane proteins tend to resist dispersion, even when the vesicle membranes are solubilized (*13*); if so, little lateral regrouping may normally be needed during recycling. On the other hand, clathrin is transported along axons to terminals (*25*), and the terminals of normal neurons apparently do contain quite an abundant supply (*181*).

It is not hard to find coated vesicles in the terminals (though they can easily be overlooked unless the microscopic preparations are suitably stained). But "bookkeeping" experiments aimed at determining whether these vesicles, or other relevant endocytic structures, are of sufficient abundance at the right times to account for most of recycling, have foundered in part because of disagreements about methods and the normalcy of conditions, and, in part, because of the possibility that the vesicles lose their coating rapidly enough to confound the measurements (e.g. *66,77,146,181,231*). Nor has the matter been decided by arguments such as that clathrin must be involved because the *shibire* mutation affects both receptor mediated endocytosis and synaptic recycling (*232*); membrane retrieval in *shibire* is still incompletely understood. [For example, intermediate compartments can participate (*120*). And under some circumstances there is an unexplained transient disappearance of membrane attributable either to disassembly (*120*) or, more likely, to the existence of reservoirs—perhaps distant ones—that have yet to be identified.]

Even for receptor-dependent endocytosis in cells where clathrin-coated vesicles clearly do play a central part, there are coexisting routes that do not depend on clathrin (*75,90,188*). This point is especially germane because the relations of coated vesicles to the intermediate structures in synaptic terminals still are uncertain (*25,40,60,66,155,171,172,181,207,257*). The original notion was that the intermediates form by fusion of coated vesicles and this view still is widely held. But there may also be circumstances in which coated vesicles bud from recycling intermediates rather than (or as well as) fusing with them. And, vacuoles or other potential intermediates in recycling may sometimes form directly from the terminal's surface (*149*). Precedents for this last possibility are encountered in gland cells, in which membrane retrieval can be mediated by tubules and vacuoles as well as coated vesicles (*1,61,93,168*). Moreover, though the matter is still not completely settled, the reigning model for recycling in neurosecretory neurons envisages retrieval of membranes of the large neurosecretory granules (see below), at least partly by way of vacuoles (*153,156a,175*).

Questions of the potential roles of clathrin-coated vesicles and recycling intermediates will arise again below in our discussion of endosomes. Here we will list a few additional items particularly needing study:

1. Small endocytic sacs and vacuoles that could be recycling intermediates are seen in the synapses of many types of neurons under seemingly normal circumstances. It will be important to determine whether structures of this sort are indeed involved in synaptic vesicle cycles, or whether they pertain to other processes such as receptor mediated endocytosis or the turnover of the plasma membrane (*34*).
2. In certain cases, presynaptic terminals contain networks of endocytic tubules that are potential sites of recycling (*73,132,147*).
3. Very little is known of neuronal recycling of exocytic membranes other than those of synaptic vesicles although there are some interesting possible phenomena. For example, by retrieving excess components or molecules

needed only transiently (e.g., for packaging or transport), endocytic structures forming at the apex of the photoreceptor cell body could help modulate the overall flow of materials during the exocytic delivery of opsin to the cell surface that precedes transport of opsin to the outer segment (*84,94*).

4. Very little is known of non-clathrin coating proteins (see e.g. *131*) in nerve terminals. We have observed that both brefeldin A and aluminum fluoride, agents that can affect such coats, engender accumulation of endocytic vacuoles in the terminals of photoreceptors (*194* and Santa-Hernandez, Johnson and Holtzman, unpublished).

B. Lysosomes and Synaptic Vesicles

From general considerations (*90*), plus the very limited available data on turnover, it appears likely that the components of synaptic vesicle membranes, like those of most membranes, are degraded within a few days to a week or two after their synthesis (*64*). In gland cells there are routes by which membranes from secretion granules can be sequestered in lysosomes (*1,168*), with multivesicular bodies serving as intermediates in the process.

Autophagic or crinophagic degradation of synaptic vesicles may sometimes take place in neuronal cell bodies (*87,98*), but most synaptic vesicles presumably leave the cell body never to return as independent small vesicles. In fully grown neurons of vertebrates, few mature lysosomes—that is, secondary lysosomes containing the residua of digestion (*88,90*)—are seen in normal axons or terminals. Still, autophagic or crinophagic breakdown of secretory material has been reported for normal neurosecretory axons or terminals (*22,153*). And, in presynaptic terminals of photoreceptors (*199*) and axons of conventional neurons (*106*), components of synaptic vesicles are subject to inclusion in multivesicular structures (MVBs). These MVBs probably are the principal sites of vesicle destruction. They, or their precursors, appear to arise in the terminals and some may acquire acid hydrolases while there (*199*), though much of their degradative activity seems to take place once they have moved to the cell body (*22,106,124,196,223*). MVBs are regularly present in the processes of normal neurons, but in mature neurons they are not very abundant in axon terminals and their frequency there seems not to be markedly related to the intensity of vesicle use for neurotransmission (*94,132*). Thus, lysosomal degradation of synaptic vesicles is probably best thought of as a steady, slow process coexisting with extensive vesicle reuse (*223*).

IV. *TRANS*-GOLGI SYSTEMS

The Golgi apparatus of neurons appears to be connected as a single network (*69,158,179*). A circumnuclear distribution, with some extension into dendrites and little or none into axons, is taken as typical (*69,158,169*) but there are many exceptions, including vertebrate retinal photoreceptors, whose Golgi apparatus extends longitudinally (see Fig. 4; *16,139*), and rapidly growing or regenerating neurons, in which the Golgi apparatus is located at one side of the nucleus (*88,222*).

The neuron's *trans*-Golgi network was first identified by virtue of its cytochemically demonstrable acid phosphatase (*158*). From the presence of this enzyme, and morphology, the system was ascribed roles in the genesis of lysosomes (*98,158*). Initially it was held to be a specialized region of ER, and thus was called GERL (*158,159*). Though now the *trans*-Golgi network is thought to be separate from the ER, it remains true that neurons and related cells, such as those of the adrenal medulla, do show a variety of intimate associations between ER and Golgi stacks at both the *cis* and the *trans* face and, more rarely, laterally (*22A,52,87,94,100*). Various of these associations jibe only uncomfortably with current dogmas about ER-to-Golgi, export-related transport. They are, at least, reminders that portrayal of Golgi functioning solely in terms of stereotyped *cis-trans*, vesicular movements may oversimplify reality inordinately (see also *150,246*). Perhaps some of the associations serve in the *cis*-Golgi-to-ER movements of current interest (*9*). Conceivably there also is transport from *trans*-Golgi regions to the ER, providing, say, Golgi-glycosylated proteins needed for some of the functions of the axonal endoplasmic reticulum discussed below.

At the *trans*-Golgi face in neurons, endocytic structures and autophagic and heterophagic lysosomes closely intermingle with endoplasmic reticulum and Golgi elements. The relations of these several systems to one another and to other parts of the Golgi apparatus, are only vaguely known. As in many cell types, it is often difficult to decide whether a particular smooth surfaced sac, tubule, or vacuole seen near the *trans* face of the Golgi apparatus is best thought of as part of the *trans*-Golgi network or as part of the endocytic systems to which *trans*-Golgi elements send hydrolases, or from which the Golgi apparatus receives recycling membranes (Fig. 2; cf. *109,124*). For frog photoreceptors we have found that, as in some other cell types (*35,131*; but see also *239*), *trans* Golgi structures do not disperse into the ER when the cells are exposed to Brefeldin A (*193,194,209*). Rather, the photoreceptors retain sac-like elements and networks that resemble *trans*-Golgi systems in morphology and in the presence of characteristic cytochemically demonstrable phosphatases.

Although the distribution of glycosyl-transferases has not been directly detailed, the patterns of lectin binding in the Golgi apparatuses of conventional neurons (*134a,177,259*) and photoreceptors (*209*) suggest that the *cis-trans* organization of glycosylation within the Golgi apparatus is much like that in other cells. Mannose 6-phosphate receptors (M6PRs) have been demonstrated in the neuronal cell body (*164*) and we have found them in the Golgi region of cultured hippocampal neurons (see Fig. 5). However, we do not yet know which membrane systems in the *trans*-Golgi zone are particularly rich in such receptors.

A. Smooth Membrane Systems in Axons

Axons contain smooth surfaced sacs and tubules (Fig. 3) collectively referred to by terms such as the "axonal agranular reticulum". These structures clearly are

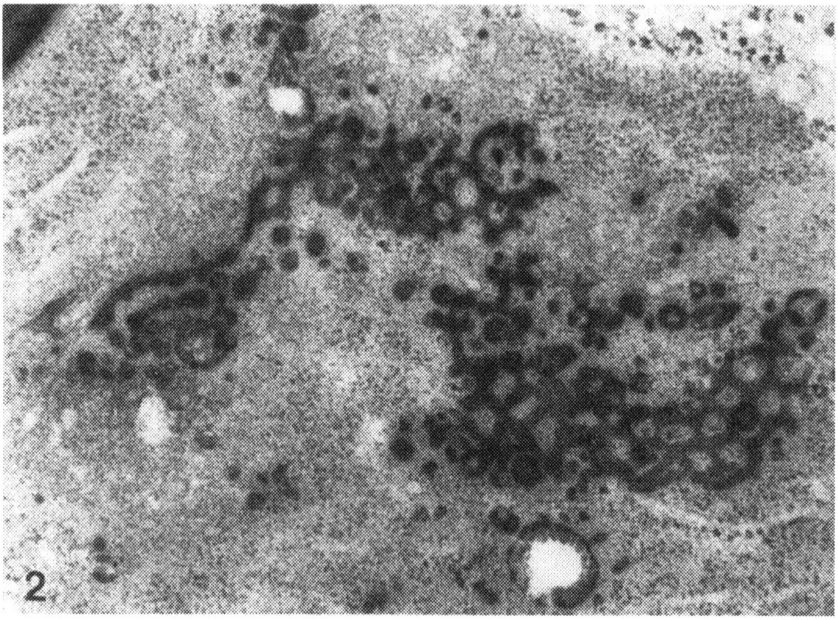

Figure 2. Portion of a frog (*Rana pipiens*) retinal photoreceptor, showing a *trans*-Golgi network containing endocytosed wheat germ agglutinin-horseradish peroxidase; the presence of the tracer has been revealed by the cytochemical reaction for peroxidase. (From work of Santa-Hernandez and Holtzman.) × 35,000.

heterogeneous in origin and function (*89,99,128,169*). Some of them pertain to retrograde transport, representing endocytic structures on their way to the cell body. Others participate in the rapid anterograde transport that carries new membranes and other materials down the axon; the vehicles of such transport are small sacs, tubules and vesicles (see below and *51,57,208,236*).

Endoplasmic Reticulum

The axonal structures also include true endoplasmic reticulum (ER), attached to the rough ER of the cell body (Fig. 4). In frog photoreceptors there is a continuous ER network, mostly smooth, from one end of the axon to the other (*147*), but these axons are unusually short and in other neurons there may be some interruptions in the continuity (discussed in *147*). The axonal ER seems not to move rapidly within the axon (*51*) and it differs from the rough ER of the cell body in several cytochemical properties (*128,147*), including the paucity of its acetylcholinesterase (*224*). Nonetheless, the axonal ER may contain BiP (*251a*) and is a likely route for movement of membrane-associated enzymes such as glucose 6-phosphatase

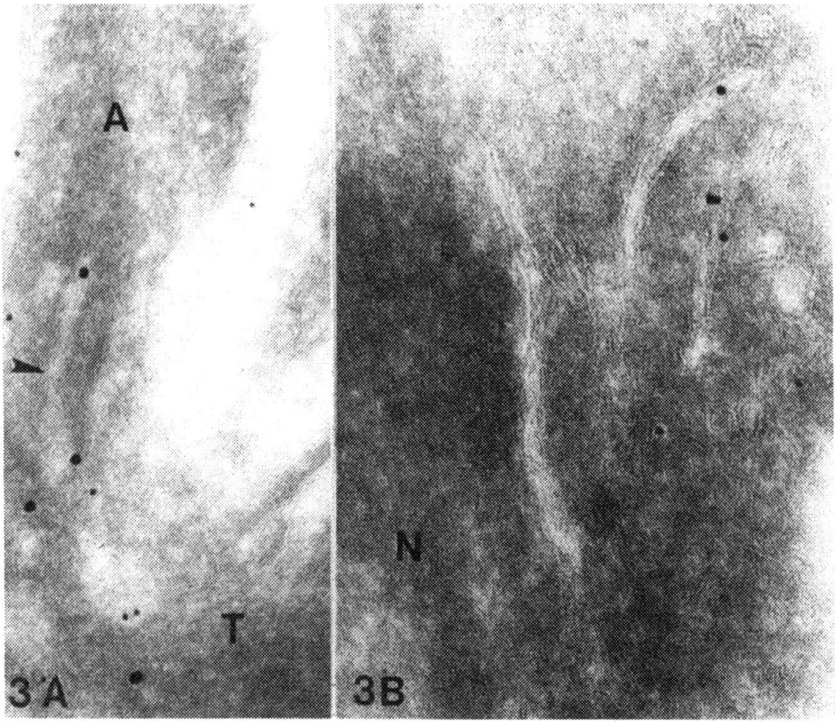

Figure 3. Regions of frog photoreceptors in frozen semithin sections immunostained with an antiserum against the synaptic vesicle protein synaptophysin from frog (*243*) followed by protein A-gold. In panel **A**, the larger gold particles, corresponding to synaptophysin, are seen associated with a sac or tubule (arrowhead) in the axon (**A**) of a rod photoreceptor. The edge of the corresponding terminal is seen at T. Panel **B** shows comparable sacs or tubules from the region near a photoreceptor nucleus (N) where extensions from the Golgi zone (cf. Fig. 11 in ref. *147*) intermingle with elements of ER and with vesicles, some of which resemble synaptic vesicles. (Work of St. Jules, Augenbraun and Holtzman.) A: × 70,000; B × 100,000.

(*22a,222*) and Ca^{2+}-accumulation machinery (*221,240,249,254*) from the cell body's ER into the axon. Indeed, from the fact that both the rough ER and the axonal ER of photoreceptors can accumulate calcium (*240*), and by analogy with the sarcoplasmic reticulum of muscle cells (*134*), one might expect the Ca^{2+}-ATPase to enter the axonal ER directly from the rough ER, without necessary involvement of the Golgi apparatus (see also *180*). On the other hand, some of the ER's Ca^{2+}-binding proteins (*249*) may follow a more circuitous route, dependent on Golgi participation (*134,226*). The presence of such binding proteins in the axonal smooth ER may account for certain of the smooth ER's distinctive staining properties (*94,147*). But what are the mechanisms that enable the cell to maintain differences between systems of rough ER and smooth ER that are directly continu-

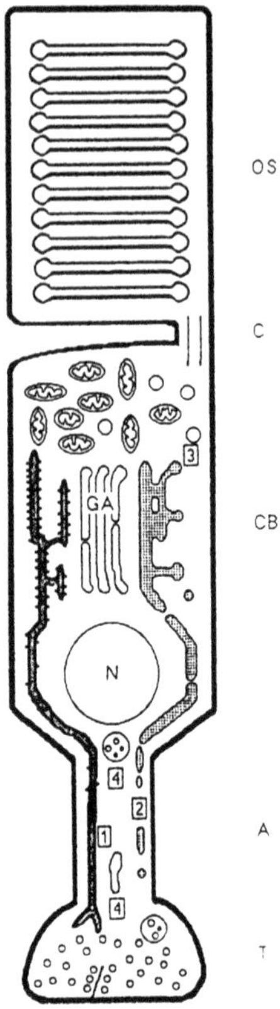

Figure 4. Diagram schematizing a photoreceptor to indicate several possible transport routes for components of the membrane delimited organelles. OS, outer segment; C, cilium that connects the cell body (CB) to the outer segment; A, axon; T, presynaptic terminal. (**1**) Certain components may move from the rough ER in the cell body to the axonal smooth ER through direct continuities. (**2**) and (**3**) For components that pass through the Golgi apparatus (GA) to the trans side, two types of subsequent routes are indicated: (**2**) Diagrams the possibility that the vesicles or other relatively small transport vehicles in which components reach their final destination, are generated at a distance from the Golgi stack, from larger, more elongate compartments. (**3**) Diagrams the likelihood that some components are packaged into their transport vehicles at, or very near the Golgi apparatus. (**4**) Included as a reminder of retrograde transport.

ous with one another (see also *221*)? [Note also the possibility that control of Ca^{2+} levels within the ER is important for the functioning of the reticulum in transporting proteins (*219*). This could bear on the proposed divisions of Ca^{2+}-storing labor between ER and "calciosomes" (*249*). A potentially related aspect of the residual controversies about mechanisms of neurotransmission is the proposal that synaptic vesicles accumulate calcium and release it from nerve terminals by exocytosis (*163*).]

Some axonal ER in regions near the cell body shows attached ribosomes (*68,147,169*). More rarely, mRNAs, ribosomes, or protein synthesis not attributable to mitochondria have been observed in distal regions of axons (including presynaptic terminals) of vertebrate and invertebrate neurons (*68,107,139,147*; Fig. 1 in *217*). Yet to be decided is the extent to which such observations reflect biologically important synthetic capacities of the axoplasm as opposed to "accidental" escapes from the cell body.

Trans-Golgi Elements

Do Golgi-associated systems extend into the axon? The stacked Golgi sacs seem not to but it may be that *trans*-Golgi elements, or very closely related systems, in some sense do. We have seen thiamine pyrophosphatase (TPPase), a cytochemically-demonstrable *trans*-Golgi enzyme, in sparsely distributed sacs or tubules in photoreceptor axons and terminals (*54,200*). Several other cytochemical techniques have revealed similarities between axonal sacs or tubules and *trans*-Golgi elements in conventional neurons (*177*). The best present guess is that these systems need not form a single, continuous network that extends the length of the axon. Instead, elongate tubules and sacs may separate from the *trans*-Golgi network in the cell body or proximal axon, and move subsequently as modest sized elements or more extensive sheets, tubules, or "rafts". Such possibilities fit with our observations of membrane systems extending away from the photoreceptor Golgi apparatus in the direction of the axon (*147*) and with observations on tubules moving away from the Golgi stacks of non-neuronal cells (*39*). As they move, such structures could fragment into smaller structures that participate in rapid anterograde transport in the distal axon.

Acid Hydrolases

One indication that this line of reasoning may have merit comes from the presence of acid hydrolases within axonal sacs and tubules, as might be expected for *trans*-Golgi elements. Such hydrolases are only minimally evident, though not completely absent, under normal conditions (*65*), but they can be seen readily after injury (*87,90,95,99,192*). Importantly, at least one such enzyme, acid phosphatase, accumulates appreciably in sacs and tubules distal to axonal interruptions (*95*) suggesting that this enzyme must normally be present along much of the axon.

The overreliance on acid phosphatase for cytochemistry (cf. *88*) is an especially serious weak spot in these connections because acid phosphatase can exhibit membrane associations that make its transport different from some of the other acid hydrolases (*251*). β-glucuronidase—another enzyme which our immunocytochemical observations on hippocampal cultures suggest to be present in axons to a significant extent (Augenbraun and Holtzman, unpublished; see also *253*)—also has atypical features (*157*). Thus, we still do not know, for example, whether the mannose 6-phosphate-dependent hydrolase transport systems operate in axons. Though there does not normally seem to be much M6PR in the axons of the very few systems studied thus far, mistreatment (lowering the axonal pH) leads to an apparent redistribution from the cell bodies into neuronal processes of cultured hippocampal neurons (*164*).

Are the hydrolases in the axonal sacs and tubules used for lysosomal functions, and if so, how? Autophagic vacuoles of the usual sort can form in axons (*86,87*), as do the acid phosphatase-containing multivesicular bodies described above and other potentially degradative structures, including ones resembling the tubular lysosomes of macrophages (*22,87,99,220*). Some of the latter may correspond to endocytic sacs and tubules carrying tracers up the axon; these, assertedly (*22*), can acquire acid hydrolases while in transit. For none of the structures that form in axons, however, is it known how the hydrolases are delivered, but one can imagine routes comparable to the ones operating in *trans*-Golgi zones and other regions of non-neuronal cell types.

B. Golgi Apparatus and Synaptic Vesicles

Though specific documentation is scanty, no one doubts that, as in most cell types, *trans*-Golgi-associated systems of neurons participate in packaging materials for exocytosis or incorporation in the cell surface. The neuronal Golgi apparatus is known to handle materials destined for transport down the axon (*74*), probably including constituents of the structures responsible for neurotransmission. Among the earliest observations of such a Golgi role were findings on neurosecretory cells, which package relatively large molecules, including peptides and proteins, into distinctive looking secretion granules with diameters of hundreds of micrometers. These granules can be seen forming in the Golgi region before they move into the axon (*24,63,153*). Most of the macromolecular constituents of the much smaller (50–100 nm diameter) synaptic vesicles that accumulate low molecular weight neurotransmitters probably also pass through the Golgi apparatus before moving down the axon. Evidence for this includes: (1) the presence, in some types of the vesicles, of glycosaminoglycans and other saccharide-rich molecules of sorts that usually are generated in the Golgi apparatus (*122,125,135,183,256,257a*); (2) the presence of the *trans*-Golgi enzyme, thiamine pyrophosphatase, in some synaptic vesicles (*200*; see *54* for review); and recent demonstrations of synaptophysin, a synaptic vesicle membrane protein, in *trans*-Golgi elements of photoreceptors (*200*) and other neurons (*106,230*). (Without the other evidence, the findings related to synaptophysin are not decisive, given the possibilities for membrane cycling via the Golgi apparatus discussed below.) However, while such information confirms the likelihood of a central role for the Golgi apparatus in the genesis of synaptic vesicles, it is by no means certain that vesicles like those that predominate at axon terminals arise as such chiefly at the *trans* face of the Golgi apparatus.

Endocytosis and Vesicle Origins

For one thing, the vesicle population of most terminals is heterogeneous. Commonly, a few vesicles ("large, dense-cored vesicles") are somewhat larger than the rest, and have an electron opaque content, whereas most of the vesicles are "clear" in ordinary electron microscope preparations (*169*). This situation, coupled with the fact that most of the vesicles seen at a given moment owe their most recent origin to the endocytic-like recycling described above, has led to proposals that what the Golgi apparatus produces is a vesicle whose contents (the vesicle's dense "core"?) are more complex than those of the "ordinary" vesicles that predominate in terminals (*18,45,94,206,212,227*). Some of these contents—pharmacologically active peptides, materials destined for the cell surface or extracellular matrix (*26*), and the like—are presumed to be released irretrievably during the first exocytosis. The membrane, however, is recycled to generate the "ordinary" vesicles, after a retrieval process that might concentrate membrane proteins essential for vesicle functions.

The present status of this hypothesis is uncertain. Many investigators are convinced that the scheme is ruled out by demonstrations that certain "larger vesicles" that are candidates for pertinent Golgi products (e.g., *111,119*) lack membrane proteins which characterize the "ordinary" vesicles, and also differ from the latter vesicles in sites and controls of exocytosis (*42,140,212,213,227-230,262*). Others argue that the lack of the proteins has not been definitively established and that adequate methods actually do detect some of them in the larger vesicles (see *18,151,251*). Still others point out that before we can decide whether the large vesicles studied thus far have been selected appropriately, more must be learned about vesicle heterogeneity, especially as regards the distribution of membrane proteins of known function, such as the enzymes and transporters of catecholamine accumulation (*45,105,163,206,212,260,261*).

Transport of Vesicle Precursor Structures?

Overlapping with the hypothesis just considered is the proposition that what the Golgi apparatus produces are precursor structures which generate synaptic vesicles as they move away from the Golgi region, as they enter or pass down the axon, or once they reach the terminals.

Maturation of neurosecretory granules during axonal transport probably includes ongoing proteolytic processing of the granule's content (*63,110,212,257a*). In neurons of electric fish that use acetylcholine for their transmitter, materials slated for synaptic vesicles are transported down the axon in membrane-associated form, but the vehicles of such transport do not accumulate neurotransmitter until they have moved a considerable distance from the cell body (*117*). During this transport, a vesicle-associated glycosaminoglycan undergoes processing, perhaps even a diminution of its size. Though the details of such changes are still being studied, this observation plus our finding of TPPase both in photoreceptor *trans*-Golgi sacs and in synaptic vesicles, especially when viewed in the light of the processing of neurosecretory materials, suggest the likelihood of some transport of processing machinery along with the constituents of the vesicles.

The small size and relatively undistinguished appearance of "ordinary" synaptic vesicles complicate microscopic searches for the vesicles' origins. Vesicles of the dense-cored variety are occasionally seen in axons and one can find bodies looking like ordinary synaptic vesicles in the axons of some neurons (reviewed in *89*). Experimentally introduced synaptic vesicles can move down axons (*133,202*). Still, the population of discernible synaptic vesicles is often quite sparse near the Golgi apparatus and in axons (*229,230,236*); we lack the requisite information about the details of membrane turnover to decide whether there are enough to replenish the terminals. Furthermore, along normal axons, near terminals, and at experimental interruptions in axons, vesicles with the appearance and cytochemical properties of synaptic vesicles seemingly can arise by budding from larger tubules or sacs (cf. *89,119,147,222*).

Overall, one can reasonably suppose that synaptic vesicles arise at more than one locale, even in a given cell. Some, such as certain large dense-cored vesicles, may form at or very near the Golgi apparatus while others may bud from structures larger than synaptic vesicles, which move away from the apparatus, giving rise to synaptic vesicles in the cell body, the axon hillock, the axon, and the terminal. As would be expected from such suppositions, some axonal sacs or tubules of photoreceptors (Fig. 3) and other neurons (*106*) contain the synaptic vesicle membrane protein, synaptophysin. Images like that in Figure 3 do not, of course, reveal the direction in which these structures are moving, but observations like the presence of TPPase in similar sacs (cf. *200*) suggest that some are likely to be enroute to the terminal. (Refs. *151* and *212* report biochemical studies on cell fractions that may also be apposite.) Bodies of this sort are among the candidates for counterparts of the "constitutive secretory structures" recently identified as likely precursors of the structures thought to be comparable to synaptic vesicles that are present in cultured PC12 cells (*182*). The PC12 cell precursors seemingly generate the smaller vesicles by steps that include passage through endocytic compartments.

Recycling of Synaptic Vesicles via the Golgi Apparatus?

Endocytosed tracers arriving from distant regions by retrograde transport accumulate near the neuronal Golgi apparatus. Though tracers like HRP wind up predominantly in lysosomes and endosomes (*87,96,124,223,225,255*) *trans*-Golgi elements do sometimes acquire HRP (*100,205*) and these elements regularly are labeled strikingly by tracers that bind to membranes [e.g., microscopically visualizable conjugates of cholera toxin or of wheat germ agglutinin; (*109,233*); Fig. 2]. Do these findings imply that synaptic vesicle-related membranes can recycle through the Golgi apparatus? In non-neuronal cells, synaptophysin seems able to recycle via Golgi associated structures, and it is possible that comparable cycling occurs locally in the cell bodies of neurons (see *71a,105*; and the sections below on "sorting" and "endosomal recycling of synaptophysin"). For neurosecretory neurons, whose secretory granules probably cannot be recharged with their peptides and proteins through synthesis or import in the terminals, recycling of membrane retrieved after exocytosis, via the Golgi apparatus, would seem to be the simplest means for reusing the membrane (*23*). The extent of such reuse is, however, not known. For conventional neurons, intersections of recycled vesicle membranes with recently assembled ones have been little studied, in part because the synaptic vesicles can reacquire transmitters without evident need for the involvement of new macromolecules.

Synaptic vesicles labeled with endocytic tracers are not extensively transported, as such, from terminals to cell bodies. The vesicles can, however, contribute membrane to the larger bodies that do undergo such transport; these bodies could conceivably mediate passage of vesicle components back to Golgi systems, perhaps after first paying a visit to the degradative apparatus. An alternative possibility, arising from the information about possible vesicle origins discussed above, is that Golgi-derived systems or other related special transport vehicles (*151*) interact in the axon or terminal with recycling vesicles. However, though sacs and tubules in terminals and axons are commonly found to become labeled with endocytic tracers, it has yet to be determined whether any of these structures can legitimately be regarded as Golgi extensions, or as comparable transport elements recently arrived from the cell body.

Sorting

Sorting and targeting the membrane constituents, lysosomal enzymes, and other materials made in the cell body is of obvious importance for neurons. In photoreceptors, immunocytochemistry, and enzyme cytochemistry (*200*), plus the effects of monensin (*137–139*), indicate that sorting of major membrane components destined for opposite ends of the cell—opsin and synaptic vesicle components—occurs in *trans*-Golgi systems. The suggestion that such systems can generate only a very limited variety of packages (*18*) may be overly "pessimistic" given, for example, the reports that *Aplysia* neurons can ship at least two different classes of secretory bodies to their processes (*110,232*). Sorting of viral membrane proteins and of lipid-linked proteins in cultured neurons follows patterns suggesting that the neuronal axon is "equivalent" to the apical surface domain of the usual model epithelial cells, and the dendrites, to the basal domain (*47,48*). Though questions remain as to how to interpret the polarity of viral budding (*263*), equivalencies of this sort would be particularly striking given the complexities of the morphogenesis of neurons (e.g., the relations of the "basal" and "apical" domains of mature neurons to the surfaces of the cells from which they originate may be complex), and the great differences between neurons and more conventional epithelia as regards cell-cell junctions and other devices that contribute to maintenance of polarity. In photoreceptors, as in other neurons, there are signs of involvement of the cytoskeleton in helping maintain differentiated cell surface domains (*201*), and beginnings have been made in analyzing the "ballet" of different proteins that localize to specific subregions of membrane surfaces; for example, the different zones of the photoreceptors' light responsive disks (*6*). On the other hand, one attractive explanation for the differences in behavior of certain membrane lipids from that of membrane proteins in photoreceptors (*58a,69a,100,116,148,255a*) is that some of the lipids can pass to the plasma membrane without the necessary involvement of the Golgi apparatus and, once at the surface of the cell body, can diffuse rapidly to other portions of the cell.

In speculating about sorting mechanisms in the *trans*-Golgi region, one can invoke the usual possibilities (*91*), such as differences in membrane association (*110*), differential mobility, or aggregation of molecules in the membrane or the lumen of Golgi-related compartments (*102*). A frequent implicit assumption is that sorting generates different populations of vesicular transport structures which the cytoskeleton then carries away from the Golgi region to appropriate final destinations. However, other possibilities deserve attention. We already have referred to the existence of smooth-surfaced sacs or tubules that extend from the Golgi apparatus toward the axon. In certain types of photoreceptors (cone cells) we also see sacs that resemble *trans*-Golgi systems in being acid-phosphatase positive, toward the other end of the cell body (in regions between the Golgi apparatus and the outer segment; *200*). Observations on non-neuronal cell types also suggest that *trans*-Golgi-related elements may extend quite a distance from the stacked Golgi sacs (*39,187*; see also the studies on effects of Brefeldin A in *131*). Can such elements provide routes for movements away from the Golgi zone that enable the neuron to "delay" generating discrete transport vehicles until the materials to be transported had already been carried at least part of the distance to their destinations (*46,54,147,200*)? Together with the movement of some axonal materials directly within the ER, this could simplify the loading and sorting tasks confronted by the cytoskeleton (cf. Fig. 4).

As in other cell types, the sorting of neuronal materials with different predominant target locations is not "perfect". For example, in photoreceptors (*200*) and other neurons (*31*), some membranes containing synaptic vesicle proteins seem to move toward the "wrong" end of the cell. In neurosecretory cells, such "missorting" can lead even to the exocytosis of secretory materials from dendrites (*174*). To what extent are such phenomena truly "mistakes", as contrasted with "intentional" movements of materials to multiple sites, or with "indirect" targeting mechanisms through which membrane molecules added first to one cell surface domain are subsequently retrieved for final delivery to another domain (*11,154*)? And to what extent do they relate to phenomena such as the occurrence of endocytosis and the presence of multivesicular bodies along dendrites or at the tip of the photoreceptor cell body? For example endocytosis of tracers into small vesicles is readily observable at the apex of the photoreceptor cell body (*84,94*) where we also have observed "missorted" synaptophysin (*200*); double-label experiments (*194*) indicate that at least some of the synaptophysin containing structures in the cell body are of endocytic origin. We have also noted that Golgi-associated synaptophysin is more heavily concentrated in *trans* elements as compared to the Golgi stack than is Golgi associated opsin (*193,200*). These findings could signify the local recycling of synaptophysin-containing vesicles in the cell body at least partly independent of the cycling in the nerve terminal. (As already mentioned, endocrine cells, including PC12 cells, are claimed to exhibit a recycling population of synaptophysin-containing vesicles along with their secretory bodies; see *71a,105,160* and below.)

V. ACIDIFIED COMPARTMENTS

Our studies with fluorescein probes on cultured hippocampal neurons (*7*) and work with HPTS on cultured chick sympathetic neurons (P. Hollenbeck, personal communication) indicate that the pHs of the lysosomes and endosomes of neuronal cell bodies fall in much the same range (roughly pH 5–7) as those of other cell types. The situation in axons and dendrites is much less clear.

A. Transmitter Accumulation

Synaptic vesicles possess proton transport machinery of the "vacuolar" type. The vesicles use this machinery to generate pH gradients or transmembrane potentials that motivate the transmembrane passage of transmitters and other materials via carriers, channels, or comparable paths (*59,108,143,216,238,261*). However, the degree to which the vesicles are acidified *in situ* is not known. In some neurons, the vesicles accumulate catecholamines or other weakly basic transmitters by mechanisms analogous to the ones responsible for the uptake of acidotropic weak bases by endosomes and lysosomes (*108,216*). At equilibrium, the vesicle interior may be substantially buffered by such transmitters. The transmitters for other neurons, including vertebrate photoreceptors, are acidic amino acids. Accumulation of such molecules presumably depends on the potential difference generated by the proton pump, rather than on the pH gradient (*37,143,238*). The intravesicular pH in this situation would depend on the interplay between proton pumping and the movement of the transmitters and of other anions. Circumstantial evidence suggests the existence of pathways through which chloride can cross the membranes of some of the vesicles (*37,143,210*); however, whether the vesicles possess chloride channels like those thought to influence the balances between pH gradients and potential differences in lysosomes or endosomes remains uncertain (*213*). Synaptic vesicles also contain internal anions, such as nucleotides and charged macromolecules (*119,214,217,218,256*), that could influence the distribution of cations across the vesicle membranes.

B. Studies with Weak Bases

We find that the chief sites at which photoreceptors and conventional neurons accumulate the weakly basic dyes used to stain acidified compartments for light microscopy are axonal varicosities and presynaptic terminals, as well as granule-like structures in the cell bodies and dendrites (*7a,8,214,215*; Figs. 5, 6, and 7). Similar sites accumulate the weak base DAMP, though in order to demonstrate such

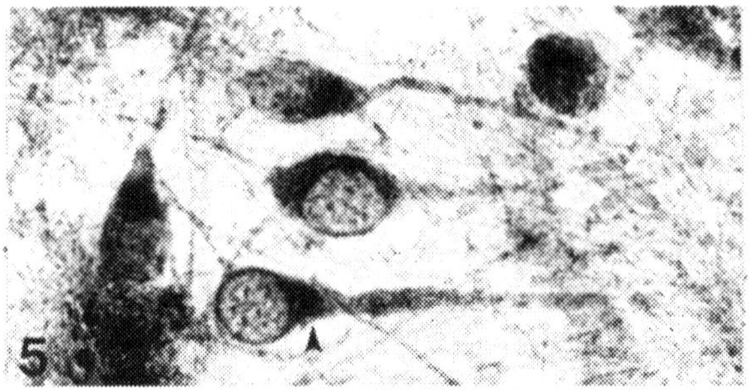

Figure 5. Light micrograph of a culture of rat hippocampal neurons stained, by an immunoperoxidase technique, to reveal the location of mannose 6-phosphate receptors (antibody against bovine receptor; courtesy of P. Lobel). Reaction product due to immunostaining is seen in the region of the Golgi apparatus (e.g. at arrowhead), which at this stage in the cells' development is located toward one pole of the nucleus. (From work of Augenbraun and Holtzman.) × 700.

accumulation reliably, we find it necessary to include endocytosable proteins in the medium used to administer DAMP (*8*); for a broader discussion of DAMP see (*5*). Apparently the endocytosed proteins provide sites, in acidified endocytic compartments, to which DAMP can be linked during fixation. Electron microscopy of DAMP preparations of frog photoreceptors and cultured hippocampal neurons indicates that the acidified sites are multivesicular bodies and other lysosome-related structures in the cell bodies and dendrites (Fig. 8), and probably small vacuoles in axonal varicosities and terminals plus some structures with the size and appearance of synaptic vesicles (Fig. 6).

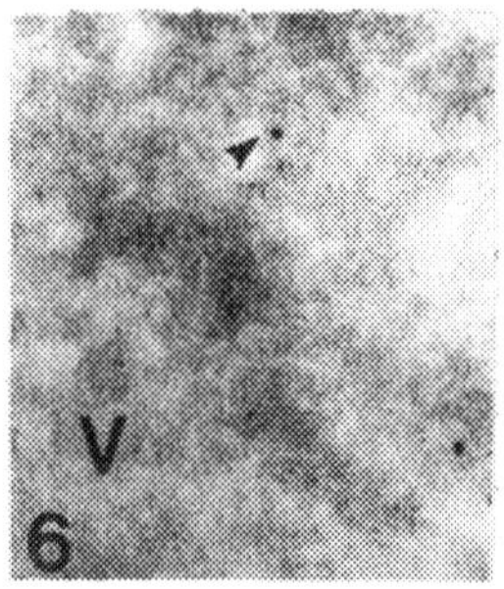

Figure 6. From a frozen semithin section of a presynaptic terminal of a frog photoreceptor in a retina incubated in DAMP and HRP by methods like those described in ref. 8 and then prepared using protein A gold as the detecting immunoreagent. The arrowhead indicates a structure with the size and appearance of a synaptic vesicle, showing a gold particle; the gold-containing structure near the lower right corner of the field probably is a similar vesicle. A synaptic vesicle exhibiting no immunostaining is present just above V. (From work of Augenbraun, St. Jules and Holtzman.) × 85,000.

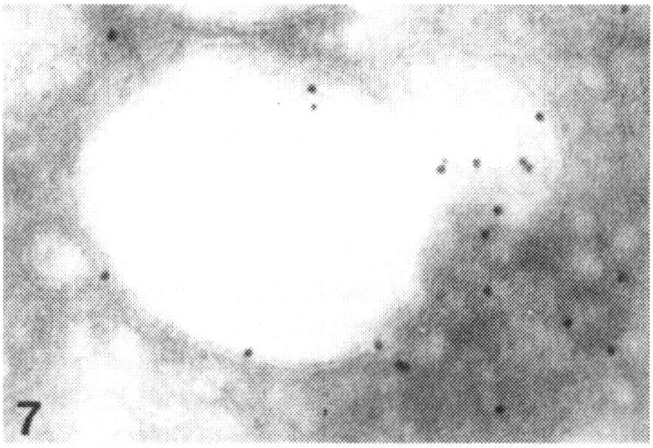

Figure 7. From a frozen semithin section of a presynaptic terminal of a frog photoreceptor in a retina incubated in NH_4Cl by methods similar to those in ref. 215 and then prepared for microscopy using anti-synaptophysin antibody and other procedures like those employed for Figure 3. Label is seen associated with the membrane of the vacuole that occupies the center of the field; synaptic vesicles lying nearby also are labeled. (From work of Augenbraun, St. Jules and Holtzman.) × 100,000.

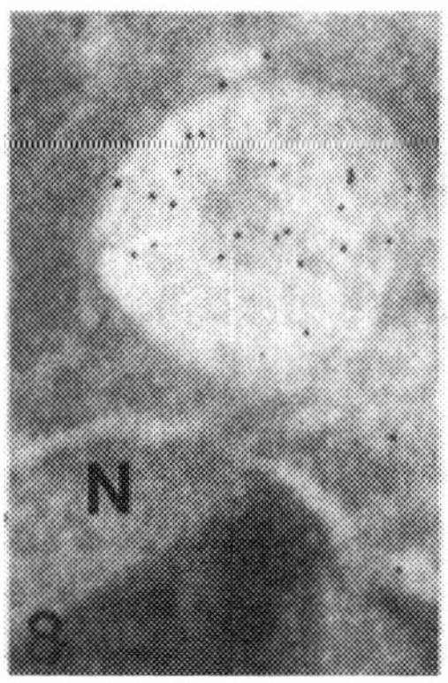

Figure 8. From the cell body of a photoreceptor in a DAMP preparation of a frog retina like that used for Figure 6. The multivesicular body occupying most of the upper part of the field, shows a concentration of gold particles. The edge of the nucleus is seen at N. As is often the case in our DAMP preparations, there are a few gold particles in the cytoplasm, not obviously associated with membrane-delimited structures. This label seems not to be accounted for by nonspecific background but we do not know whether it reflects a degree of acidification of the cytoplasm or rather represents DAMP that has escaped from the membrane bounded compartments during tissue preparation (cf. 8). (From work of Augenbraun, St. Jules and Holtzman.) × 45,000.

These last observations suggest that at least a few of the synaptic vesicles can maintain a low internal pH (if pH is a meaningful notion for structures as small as the vesicles), but we cannot yet decide whether this is true only during the period shortly after retrieval, or remains the case later when the vesicles are fully mature. An added complication is that neurons utilizing acidic amino acids as transmitters might show some degree of acidification of their cytosol during the genesis or accumulation of these molecules (see Fig. 8).

Ammonium chloride and chloroquine, two of the "vacuologenic" weak basic agents used to "swell" acidified compartments in other cell types (*30,162*), and the weak-base vital dye, Acridine Orange, produce vacuoles of a spectrum of sizes larger than the synaptic vesicles in the presynaptic terminals of photoreceptors (*211*; for comparable effects of monensin see *137*). In initial immunocytochemical studies we have found the membranes of many of these vacuoles to contain synaptophysin (Fig. 8). The bases also affect the retrieval of membrane from the cell surface. For example, in retinas treated with NH_4Cl in high-potassium media that maximize vesicle exocytosis, the photoreceptors exhibit the marked bulging of presynaptic terminal plasma membrane expected when retrieval has been unable to keep up with transmitter release (*215*). Many of the vacuoles in the photoreceptor terminals, normal or perturbed, are endocytic, and in this and other regards, resemble the intermediate compartments described above in our discussion of recycling of synaptic vesicles. Nonvacuologenic weak bases [benzylamine; tributylamine (*30*)] do not generate vacuoles as large as those seen with NH_4Cl or chloroquine, though they may sometimes retard retrieval from the cell surface or increase the abundance of smaller, endocytically derived vacuoles, somewhat larger than synaptic vesicles and resembling structures seen in untreated terminals.

Our observations suggest that there is a stage in the recycling of synaptic vesicles in photoreceptors that is responsive to changes in transmembrane pH gradients and to other perturbations in the ionic environment (*132*). Certain changes in this environment can drastically alter the recycling of synaptic vesicles from the cell surface or from intermediate compartments in the terminals (e.g. *132*). However, in the photoreceptors, and in cultured hippocampal neurons, weak bases that largely abolish acidification of the terminals do not prevent substantial endocytic labeling of synaptic vesicles, suggesting that passage through a markedly acidified compartment is not obligatory for synaptic vesicle recycling (*7,92,215*). Thus, what our findings on weak base-induced "swelling" imply for normal recycling mechanisms will only be evident when more is known about the possible coexistence of multiple routes for vesicle recycling and about: (1) possible toxic effects of the weak bases and other agents we use; (2) the heterogeneity of the synaptic compartments that respond to weak bases (*117,215*); and (3) the details of how weak bases actually affect membrane phenomena. Usual accounts of these effects center around presumed impacts on osmotic relations ("swelling"), on ligand-receptor binding, and on other factors that could affect lateral mobilities of membrane components. But simple osmotic swelling of actively acidified compartments by agents like NH_4Cl, might be expected to burst some of the compartments were it not limited by presently unknown factors (*44*), or by compensatory effects on the flux of membrane through the compartments' surfaces. Perhaps swelling is better conceptualized as a failure to shrink normally (*90*; note also that some of the "swelling" we see in nerve terminals and that others observe in other systems may owe its origins partly to intracellular fusions; cf. 215). In these connections, it is interesting that many of the mature synaptic vesicles in the terminals we study do not swell dramatically in the presence of weak bases. This might reflect resistance of the vesicles to osmotic effects (discussed in *38,190,215*). Or it might imply important differences—in proton "pumping", ion channels (*210*), internal binding sites (*151,217*), or the aggregation of the vesicles' contents—between mature vesicles and newly recycled vesicles or recycling intermediates. Perhaps the smaller of the "swollen" endocytic structures we see are acidified, early recycling compartments whose normal size is not very much larger than the synaptic vesicles (see also *53,60,149*). Changes in the proton gradient, and resistance to weak bases, might accompany the accumulation of transmitters as these compartments mature into functional vesicles.

Initial findings by Santa-Hernandez, Johnson and Holtzman (*193,194*) suggest that, under some circumstances, endocytosed sucrose or gluconate can help generate vacuoles within photoreceptor terminals, comparable to the vacuoles seen with weak bases. If so, this would seem to imply that early in their life history, the recycling compartments in the terminals are susceptible to direct perturbation of their osmotic relations as well as to the indirect perturbations produced by the bases.

"Proton pumps" from the recycling synaptic vesicle membranes, logically, are prime suspects for the enzymes producing low pHs in the endocytic structures of nerve terminals. More speculative is the possibility that operation of such pumps, or the release of protons, during the sojourn of synaptic vesicle membranes in the cell surface, acidifies nearby extracellular spaces. This type of process could help account for the perturbed retrieval seen with weak bases. Also, bear in mind that some of the effects we see could conceivably reflect alterations in the pH of the cytosol, especially if the cytosol in the terminals normally maintains a pH lower than the extracellular medium. Our findings indicate only that there probably are compartments, within the terminals, whose pH is lower still.

Traces of DAMP turn up in *trans*-Golgi associated structures in the photoreceptors, but we have not yet studied this systematically enough to establish which structures are significantly acidified. NH_4Cl and chloroquine do perturb the *trans*-Golgi systems, which would be expected were the systems at a lower pH than their surroundings. Such weak base effects, and the effects of monensin (*137–139*) are, however, only suggestive, since the agents can influence secretory pathways in more than one way (*189*).

V. ENDOSOMES

Where does the body of information discussed so far leave us with respect to the existence and roles of endosomes in nervous tissue? To some extent, as will be clear from other contributions to this book, the answer depends on what one wants to call an endosome, but little is to be gained at present in demanding a precise definition (cf. *90*). It seems useful enough to think of endosomes as a somewhat heterogeneous class of intermediate compartments which acquire endocytosed materials very soon after such materials enter the cell, and then evolve relatively rapidly through a series of stages, during which the extent of their acidification may change and they also sometimes acquire (limited) degradative capacities. By transformation, transport, or fusion, endosomes can distribute portions of their contents or membranes to other endosomes, lysosomes, the cell surface, or Golgi-associated structures.

Some "traditional" endosomes fulfilling many of these criteria almost certainly do exist in neurons, functioning in much the same way as elsewhere. The best candidates are the tubules, vacuoles, and multivesicular structures found in many regions of the neuron that take part in the accumulation of endocytic tracers which, sooner or later, are deposited mostly in lysosomes in the cell body. These structures in neurons resemble the ones in other cell types. For example, as mentioned above, bodies of the category, now called "tubular" endosomes and lysosomes (cf. *184*), are common in axons (*89*), dendrites (*166*), and cell bodies. As in other cell types (*131*), Brefeldin A increases the abundance of such tubular endocytic structures in photoreceptors (*194*). It will be important to determine directly how the properties of these and other endosomes influence neuronal handling of nutrient materials, receptors, growth factors, transcytosed toxins, and other molecules. Tetanus toxin, for example, seems to show pH- dependent interactions with membranes reminiscent of the biologically relevant interactions of other toxins with endosomal membranes (*83,161*). And the fact that at least one growth factor is fragmented by neurons (*252*) calls to mind the endosomal "processing" of viruses and proteins (cf. *90*).

A. Recycling of Synaptic Vesicles

Is it appropriate to think of the recycling of synaptic vesicles as proceeding, normally, through an endosome-dependent route? Believers in "rapid" retrieval might argue that the specialized nature of the membranes involved—the synaptic vesicle membrane, and the plasma membrane at release sites—obviates the need for large intermediates. Moreover, since synaptic vesicles can acidify themselves, intermediates could be superfluous insofar as the generation of a proton gradient is concerned. Still, intermediate compartments considered at several points above, do have certain of the characteristics of early endosomes; they are labeled by endocytic tracers, seem to be acidified, and can regenerate the smaller synaptic vesicles that eventually undergo new exocytic fusions with the cell surface.

Conceivably nerve terminals make use of bodies of diverse sizes to fulfill endosomal functions, sometimes making do with synaptic vesicle-sized structures, and sometimes needing larger bodies. Comparable diversity has been claimed for endosomes of non-neuronal cells (*53*). Perhaps the questions of vesicle coating raised earlier will be resolved along similar lines. One could imagine, for example, that synaptic vesicles can arise either from the cell surface or from endosome-like compartments by mechanisms that may not always require clathrin. Such mechanisms could resemble the suspected budding of vesicles or tubules, not coated with clathrin, from endosomes in non-neuronal cells. Clathrin might, however, be required to generate endosome-like intermediates (large and small) when retrieval from the cell surface is "delayed" (see above).

One unknown in these connections is whether the same endosomes that recycle "old" synaptic vesicles participate in the biogenesis of new ones (*182*). A related central question is whether the intermediate compartments, that can take part in recycling of synaptic vesicles, carry out differential sorting of the type associated with endosomes. Are they, for example, sites at which synaptic vesicle membrane components reaggregate before the vesicles bud off? If so, do they leave behind "unwanted" components that either return to the cell surface, or eventually are transported away from the terminals? The fact that tracers like HRP pass into apparently mature synaptic vesicles, and that unlike "traditional" endosomes, the intermediates in synaptic recycling have not been found to contribute extensively to lysosome-related compartments, might conflict with such sorting. On the other hand, even "traditional" endosomes do cycle some of their contents into structures that eventually return to the cell surface. Certain intermediates in nerve terminals exhibit a more dispersed distribution of intramembrane particles than do the synaptic vesicles, raising the possibility that the particles redistribute during vesicle genesis (*149*). Cultured neurons are able to retain endocytosed antisynaptic vesicle antibodies at sites of vesicle cycling, while sending tracers such as wheat germ agglutinin up the axons to the cell body (*142*). And, sorting is one of several possible interpretations of observations that different endocytic tracers can label different proportions of vacuole-like structures versus synaptic vesicles (*211*).

Key questions of endosome involvement also remain open for membrane retrieval in gland cells. Perhaps the lysosome-related route in gland cells (mentioned above) permits endosomal sorting of wanted from unwanted materials, both soluble and membrane-associated. Even assuming the correctness of the conclusion that secretory granule membranes remain as patches in the cell surface until they are retrieved, this may not imply that the patches are inviolable as regards diffusion of components in from the adjacent plasma membrane (*190*). One problem, applying both to gland cells and to synapses, is that we lack detailed information about the composition of the plasma membrane domains at the sites where exocytosis takes place. One could argue, for example, that nerve terminals do not ordinarily need to sort synaptic vesicle components from the receptors with which "traditional" endosomes must deal because the geometry of organization of the neuronal surface minimizes intermixing of these two classes of membrane molecules.

Recall also that membrane recycling via "traditional" endosomes in non-neuronal cells has been studied largely by following components that return rapidly to the plasma membrane or go to mature lysosomes. Much less attention has been devoted to components (e.g., elements of the acidification system; *78a*) that might have other fates, such as remaining within the cell and participating in the intracellular phases of the formation of "new" endosomes (*90*). In some respects, recycled synaptic vesicles would more closely resemble this last class of components than the others.

B. Recycling of Synaptophysin

The membrane protein, synaptophysin mentioned several times already, is prominent in the synaptic vesicles of diverse types of neurons (*105,200,230,244*). It also is found in a recycling vesicle population, still only partly characterized, that is present in endocrine gland cells along with the cells' secretory granules (*31,105,160*). When expressed, after transfection, in cells that ordinarily do not make it, the protein can enter membranes that recycle through the cell surface (*31,127,129*).

One report suggested that, when transfected, cells that are neither neuronal or endocrine in origin can generate a class of vesicles in which synaptophysin is the principal membrane protein (*127*). This might imply that the genesis and sorting of synaptic vesicle membrane depends relatively simply upon information built into the membrane components, related perhaps to features responsible for the assembly of the oligomers in which the synaptophysin of synaptic vesicles is found.

However, laboratories working on other cell types have failed to confirm such segregation. They find instead that only neurons and endocrine cells can segregate synaptophysin into vesicles lacking recyclable membrane proteins which are not synaptic vesicle constituents, such as transferrin receptors (*31,129*). Of special interest in the context of this chapter is that the cells capable of such segregation also form compartments in which synaptophysin and the nonsynaptic vesicle proteins do colocalize. This could mean that recycling synaptophysin and the other proteins are first retrieved together into endosomes and then are separated. Kinetic data, and studies on the details of localization of the proteins at the cell surface and elsewhere, are needed to definitively rule out alternative possibilities, such as the existence of more than one, independently cycling populations of membranes with different composition.

Transfection experiments with modified synaptophysin DNA suggests that synaptophysin's cytoplasmic, C-terminal domain is necessary for efficient endocytic internalization of the protein in 3T3 cells (*130*). Additional transmembrane or lumenal domains, as well as an appropriate context provided by specific types of cells, may be needed for sorting into synaptic vesicle-like packages (*112*).

VII. CYTOSKELETAL PARTICIPATION

The cast of cytoskeletal characters that controls the behavior of the structures we have been considering overlaps that in other cell types, though the nervous system may contain its own forms of some of the proteins (*25,198,258*). Oriented arrays of microtubules and filaments are prominent in neurons. Potential "motors" for membrane-delimited compartments have been identified (*80–82*) including proteins that may help move synaptic vesicles or their precursors (*72*). Kinesin is among the likely anterograde motors, and dynein among the probable retrograde motors. Certain of the other proteins that have been tentatively implicated in vesicle movements or in exocytosis, notably synapsins (*241,244*) and rab3A (*141,213*), may regulate or otherwise participate in the interactions of the vesicle membranes with the cytoskeleton (*43,232*) and the plasma membrane; probable controls include phosphorylations (*241*), GTP-GDP cycles (*213*), and the like. Related contributions of calcium have long been suspected from electrophysiology and the effects of toxins (e.g., *32,35,43,108,189,242,261*); suspected participation by particular membrane proteins is receiving considerable current attention (e.g., *92,112,170*).

Much more is known about the involvement of the cytoskeleton in delivery of molecules to terminals, in retrograde transport, and in exocytosis (*3,80,232*) than about endocytosis and subsequent events. [The recent observation that the *shibire* mutant of *Drosophila* affects a dynamin-like protein may portend a change in this situation (*35a,198,245*).] For our present interests, one important general conclusion is that the long-range migrations of synaptic vesicle components and endocytic structures, and facets of the organization of elongate intraaxonal membrane systems (*176*), are governed significantly by formed cytoskeletal elements such as the microtubules (*3,80–82*). Local circulation of synaptic vesicles in terminals probably involves more specialized components including synapsins (*43,105,244*). Neurons, such as the photoreceptors that lack certain synapsins (*43,134b*), may substitute the still mysterious components of the photoreceptors' synaptic ribbons along which synaptic vesicles—or, sometimes, recycling intermediates (*132*)—accumulate before exocytosis. But it still not clear why, for example, synaptic vesicles circulate principally within circumscribed regions of the nerve terminal (*15*), even though larger bodies thought to participate in degradation of the vesicles can be transported up the axon. Synaptic vesicle-rich cell fractions have been found to contain cytoplasmic dynein (*123*). Although the distribution of this protein in the fractions has yet to be definitively determined, these findings could imply that many of the vesicles possess a "retrograde motor" for transport along axonal microtubules. Must one then invoke geometric factors, such as restricted access to microtubules, or the participation of agents that regulate dynein (e.g., *67*), to explain the vesicles' behavior?

In trying to understand the regulation of migration of newly made membrane components, endocytic structures, and lysosomes in neurons, attention has begun to focus on differences between axons and dendrites. In at least some cases, these regions differ in microtubule associated proteins (*3,68,80–82*). Moreover, in axons (of the very few neurons studied), microtubules mostly have their (+) ends oriented away from the cell body (toward the terminals), whereas in dendrites microtubules with both orientations are intermixed. Thus, for example, one might expect structures whose movements are biased toward the (–)-end direction to be scarce in axons relative to dendrites. Correspondingly, rather than depending on selective restrictions in anterograde axonal transport, the paucity of mature lysosomes in axons as compared with dendrites, and the abundance of lysosomes near the neuron's Golgi region, could reflect tendencies of lysosomes and maturing endocytic structures to move toward the (–) ends of microtubules (*3,80–82*). [Concentrations of (–) ends would be expected at the microtubule organizing centers that often seem to lie close to the Golgi apparatus.] Similarly, imprecisions in sorting of materials between the opposite poles of the neuron might partly be due to the availability in dendrites, as well as in axons, of microtubules with their (+) ends pointing away from the cell body. And, the seeming insensitivity of opsin transport in photoreceptors to "antimicrotubule" agents (*248*; but see also *118*) might reflect distinctive transport mechanisms related to the fact that the microtubules in the photoreceptor cell body are oriented with their (+) ends pointing away from the outer segment (*234*).

VIII. PERSPECTIVES

The information presented above does not tie neatly into a simple package. We will end, therefore simply with the cautionary note that it is probably wisest to keep an open mind as to the degree to which the details of phenomena in neurons are shadowed in other cells. No doubt there are fundamental similarities among comparable processes in different cell types, both at the molecular level and at more complex levels. At times, however, the differences prove as interesting as the similarities. The issue of the involvement of endosomes in synaptic vesicle recycling could provide an example of this. Even, for instance, if one assumes that the recycling process is closely tied evolutionarily, to receptor-mediated endocytosis, we will still be left puzzled by such issues as the absence of evident ligand binding during retrieval of synaptic vesicle membranes. It will also be important to establish whether such phenomena as the use of clathrin-coated endocytic vesicles and endosomes to help alter CAM-dependent interactions among *Aplysia* neurons (*8a*) are unique to the nervous system or rather represent particular employment of general mechanisms.

ACKNOWLEDGMENTS

The work of our laboratory is supported by NEI grant 03168. R. St. Jules has received support from Training Grant NS07258, and M. Santa-Hernandez has received support from an NIH-funded American Psychological Association Minority Fellowship program and from a Danforth Foundation program administered by Columbia University. We thank Wanda Setlik for extraordinary technical assistance, and Drs. R.G.W. Anderson, E. Fujimoto, P. Lobel, M. Rosenthal, and F. Valtorta for antibodies or other reagents. We are grateful for preprints and other information provided to us by Drs. W. Betz, P. De Camilli, N.K. Gonatas, P. Greengard, W.B. Huttner, R. Kelly, B.G. Schneider, K. Simons, F. Valtorta, H. Zimmermann, and L. Zipursky.

REFERENCES

1. Abrahams, S.J. and Holtzman, E. (1973). Secretion and endocytosis in insulin-stimulated rat adrenal medulla. *J. Cell Biol.* 56: 540–558.
2. Aizenman, Y., Wechsel, M.E., and Vellis, J. (1986). Changes in insulin and transferrin requirements of pure brain cultures during embryonic development. *Proc. Natl. Acad. Sci. USA* 83: 2263-2266.
3. Allan, V.J., Vale, R.D., and Navone, F. (1991). Microtubule based organelle transport in neurons. In: *The Neuronal Cytoskeleton*, pp. 257-282 (Burgoyne, R.D., Ed.). New York: Wiley-Liss.
4. Almers, W. and Tse, F.W. (1990). Transmitter release from synapses: does a presassembled fusion pore initiate exocytosis? *Neuron* 4: 813–818.
5. Anderson, R.G.W. and Orci, L. (1988). A view of acidic intracellular compartments. *J. Cell Biol.* 106:539–544.
6. Arikawa, K., Molday, L.L., Molday, R.S., and Williams, D.S. (1992). Localization of peripherin/rds in the disk membranes of cone and rod photoreceptors: relationship to disk membrane morphogenesis and retinal degeneration. *J. Cell Biol.* 116: 659–668.
7. Augenbraun, E., Maxfield, F.R., St. Jules, R., Setlik, W., and Holtzman, E. (1993). Properties of acidified compartments in hippocampal neurons. *Eur. J. Cell Biol.* (In press).
7a. Augenbraun, E. and Holtzman, E. (1990). Microscopic studies of low pH compartments in neurons and photoreceptors. *Abs. Soc. Neurosci.* 16: 500.
8. Augenbraun, E., Sulzer, D., Rayport, S., Setlik, W., and Holtzman, E. (1990). Experiments on the use of DAMP to study retina and cultured neurons. *J. Histochem. Cytochem.* 38: 1927–1931.
8a. Bailey, C.H., Chen, M., Keller, F., and Kandel, E.R. (1992). Serotonin-mediated endocytosis of apCAM: an early step of learning related synaptic growth in *Aplysia. Science* 256: 645–649.
9. Balch, W.E. (1990). Molecular dissection of early stages of the eucaryotic secretory pathway. *Curr. Op. Cell Biol.* 2: 634–641.
10. Barde, Y-A. (1989). Tropic factors and neuronal survival. *Neuron* 2: 1525–1534.
11. Bartles, J.R. and Hubbard, A.L. (1988). Plasma membrane protein sorting in epithelial cells: do secretory pathways hold the key? *TIBS* 13: 181–184.
12. Baskin, D.G., Wilcox, B.J., Figlewicz, D.P., and Dorsa, D.M. (1988). Insulin and insulin-like growth factors in the central nervous system. *Trends Neurosci.* 11: 106–111.
13. Bennett, M.K., Calakos, N., Kreiner, T., and Scheller, R.H. (1992). Synaptic vesicle membrane proteins interact to form a multimeric complex. *J. Cell Biol.* 116: 761–775.
14. Besharse, J.C. (1986). Photosensitive membrane turnover: differentiated membrane domains and cell-cell interaction. In: *The Retina: A Model for Cell Biology Studies*, pp. 297–352 (Adler, R. and Farber, D., Eds.). Academic Press, Orlando FL.
15. Betz, W.J., Mao, F., and Bewick, G.S. (1992). Activity dependent fluorescent staining and destaining of living vertebrate motor nerve terminals. *J. Neurosci.* 12: 363–373.
16. Bird, A.C., Flannery, J.G., and Bok, D. (1988). A diurnal rhythm in opsin content of *Rana pipiens* rod inner segments. *Invest. Ophthalmol.* 29: 1028–1039.
17. Blessing, W.W., Li, Y-W., and Wesselingh, S.L. (1991). Transneuronal transport of Herpes simplex virus from the cervical vagus to brain neurons with axonal inputs to central vagal sensory nuclei in the rat. *Neurosci.* 42: 261–274.
18. Boarder, M.R. (1989). Presynaptic aspects of cotransmission: relationship between vesicles and neurotransmitters. *J. Neurochem.* 53: 1–11.
19. Bok, D. (1985). Retinal photoreceptor-pigment epithelium interactions. *Invest. Ophthalmol.* 26: 1659–1694.
20. Bonzelius, F. and Zimmermann, H. (1990). Recycled synaptic vesicles contain vesicle but not plasma membrane marker, newly synthesized acetylcholine and a sample of extracellular medium. *J. Neurochem.* 55: 1266–1273.
21. Boyle, D., Tien, L., Cooper, N.G.F., Shepherd, V., and McLaughlin, B.J. (1991). A mannose receptor is involved in retinal phagocytosis. *Invest. Ophthalmol.* 32: 1464–1470.

22. Broadwell, R.D. and Brightman, M.W. (1979). Cytochemistry of undamaged neurons transporting exogenous proteins *in vivo*. *J. Comp. Neurol.* 185: 31–74.

22a. Broadwell, R.D. and Cataldo, A.M. (1983). The neuronal endoplasmic reticulum: its cytochemistry and contributions to the endomembrane system. *J. Histochem. Cytochem.* 31: 1077–1088.

23. Broadwell, R.D., Cataldo, A.M., and Balin, B.J. (1984). Further studies of the secretory process in hypothalamo-neurohypophysial neurons: an analysis using immunocytochemistry, wheat germ agglutinin peroxidase and native peroxidase. *J. Comp. Neurol.* 228: 155–167.

24. Broadwell, R.D., Oliver, C., and Brightman, M.W. (1979). Localization of neurophysin within organelles associated with protein synthesis and packaging in the hypothalamo-hypophyseal system: an immunocytochemical study. *Proc. Natl. Acad. Sci. USA* 76: 5999–6003.

25. Brodsky, F.M., Hill, B.L., Acton, S.L., Näthke, I., Wong, D.H., Ponnambalam, S., and Parham, P. (1991). Clathrin light chains: arrays of protein motifs that regulate coated-vesicle dynamics. *Trends Biochem. Sci.* 16: 208–213.

26. Buckley, K.R., Schwietzer, E.S., Miljanisch, G.P., Clift-O'Grady, L., Kushner, P.D., Reichardt, L.F., and Kelly, R.B. (1983). A synaptic vesicle antigen is restricted to the junctional region of the presynaptic plasma membrane. *Proc. Natl. Acad. Sci. USA* 80: 7342–7346.

27. Bunge, M.B. (1977). Initial endocytosis of peroxidase or ferritin by growth cones of cultured nerve cells. *J. Neurocytol.* 6: 407–439.

28. Bursztajn, S. (1984). Coated vesicles are associated with acetylcholine receptor at nerve-muscle contact. *J. Neurocytol.* 13: 503–518.

29. Cabot, J.B., Mennone, A., Bogan, N., Carroll, J., Evinger, C., and Erichsen, J.T. (1991). Retrograde, transsynaptic and transneuronal transport of fragment C of tetanus toxin by sympathetic preganglionic neurons. *Neurosci.* 40: 805–823.

30. Cain, C.C. and Murphy, R.F. (1986). Growth inhibition of 3T3 fibroblasts by lysosomotrophic amines: correlation of effects on intravesicular pH but not vacuolation. *J. Cell Physiol.* 129: 65–70.

31. Cameron, P.L., Sudhof, T.C., Jahn, R., and De Camilli, P. (1991). Colocalization of synaptophysin with transferrin receptors: implications for synaptic vesicle biogenesis. *J. Cell Biol.* 115: 151–164.

32. Ceccarelli, B. and Hurlburt, W.P. (1980). Vesicle hypothesis of the release of quanta of acetylcholine. *Physiol. Rev.* 60: 396–441.

33. Ceccarilli, B., Hurlburt, W.P., and Mauro, A. (1973). Turnover of transmitter and synaptic vesicles at the frog neuromuscular junction. *J. Cell Biol.* 57: 499–524.

34. Chang, T.P.O. and Reese, T.S. (1987). Recycling of plasmalemma in chick tectal growth cones. *J. Neurosci.* 7: 1752–1759.

35. Chege, N.W. and Pfeffer, S.R. (1990). Compartmentation of the Golgi complex: Brefeldin A distinguishes *trans*-Golgi cisternae from the *trans*-Golgi network. *J. Cell Biol.* 111: 893–899.

35a. Chen, M.S., Obar, R.A., Schroeder, C.S., Austin, T.W., Poodry, C.A., Wadsworth, S.C., and Vallee, R.B. (1991). Multiple forms of dynamin are encoded by *shibire*, a *Drosophila* gene involved in endocytosis. *Nature* 351: 583–586.

36. Chow, R.H., van Rüden, L., and Neher, E. (1991). Delay in vesicle fusion revealed by electrochemical monitoring of single secretory events in adrenal chromaffin cells. *Nature* 356: 60–63.

37. Cidon, S. and Sihra, T.S. (1989). Characterization of a H^+-ATPase in rat brain synaptic vesicles. *J. Biol. Chem.* 264: 8281–8288.

38. Clementi, F. (Ed). (1989). Neurotransmitter release: the neuromuscular junction. *Cell Biol. Int. Reps.* 13: 981–1075.

39. Cooper, M.S., Cornell-Bell, A.M., Chernjavsky, A., Dani, J.W., and Smith, S.O. (1990). Tubulovesicular processes emerge from *trans* Golgi cisternae, extend along microtubules and interlink adjacent *trans* Golgi elements into a reticulum. *Cell* 61: 135–145.

40. Cooper, N.G.F. and McLaughlin, B.J. (1983). Tracer uptake by photoreceptor synaptic terminals. *J. Ultrast. Res.* 84: 252–267.

41. Cull-Candy, S.G., Fohlman, J., Gustavsson, D., Lullman-Rausch, R., and Thesleff, S. (1976). The effects of taipoxin and notexin on the function and fine structure of the murine neuromuscular junction. *Neurosci.* 1: 175–180.
42. Cutler, D.F. and Cramer, L.P. (1990). Sorting during transport to the surface of PC12 cells: divergence of synaptic vesicle and secretory granule proteins. *J. Cell Biol.* 110: 721–730.
43. DeCamilli, P. and Jahn, P. (1990). Pathways to regulated exocytosis in neurons. *Ann. Rev. Physiol.* 52: 625–645.
44. DeCourcy, K. and Storrie, B. (1991). Osmotic swelling of endocytic compartments induced by internalized sucrose is restricted to mature lysosomes in cultured mammalian cells. *Exp. Cell Res.* 192: 52–60.
45. DePotter, W.P. and Chubb, I.W. (1977). Biochemical observations on the formation of small adrenergic vesicles in the splenic nerve of the dog. *Neurosci.* 2: 167–174.
46. Deretic, D. and Papermaster, D.S. (1991). Polarized sorting of rhodopsin on post Golgi membranes in frog retinal photoreceptor cells. *J. Cell Biol.* 113: 1281–1293.
47. Dotti, C.G. and Simons, K. (1990). Polarized sorting of viral glycoproteins to the axon and dendrites of hippocampal neurons in culture. *Cell* 62: 63–72.
48. Dotti, C.G., Parton, R.G., and Simons, K. (1991). Polarized sorting of glypiated proteins in hippocampal neurons. *Nature* 349: 158–161.
49. Dumas, M., Schwab, M.C., and Thoenen, H. (1979). Retrograde axonal transport of specific macromolecules as a tool for characterizing nerve terminals and membranes. *J. Neurobiol.* 10: 179–197.
50. Eguchi, E. and Waterman, T.M. (1976). Freeze-etch and histochemical evidence for cycling in crayfish photoreceptor membranes. *Cell Tissue Res.* 169: 418–434.
51. Ellisman, M.H. and Lindsey, J.D. (1983). The axonal endoplasmic reticulum within myelinated axons is not transported rapidly. *J. Neurocytol.* 12: 393–411.
52. Ellisman, M.H., Lindsey, J.D., Carragher, B.O., Kiyonaga, S.H., McEwan, L.R., and McEwen, B.F. (1990). Three-dimensional tomographic reconstructions of components of the Golgi apparatus imaged by selective staining and high voltage electron microscopy. *J. Cell Biol.* 11: 199A.
53. Eskelinen, S., Kok, J.W., Sormunen, R., and Hoekstra, D. (1991). Coated endosomal vesicles: sorting and recycling compartments for transferrin in BHK cells. *Eur. J. Cell Biol.* 56: 210–222.
54. Evans, J.A. and Holtzman, E. (1982). Thiamine pyrophosphatase localization in neuronal axons and terminals. *J. Histochem. Cytochem.* 40: 489–490.
55. Evans, J., Hood, D., and Holtzman, E. (1978). Differential effects of cobalt ions on rod and cone synaptic activity in the isolated frog retina. *Vis. Res.* 18: 145–151.
56. Evans, J., Liscum, L., Hood, D.C., and Holtzman, E. (1981). Uptake of horseradish peroxidase by presynaptic terminals of bipolar cells and photoreceptors of the frog retina. *J. Histochem. Cytochem.* 29: 511–518.
57. Fahim, M.A., Lasek, R.J., Brady, S.T., and Hodge, A.J. (1985). AVEC-DIC and electron microscopic analyses of axonally transported particles in cold block squid giant axons. *J. Neurocytol.* 14: 689–704.
58. Farquhar, M.G. (1981). Membrane recycling in secretory cells. *Meth. Cell Biol.* 23: 400–428.
58a. Fliesler, S.J. and Basinger, S.F. (1987). Monensin stimulates glycerolipid incorporation into rod outer segment membranes. *J. Biol. Chem.* 262: 17516–17523.
59. Floor, E., Leventhal, P.S., and Schaeffer, S.F. (1990). Partial purification and characterization of the vacuolar ATPase of mammalian synaptic vesicles. *J. Neurochem.* 55: 1663–1670.
60. Fried, R.C. and Blaustein, M.P. (1978). Retrieval and recycling of synaptic vesicle membrane in pinched off nerve terminals (synaptosomes). *J. Cell Biol.* 78: 685–700.
61. Fujimoto, T. and Ogawa, K. (1989). Retrieving vesicles in secretion-induced rat chromaffin cells contain fodrin. *J. Histochem. Cytochem.* 37: 1589–1599.
62. Gainer, H. (1978). Intercellular transfer of proteins from glial cells to axons. *Trends Neurosci.* 1: 93–96.

63. Gainer, H., Russell, J.T., and Brownstein, M.J. (1982). Axonal transport, neurosecretory vesicles and the endocrine neuron. In: *Axoplasmic Transport in Physiology and Pathology*, pp. 44–50 (Weiss, D.G. and Gorio, A., Eds.). Springer-Verlag, Berlin.
64. Garner, J.A. and Mahler, H.R. (1987). Biogenesis of presynaptic terminal proteins. *J. Neurochem.* 49: 905–915.
65. Gatzinksy, K.P., Berthold, C-H., and Corneliuson, O. (1988). Acid phosphatase activity at nodes of Ranvier in alpha motor neurons and dorsal ganglion neurons of the cat. *J. Neurocytol.* 17: 531–544.
66. Gennaro, J., Nastuk, W., and Rutherford, D. (1978). Reversible depletion of synaptic vesicles induced by application of high external potassium to frog neuromuscular junctions. *J. Physiol. (Lond.)* 280: 237–247.
67. Gill, S.R., Schroer, T.A., Szilak, I., Steuer, E.R., Sheetz, M.P., and Cleveland, D.W. (1991). Dynactin, a conserved, ubiquitously expressed component of an activator of vesicle motility mediated by cytoplasmic dynein. *J. Cell Biol.* 115: 1639–1650.
68. Ginzburg, I. (1991). Neuronal polarity: targeting of microtubule components into axons and dendrites. *Trends in Biochem. Sci.* 16: 257–261.
69. Goldfischer, S. (1982). The internal reticular apparatus of Camillo Golgi. *J. Histochem. Cytochem.* 30: 717–733.
69a. Gordon, W.C. and Bazan, N.G. (1990). Dorosahexaneonic acid utilization during rod photoreceptor cell renewal. *J. Neurosci.* 10: 2190–2202.
70. Grafstein, B. and Forman, D.S. (1980). Intracellular transport in neurons. *Physiol. Rev.* 60: 1167–1283.
71. Green, S.A. and Kelly, R.B. (1990). Endocytic membrane traffic to the Golgi apparatus of a regulated secretory cell line. *J. Biol. Chem.* 265: 21269–21275.
71a. Green, S.A. and Kelly, R.B. (1992). Low density lipoprotein receptors and cation-independent mannose-6-phosphate receptors are transported from the cell surface to the Golgi apparatus at equal rates in PC12 cells. *J. Cell Biol.* 117: 47–56.
71b. Haass, C., Koo, E.H., Mellon, A., Hung, A.Y., and Selkoe, D.J. (1992). Targeting of cell surface β-amyloid precursor protein to lysosomes: alternative processing into amyloid-bearing fragments. *Nature* 357: 500–503.
72. Hall, D.H. and Hedgecock, E.M. (1991). Kinesin-related gene unc104 is required for axonal transport of synaptic vesicles in *C. elegans*. *Cell* 65: 837–847.
73. Hama, K. and Saito, K. (1977). Fine structure of the afferent synapse of the hair cells in the saccular macula of the goldfish with special reference to the anastomosing tubules. *J. Neurocytol.* 6: 361–373.
74. Hammerschlag, R., Stone, G.C., Bolen, F.A., Lindsey, J.D., and Ellisman, M.H. (1982). Evidence that newly synthesized proteins destined for fast axonal transport pass through the Golgi apparatus. *J. Cell Biol.* 93: 568–575.
75. Hansen, S.H., Sandvig, K., and van Deurs, B. (1991). The preendosomal compartment comprises distinct coated and noncoated endocytic vesicle populations. *J. Cell Biol.* 113: 731–741.
76. Herzog, V. (1981). Pathways of endocytosis in secretory cells. *Trends. Biochem. Sci.* 6: 319–322.
77. Heuser, J.E. (1989). The role of coated vesicles in recycling of synaptic vesicle membrane. *Cell Biol. Int. Repts.* 13: 1063–1075.
78. Heuser, J. (1989). Review of electron microscopic evidence favoring vesicle exocytosis as the structural basis for quantal release during synaptic transmission. *Quart. J. Exp. Physiol.* 74: 1051–1069.
78a. Heuser, J.E., Padh, H., and Steck, T.L. (1990). Deep etching EM of proton pumps in the acidosomes of Dictyostelium amebae. *J. Cell Biol.* 111: 315A.
79. Heuser, J.E. and Reese, T.S. (1973). Evidence for recycling of synaptic vesicle membrane during transmitter release at the frog neuromuscular junction. *J. Cell Biol.* 57: 315–344.
80. Hirokawa, N. (1991). Molecular architecture and dynamics of the neuronal cytoskeleton. In: *The Neuronal Cytoskeleton*, pp. 5–74 (Burgoyne, R.D., Ed.). Wiley-Liss, New York.

81. Hirokawa, N., Sato, Y., Yoshida, T., and Kawashima, T. (1990). Brain dynein localizes on both anterogradely and retrogradely transported membranous organelles *in vivo*. *J. Cell Biol.* 111: 1027–1037.
82. Hirokawa, N., Sato-Yoshitake, R., Kobayashi, N., Pfister, K.K., Bloom, G.S., and Brady, S.T. (1991). Kinesin associates with anterogradely transported membranous organelles *in vivo*. *J. Cell Biol.* 114: 295–302.
83. Hoch, D.H., Romero-Mira, M., Ehrlich, B.E., Finklelstein, A., DasGupta, B.R., and Simpson, L.L. (1985). Channels formed by botulinum, tetanus and diphtheria toxins in planar lipid bilayers: relevance to translocation of protein across membranes. *Proc. Natl. Acad. Sci. USA* 82: 1692–1696.
84. Hollyfield, J.G. and Rayborn, M.E. (1987). Endocytosis in the inner segment of rod photoreceptors: analysis of *Xenopus* laevis retinas using horseradish peroxidase. *Ext. Eye Res.* 45: 703–720.
85. Hollyfield, J.G., Varner, H.H., Rayborn, M.E., Liou, G.I., and Bridges, C.D. (1985). Endocytosis and degradation of interstitial retinol-binding protein: differential capabilities of cells that border the interphotoreceptor matrix. *J. Cell Biol.* 100: 1676–1681.
86. Holtzman, E. (1969). Lysosomes in the physiology and pathology of neurons. In: *Lysosomes in Biology and Pathology*, Vol. 1, pp. 192–216 (Dingle, J.T. and Fell, H., Eds.). North Holland, Amsterdam.
87. Holtzman, E. (1971). Cytochemical studies of protein transport in the nervous system. *Phil. Trans. Roy. Soc. Lond. B.* 261: 407–421.
88. Holtzman, E. (1976). *Lysosomes, A Survey*. Springer-Verlag, Vienna.
89. Holtzman, E. (1977). The origin and fate of secretory packages, especially synaptic vesicles. *Neurosci.* 2: 327–355.
90. Holtzman, E. (1989). *Lysosomes*. Plenum Press, New York.
91. Holtzman, E. (1992). Intracellular targeting and sorting. *Biosci.* 42: 608–620.
92. Holtzman, E. (1992). Neuronal membrane traffic. *Curr. Op. Neurobiol.* 2: 607–612.
93. Holtzman, E. and Dominitz, R. (1968). Cytochemical studies of lysosomes, Golgi apparatus and endoplasmic reticulum in secretion and protein uptake by adrenal medulla cells of the rat. *J. Histochem. Cytochem.* 16: 320–336.
94. Holtzman, E. and Mercurio, A. (1980). Membrane circulation in neurons and photoreceptors: some unresolved issues. *Int. Rev. Cytol.* 67: 1–67.
95. Holtzman, E. and Novikoff, A.B. (1965). Lysosomes in the rat sciatic nerve following crush. *J. Cell Biol.* 27: 651–669.
96. Holtzman, E. and Peterson, E.R. (1969). Protein uptake by mammalian neurons. *J. Cell Biol.* 40: 863–869.
97. Holtzman, E., Freeman, A.R., and Kashner, L.A. (1971). Stimulation dependent alterations in peroxidase uptake by lobster neuromuscular junctions. *Science* 173: 733–736.
98. Holtzman, E., Novikoff, A.B., and Villaverde, H. (1967). Lysosomes and GERL in normal and chromatolytic neurons of the rat ganglion nodosum. *J. Cell Biol.* 33: 419–436.
99. Holtzman, E., Schacher, S., Evans, J., and Teichberg, S. (1977). Origin and fate of the membranes of secretion granules and synaptic vesicles: membrane circulation in neurons, gland cells and retinal photoreceptors. In: *Cell Surface Reviews 4: The Synthesis, Assembly and Turnover of Cell Surface Components*, pp. 165–246 (Poste, G. and Nicholson, G.L., Eds.). Associated Scientific Publishers (ASP: Elsevier), Amsterdam.
100. Holtzman, E., Teichberg, S., Abrahams, S.J., Citkowitz, E., Crain, S.M., Kawai, N., and Peterson, E.R. (1973). Notes on synaptic vesicles and related structures, endoplasmic reticulum, lysosomes and peroxisomes in nervous tissue and the adrenal medulla. *J. Histochem. Cytochem.* 21: 349–385.
101. Hurlburt, W.P. (1989). The correlation between vesicle loss and quantal secretion at the frog neuromuscular junction. *Cell Biol. Int. Repts.* 13: 1053–1062.
102. Huttner, W.B., Gerdes, H-H., and Rosa, P. (1991). The granin (chromogranin/secretogranin) family. *Trends Biochem. Sci.* 16: 27–30.

103. Hyndman, A.G. and Zeevalk, G.D. (1988). Transferrin and iron in cultured chick embryonic neurons: a comparison between human and chick transferrins. *J. Cell Physiol.* 134: 238–244.

104. Ignatius, M.J., Shooter,, E.M. Pitas, R.E., and Mahley, R.W. (1987). Lipoprotein uptake by neuronal growth cones *in vitro*. *Science* 236: 959–962.

105. Jahn, R. and De Camilli, P. Membrane proteins of synaptic vesicles: markers for neurons and neuroendocrine cells and tools for the study of neurosecretion. In: *Markers for Neurons and Endocrine Cells* (Gratzl, M. and Langley, K., Eds.). Verlaggesellschaft Weinheim, Germany. (In press, (1991).

106. Janetzko, A., Zimmermann, H., and Volknandt, W. (1989). Intraneuronal distribution of a synaptic vesicle membrane protein; antibody binding sites at axonal membrane compartments and *trans* Golgi network and accumulation at nodes of Ranvier. *Neurosci.* 32: 65–77.

107. Jirikowski, G.F., Sanna, P.P., and Bloom, F.E. (1990). mRNA coding for oxytocin is present in axons of the hypothalamo-neurohypophyseal tract. *Proc. Natl. Acad. Sci. USA* 87: 7400–7404.

108. Johnson, R.G. (Ed). *Cellular and Molecular Biology of Hormone and Neurotransmitter-containing Secretory Vesicles.* New York: N.Y. Acad. Sciences. (N.Y. Acad. Sci. Annals, Vol. 493), 1987.

109. Joseph, K.C., Stieber, A., and Gonatas, N.K. (1979). Endocytosis of cholera toxin in GERL-like structures of murine neuroblastoma cells pretreated with GM1 ganglioside: cholera toxin internalization into neuroblastoma GERL. *J. Cell Biol.* 81: 543–554.

110. Jung, L.J. and Scheller, R.H. (1991). Peptide processing and targeting in the neuronal secretory pathway. *Science* 251: 1330–1335.

111. Kagotani, Y., Picart, R., Barret, A., Wiedenmann, B., Huttner, W.B., and Tixier-Vidal, A. (1991). Subcellular localization of secretogranin II and synaptophysin by immunoelectron microscopy in differentiated hypothalamic neurons in culture. *J. Histochem. Cytochem.* 39: 1507–1518.

112. Kelly, R.B. (1993). Storage and release of neurotransmitters. *Cell 72/Neuron 10* (Suppl.): 43–53.

113. Kartenbeck, J., Schmelz, M., Franke, W.W., and Geiger, B. (1991). Endocytosis of junctional cadherins in bovine kidney epithelial (MDBK) cells cultured in low calcium medium. *J. Cell Biol.* 113: 881–892.

114. Keller, R.K. and Fliesler, S.J. (1990). Brefeldin A blocks transport of proteins, but not lipids, to the rod outer segment. *Invest. Ophthalmol.* 31: 470 (ARVO Absts.).

115. Keller, R.K. and Fliesler, S.J. (1991). Intracellular transport of phospholipids, squalene and protein to the rod outer segment is differentially affected by temperature and energy inhibitors. *Invest. Ophthalmol.* 32: 1009 (ARVO Abstrs.)

116. Kelly, R.B. (1988). The cell biology of the nerve terminal. *Neuron* 1: 431–438.

117. Kiene, M-L. and Stadler, H. (1987). Synaptic vesicles in electromotoneurons: I. Axonal transport, site of transmitter uptake and processing of a core proteoglycan during maturation. *EMBO J.* 6: 2209–2215

118. Kirsch, M. and Burnside, B. (1992). Opsin transport in fish rod photoreceptors. *Invest. Ophthalmol.* 33: 738 (ARVO Absts).

119. Klein, R.L. and Thureson-Klein, Å.K. (1990). Neuropeptide costorage and exocytosis by neuronal large dense-cored vesicles: how good is the evidence? *Curr. Asp. Neurosci.* 2: 219–258.

120. Koening, J.H. and Ikeda, K. (1989). Disappearance and reformation of synaptic vesicle membrane upon transmitter release observed under reversible blockage of membrane retrieval. *J. Neurosci.* 9: 3844–3860.

121. Kramer, H., Cagan, R.L., and Zipursky, S.L. (1991). Interaction of bride of sevenless and the sevenless tyrosine kinase receptor. *Nature* 352: 207–215. (Also *Cell* 69: 393.)

122. Kuhn, P.M., Volknandt, W., Stadler, H., and Zimmermann, H. (1988). Cholinergic vesicle specific proteoglycan stability in isolated vesicles and in synaptosomes during induced transmitter release. *J. Neurochem.* 50: 11–16.

123. Lacey, M.L. and Haimo, L.T. (1992). Cytoplasmic dynein is a vesicle protein. *J. Biol. Chem.* 267: 4793–4798

124. LaVail, J.H. and LaVail, M.M. (1974). The retrograde transport of horseradish peroxidase in the chick visual system. *J. Comp. Neurol.* 157: 303–358.

125. Ledeen R.W., Parsons, S.M., Diebler, M.F., Sbashnig-Agler, M., and Lazereg, S. (1988). Ganglioside composition of synaptic vesicles from Torpedo electric organ. *J. Neurochem.* 51: 1465–1469.

126. Lentz, T.L. (1983). Cellular membrane reutilization and synaptic vesicle recycling. *Trends Neurosci.* 6: 48–53.

127. Leube, R.E., Wiedenmann, B., and Franke, W.W. (1989). Topogenesis and sorting of synaptophysin: synthesis of a synaptic vesicle protein from a gene transfected into non-neuroendocrine cells. *Cell* 59: 433–446.

128. Lindsey, J.D. and Ellisman, M.H. (1985). The neuronal endomembrane system III. The origin of the axoplasmic reticulum and discrete axonal cisternae at the axon hillock. *J. Neurosci.* 5: 3135–3144.

129. Lindstedt, A.D. and Kelly, R.B. (1991). Synaptophysin is sorted from endocytotic markers in neuroendocrine PC12 cells but not transfected fibroblasts. *Neuron* 7: 309–317.

130. Lindstedt, A.D. and Kelly, R.B. (1991). Endocytosis of the synaptic vesicle protein, synaptophysin, requires the COOH-terminal tail. *J. Physiol.* 85: 90–96.

131. Lippincott-Schwartz, J., Yuan, L., Tipper, C., Amherdt, M., Orci, L., and Klausner, R.D. (1991). Brefeldin A's effects on endosomes, lysosomes and the TGN suggest a general mechanism for regulating organelle structure and membrane traffic. *Cell* 67: 601–616.

132. Liscum, L., Hauptman, P.J., Hood, D.C., and Holtzman, E. (1982). Effects of barium and tetraethylammonium on membrane circulation in frog retinal photoreceptors. *J. Cell Biol.* 95: 296–309.

133. Llinas, R., Sugimori, M., Lin, J.W., Leopold, P.L., and Brady, S.T. (1989). ATP dependent directional movement of rat synaptic vesicles injected into the presynaptic terminal of squid giant synapses. *Proc. Natl. Acad. Sci. USA* 86: 5656–5660.

134. MacLennan, D.H. and Campbell, K.P. (1979). Structure, function and biosynthesis of sarcoplasmic reticulum proteins. *Trends. Bioch. Sci.* 4: 148–150.

134a. Malchiodi, F., Rambourg, A., Clermont, Y., and Caroff, A. (1986). Ultrastructural localization of Concanavalin A binding sites in the Golgi apparatus of various types of neurons in dorsal root ganglia. *Am. J. Anat.* 177: 81–95.

134b. Mandell, J.W., Czernik, A.J., De Camilli, P., Greengard, P., and Townes-Anderson, E. (1992). Differential expression of synapsin I and II among rat retinal synapses. *J. Neurosci.* (In press).

135. Margolis, R.U., Ledeen, R.W., Sbashnig-Agler, M., Byrne, M.C., Klein, R.L., Douglas, B.M., and Margolis, R.K. (1987). Complex carbohydrate composition of large dense-cored vesicles from sympathetic neurons. *J. Neurochem.* 49: 1839–1844.

136. Markelonis, G.J., Oh, T.H., Park, L.P., Cha, C.Y., Sofia, C.A., Kim, J.W., and Azari, P. (1985). Synthesis of the transferrin receptor by cultures of embryonic chick spinal neurons. *J. Cell Biol.* 100: 8–17.

137. Matheke, M.L., Kalff, M., and Holtzman, E. (1983). Effects of monensin on photoreceptors of isolated frog retinas. *Tissue and Cell* 15: 509–513.

138. Matheke, M.L., Fliesler, S.J., Basinger, S.F., and Holtzman, E. (1984). The effects of monensin on transport of membrane components in the frog retinal photoreceptor I. Light microscopic autoradiography and biochemical analysis. *J. Neurosci.* 4: 1086–1092.

139. Matheke, M.L. and Holtzman, E. (1984). Effects of monensin and of puromycin on transport of membrane components in the frog retinal photoreceptor II. Electron microscopic autoradiography of proteins and glycerolipids. *J. Neurosci.* 4: 1093–1103.

140. Matteoli, M., Navone, F., Haimann, C., Cameron, P.L., Solimena, M., and De, P. Camilli (1989). Secretory organelles of neurons and their relationship to organelles of other cells. *Cell Biol. Int. Repts.* 13: 981–992.

141. Matteoli, M., K. Takei, R. Cameron, P. Hurlbut, P.A. Johnston, T.C. Südhof, R. Jahn and P. De Camilli. (1991). Association of Rab3A with synaptic vesicles at late stages of the secretory pathway. *J. Cell Biol.* 115: 625–633.
142. Matteoli, M., Takei, K., Perin, M.S., Südhof, T.C., and De Camilli, P. (1992). Exo-endocytotic recycling of synaptic vesicles in developing processes of cultured hippocampal neurons. *J. Cell Biol.* (In press).
143. Maycox, P.R., Deckwerth, T., Hell, J.W., and Jahn, R. (1988). Glutamate uptake by brain synaptic vesicles. Energy dependence of transport and functional reconstitution in proteoliposomes. *J. Biol. Chem.* 263: 15423–15428.
144. Melanby, J. and Green, J. (1981). How does tetanus toxin act? *Neurosci.* 6: 281–300.
145. Mercola, M. and C.D. Stiles (1988). Growth family superfamilies and mammalian embryogenesis. *Development* 102: 451–460.
146. Mershul, C.E. and Pappas, G.D. (1984). The relationship of pinocytosis and synaptic vesicles at the frog neuromuscular junction. *Brain. Res.* 290: 1–18.
147. Mercurio, A. and Holtzman, E. (1982). Smooth endoplasmic reticulum and other agranular reticulum in frog retinal photoreceptors. *J. Neurocytol.* 11: 263–293.
148. Mercurio, A. and Holtzman, E. (1982). Ultrastructural localization of glycerolipid synthesis in rod cells of the isolated frog retina. *J. Neurocytol.* 11: 295–322.
149. Miller, T.M. and Heuser, J.E. (1984). Endocytosis of synaptic vesicle membrane at the frog neuromuscular junction. *J. Cell Biol.* 98: 685–98
150. Moore, P.J., Swords, K.M.M., Lynch, M.A., and Staehelin, L.A. (1991). Spatial organization of the assembly pathways of glycoproteins and complex polysaccharides in the Golgi apparatus of plants. *J. Cell Biol.* 112: 589–602.
151. Morin, P.J., Liu, N., Johnson, R.J., Leeman, S.E., and Fine, R.E. (1991). Isolation and characterization of rapid transport vesicle subtypes from rabbit optic nerve. *J. Neurochem.* 57: 415–427.
152. Moriyama, Y. and Nelson, N. (1987). Internal anion binding sites and membrane potential dominate the regulation of proton pumping by the chromaffin granule ATPase. *Biochem. Biophys. Res. Comm.* 149: 140–144.
153. Morris, J.F., Nordmann, J.J., and Dyhall, R.E.J. (1978). Structure function correlation in mammalian neurosecretion. *Int. J. Exp. Path.* 18: 1–28.
154. Mostov, K., Apodaca, G., Aroeti, B., and Okamoto, C. (1992). Plasma membrane sorting in polarized epithelial cells. *J. Cell Biol.* 116: 577–583.
155. Newman, T.M. and Severs, N.J. (1990). Arrested exocytosis of synaptic vesicles at a glutamatergic synapse: the insect fast excitatory neuromuscular junction. *Cell Biol. Int. Repts.* 14: 849–863.
156. Nielson, F.C., Wang, E., and Gammeltoft, S. (1991). Receptor binding, endocytosis and mitogenesis of insulin-like growth factors I and II in fetal rat brain neurons. *J. Neurochem.* 56: 12–21.
156a. Nordmann, J.J. and Artault, J.-C. (1992). Membrane retrieval following exocytosis in isolated neurosecretory nerve endings. *Neurosci.* 49: 201–207.
157. Novack, E.K., Baumann, H., Ovnic, M., and Swank, R.T. (1991). Expression of egasyn-esterase in mammalian cells: sequestration in the endoplasmic reticulum and complexation with β-glucuronidase. *J. Biol. Chem.* 266: 6377–6380.
158. Novikoff, A.B. (1967). Enzyme localization and ultrastructure of neurons. In: *The Neuron*; pp. 255–318. (Hyden, H., Ed.). Elsevier, New York.
159. Novikoff, P.M., Novikoff, A.B., Quintana, N., and Hauw, J.J. (1971). Golgi apparatus, GERL and lysosomes of neurons in rat dorsal root ganglia studied by thick section and thin section cytochemistry. *J. Cell Biol.* 50: 859–886.
160. O'Grady, L.C., Lindstedt, A.D., Lowe, A.W., Grote, E., and Kelly, R.B. (1990). Biogenesis of synaptic vesicle like structures in a pheochromocytoma cell line, PC12. *J. Cell Biol.* 110: 1693–1703.

161. Oh, T.H., Markelonis, G.I., Royal, G.M., and Bregman, B.J. (1986). Immunocytochemical distribution of transferrin and its receptor in the developing chick nervous system. *Dev. Brain Res.* 30: 207–220.
162. Okhuma, S. and Poole, B. (1981). Cytoplasmic vacuolation of mouse peritoneal macrophages and the uptake into lysosomes of weakly basic substances. *J. Cell Biol.* 90: 656–664.
163. Oorschot, D.E. and Jones, D.G. (1987). The vesicle hypothesis and its alternatives: a critical assessment. *Curr. Top. Res. Synapses* 4: 85–153.
164. Parton, R.G., Dotti, C.G., Bacallao, R., Kurtz, I., Simons, K., and Prydz, K. (1991). pH-induced microtubule dependent redistribution of late endosomes in neuronal and epithelial cells. *J. Cell Biol.* 113: 261–274.
165. Parton, R.G., Ockleford, C.D., and Critchley, D.R. (1987). A study on the mechanism of internalization of tetanus toxin by primary mouse spinal cord cultures. *J. Neurochem.* 49: 1057–1068.
166. Parton. R.G., Simons, K., and Dotti, C.G. (1992). Axonal and dendritic endocytic pathways in cultured neurons. (In press).
167. Patzak, A. and Winkler, H. (1986). Exocytotic exposure and recycling of membrane antigens of chromaffin granules: ultrastructural study after immunolabeling. *J. Cell Biol.* 102: 510–515.
168. Patzak, A., Aunis, D., and Langley, K. (1987). Membrane recycling after exocytosis: an ultrastructural study of cultured chromaffin cells. *Exp. Cell Res.* 171: 346–356.
169. Peters, A., Palay, S.L., and deF. Webster, H. (1991). *The Fine Structure of the Nervous System*, 3rd ed. Oxford University Press, New York. (See also 2nd ed., 1976.)
170. Petrenko, A.G., Perin, M.S., Davietov, B.A., Ushkaryov, Y.A., Geppert, M., and Südhof, T.C. (1991). Binding of synaptogamin to the α-latrotoxin receptor implicates both in synaptic vesicle exocytosis. *Nature* 353: 65–68.
171. Pfeffer, S. and Kelly, R.B. (1985). The subpopulation of brain coated vesicles that carries synaptic vesicle proteins contains two unique polypeptides. *Cell* 40: 949–957.
172. Pocock, J.M., Nicholson, R.A., and Osborne, M.P. (1990). Effects of colchicine on central and peripheral nerve terminals of the locust *Schistocerca gregoria. J. Neurocytol.* (19: 574–582).
173. Pomerat, C.M., Hendelman, W.J., Raiborn, C.W., and Massey, J.R. (1967). Dynamic activities of nervous tissue *in vitro*. In: *The Neuron*, pp. 119–178 (Hyden, H., Ed.). Elsevier, New York.
174. Pow, D.V. and Morris, J.F. (1989). Dendrites of hypothalamic magnocellular neurons release neurohyophyseal peptides by exocytosis. *Neurosci.* 32: 435–439.
175. Pow, D.V. and Morris, J.F. (1991). Membrane sorting during exocytosis and endocytosis in neuroendocrine neurons and endocrine cells: use of colloidal gold particles and immunocytochemical discrimination of membrane compartments. *Cell Tiss. Res.* 264: 299–316.
176. Price, R.L., Lasek, R.J., and Katz, M.J. (1991). Microtubules have special physical association with smooth endoplasmic reticulum and mitochondria in axons. *Brain Res.* 540: 209–216.
177. Quatacker, J.R., Annaert, W.G., and de Potter, W.P. (1990). Detection of sialic acid residues in the axonal reticulum of rat superior cervical ganglion cells by lectin-gold cytochemistry. *J. Histochem. Cytochem.* 38: 1445–1449.
178. Ragsdale, C. and Woodgett, J. (1991). *trking* neurotrophic receptors. *Nature* 350: 660–661.
179. Rambourg, A. and Clermont, Y. (1990). Three-dimensional electron microscopy: structure of the Golgi apparatus. *Eur. J. Cell Biol.* 51: 189–200.
180. Rambourg, A. and B. Droz (1980). Smooth endoplasmic reticulum and axonal transport. *J. Neurochem.* 35: 16–25.
181. Rash, J.E., Walrond, J.P., and Morita, M. (1988). Structural and functional correlates of synaptic transmission in the vertebrate neuromuscular junction. *J. Elec. Mic. Tech.* 10: 153–186.
182. Regnier-Vigouroux, A., Tooze, S.A., and Huttner, W.B. (1991). Newly synthesized synaptophysin is transported to synaptic-like microvesicles via constitutive secretory vesicles and the plasma membrane. *EMBO J.* 10: 3589–3601.
183. Reichardt, L.F. and Kelly, R.B. (1983). A molecular description of nerve terminal function. *Ann. Rev. Bioch.* 52: 871–926.

184. Ripps, H., Shakib, M., and Mac Donald, E.D. (1976). Peroxidase uptake by photoreceptor terminals of the skate retina. *J. Cell Biol.* 70: 86–96.
185. Ronnevi, L. (1979). Spontaneous phagocytosis of c type synaptic terminals by spinal motor neurons in kitten: an electron microscopic study. *Brain Res.* 162: 189–199.
186. Rosenbluth, J. and Wissig, S.I. (1964). The distribution of exogenous ferritin in toad spinal ganglia and the mechanisms of its uptake by neurons. *J. Cell Biol.* 23: 307–325.
187. Roth, J., Taatjes, D.J., Lucocq, J.M., Weinstein, J., and Paulson, J.C. (1985). Demonstration of an extensive transtubular network continuous with the Golgi apparatus stack that may function in glycosylation. *Cell* 43: 287–295.
188. Rothberg, K.G., Ying, Y., Kolhouse, J.F., Kamen, B.A., and Anderson, R.G.W. (1990). The glycosphingolipid-linked folate receptor internalizes folate without entering the clathrin-coated pit endocytic pathway. *J. Cell Biol.* 110: 637–649.
188a. Rothberg, K.G., Heuser, J.E., Donzell, W.C., Ying, Y.S., Glenner, J.R., and Anderson, R.G.W. (1992). Caveolin: a protein component of caveolae membrane coats. *Cell* 68: 673–682.
189. Sabban, E.L., Schwartz, J., and McMahon, A. (1990). Effects of compounds which disrupt proton gradients on secretion of neurosecretory proteins from PC12 pheochromocytoma cells. *Neurosci.* 38: 561–570.
190. Sahara, N. and Suzuki, K. (1990). Lateral diffusion of luminal membrane components during secretion in parotid acinar cells of the rat. *Cell Tissue Res.* 261: 461–466.
191. Sahenk, Z. and Brown, A. (1991). Weak-base amines inhibit the anterograde to-retrograde conversion of axonally transported vesicles in nerve terminals. *J. Neurocytol.* 20: 365–375.
192. Sahenk, Z., Whitaker, J.N., and Mendell, J.R. (1990). Immunocytochemical evidence for the retrograde transport of intraaxonal cathepsin D: possible relevance to the dying back process. *Brain Res.* 510: 1–6.
193. Santa-Hernandez, M., St. Jules, R., Augenbraun, E., and Holtzman, E. (1991). *trans*-Golgi systems in frog (Rana) photoreceptors. *J. Cell Biol.* 115: 210A.
194. Santa-Hernandez, M., Augenbraun, E., St. Jules, R., and Holtzman, E. (1992). Membrane cycling and endosomes in cell bodies and terminals of frog retinal photoreceptors. *Neurosci. Abstr.* 18: 1338.
195. Schacher, S., Holtzman, E., and Hood, D.C. (1974). Uptake of horseradish peroxidase by frog photoreceptor synapses in the dark and the light. *Nature* 249: 261–263.
196. Schacher, S., Holtzman, E., and Hood, D.C. (1976). Synaptic activity of frog retinal photoreceptors: a peroxidase uptake study. *J. Cell Biol.* 70: 178–192.
197. Schaeffer, S.F. and Raviola, E. (1978). Membrane recycling in the cone cell endings of the turtle retina. *J. Cell Biol.* 79: 802–825.
198. Scaife, R. and Margolis, R.L. (1990). Biochemical and immunochemical analysis of rat brain dynamin interaction with microtubules and organelles *in vivo* and *in vitro*. *J. Cell Biol.* 111: 3023–3033
199. Schmied, R. and Holtzman, E. (1987). A phosphatase activity and a synaptic vesicle antigen in multivesicular bodies of frog retinal photoreceptor terminals. *J. Neurocytol.* 16: 627–637.
200. Schmied, R. and Holtzman, E. (1989). Involvement of the Golgi apparatus in sorting of materials to opposite ends of frog retinal photoreceptors. *J. Neurobiol.* 20: 115–138.
201. Schneider, B.G. (1992). Na^+, K^+–ATPase isoforms in the retina. *Int. Rev. Cytol.* (In press).
202. Schroer, T.A., Brady, S.T., and Kelly, R.B. (1985). Fast axonal transport of foreign synaptic vesicles in squid axoplasm. *J. Cell Biol.* 101: 568–572.
203. Schwab, M.E., Suda, K., and Thoenen, H. (1979). Selective retrograde transsynaptic transfer of a protein, tetanus toxin, subsequent to its retrograde axonal transport. *J. Cell Biol.* 82: 798–810.
204. Shepherd, V.L., Tarnowski, B.I., and McLaughlin, B.J. (1991). Isolation and characterization of a mannose receptor from human pigment epithelium. *Invest. Ophthalomol.* 32: 1779–1784.
205. Siegel, J.H. and Brownell, W.E. (1986). Synaptic and Golgi membrane recycling in cochlear hair cells. *J. Neurocytol.* 15: 311–328.

206. Smith, A.D. (1973). Mechanisms involved in the release of noradrenaline from sympathetic nerves. *Brit. Med. Bull.* 29: 123–129.
207. Smith, J.E. and Smith, D.O. (1984). Coated vesicle morphology and subpopulations at the neuromuscular junction. *Brain Res.* 299: 383–388.
208. Smith, R.S. (1980). The short term accumulation of axonally transported organelles in the region of localized lesions of single myelinated axons. *J. Neurocytol.* 9: 39–65.
209. St. Jules, R., Santa-Hernandez, M., Ungar, F., and Holtzman, E. (1993). Features of the polarity of the Golgi apparatus of frog photoreceptors: studies with lectins and with Brefeldin A. *Exp. Eye. Res.* (In press).
210. Stanley, E.F., Ehrenstein, G., and Russel, J. (1988). Evidence for anion channels in secretory vesicles. *Neurosci.* 25: 1035–1039.
211. Stieber, A., Erulkar, S.D., and Gonatas, N.K. (1989). A hypothesis for the superior sensitivity of wheat germ agglutinin as a neuroanatomical probe. *Brain Res.* 495: 131–139.
212. Studelska, D.R. and Brimijoin, S. (1989). Partial isolation of two classes of DBH containing particles undergoing rapid axonal transport in rat sciatic nerves. *J. Neurochem.* 53: 622–630.
213. Südhof, T.C. and Jahn, R. (1991). Proteins of synaptic vesicles involved in exocytosis and membrane recycling. *Neuron* 6: 665–677.
214. Sulzer, D. and E. Holtzman (1986). Alcian Blue and Neutral Red staining of retinal synaptic layers. *J. Histochem. Cytochem.* 34: 1513-1515.
215. Sulzer, D. and Holtzman, E. (1989). Acidification and an endosome-like compartment in the presynaptic terminals of frog retinal photoreceptors. *J. Neurocytol.* 18: 529–540.
216. Sulzer, D. and Rayport, S. (1990). Amphetamine and other psychostimulants reduce pH gradients in midbrain dopaminergic neurons and chromaffin granules: a mechanism of action. *Neuron* 5: 797–808.
217. Sulzer. D., Piscopo, I., Ungar, F., and Holtzman, E. (1987). Lead dependent deposits in diverse synaptic vesicles: suggestive evidence for the presence of anionic binding sites. *J. Neurobiol.* 18: 467–484.
218. Sulzer, D., Ungar, F., and Holtzman, E. (1990). Further studies on the cationic staining of retinal photoreceptors. *J. Histochem. Cytochem.* 38: 743–745.
219. Suzuki, E.K., Bonifacino, J.S., Lin, A.Y., Davis, M.M., and Klausner, R.D. (1991). Regulating the retention of T cell receptor α-chains within the endoplasmic reticulum: Ca^{2+} dependent association with BiP. *J. Cell Biol.* 114: 189–206.
220. Swanson, J., Bushnell, A., and Silverstein, S.C. (1987). Tubular lysosome morphology and distribution within macrophages depend on the integrity of cytoplasmic microtubules. *Proc. Natl. Acad. Sci. USA* 84: 1921–1925.
221. Takei, K., Stukenbrok, H., Metcalf, A., Mignery, G.A., Südhof, T.C., Volpe, P., and De Camilli, P. (1992). Ca^{2+} stores in Purkinje neurons: endoplasmic reticulum subcompartments demonstrated by the heterogeneous distribution of the $InsP_3$ receptor, Ca^{2+}-ATPase and calsequestrin. *J. Neurosci.* 12: 489–505.
222. Teichberg, S. and Holtzman, E. (1973). Axonal agranular reticulum and synaptic vesicles in cultured embryonic chick sympathetic neurons. *J. Cell Biol.* 57: 88–108.
223. Teichberg, S., Holtzman, E., Crain, S.M., and Peterson, E.R. (1975). Circulation and turnover of synaptic vesicle membranes in cultured mouse spinal cord neurons. *J. Cell Biol.* 67: 215–230.
224. Tennyson, V.M. and Brzin, M. (1970). The appearance of acetylcholinsterase in the dorsal root neuroblast of the rabbit embryo: a study by electron microscope cytochemistry and microgasimetric analysis with the magnetic diver. *J. Cell Biol.* 46: 64–80.
225. Thoenen, H. and Barde, Y-A. (1980). Physiology of nerve growth factor. *Physiol. Rev.* 60: 1284–1335.
226. Thomas, K., Navarro, J., Benson, R.J.J., Campbell, K.P., Rotundo, R.L., and Fine, R.E. (1989). Newly synthesized calsequestrin, destined for the sarcoplasmic reticulum is contained in early/intermediate Golgi-derived clathrin coated vesicles. *J. Biol. Chem.* 264: 3140–3145.

227. Thureson-Klein, A.K. and Klein, R. (1990). Exocytosis from neuronal large dense cored vesicles. *Int. Rev. Cytol.* 121: 67–126.
228. Thureson-Klein, A., Klein, R.L., and S-H. C. Yen (1975). Morphological effects of osmolarity on purified noradrenergic vesicles. *J. Neurocytol.* 4: 609–627.
229. Tixier-Vidal, A., Barret, A., Faivre-Bauman, A., Huttner, W., and Wiedenmann, B. (1992). Differential expression and subcellular localization of secretogranin II and synaptophysin during early development of mouse hypothalamic neurons in culture. *Neurosci.* 47: 967–978.
230. Tixier-Vidal, A., Faivre-Bauman, A., Picart, R., and Wiedenmann, B. (1988). Immunoelectron microscopic localization of synaptophysin in a Golgi subcompartment of developing hypothalamic neurons. *Neurosci.* 26: 847–861.
231. Torri-Tarelli, F., Haimann, C., and Ceccarelli, B. (1987). Coated pits and coated vesicles during enhanced quantal release of acetylcholine at the neuromuscular junction. *J. Neurocytol.* 16: 205–214.
232. Trimble, W.S., Linial, M., and Scheller, R.H. (1991). Cellular and molecular biology of the presynaptic nerve terminal. *Ann. Rev. Neurosci.* 14: 93–122.
233. Trojanowski, J.O., Gonatas, J.O., and Gonatas, N.K. (1981). A light and electron microscopic study of the intraneuronal transport of horseradish peroxidase and wheat germ agglutinin-peroxidase conjugates in the rat visual system. *J. Neurocytol.* 10: 441–456.
234. Troutt, L.L., Wang, E., Bagh-Roehl, K., and Burnside, B. (1990). Microtubule nucleation and organization in teleost photoreceptors: microtubule recovery after elimination by cold. *J. Neurocytol.* 19: 213–223.
235. Tsukita, S. and Ishikawa, H. (1976). Three-dimensional distribution of smooth endoplasmic reticulum in myelinated nerves. *J. Elect. Micr.* 25: 141–149.
236. Tsukita, S. and Ishikawa, H. (1980). The movement of membranous organelles in axons. *J. Cell Biol.* 84: 513–30.
237. Tytell, M. and Lasek, R.J. (1985). Glial polypeptides transferred into the squid giant axon. *Brain Res.* 324: 223–229.
238. Ueda, T. and Shigetaka, N. (1983). Adenosine-triphosphate dependent uptake of glutamate into protein I-associated synaptic vesicles. *J. Biol. Chem.* 258: 698–699.
239. Ullmer, J.B. and Palade, G.E. (1991). Effects of brefeldin A on the Golgi complex, endoplasmic reticulum and viral envelope proteins in murine erythroleukemia cells. *Eur. J. Cell Biol.* 54: 38–54.
240. Ungar, F., Piscopo, I., Letizia, J., and Holtzman, E. (1984). Uptake of calcium by the endoplasmic reticulum of the frog photoreceptor. *J. Cell Biol.* 98: 1645–1655.
241. Valtorta, F., Befenati, F., and Greengard, P. (1992). Structure and function of the synapsins. *J. Biol. Chem.* 267: 7195–7198.
242. Valtorta, F., Fesce, R., Grohovaz, F., Haimann, C., Hurlbut, W.P., Iezzi, N., Torri Tarelli, F., Villa, A., and Ceccarelli, B. (1990). Neurotransmitter release and synaptic vesicle recycling. *Neurosci.* 35: 477–489.
243. Valtorta, F., Jahn, R., Fesce, R., Greengard, P., and Ceccarelli, B. (1988). Synaptophysin (p38) at the frog neuromuscular junction: its incorporation into the axolemma and recycling after intense quantal secretion. *J. Cell Biol.* 107: 2717–2728.
244. Valtorta, F., Torri Tarelli, F., Campanati, L., Villa, A., and Greengard, P. (1989). Synaptophysin and synapsin I as tools for the study of the exo-endocytotic cycle. *Cell Biol. Int. Repts.* 13: 1023–1052.
245. Van der Bliek, A.M. and Myerowitz, E.M. (1991). Dynamin-like protein encoded by the *Drosophila shibire* gene associated with vesicular traffic. *Nature* 351: 411–414.
246. Valterson, C., Dutton, A.M., and Singer, S.J. (1990). Transfer of secretory protein from the endoplasmic reticulum to the Golgi apparatus: distinction between homologous and heterologous transport in intact heterokaryons. *Proc. Natl. Acad. Sci. USA* 87: 8175–8179.
247. Vance, J.E., Pan, D., Vance, D.E., and Campenot, R.B. (1991). Biosynthesis of membrane lipids in rat axons. *J. Cell Biol.* 115: 1061–1068.

248. Vaughan, D.K., Fisher, S.K., Bernstein, S.A., Hale, I.L., Linberg, K.A., and Matsumoto, B. (1989). Evidence that microtubules do not mediate opsin vesicle transport in photoreceptors. *J. Cell Biol.* 109: 3053–3062.

249. Villa, A., Podini, P., Clegg, D.O., Pozzan, T., and Meldolesi, J. (1991). Intracellular Ca^{2+} stores in chicken Purkinje neurons: differential distribution of the low affinity, high capacity Ca^{2+} binding protein, calsequestrin, of Ca^{2+}-ATPase, and of the ER lumenal protein, BiP. *J. Cell Biol.* 113: 779–791.

250. Wagner, J.A. and Kostyk, S.K. (1990). Regulation of neural cell survival and differentiation by peptide growth factors. *Curr. Op. Cell Biol.* 2: 1050–1057.

251. Waheed, A., Gottschalk, S., Hille, A., Krentler, C., Pohlmann, R., Braulke, T., Hauser, H., Geuze, H., and von Figura, K. (1988). Human lysosomal acid phosphatase is transported as a transmembrane protein in transformed baby hamster kidney cells. *EMBO J.* 7: 2351–2358.

251a. Villa, A., Sharp, A.H., Racchetti, G., Podini, P., Bole, D.G., Dunn, W.A., Pozzan, T., Snyder, S.H., and Meldolesi, J. (1992). The endoplasmic reticulum of Purkinje neuron cell body and dendrites: molecular identity and specializations for Ca^{2+} transport. *Neurosci.* 49: 467–477.

252. Walicke, P.A. and Baird, A. (1991). Internalization and processing of basic fibroblast growth factor by neurons and astrocytes. *J. Neurosci.* 11: 249–2258.

253. Waltino, O. and Talanti, S. (1965). Histochemical localization of a-glucuronidase in rat brain. *Nature* 205: 499–500.

254. Walz, B. and Baumann, O. (1989). Calcium sequestering cell organelles. *Prog. Histochem. Cytochem.* 20: 1–45.

255. Weiss, D.G. and Gorio, A. (Eds.). (1982). *Axoplasmic Transport in Physiology and Pathology.* Springer-Verlag, Belrin.

255a. Wetzel, M.G., Tulan, E.B., and Besharse, J.C. (1992). Tunicamycin does not inhibit transport of phospholipid to *Xenopus* rod photoreceptor outer segments. *Invest. Opthalmol.* 33: 1008 (ARVO Abs.).

256. Whittaker, V.P. (1987). Cholinergic synaptic vesicles from the electromotor nerve terminals of Torpedo. *Ann. N.Y. Acad. Sci.* 493: 77–91.

257. Wiley, R.G., Spencer, C., and Pysh, J.J. (1987). Time course and frequency dependency of synaptic vesicle depletion and recovery in electrically stimulated sympathetic ganglia. *J. Neurocytol.* 16: 359–372.

257a. Winkler, H. and Fischer-Colbrie, R. (1990). Common membrane proteins of chromaffin granules, endosomes and synaptic vesicles: properties, tissue distribution, membrane topopgraphy and regulation of synthesis. *Neurochem. Int.* 7: 245–262.

258. Wong, D.H., Ignatius, M.J., Proskky, G., Parham, P., Trojanowski, J.Q., and Brodsky, F.M. (1990). Neuron-specific expression of high molecular weight clathrin light chain. *J. Neurosci.* 10: 3025–3031.

259. Wood, J.G., Byrd, F.I., and Gurd, J.W. (1981). Lectin cytochemistry on cell membranes of rat cerebellum. *J. Neurocytol.* 10: 149–159.

260. Zimmermann, H. (1990). Neurotransmitter release. *FEBS Letts.* 268: 394–399.

261. Zimmermann, H. (1991). Cellular and molecular basis of neurotransmitter release. In: *Synapse-Transmission Modulation,* pp. 97–110. (Elsner, N. and Penzlin, H., Eds.). George Thieme-Verlag, Stuttgart.

262. Zimmermann, H., Volknandt, W., Henkel, A., Bonzelius, F., Janetzko, A., and Kanaseki, T. (1989). Synaptic vesicle membrane: origin, axonal distribution, protein components, exocytosis and recycling. *Cell Bio. Int. Reps.* 13: 993–1006.

263. Zurzolo, C., Polistina, C., Saini, M., Gentile, R., Aloj, L., Migliaccio, G., Bonatti, S., and Nitsch, L. (1992). Opposite polarity of virus budding and of viral envelope glycoprotein distribution in epithelial cells derived from different tissues. *J. Cell Biol.* 117: 551–564.

THE ROLE OF ENDOCYTOSIS IN EPIDERMAL GROWTH FACTOR SIGNALING

Bryan K. McCune, William R. Huckle, and H. Shelton Earp

Advances in Cell and Molecular Biology of Membranes
Volume 1, pages 269–305

ISBN: 1-55938-362-3

ABSTRACT

Within five seconds of addition to cultured cells, the epidermal growth factor (EGF) binds to the EGF receptor and activates the receptor's intrinsic tyrosine kinase. If EGF remains present for 6 to 8 hours, DNA synthesis and cell proliferation occur. Activation of the receptor tyrosine kinase activity is essential for these events. EGF also produces rapid internalization of the EGF receptor, resulting in the formation of endocytotic vesicles bearing activated receptor tyrosine kinase. Several functions for EGF internalization have been postulated. This first is to terminate the receptor signal by degrading EGF:EGF receptor complexes, and numerous investigations have substantiated receptor down-regulation by the degradative pathway. Activation of the receptor tyrosine kinase appears to increase the rate of receptor entry into a high-affinity internalization pathway, but the mechanism of kinase-enhancement of internalization is unknown. Whether the receptor's subsequent choice of intracellular pathway (recycling vs. degradation) is influenced by tyrosine phosphorylation remains controversial and may differ among cell types. Several lines of evidence indicate that EGF receptor signaling occurs while the receptor is on the cell surface, and this may be one reason that coordinate removal of activated surface receptor must occur. The importance of receptor down-regulation is demonstrated by studies of EGF receptor mutants that have lost their capacity to be internalized. When cells bearing such receptors are treated with EGF, proliferation not only occurs, but the persistent growth signal can result in neoplastic transformation.

A second, more speculative, function for internalization has been proposed: to propagate the growth signal by moving an active tyrosine kinase to an intracellular position from which it could extend its signaling function. At first examination, this would seem incompatible with the finding that noninternalizable receptors can signal growth. However, the noninternalizing mutant receptors have additional properties that differentiate them from wild-type receptors; the mutants produce a much more dramatic increase in tyrosine phosphorylation and, thus, may bypass normal growth control mechanisms. Wild-type receptors may require the spatial arrangements provided by the internalization pathway to efficiently send a growth signal. These spatial arrangements may include the path of lateral movement toward the site of internalization as well as the potential function from an intracellular endosome. Precise placement of kinase and substrates may not be necessary under experimental conditions featuring mutant receptors or a high level of receptor occupancy. However, at physiologic concentration of ligands that produce only partial receptor occupancy, efficient signal transduction may be needed.

The following points about EGF receptor internalization and signaling are salient. First, autophosphorylated EGF receptors form multimeric complexes with proteins. This occurs when enzymes and structural proteins bearing *src* homology 2 (SH2) regions bind specifically to phosphorylated tyrosine residues on activated EGF receptors. Second, additional enzymes bind to other parts of the receptor's cytoplasmic domain. Third, the EGF receptor tyrosine kinase remains activated in internalized vesicles. Thus, when an activated, autophosphorylated EGF receptor moves, it may carry with it a phosphoprotein complex with multiple enzymatic activities. The lateral movement of this complex as it approaches the internalization site may deliver receptor and associated proteins to specified locales in the membrane, in effect, physically orchestrating signal transduction components to a membrane environment rich in substrates. The action of the receptor's multimeric complex may change the membrane environment itself, releasing signals from lipids, modulating the binding of cytoskeletal proteins to lipids, or directly associating with or altering cytoskeletal elements. Whether this complex of receptor and associated signaling molecules remains intact in early or even late endosomes and what function the complex plays beyond simply directing internalization are questions for future studies.

I. INTRODUCTION

A. Overview

The interaction of a peptide hormone or growth factor with its cell surface receptor is only the beginning of a complex cascade of physiochemical events. Ligand receptor interactions control a series of enzymatic reactions and ion fluxes which comprise signal transduction, and often result in subsequent receptor internalization and degradation. Each hormone receptor has a signaling function. Many act by influencing GTP-binding regulatory proteins. Several have been recently postulated to be ligand-activated serine/threonine protein kinases, and many, particularly the cytokine receptors, have unknown signal transduction mechanisms (*3,34,55,106,145*). Receptor internalization following ligand binding has been described for each class; for example, the G-protein-linked vasopressin receptors (*100*), members of the activin/transforming growth factor beta (TGFβ) receptor family (*105*) and receptors for the cytokines IL-2 (*110*) and tumor necrosis factors (*74*). One additional large class of receptors, the class on which this review will focus, are ligand-activated tyrosine-specific protein kinases that internalize upon activation; for example, receptors for epidermal growth factor (EGF), platelet-derived growth factor (PDGF) (*78,141*), insulin (*54,82,103,131*), colony-stimulating factor-1 (CSF-1) (*42,60*), and nerve growth factor (NGF) (*15,50,67*). Their down-regulatory mechanisms, which often entail uptake of the receptor into the cell and the subsequent degradation of both receptor and ligand, prevent a high level of persistent cell stimulation. The process by which ligand initiates movement of the hormone-receptor complex through a maze of intracellular membrane compartments is referred to as receptor-mediated endocytosis (RME). In this chapter, we will examine the mechanism of growth factor RME and discuss the possible relationships between endocytosis and receptor signaling.

RME is a general phenomenon, and a variety of macromolecules gain entrance to the cell interior through interaction with specific surface receptors. The process has a number of functions. For example, RME facilitates clearance of proteins from the extracellular space (e.g., the asialoglycoprotein-receptor system), delivery of certain nutrients to the cell interior [the transferrin-transferrin receptor system for iron and the low density lipoprotein (LDL) receptor system for cholesterol], and transport of immunoglobulins across epithelial cells (the IgA-receptor system)

(*13,137,154*). However, the largest category of ligands that enter the cell by RME are regulatory polypeptides, including the growth factors, hormones, and other cytokines. In this chapter, we will confine our in-depth discussion to one example: ligands internalized by the EGF receptor. This is, in part, for historical reasons. The chemical nature of the EGF molecule allows it to be modified (iodinated or labeled with fluorescent tags) with little loss of biologic function. This allowed early studies of RME to be conducted with a biologically active ligand. These experiments were some of the first to analyze the intracellular fate of peptides bound to surface receptors (see *21,81,124* for reviews). Extensive studies of EGF:EGF receptor complexes have also been possible because the EGF receptor was the first growth factor receptor cloned and sequenced. EGF receptor cDNA expression and site-directed mutagenesis have led to a plethora of sophisticated studies that have provided insight into EGF receptor internalization. To initiate our discussion, we will present a brief review of RME. The general aspects of RME have been the subject of a number of recent reviews (see Refs. *13,136,137,154* and chapters in this volume).

B. Endocytotic Pathways of Growth Factors, Hormones, and Cytokines

Following binding of ligand to cell surface receptors, receptors cluster in surface membrane invaginations called "coated pits" (*58*). While certain types of unoccupied surface receptors are concentrated in preexisting coated pits (e.g., LDL receptors), most unoccupied receptors are evenly distributed throughout the plasma membrane, and become concentrated in coated pits only after ligand binding (*58,154*). The "coat" of the coated pit is comprised of a basket-like lattice of trimers of the 180-kDa protein clathrin which reside on the cytoplasmic surface of the invaginated membrane (*79,120*). The formation of clathrin coats is a spontaneous, ATP-independent process (*120*). The clathrin coat appears to be essential for the early stages of receptor-mediated endocytosis; interference with the clathrin coat by biochemical means or using microinjected antibodies severely impedes internalization of ligand-receptor complexes (*45,94*). As endocytosis proceeds, the coated invagination is pinched off, forming a membrane-bound coated vesicle with the ligand oriented toward the lumen and intracellular signaling domain (e.g., the EGF receptor tyrosine kinase domain and C terminal autophosphorylation sites) remaining toward the cytoplasm. Through the action of an uncoating ATPase, the clathrin coat soon is disassembled and recycled. Vesicles then fuse to form larger endosomes, and the vesicle lumen is acidified to an increasing degree over time by the action of an ATP-driven proton pump (*13,58,79,120,136,137,154*).

The role of various compartments (early and late endosomes, multivesicular bodies, and tubular systems) in the decision to recycle or degrade receptors or ligands is the subject of much recent interest. Receptor/ligand systems have been grouped by several authors into four classes based on their intracellular fate (*13,137*). In class I systems, the receptor is recycled to the plasma membrane and the ligand is degraded in lysosomes. In class II systems, both receptor and ligand recycle back to the surface. In class III systems, both receptor and ligand proceed to lysosomes where they are degraded. In class IV systems, ligand-receptor complexes are transported across cells through a process referred to as transcytosis. The major and minor pathways of the EGF receptor are discussed below.

The Degradative Pathway (Class III)

In normal cells, the most common pathway taken by internalized growth factor: receptor complex is one resulting in lysosomal degradation of both ligand and receptor. This is the pathway typically taken by EGF:EGF receptor complex. Although exceptions will be noted, in many cell types multiple early endosomes coalesce to form a heterogeneous population of intermediate endosomes. As the endosomes enlarge, secondary invaginations occur resulting in multivesiculated endosomes or "multivesicular bodies." At some later point in time, endosome contents, including both ligand and receptor, are degraded in the lumen of the lysosome.

The Receptor Recycling Pathway (Class I and II)

Progressive acidification of the endosomal compartment during RME favors dissociation of ligand/receptor complexes (*165*). Since the receptor remains associated with the endosomal membrane, ligands and receptors *may* be routed separately. In the receptor recycling pathway, endosomes enriched in unoccupied receptors are routed back to the plasma membrane and assume their original orientation, while the dissociated ligand continues on to an intracellular destination. The receptor recycling pathway was first described for the LDL-receptor system (*58*) and is most common in receptors involved in protein clearance or nutrient delivery. Both transferrin and the transferrin receptor are recycled to the membrane, leaving only their iron within target cells. In addition, many ligand-receptor complexes have several possible fates depending on the cell type and other, often unknown, conditions. For example, even though the principal fate of EGF:EGF receptor complexes is degradation, in many cells, especially tumor cell lines with high numbers of EGF receptors, receptor is recycled to the plasma membrane, and undergraded EGF or transforming growth factor alpha (TGFα) can be found in the medium after recycling. In addition, NGF and NGF receptors have been observed to recycle in PC12 pheochromocytoma cells (*15,50*). This variability in the recycling/degradation decision from cell-type to cell-type sometimes makes it difficult to directly compare studies of ligands and receptors performed in different laboratories.

The Transcytotic Pathway (Class IV)

The transcytotic pathway, originally described for the IgA-receptor system, functions to transport lymphocyte-derived IgA across epithelial cells for release onto epithelial surfaces (2). In this pathway, both receptor and ligand are transported in endosomes to the apical plasma membrane surface where the receptor is cleaved, releasing the IgA with attached secretory component (*2*). There is evidence for epithelial cell transcytosis of EGF, including vectorial transport across the hepatocyte *in vivo* and MDCK cells *in vitro*, but the role of the EGF receptor beyond the initial internalization of ligand is not clear (*12,14,36,101,143*).

C. Potential Roles for Endocytosis in Receptor Signaling

For most ligands and receptors, components of the signal transduction cascade remain incompletely characterized. For growth factors, the receptor is the initial component of the signaling cascade encountered, and has a role beyond merely mediating ligand internalization. Indeed, in most ligand/receptor systems, the ligand functions principally to activate its cognate receptor. This point is demonstrated by a number of studies in which antireceptor antibodies succeed in mimicking the effect of the natural ligand (*119,134*). Thus, the biochemical action of the activated receptor, and its intracellular pathway, are both important in considering the function of endocytosis.

A number of growth factor receptors (including the insulin receptor) have in common an intrinsic tyrosine kinase domain which is essential for growth signal-

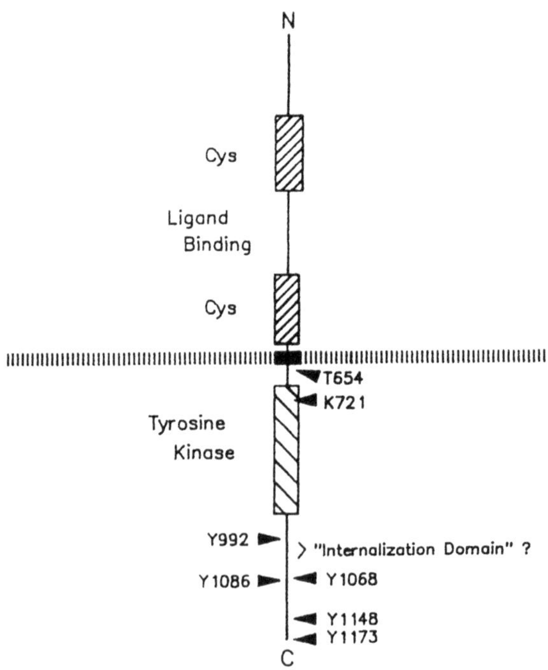

Figure 1.

ing. The tyrosine kinase-containing receptors have been intensively studied, in part because full-length cDNAs have been available for several years, facilitating structure/function studies. In addition, there is clear association of these receptors' enzyme activity with cell transformation (*1,149*). Initial studies for virtually all of these receptors have concluded that tyrosine kinase activity is necessary for signal transduction. Receptors for other cytokines, such as interleukin-2 (IL-2), IL-3, and growth hormone also have single transmembrane domains but lack a tyrosine kinase (*1*). Their signal transducing mechanisms are poorly characterized, but appear also to involve increases in tyrosine phosphorylation, regulated perhaps by intracellular tyrosine kinases or phosphatases. However, these receptors do not carry an intrinsic tyrosine kinase into the cytoplasm. The EGF receptor does, but the functional consequences of this act are still undefined.

II. THE EGF RECEPTOR AS A MODEL FOR RECEPTOR ENDOCYTOSIS

A. Introduction: The EGF Family of Growth Factors and the EGF Receptor

EGF and Other Members of the EGF Family

EGF was originally extracted from mouse salivary glands, an organ where the mature 53 amino acid growth factor is cleaved by a specific protease from a large precursor and stored (reviewed in *19,20,21*). In bioassays, this peptide caused growth and maturation of epidermal tissue and hence its name. EGF is found in high concentrations in many exocrine secretions (e.g., saliva and breast milk) but only at low concentrations in the blood. The 1200-amino acid EGF precursor is also expressed as a transmembrane protein on the surface of some cells; for example, in the kidney and uterus. Again, after cleavage, EGF precursor fragments appear in urine or in uterine fluids. It appears likely that growth stimulation via the EGF receptor is mediated by EGF or EGF family members (see below) that are either present in exocrine secretions, or that are made and released locally as autocrine or paracrine growth agents (*20*).

Three other members of the EGF family have been discovered in mammalian tissues. All have homology to EGF in the placement of the six cysteine residues that form disulfide bridges. These are the primary determinants of the folded structure which allow four distinct growth factors to each bind to and activate the EGF receptor (*20*). TGFα, the second family member sequenced, is a 50-amino acid peptide which, like EGF, is cleaved from a transmembrane precursor (159 and 160 amino acids in the rat and human, respectively) (*40,97*). TGFα expression is kept at very low levels in normal tissues but can be induced by growth stimuli (*32*). In contrast, TGFα is derepressed in many carcinoma cells, with as many as 75% of certain tumor types producing detectable levels of TGFα (*41*). TGFα production in cancer cells may result in an autocrine loop that enhances tumorigenicity *in vivo*. The unproteolyzed TGFα precursor embedded in the membrane of one cell type has been shown to stimulate the tyrosine kinase activity and function of the EGF receptor on an adjacent cell (*11,163*). Whether the stimulated EGF receptor is able to loosen the transmembrane TGFα from its site of membrane insertion and internalize the ligand-receptor complex into the receptor-bearing cell is unclear, but such a possibility has precedent: the *sevenless* receptor from the R7 cell of the *Drosophila* compound eye is capable of internalizing its ligand, *boss*, which is a membrane-embedded ligand found on an adjacent cell (*88*).

The recent discoveries of amphiregulin (*122,138,139*) and heparin binding (HB)-EGF (*9,68*) have doubled the size of the known EGF family. Both of these molecules are larger than EGF or TGFα, containing N-terminal extensions encoding basic regions that appear to allow these two factors to bind to heparin, heretofore a defining characteristic of the fibroblast growth factor (FGF) family (*83*). Recent studies of heparin binding EGF are particularly exciting in that, in certain cell types, HB-EGF is 40 times more potent than EGF (*56*). It has been postulated that the co-binding of HB-EGF to both heparan-containing glycosaminoglycans and to the EGF receptor vastly enhances signal generation. TGFα, amphiregulin and HB-EGF have all been shown to differ in their biological activity or potency when compared to EGF (*4,68,73,129,133,135,139*). In most cases, these differences can neither be ascribed to a unique receptor subtype or attributed simply to a higher affinity for the receptor. These differences, may, in part, be explained by subtle alterations in how the individual ligands bring about or sustain endocytosis or receptor degradation. For example, evidence has been presented that TGFα dissociates from the EGF receptor at a higher pH than does EGF (*49*). This should result in dissociation of TGFα from the EGF receptor at an earlier stage in the endosome acidification process. In turn, this might inactivate the internalized receptor, change the fate of the ligand receptor complex and result in the recycling of TGFα to the cell surface at a higher rate than EGF (*37,49*). The consequence would be to create a more prolonged biological signal because consistent recycling and rebinding of TGFα to the EGF receptor would prolong surface EGF receptor signal. As noted above, another way of obtaining a prolonged signal might be seen with HB-EGF. The dual binding of HB-EGF to heparan-containing glycosaminoglycans and the EGF receptor may prolong the EGF receptor's stay on the surface. As will be discussed in Sections III and V, persistence of the EGF receptor surface signal by mutant receptors incapable of downregulation has been shown to lower the dose of EGF necessary to stimulate proliferation. The vast majority of studies of EGF receptor endocytosis and structure function have been performed using EGF as the test ligand. Detailed studies must be performed with the other ligands (e.g.,TGFα and HB-EGF) if we are to understand whether EGF receptor endocytosis, downregulation and signaling are modified by distinct properties of specific EGF family members.

EGF Receptor: Structure-Function Relations

The EGF receptor has served as a prototype for the study of polypeptide receptors that bear intrinsic tyrosine kinase activity; it was the first receptor shown to have a ligand-activated protein tyrosine kinase (*22,151*), and the first receptor cloned and sequenced (*44,150*). Combined biochemical, molecular, and cellular studies have enhanced our knowledge of the functional domains of the EGF receptor (Fig. 1). The external domain comprises 50% of the receptor and is characterized by two regions that are rich in cysteine residues. There are 10 to 11 sites of N-linked glycosylation and at least 40-kDa M_r is added to the receptor protein backbone by the addition of carbohydrate (*19*). The receptor also contains mannose phosphate, a sugar that may be important in intracellular routing of the receptor (*147*). Data suggest that the external domain ligand-binding site is localized just carboxyl to the first cysteine-rich domain (*96,164*; Bishop and Earp, unpublished results). A preponderance of evidence suggests that EGF binding changes the conformation of the external domain, leading to dimerization and oligomerization of occupied receptors. Dimerization is thought to be an initial step in transducing the activation signal across the transmembrane region. As will be discussed below, EGF receptor dimers have been shown to persist even after the receptor appears in endosomes.

The predicted transmembrane region is a hydrophobic amino acid peptide in the center of the EGF receptor molecule. Recent experiments have shown that shortening of the transmembrane region by several amino acids, or insertion of prolines have little or no effect on EGF receptor signal transduction (*18*). In addition, insertion of an activating mutation similar to that of the *neu* oncogene into the EGF receptor transmembrane domain does not alter or activate signaling (*18*). The *neu* oncogene, a protein highly homologous to the EGF receptor, is created from the cellular *neu* proto-oncogene by a point mutation in the transmembrane region. This mutation activates this receptor's cytoplasmic tyrosine kinase, presumably by causing dimers to form spontaneously in the absence of the ligand (*158*). No such mutation for the EGF receptor has been found in tumors nor has one been created by recombinant means. In summary, the EGF receptor transmembrane region can be altered substantially without influencing signal transduction.

The cytoplasmic domain of the receptor can be divided roughly into three functional areas. The first, just below the plasma membrane, is a regulatory region which, when modified by phosphorylation, can block signal transduction across the membrane (see Section IV). The second domain is a region of 290 amino acids that is highly homologous to the more than 70 other tyrosine kinase domains which have been sequenced (*62*). A number of highly conserved residues in this region identify the EGF receptor as a tyrosine kinase. As with other kinases, residues which bind ATP (e.g., Lys 721) and participate in its hydrolysis have been identified. This knowledge has been used to make point mutations that abolish ATP binding and result in a kinase inactive receptor mutants (see Section III). The third cytoplasmic domain is the C terminal, 15 to 20 kDa of the receptor, that contains three major and at least two minor tyrosine autophosphorylation sites. In the inactive state, the C terminus is thought to fold into the catalytic domain, denying access of potential substrates to the catalytic site (*20*). In addition, the folded C terminus encloses specific sequences that, when exposed, may associate with proteins in the internalization pathway. Upon activation, three C terminal tyrosines can be autophosphorylated (Tyr 1068, Tyr 1148, Tyr 1173), and the C terminus is thought to swing away from the catalytic region exposing sequences that associate with the high affinity internalization apparatus and help select substrates. It is now clear that certain receptor substrates can complex to the phosphorylated EGF receptor through specific regions referred to as *src* homology 2 (SH2) domains (*16,85*). This 100-amino acid sequence was originally defined in the *src* oncogene as an area outside the tyrosine kinase domain (*src* homology 1) that was homologous to regions in other nonreceptor tyrosine kinases (*85*). SH2 groups show a pronounced preference for binding to phosphorylated forms of EGF receptor. The presence of SH2 groups in substrates, such as the *ras* GTPase-activating protein or phospholipase-C-γ, dictates that when the autophosphorylated receptor C terminus is exposed (*85*), it will attract proteins with SH2 groups that can bind to and move with the receptor.

B. Ligand-Dependent Internalization of the EGF Receptor

Endocytotic Pathways of the Complex

The metabolic fate of the EGF/receptor complex has been extensively studied in a number of cell types using biochemical and histochemical techniques. We have sketched a hypothetical internalization schema in Figure 2. The schema synthesizes the work of a number of investigations using different model systems and is not thought to be accurate under all conditions in all cell types. The unactivated receptor is folded upon itself, covering the tyrosine kinase domain and sequences that dictate internalization. EGF activates the receptor kinase, initiating movement into "coated pits" and entry into a series of endosomal compartments. The kinase remains active, at least in early endosomal compartments. Eventually, the kinase is inactivated, and the receptor and ligand are degraded. Choices are made about whether to recycle, transcytose, or enter lysosomes. The role of intracellular kinase in signaling and directing the receptor's choice of an intracellular route will be discussed.

The unoccupied EGF receptor is diffusely distributed throughout the plasma membrane both *in vivo* and *in vitro* (*6,46,47,113*). In an epidermoid carcinoma cell line (A431) however, almost half of the surface receptors were present in clusters of two or more in the absence of EGF, as determined by quantitative analysis of immunogold labeling (*152*). It was not reported whether the preclustered receptors were in coated pits. It must be borne in mind that EGF receptor expression in A431 cells is aberrant, with 2 to 6 million receptors per cell, 10 to 100 times the level found in normal cells (*33*). In EGF receptor-containing cells, the binding of EGF induces rapid clustering in coated pits (*6,46,47,63,64,113*). A detailed kinetic analysis of EGF internalization by Lund et al. supported a model in which occupation of receptors increased their affinity for coated pit proteins, thereby promoting clustering and internalization (*99*). They hypothesized that the occupied receptor's activated tyrosine kinase might be responsible for this change in affinity for the internalization pathway, although other studies suggest that the tyrosine

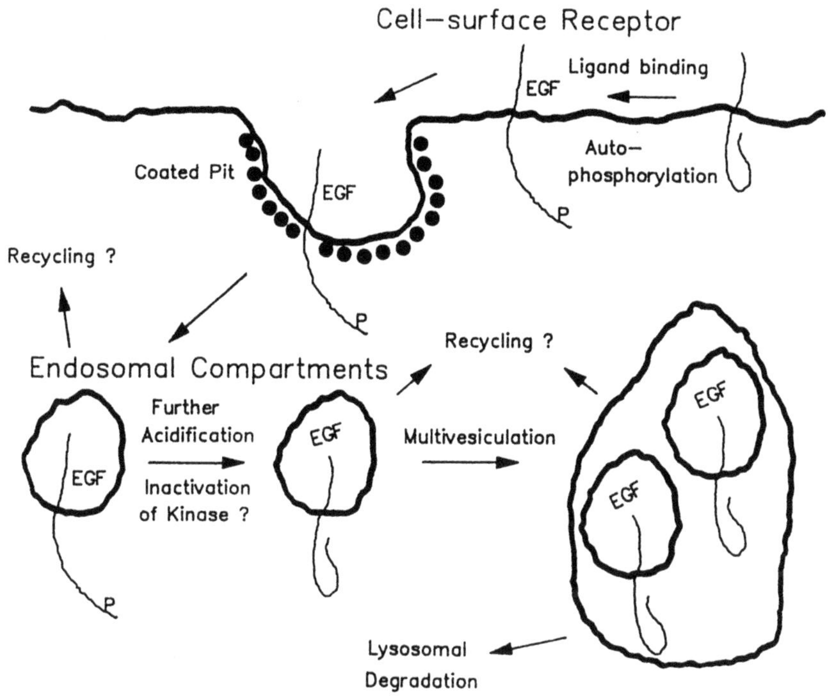

Figure 2.

kinase is not necessary for clustering in and internalization via coated pits (see Section III). It is clear that the tyrosine kinase is not needed for dimerization, the postulated first step in receptor activation. It has been shown that the truncated external receptor domain can dimerize in an EGF-dependent manner (*95*).

After clustering, ligand/receptor complexes are internalized in coated vesicles, then appear in the noncoated endosomal compartment (*6,46,47,63,64,113,114*). In some cells, EGF-EGF receptor complexes have been observed to pass through three types of endosomes: Type I, early endosomes of 100 to 200 nm in diameter, located near the cell periphery; Type II, late endosomes located in the Golgi-lysosome region of the cell; Type III, multivesiculated endosomes (multivesicular bodies) also located in the Golgi-lysosome region (*46,47,81,113,114*). Recent studies using video microscopy have shown that, in some cells, the distinction between the individual compartments—early, late, multivesicular body—may be overstated. In these studies, recycling transferrin receptors and EGF receptors targeted for degradation in lysosomes both were observed within one contiguous internal tubular structure that appears to separate, sort, and redirect receptors to their various fates (*71*). These authors have raised the possibility that the EGF receptor tyrosine kinase may be involved in routing the receptor to the pole of the tubule that targets receptor to the degradative pathways (*52*).

Association with the internalization system can be achieved in at least one ligand-independent manner. Infection of cells with adenovirus or expression of the E3 adenoviral proteins results in profound decreases in EGF receptor surface number. The biosysnthetic rate of EGF receptor synthesis was not altered by E3 expression, and therefore, it seems that expression of the E3 complex (a 10-kDa and a 14-kDa protein) results in a greatly increased rate of EGF receptor internalization (*17,148*).

Some form of proteolysis appears to proceed in the internalized pre-lysosomal vesicles. Renfrew and Hubbard have demonstrated that proteolytic processing of EGF begins in the endosomal compartment (*128*), probably before receptor degradation commences (*108,121,132,162*). In liver, different proteases may selectively degrade EGF in early and late endosomes. Not all the EGF is proteolyzed, because ultrastructural studies have demonstrated that EGF and EGF receptors continue to be colocalized within the internal vesicles of multivesicular bodies immediately preceding the appearance of receptor in lysosomes (*114*). Pastan and co-workers provided evidence that endocytosed EGF became associated with coated pits in the Golgi apparatus of cultured KB cells just prior to degradation in lysosomes (*64*). In contrast, transferrin, which is recycled along with its receptor, showed no such localization in Golgi-associated coated pits (*64*).

Transcytosis and Receptor Recycling

At least two alternatives to receptor degradation in lysosomes exist: recycling to the surface; and routing across the cell. Transcytosis of EGF has been demonstrated in the rat liver and in cultured Madin-Darby canine kidney (MDCK) cells. Jones and co-workers studied the fate of iodinated EGF injected into rat portal vein (*14,143*). At nonsaturating doses, almost all of the EGF administered was endocytosed by hepatocytes in a specific manner, and most of this was degraded in lysosomes. A small fraction (4% of the injected dose), however, was secreted into the bile intact and bioactive. This fraction was not affected by lysosomal inhibitors, suggesting a receptor-mediated endocytic pathway independent of lysosomes (*14*). Several groups have studied transport of EGF in polarized MDCK cells grown on dual-sided chambers (*101*). They found EGF receptors only on the basal surface, and observed unidirectional (basal to apical) transcytosis of 50% of surface-bound EGF. The receptor may not be involved in this transport to the apical surface (*12*).

In most cultured cell lines, including epithelial and fibroblast cells, both receptor and EGF are degraded in lysosomes with no evidence of substantial receptor recycling (*6,63,112,113,114,144*). However, under some circumstances in normal cells, and often in neoplastically transformed cells that overexpress receptor, or in receptor negative cells, that have been transfected with EGF receptor expression vectors, recycling of receptor occurs (*116,161*). This fundamental difference has complicated comparison of results from investigators who use different cell types.

The problems of assessing recycling of ligand-bound EGF receptors are typified by divergent results in studies of intact liver and liver cells. Using liver perfusion, Dunn et al. observed that rat liver had a cycloheximide-resistant capacity for internalization and degradation of EGF that far exceeded its surface receptor content (*46*). Similar results were obtained by another group studying primary hepatocyte cultures (*56*). Dunn originally suggested that receptors were recycled, but a later study suggested the existence of a large latent pool of receptors that was recruited to the cell surface (*47*). Another group, using a portal vein injection approach *in vivo*, studied the fate of hepatocyte EGF receptors as a function of EGF dose (*91*). After injection of low doses of EGF (resulting in less than 50% receptor occupancy), cycloheximide-resistant reappearance of the lost surface receptors occurred. EGF binding studies and immunoblotting using EGF receptor antibodies, however, failed to identify latent or internal pools of receptors that could account for the replenished surface receptor. These investigators concluded that the internalized receptors were recycled. Most of the injected dose of ligand was degraded, but a small proportion appeared in the circulation intact. This implies differential handling of ligand and receptor. In contrast, higher doses of injected EGF (resulting in greater than 50% receptor occupancy) induced loss of receptors and degradation of the entire injected dose of ligand (*91*). This suggested that a high level of receptor occupancy somehow dictated that all receptors enter the degradative rather than the recycling pathway. This group has also demonstrated that insulin injection internalized insulin and its receptor (*80*).

A third group performed a similar study with high-dose injection of EGF into the portal circulation (*75*). They subsequently isolated three intracellular vesicular compartments and assessed EGF and transferrin receptor levels in these compartments. In control liver, EGF receptor was found in all three compartments (compartment of uncoupling receptor and ligand, multivesicular bodies, and receptor-recycling compartment). This indicated a constitutive flow of EGF receptor into a recycling compartment. Upon maximal EGF challenge, *no* EGF receptor was found in the receptor-recycling compartment which continued to ferry transferrin receptor back to the surface (*75*). One interpretation of this series of reports is that the liver is a high capacity system for clearing EGF from the blood; in effect, reducing the blood levels of a powerful growth factor. Recycling of receptor may occur at low levels of ligand to amplify EGF uptake; high receptor occupancy may produce a sufficient EGF receptor tyrosine kinase signal to clear the receptor from the surface by placing all receptors into a noncycling pathway.

Recycling of EGF also was reported in a pancreatic carcinoma cell line (*87*). In this study, internalized EGF was released both in an intact form and as a processed species which retained its binding activity. The fate of the receptor was not investigated. This observation probably represents a transcytotic pathway, since agents such as methylamine, which block true recycling pathways, did not interfere with the release of EGF (*87*).

EGF Receptor Tyrosine Kinase Activity during Endocytosis

The tyrosine kinase is essential for EGF receptor growth signaling and appears to have a role in bringing the receptor into the endocytotic pathway (see Section III). In addition, the endocytic pathway may serve to translocate an active kinase to a position where it can interact more efficiently with intracellular substrates that are not as easily accessible from the cell surface. Lastly, the kinase activity may modulate intracellular receptor traffic and, in turn, may be modulated by endocytotic mechanisms such as cell-surface clustering, endosome acidification, and prelysosomal proteolytic activity. In early experiments, several groups approached this problem by determining the activity of EGF receptor kinase in isolated endosomes following EGF-induced internalization. They found that the EGF receptor tyrosine kinase activity is readily measurable *in vitro* by determining rates of receptor autophosphorylation and tyrosine phosphorylation of exogenous substrates (*33,77,111*). Cohen and Fava demonstrated the appearance of activated EGF receptor tyrosine kinase in a population of internal vesicles in EGF-exposed A431 cells (*33*). The receptor kinase was active *in vitro* in autophosphorylation and phosphorylation of added substrate. The partially-purified vesicle population was shown by electron microscopy to contain small coated vesicles, small and large endosomes, and Golgi elements; it was not determined specifically which vesicles contained the active kinase.

Herman and Earp have made similar observations using endosomal fractions isolated from EGF-treated A431 cells by percoll density gradient centrifugation. In untreated cells, the receptor resided in vesicles with plasma membrane characteristics. When cells were treated with EGF at 16 °C, the EGF receptor shifted to membrane fraction with the characteristics of endosomes. In addition, the autophosphorylation of the receptor in endosomal fractions from EGF-treated cells was no longer EGF-dependent. It had been and remained activated by the EGF added to the cell which had been internalized within the endosomes (unpublished results). The EGF receptor tyrosine kinase in plasma membrane fraction from untreated cells was only activated by adding of exogenous EGF. Kay et al. made similar observations in EGF-injected rat liver (*77*). Increased EGF receptor kinase activity, as measured by autophosphorylation, was observed in Golgi-endosome fractions as early as 30 seconds after injection of EGF. The peak activity was detected at 5 to 15 minutes; the rise in internalized EGF receptor kinase activity corresponded to the loss of surface receptors (*77*). This same group demonstrated that insulin injection produced an activated insulin receptor tyrosine kinase isolatable from liver endosomes (*80*). Using electron microscopic radioautography, these workers showed that injected EGF became associated with the endosomal and not the Golgi elements of the isolated internal membrane fractions (*77*).

The problem with each of the above observations was that the "active" EGF receptor kinase may, in fact, have been receptor reactivated during endosomal isolation. In other words, it is possible that EGF dissociated from the receptor by an acid environment of the endosomal compartment may have rebound to the receptor as the pH gradient was dissipated following cell homogenization. Additionally, these studies support the conclusions of the earlier reports that endosomes from EGF-treated cells carried active EGF receptor kinases. For example, recent reports indicate that EGF remains bound to the receptor, at least in early endosomes (*23,90,92,113,140*). Moreover, Nesterov et al. permeablized A431 cells after EGF-directed EGF receptor internalization, and showed that an exogenous peptide substrate could enter the cell postpermeablization. This substrate could be tyrosine phosphorylated by an internal tyrosine kinase, presumably the activated endosomal EGF receptor kinase (*117*).

Our laboratory used a somewhat different approach to examine the relationship between EGF-induced endocytosis and EGF receptor kinase activity in a rat liver epithelial cell line with an EGF receptor content similar to rat hepatocytes (*112*). After incubation in culture with EGF for various times, cells were analyzed for: (1) EGF surface binding; (2) EGF receptor immunolocalization; (3) total EGF receptor content; (4) EGF receptor kinase activity in cell homogenates, and (5) tyrosine phosphorylation of cellular proteins. EGF-induced rapid (5 min) internalization of >90% of surface receptors without evidence of significant recycling. Control cells exhibited little basal EGF receptor kinase activity, as assessed by autophosphorylation in cell homogenates; preincubation of these homogenates with EGF markedly stimulated the kinase activity. Maximally-increased EGF receptor kinase activity was detected in homogenates made 15 minutes after the addition of saturating doses of EGF to cell cultures, a period during which internalization of surface receptors was complete but significant receptor degradation had not yet occurred. Preincubation of the homogenates from EGF-treated cells with additional EGF did *not* further increase the kinase activity, indicating that receptor occupancy was complete and that the receptor population had been maximally activated by the EGF added to the intact cell. The kinase activity measured in homogenates gradually decreased, paralleling the degradation of ^{35}S-labeled receptor. If receptor trafficking and, thus, degradation was blocked by treatment of cells with methylamine or at low temperature (16 °C), kinase activity in homogenates made at times up to 2 hours after EGF addition to intact cells persisted at high levels. For at least 2 hours, methylamine-treatment of cells allowed large patches of immunoreactive EGF receptor to accumulate intracellularly. Even 2 hours after EGF treatment, stimulated receptor kinase activity could be detected *in vitro* when detergent was included in the assay. Detergent was not necessary to detect stimulation of kinase when EGF was added directly to homogenates of control cells (in which EGF receptors were still in the plasma membrane). Analysis of phosphotyrosine substrates by anti-p-tyr immunoblotting showed a gradual disappearance of p-tyr substrate at 1 to 2 hours in methylamine or 18 °C treated cells. Thus, we postulated that under these circumstances internalized EGF receptors were *potentially* kinase active, but that the kinase domains became sequestered in a membrane-bound intracellular compartment that inhibited the intracellular kinase activity. This inhibition occurred even though there was *no* acidification in the vesicles of methylamine-treated cells. The kinase was either constrained physically or biochemically, or was placed in a compartment in which ATP was not available. The multivesicular body is a prime candidate for such a late compartment. EGF receptors, and specifically kinase domains, have been found in the lumen of these structures, where inhibitors of lysosomal degradation cause their accumulation (*52,114*).

In the studies by Cohen and Fava and Kay et al. described earlier, detection of translocated EGF receptor kinase activity in Golgi-endosome fractions did not depend on the presence of detergent (*33,77*). This suggested that the receptor was oriented with the kinase (cytoplasmic) domain on the outside of the endosome, as would be predicted. Cohen and Fava further confirmed this orientation by demonstrating that the kinase was trypsin-sensitive. When we studied internalized kinase activity, we detected trypsin-resistant kinase activity after receptor was allowed to accumulate in methylamine-treated cells (McCune and Earp, unpublished results). This indicates that under these conditions, the cytoplasmic kinase and autophosphorylation sites were protected from trypsin by their inclusion within another vesicular wall. As noted above, our assay system also measured significant EGF receptor kinase activity at early times (5–15 min) after >90% internalization of the receptor. At this time, we did not observe the disappearance of autophosphorylated EGF receptor and substrates. However, the EGF-independent phosphorylation still required detergent for full expression of high level *in vitro* kinase activity, suggesting that in these cells some form of biochemical constraint still exists when the receptor is fully active inside the cell.

The Role of Endosome Acidification in Terminating the Activity of Intracellular Receptor Kinase

The relation of EGF activation of the receptor kinase and the EGF-induced appearance of tyrosine phosphoproteins was also studied in cultured rat liver epithelial cells (*112*). Within 5 seconds of EGF addition to the culture media, tyrosine phosphorylation of at least seven proteins was detected by immunoblotting with antiphosphotyrosine antibodies. If incubations were carried out at 0 °C to prevent endocytosis, phosphorylation of these same seven proteins still occurred, suggesting that they were accessible to activated plasma membrane receptor kinase. At least three substrates showed delayed increases, appearing at 45 to 90 seconds after the addition of EGF at 37 °C; these phosphoproteins did not appear even with prolonged incubation of cells with EGF at 0 °C. Although other explanations exist, this time frame of the secondary phosphorylation is compatible with entry of the kinase into the early endosome compartment, raising the possibility that some substrate phosphorylation is internalization-dependent.

Bergeron and co-workers have extensively studied a 55-kDa phosphotyrosine substrate in rat liver after injection of EGF in the portal vein (*155*). This substrate copurifies with the receptor from endosomes in EGF-treated livers, suggesting that it may be continually phosphorylated in endosomes by the activated receptor. Whether this substrate is important for EGF receptor internalization in liver remains to be determined. The phosphotyrosine content of the EGF receptor and other substrates declined after 5 to 15 minutes in our liver epithelial cells. However, the fact that phosphorylation remained at maximal levels beyond 5 minutes argues for effective and continued phosphorylation by internalized receptors in early endosomal compartments. First, greater than 90% of the receptor is internal at this time. Second, when receptor is stripped of ligand at low pH (*61*), the receptor tyrosine kinase activity is inhibited within a very short time (1–2 min), and tyrosine phosphatases rapidly dephosphorylate the receptor and substrates. This rapid loss of phosphotyrosine has been shown by studying activation of EGF receptor tyrosine phosphorylation at 0 °C in intact cells and its reversal by a low pH treatment of the cells (*146*). Under conditions which prevented endosome acidification and receptor degradation, the loss of p-tyr EGF receptor and EGF-dependent p-tyr substrates is slowed by 50 to 100%. This observation indicates that the intracellular EGF receptor kinase activity is normally inhibited as endosomes acidify and ligand is increasingly dissociated. The timing indicates that the acidification in the earliest vesicles internalized is *not* sufficient to acid-dissociate EGF from the receptor. However, the fact that the receptor and its substrates are dephosphorylated at a defined rate, even without vesicular acidification, indicates that some other mechanism exists for terminating the intracellular receptor signals. As suggested above, this may be accomplished by constraints placed upon the kinase or its substrates (including ATP) by entry into the multivesicular body.

In the intact cell, the acidic environment achievable in some endosomal compartments (about pH 5–6) should substantially reduce the affinity of most ligands for their receptors (*13*). As noted above, removal of cell surface-bound EGF by acid washing dramatically reduces the phosphotyrosine content of the receptor and reverses surface receptor EGF-dependent dimerization (*140*), indicating that continued receptor occupation is necessary to maintain kinase activity. However, the ligand concentration in the early endosome may be sufficiently high to overcome a slight acid pH-induced reduction in affinity and permit continued binding. Several *in situ* ultrastructural studies support the hypothesis of continued binding of EGF to the receptor in endosomes. These studies have localized endosomal EGF to the inner luminal surface rather than free in the lumen (*23,90,92,113*). Recent studies showing a relatively high level of persistent EGF receptor dimers in internalized vesicles of A431 and NIH 3T3 cells also indicate that EGF remains bound to the receptor in at least some internal compartments (*140*). The NIH 3T3 cells stably expressing a moderate number of human EGF receptors retain a much higher proportion of their internalized EGF receptors as dimers than do A431 cells (*140*). These results again point to cell-type specificity in EGF and EGF receptor metabolism.

The Role of EGF Receptor Clustering and Dimerization

The mechanism by which EGF binding activates the receptor's kinase is still not completely resolved. EGF can induce receptor oligomerization in detergent extracts and in isolated membranes, and this event has been correlated with increased kinase activity (*10,149,167,168*). Rapid EGF-dependent receptor dimerization and oligomerization also have been demonstrated in intact cells by biochemical methods (*31,51,140*). In cells, dimerized receptors were generally phosphorylated *but* not all phosphorylated receptors were dimerized (*140*). Thus, correlational data has led to the proposal that receptor dimerization is necessary for kinase activation and, consequentially, for signal transduction in the living cell (*149*). Under many circumstances, EGF-induced receptor clustering immediately follows ligand binding and precedes internalization in living cells. In a sense, clustering has been considered an extension of dimerization that results in the large molecular weight oligomers seen upon chemical cross-linking of ^{125}I-EGF to receptor (*51*).

In the usual case, dimerization and clustering may both precede endocytosis, but are they obligatory for activation of signal transduction? Several observations argue against their absolute necessity. Activation of EGF receptor tyrosine kinase activity without dimerization has been achieved by incubation with sphingosine (*118*), F_{ab} fragments of monoclonal antibodies (*134*), or EGF itself (*86*). Carraway and Cerione recently observed that significant receptor microaggregation did not occur in A431 cells within the time frame of early EGF-stimulated events such as calcium flux (*24*). We have observed marked EGF-dependent stimulation of tyrosine phosphorylation of EGF receptors and most other cellular substrates in cells incubated with EGF at 0 to 4 °C (*112*). Many studies have demonstrated morphologically and biochemically that receptor clustering and oligomerization are prevented at this temperature (*6,46,51,113*). Thus, kinase activation and receptor oligomerization/clustering appear to be dissociable events.

Desensitization of Receptor Activity after Ligand Binds

Initial events in signaling and in receptor-mediated endocytosis may modulate EGF receptor kinase activity by producing forms of receptor desensitization. The forms of desensitization to be discussed occur rapidly (within 1–3 min), probably while the receptor is on the plasma membrane. Activation of phospholipase-C-γ can activate protein kinase C which, in some cells, prevents further EGF signaling by downmodulating tyrosine kinase activity (see Section IV-A). Davis and coworkers (*35*) have suggested that phosphorylation of Ser 1046/7 is also capable of downmodulating kinase activity. Their studies demonstrate that receptor fragments containing these residues can be phosphorylated by calmodulin-dependent protein kinase II *in vitro*.

Our laboratory has observed a transient suppression of EGF receptor kinase activity measured in homogenates 1 to 2 minutes after addition of EGF to cells (*111*). This inhibition was not altered in cells depleted of >95% of their protein kinase C, indicating that inhibition of receptor kinase via activation of protein kinase C was *not* responsible for our observations. In addition, the strict EGF dose-dependence of this phenomenon suggested that receptor occupancy of individual receptors was necessary for the effect. Since the time frame of suppressed activity corresponds to the clustering of EGF receptors, we hypothesize that association of surface EGF receptor with proteins of the internalization apparatus results in kinase inhibition, thus limiting its action at the cell surface.

Similar observations were made by Lai et al. in rat liver (*92*). In their study, subcellular fractionation demonstrated that the suppressed kinase activity was associated with the plasma membrane and not Golgi-endosome fractions. Several other reports suggest that EGF stimulates association of the receptor with the cytoskeleton (*93,153*). Thus, we favor the explanation that EGF drives an occupied receptor into an association with elements that transiently prevent detection of activated receptor kinase. Continuation of the internalization process relieves this constraint. However, the findings of Davis and colleagues (*35*) indicate that Ser/Thr kinase-mediated forms of desensitization cannot be excluded. It may be that activation of an individual receptor by EGF exposes a site for phosphorylation by a Ser/Thr kinase, allowing specific desensitization of occupied receptor in the same way that exposure of internalization sequence promotes entry of ligand-occupied receptor into the high affinity internalization pathway (*161*). The effect of transient desensitization processes may be relieved as the receptor is internalized and the inhibitory Ser/Thr phosphate is removed by phosphatase.

III. THE STUDY OF EGF RECEPTOR ENDOCYTOTIC MECHANISMS BY SITE-DIRECTED MUTAGENESIS

The availability of cDNA encoding the human EGF receptor (*150*), and the ability to express wild-type and specifically mutated receptors in cell culture model systems, have provided a powerful means of examining receptor structure-function relationships in the endocytotic process. Investigations using receptor mutants have addressed three major aspects of ligand-dependent internalization: (1) the role of receptor kinase activity and substrate phosphorylation; (2) the role of receptor autophosphorylation; and (3) the role of the carboxyl terminus, including regions distinct from the kinase domain and phosphorylation sites.

A. Receptor Kinase Mutants

Kinase-negative congeners of the EGF receptor have been prepared in several laboratories by mutating Lys721, a residue involved in ATP binding that is conserved throughout the large family of protein kinases (*62*). When expressed in appropriate host cells, mutant receptors containing Met721 or Ala721 bind EGF with high affinity, but fail to elicit typical EGF-stimulated cellular responses, including phosphoinositide hydrolysis, Ca^{2+} mobilization, proto-oncogene induction, DNA synthesis, or cell proliferation (*28,70,115*). Studies such as these have provided crucial support for the hypothesis that EGF signal transduction requires activation of the receptor kinase activity and, implicitly, phosphorylation of specific protein substrates.

Kinase-negative receptors also show aberrant endocytotic behavior, but the precise nature of the defect has been the subject of some disagreement. One group suggests that kinase activation greatly enhances initial receptor internalization but has little effect on the choice between recycling and degradative pathways, while another group suggests that activated kinase dictates choice of the degradative pathway but does not greatly influence initial internalization. In a series of studies by the first group, Met721 mutant receptors have been compared to wild-type receptors expressed in mouse B82 cells (*27,28,161*). In these experiments, internalization (determined by measuring cell-associated ligand after removal of surface ligand at low pH) has been monitored for receptors occupied either with labeled EGF or a labeled nonactivating monoclonal antibody. These procedures allow calculation of "specific internalization rates," which may be viewed as the probability that a labeled surface receptor will internalize per unit time. These experiments have demonstrated that EGF occupancy of wild-type receptors can provoke an approximate 10-fold increase in specific internalization rate, whereas EGF-occupied Met721 receptor internalize at a rate no greater than their constitutive (antibody-occupied) rate. These results indicate that, in B82 cells, expression of tyrosine kinase activity is important for ligand-dependent internalization of EGF receptors (*161*). The same investigators found no kinase-dependent difference in the rate of receptor recycling, as measured by the return of previously internalized EGF to the cell surface. Interestingly, studies of the B82 transfectants also revealed that specific internalization rates for the EGF-occupied wild-type receptor declined as a function of increasing surface receptor occupancy, approaching the rates observed with kinase-negative or unactivated wild-type receptors. This finding is interpreted as indicative of a high-affinity, saturable endocytotic mechanism accessible to activated, kinase-competent receptors.

In contrast, other investigators, studying Ala721 receptors expressed in NIH 3T3 cells, have reported that such receptors internalize at wild-type rates (*70*). Similar findings were made by these investigators in recent studies in which EGF added to NIH 3T3 cells expressing wild-type or mutant receptors was found to enter internalized (acid-stable) compartments with similar kinetics (*52*). This delivery of EGF to internal sites by mutant receptors, however, was accompanied by a relative conservation of surface EGF binding sites, interpreted by these authors as indicative of more rapid recycling of the mutants. Similar findings were made with receptors labeled with an antibody that binds to but does not activate the receptor kinase. Immunohistochemical studies suggested that this preferential recycling of kinase-negative receptors might stem from differential sorting in a late endosomal compartment. The sorting mechanism was postulated to actively select kinase-positive receptors for routing to lysosomes (*52*).

The reasons for the discrepant findings between these two groups of studies are not entirely clear. Among the more mundane possibilities is that, in some cases, different methods of analysis have been used. Wiley et al. (*161*) have pointed out the importance of distinguishing kinetically between: (1) ligand-induced internalization by high-affinity, saturable pathways; and (2) random entrapment of receptors in coated or smooth pits, either in the absence of ligand-activated kinase or under conditions at which high-affinity pathways are saturated. Thus, finding similar rates of wild-type and kinase-negative receptor internalization may stem from their measurement at relatively high degrees of receptor occupancy (*52*). On the other hand, divergent findings may, in time, reveal important cell-type-specific differences in endocytotic capabilities, or details of endocytotic mechanisms not yet appreciated. For example, just as the saturability of high-affinity internalization might mask differences between wild-type and mutant receptors at high ligand occupancy, so might high constitutive rates of recycling in neoplastic cells or the existence of saturable sorting/lysosomal routing mechanisms mask possible kinase-dependent effects beyond the internalization event. While the evidence is strong that tyrosine kinase activation can have a pronounced effect on receptor internalization rate, the evidence for discounting a role for active receptor kinase in the endocytotic compartments is, in our opinion, not compelling. In fact, evidence from other studies in which TGFα dissociation from the receptor in endosomes leads to decreases in intracellular EGF receptor kinase activity (*49*) and increased receptor recycling argue for an effect of kinase activity in the choice of the degradative pathway. In any event, investigators would likely agree that kinase-negative EGF receptors are not subject to normal downregulation (as defined by accelerated net entry into a degradative pathway), and that kinase-dependent receptor conformations and/or interactions with kinase substrates are required for efficient endocytosis.

B. Mutation of Autophosphorylation Sites

The known sites of EGF receptor autophosphorylation include three major sites: Tyr 1068, 1148, and 1173, and a recently detected minor site at Tyr 1086. All lie in the C-terminal portion of the protein (residues 941-1186), distal to the consensus kinase domain (*43,102*). [An additional site, Tyr992, is phosphorylated when the primary sites are removed by receptor truncation (*157*).] In unstimulated cells, the nonphosphorylated C terminus appears to exert an autoinhibitory effect on the kinase domain; autophosphorylation, triggered by EGF binding, is believed to promote a conformational change that allows access to the kinase domain by protein substrates (*8,19,149*). Thus, mutation of the autophosphorylation sites of the EGF receptor provides the means to evaluate this aspect of receptor activation in the endocytotic process. In addition, a growing body of evidence suggests that the EGF receptor C terminus, apart from its proposed permissive role in kinase activation, may contain important determinants of protein substrate selection. In particular, autophosphorylated tyrosine residues appear to serve as specific binding sites for proteins (e.g., phospholipase C-γ) that contain *src*-homology domains (reviewed in *85*). It seems reasonable to postulate, then, that phosphorylation sites might also participate in interactions with the endocytotic apparatus.

Three EGF-receptor mutants, in which the major autophosphorylation sites (Tyr1068, 1148, and 1173) had been individually mutated to phenylalanine, bound

EGF with appropriate affinity and were internalized and degraded at rates similar to wild-type receptors when expressed in NIH 3T3 cells. Each mutant form transduced a mitogenic signal in response to EGF (*69*). Similarly, double Tyr-to-Phe point mutants downregulated at wild-type rates and were biologically active; however, triple point mutants were deficient in several EGF-stimulated functions, including kinase activity toward exogenous substrates, mitogenesis, and rates of internalization and degradation (*27,65,142*). No changes in receptor recycling rates were noted with the triple mutant (*142*).

In cells expressing the triple mutant, EGF-stimulated receptor autophosphorylation occurred at approximately 30% of wild-type levels (*142*), presumably owing to phosphorylation of the remaining known sites (Tyr1086 and 992). Furthermore, EGF-stimulated phosphorylation of a well-characterized endogenous receptor substrate, phospholipase C-γ_1, was decreased approximately 50% in cells expressing triple, but not single or double, mutants (when data were normalized for numbers of EGF binding sites present among various transfected lines). These results, consistent with those obtained using kinase-negative mutants, suggest that normal, ligand-directed receptor endocytosis requires that the receptor kinase be fully competent for activation. These studies alone, however, do not resolve the question of whether: (1) interaction of phosphorylated residues with endocytotic elements, or (2) relief of kinase autoinhibition is the consequence of autophosphorylation relevant to its role in endocytosis.

C. C-Terminal Truncations

To determine the relative contributions of autophosphorylated residues, per se, and kinase activation, we may consider the results of C-terminal truncation experiments. A number of such studies have indicated that the EGF receptor C terminus contains regions, distinct from sites of autophosphorylation, that are critical for interaction with the endocytotic apparatus. Receptors truncated at amino acid residue 1123, 1052, or 1022 have been shown to undergo EGF-stimulated internalization when expressed in host cells (*27,98*). Further truncation to residue 973 produced receptors that retained EGF-stimulated kinase activity but were resistant to internalization by the high-affinity pathway (*27*), suggesting that the region between residues 973 and 1022 contains elements essential for endocytosis. A kinase-negative fusion protein, composed of receptor truncated at residue 688 plus residues 944-1022, underwent EGF-stimulated internalization, although at much lower specific rates than the wild-type receptor. A recent report from the same laboratory has more precisely mapped this region that apparently directs internalization (*25*). In these studies, the sequence encoding residues 993-1022 was found to confer ligand-induced, high-affinity internalization when fused to a receptor truncated at residue 958, even though the mutant protein contains none of the known sites of autophosphorylation (*25,157*). Residues 974-991 have limited internalizing activity, since kinase-active receptors truncated at residue 991 showed partial retention of high-affinity ligand-induced internalization, whereas, as noted above, truncation to residue 973 abolished this property. Thus, these studies have delineated two peptide regions (974-991 and 993-1022) that independently direct ligand-stimulated, high-affinity receptor internalization. It should be noted that neither peptide contains the NPXY motif required for internalization of the LDL receptor (see *141* for review); this motif does occur three times in the EGF receptor C-terminal region (preceding tyrosines 1086, 1114, and 1148), but only at loci apparently dispensable for ligand-stimulated internalization (*27*).

Taken together, receptor truncation studies suggest that a measurable degree of EGF-dependent receptor endocytosis minimally requires only two factors: ligand activation, and the presence of a competent "internalization domain." In the limited context of endocytosis, the kinase domain and autophosphorylation sites might be considered important only inasmuch as they provide the wild-type receptor with a mechanism to unmask the residues involved in interaction with the endocytotic apparatus. However, a more active role for the kinase in the holoreceptor seems probable, since neither of the internalization-directing peptides described above was found to be functional in M721 kinase-negative mutants (unpublished data cited in reference *25*). In addition, the regulatory features of the receptor that lie C terminal to residue 1022 likely play key roles in both kinase substrate selection and signal termination. It is important to note that serial C-terminal truncations of the EGF receptor are accompanied by dramatic increases in the number of readily detectable tyrosine phosphorylated proteins (*27,39,157*). This increased responsiveness to EGF may reflect relaxation of substrate specificity in the absence of critical C-terminal domains, as well as the prolonged presence of activated, internalization-deficient receptors on the cell surface. In any case, the importance of normal receptor downregulation is underscored by the increased potency of EGF in transforming NR6 cells expressing noninternalizing receptors (*159*). In future studies, it is anticipated that the ability of defined receptor peptides to convey an internalization signal will expedite the discovery of the endocytotic components with which the receptor might interact.

IV. SPECIAL TOPICS

A. The Effect of Phorbol Esters on the EGF Receptor

The cytoplasmic juxtamembrane region of the receptor contains a protein kinase C phosphorylation site at threonine 654. Protein kinase C-dependent phosphorylation of Thr 654 has a remarkable effect on the receptor. First, in intact cells, Thr 654 phosphorylation reduces the affinity of the receptor for EGF by 10-fold, physically altering the state of the external domain or its ability to form high affinity sites by dimerization. In addition, Thr 654 phosphorylation abrogates EGF-dependent signal transduction by drastically reducing the ability of even high concentrations of EGF to activate the receptor's tyrosine kinase. One function of this negative regulatory site may be to attenuate EGF signaling (*20*). In many cells, EGF stimulation of its receptor results in activation of phospholipase C-γ which, in turn, creates a diacylglycerol and calcium signal that stimulates protein kinase C activity (see *66*). The end result, phosphorylation of Thr 654, significantly blunts EGF receptor signal transduction, including EGF-activated EGF receptor tyrosine kinase activity. For example, pretreatment of liver epithelial cells with phorbol ester blocks EGF-dependent tyrosine phosphorylation of phospholipase C-γ (*72*). This desensitization has recently been shown by studies of cellular morphology that examine ruffling of EGF-treated cells. Ruffling proceeds for several minutes after EGF addition but then stops, perhaps due to activation of protein kinase C and inhibition of the receptor tyrosine kinase. When a site-directed mutant changing Thr 654 to Ala is expressed in cells, the mutant receptor continues to send a ruffling signal in the presence of EGF (*160*). Others have suggested that additional ser/thr kinases phosphorylate the receptor and decrease the response to EGF (*35*). These desensitization mechanisms would occur earlier than the downregulation brought about by internalization and degradation of the receptor.

Activation of protein kinase C by phorbol esters can, in some cells, lead to a specialized form of internalization of the EGF receptor. Several groups, including our own, have clearly shown that phorbol ester activation of protein kinase C in epithelial cells decreases EGF receptor surface number by inducing EGF receptor internalization (*7,111*). However, the TPA-induced internalization is different in consequence from that produced by EGF. The maximum number of receptors internalized is much less than 90 to 100% maximum that can be achieved by EGF. More importantly, the receptor in phorbol ester-treated cells is not degraded but rather is placed into the recycling pool and returns to the surface within several hours (*7,111*). Phorbol ester-induced internalization may not take place in every cell type and is probably due to stimulation of general receptor internalization (e.g., transferrin receptor) rather than a specific result of EGF receptor Thr 654 phosphorylation.

B. Do EGF Receptor Complexes Translocate to the Nucleus?

In light of the action of steroid receptors in the nucleus, the possibility that polypeptide cytokine/receptor complexes deliver growth factor or receptor to the nucleus has been considered by several groups. Isolated reports have described translocation of EGF and/or EGF receptors to the nucleus following EGF binding to surface receptors. Raper et al. described shunting of injected EGF to the hepatocyte nucleus in regenerating rat livers in contrast to lysosomal degradation in controls (*127*). Nuclear translocation and binding of EGF to a chromatin protein was described by Rakowicz-Szulczynska et al. (*126*). Jiang and Schindler used an EGF receptor-deficient mutant 3T3 fibroblast cell line reconstituted with active EGF receptors derived from rat liver membranes to study nuclear translocation (*76*). Their data indicated that a small proportion of internalized EGF:EGF receptor complexes became associated with the nucleus. Inhibitors of the recycling/lysosomal pathway such as chloroquine have been reported to enhance nuclear accumulation of EGF (*130*), suggesting that the nuclear pathway is a prelysosomal branch of the endocytotic pathway. In contrast, many investigators have failed to see nuclear translocation of EGF receptor (*23*), although these negative studies are not often published. In contrast to EGF, FGF contains a nuclear localization signal domain which, when mutated, significantly affects the biologic action of FGF (*125*). Other studies have suggested accumulation of insulin (*123*), NGF (*166*) and IL-1 (*59*) following mediated endocytosis, but the functional consequences of these findings are not clear.

C. Generation of EGF and EGF-Receptor Degradation Products

Several investigators have described limited proteolytic processing of EGF in the prelysosomal endosome compartment that results in distinct species of growth factors (*108,121,132,162*). The initial species generated have equal or greater affinity for EGF receptors, while subsequent species have reduced affinity (*108,121,132,162*). In a report by Korc and Magun, a high molecular weight species of processed EGF released by pancreatic carcinoma cells was capable of rebinding to surface receptors (*87*). Thus, there is the potential for bioactivity of partially processed EGF. A study by Matrisian et al. suggested that EGF processing might be necessary for some late effects on cellular metabolism (*107*). Early studies by Fox and Das suggested that endosomal proteolytic processing of the EGF receptor might play a role in signal transduction (*53*). The orientation of the EGF receptor in endosomes could lend itself to processing of the extracellular (EGF-binding) domain to create an erb B-like molecule (erb B is a retroviral oncogene created by transduction or insertion promotion of the chicken EGF receptor gene). Little evidence for such processing has been found. In most cells the receptor may be protected from proteolysis in the endosome compartment until it is completely degraded (*20*). However, Decker has observed the generation of an N-terminally truncated 125-kDa receptor fragment during ligand-dependent internalization of the human receptor overexpressed in NIH 3T3 cells (*37,38*).

V. SYNOPSIS: THE BIOLOGIC IMPORT OF SURFACE SIGNALING AND RECEPTOR INTERNALIZATION

Upon ligand binding, receptor kinase activation results in autophosphorylation of the receptor's C terminus. This, in turn, exposes at least two regions whose sequences direct association with the high affinity internalization pathway (*25*). The activated kinase may also help speed entry into the internalization pathway by phosphorylating specific substrates. The activation of receptor dimerization and the internalization process probably initiates receptor clustering. After clustering and internalization via coated pits, the receptor spends a variable period of time in early and late endosomal compartments and is degraded in lysosomes. The length of time for each of these steps depends on the cell type and the number of EGF receptors expressed on the cell surface. This issue of cell type difference in receptor metabolism clearly plagues studies of EGF receptor internalization and its biologic function. One example is the A431 cell which exhibits 2 to 6 × 10^6 receptors per cell, and virtually never clears all of its surface receptors, presumably because the sheer receptor number prevents efficient entry of EGF-bound receptor into the degradation pathway. Even 4 to 6 hours after EGF addition, one can still find evidence of extensive receptor occupation (*146*). The effect of high concentrations of EGF on A431 cells is cell death, demonstrating that an overwhelming, persistent signal can have deleterious effects.

In contrast, when high EGF concentrations are added to normal cells, approximately 90% of the receptor can be internalized within 2 to 10 minutes (*112*). Yet, in untransformed cells, there is a requirement for prolonged EGF receptor signaling to achieve a proliferative signal; i.e., a 6 to 8 hour exposure to extracellular EGF is needed to induce DNA synthesis in previously quiescent cells (*19,21*). Thus, there is a paradox: greater than 90% of the EGF receptor is degraded rapidly, but important aspects of receptor signaling are incomplete. In addition, since mutant EGF receptors without kinase activity fail to stimulate DNA synthesis (*28,70,115*), the above data strongly suggest a requirement for an active EGF receptor tyrosine kinase for at least 6 to 8 hours. In normal cells the half-life of ligand-occupied receptors is 1 hour or less (due to ligand-directed receptor degradation) (*144*). Thus, a 6 to 8 hour tyrosine kinase signal can only be provided by recruitment from a pool of unactivated receptors, consisting possibly of newly synthesized receptors which reach the surface throughout the next 6 to 8 hours. We and others have demonstrated that EGF can accelerate EGF receptor synthesis during the first 6 hours of EGF action (*29,48,89*). However, occupancy of the small percentage (<10%) of receptor that escapes initial internalization appears to be sufficient to generate the persistent signal required for cell growth. In fact, once the initial receptor clearance by high concentrations of EGF is complete, the amount of DNA synthesis in cultured cells is directly proportional to the occupancy of the remaining surface receptors (*84*). At this point, there may be only 2000 to 6000 surface receptors per cell, and EGF:EGF receptor signal transduction must be relatively efficient for the requisite 6 to 8 hours of EGF action. This need for efficiency may mean that the signal transduction takes advantage of the physical attributes of the receptor environment, including the ability to select crucial substrates or deliver signal transduction complexes to the right locale through movement along a predetermined internalization pathway.

The studies above establish the need for persistent stimulation by the receptor, but have not defined *what* signal is needed. Is the critical tyrosine phosphorylation that must be generated persistently transmitted while the EGF receptor is on the external surface of the cell, *or* are some components of the proliferative signal transduced from an activated endosome-based receptor with access to a key intracellular substrate? Alternatively, signaling may occur at the surface, but only as the receptor is moving laterally towards its rendezvous with the internalization machinery. For example, the autophosphorylated EGF receptor can bind phospholipase Cγ through this enzyme's SH2 group (*85*). The receptor:PLCγ complex may move laterally towards a point of cytoskeletal attachment where PLCγ's action upon phosphatidylinositol 4,5-bisphosphate (PIP_2) removes the IP_3. This can alter the binding of profilin to PIP_2 and potentially release actin binding proteins from a site of cytoskeletal attachment (*57*), and, in turn, may have import for the cell shape changes that occur in growth states. Therefore, receptor activation, protein complexing, and movement to the internalization site may have a clearcut function. In fact, immunofluorescence studies of phospholipase Cγ localization indicates that the enzyme associates with the actin cytoskeleton to a greater degree in EGF-treated cells than in control cells (*109*).

This review details the evidence that the EGF receptor *could* signal from endosomes: (1) EGF stays bound to the EGF receptor in at least some endosome compartments; (2) EGF receptor dimers (the putative activated receptor form) are present in endosomes; (3) the EGF receptor remains autophosphorylated while being internalized; and (4) the receptor is an active endosome-based kinase. All of these constitute necessary criteria for endosomal-based signaling: the ligand-internalized receptor is in the stimulated state.

However, the presence of an activated kinase does not prove that it performs a necessary function. In fact, when one examines the available evidence, it appears that persistent signaling from surface EGF receptor is necessary and can be sufficient to transmit an EGF growth signal. Persistent expression of a large number of EGF-stimulated EGF receptors at the cell surface has been achieved using receptor mutants with deletions of their "internalization domain." EGF treatment results in disordered growth and neoplasia in cells expressing these noninternalizating receptors (*159*). The above data have been interpreted to indicate that internalization of the receptor is *not* important for growth control, and that all aspects of signaling *can* be performed by the truncated receptor at the surface of the cell. The latter statement is correct, but it is unclear that this experiment precludes a signaling role for internalized receptor in the normal cell. Examination of the data reveals that the extent of EGF-stimulated tyrosine phosphorylation is enhanced dramatically in cells expressing these mutants (*39,157*). Persistent overphosphorylation of a number of substrates may bypass the normal route of cell growth control and contribute to neoplastic transformation.

A second line of experimentation also shows that surface signaling can enhance growth. Expression of the E5 protein of the bovine papilloma virus is capable of transforming cells. Interestingly, the expression of this protein *decreases* the EGF-dependent downregulation of EGF receptor from the surface of E5 bearing cells. Thus, the E5 protein appears to partially recapitulate the phenotype of the internalization-deficient mutant, prolonging an EGF signal from a large complement of surface receptors (*104*). In addition, experiments indicate that when added to cells, TGFα and Val47-EGF (a protease-resistant EGF with low affinity) recycle more readily than EGF. This could result in rebinding of EGF receptors on the cell surface, sending a more prolonged surface growth signal. In accord with this prediction, both TGFα (*49*) and Val47-EGF (*156*) under some experimental conditions were more effective than EGF in stimulating growth (*5*). Each of these experimental observations indicate that internalization of the receptor may function to terminate the signal and control the growth response. On the other hand, inhibition of receptor downregulation does not always strengthen the growth control signal. TGFβ decreases EGF receptor internalization *but* also inhibits EGF-dependent cell growth (*5*).

In general, manipulations that prolong cell-surface signaling (as opposed to rapid internalization) enhance EGF-stimulated cell growth. Do these results end the argument? If surface signaling is both necessary and sufficient to promote growth, why postulate a role for internalized receptor signaling? The reason for continued investigation has to do with the particulars of the experiments that demonstrate the primacy of surface signaling. These experiments have used: (1) mutated receptors with relaxed substrate specificity; (2) cells which overexpress normal receptors; (3) viral gene products which perturb EGF receptor metabolism; or (4) saturating concentrations of ligand that may not be encountered by normal cells. Under these conditions, the strength of the receptor's signal may be enhanced sufficiently to bypass physiologic-signaling pathways. In other words, large-scale activation of receptors that are *not* efficiently internalized may well allow phosphorylation of a critical substrate or mobilization of a critical phosphoprotein complex by surface receptors. In the normal cell, with low but persistent receptor occupancy, efficient phosphorylation of a critical substrate or construction of phosphoproteins complexes involving SH2-containing proteins may occur efficiently only as the activated receptor moves laterally and then internally. The receptor which binds enzymes at autophosphorylated residues and other sites (*30*) brings together a specific cast of signaling molecules that proceed to their site of action using the movement engendered by entry into the internalization pathway. Whether the signal occurs at the surface or from endosomes in unknown. Ultimately, this question can only be resolved when the key secondary steps in receptor signal transduction are known, and we can study the locale at which the critical modifications occur.

ACKNOWLEDGMENTS

This manuscript contains work supported by U.S. Public Health Service Grants DK-30002 and DK-31683. We would like to thank Rockell Landry for her excellent help in manuscript preparation and Brian Herman for his critical reading of the manuscript.

REFERENCES

1. Aaronson, S.A. (1991). Growth factors and cancer. *Science* 254: 1146–1153.
2. Apodaca, G., Bomsel, M., Arden, J., Breitfeld, P.P., Tang, K., and Mostov, K.E. (1991). The polymeric immunoglobulin receptor. A model protein to study transcytosis. *J. Clin. Invest.* 87: 1877–1882.

3. Attisano, L., Wrana, J.L., Cheifetz, S., and Massague, J. (1992). Novel activin receptors: distinct genes and alternative mRNA splicing generate a repertoire of serine/threonine kinase receptors. *Cell* 68: 97–108.
4. Barrandon, Y. and Green, H. (1987). Cell migration is essential for sustained growth of keratinocyte colonies: the roles of transforming growth factor alpha and epidermal growth factor. *Cell* 50: 1131–1137.
5. Baskin, G., Schenker, S., Frosto, T., and Henderson, G. (1991). Transforming growth factor beta 1 inhibits growth factor receptor endocytosis and down-regulation in cultured fetal rat hepatocytes. *J. Biol. Chem.* 266: 13238–13242.
6. Beguinot, L., Werth, D., Ito, S., Richert, N., Willingham, M.C., and Pastan, I. (1986). Functional studies on the EGF receptor with an antibody that recognizes the intracellular portion of the receptor. *J. Biol. Chem.* 261: 1801–1807.
7. Beguinot, L., Hanover, J.A., Ito, S., Richert, N.D., Willingham, M.C., and Pastan, I. (1985). Phorbol esters induce transient internalization without degradation of unoccupied epidermal growth factor receptors. *Proc. Natl. Acad. Sci. USA* 82: 2774–2778.
8. Bertics, P.J. and Gill, G.N. (1985). Self-phosphorylation enhances the protein-tyrosine kinase activity of the epidermal growth factor receptor. *J. Biol. Chem.* 260: 14642–14647.
9. Besner, G., Higashayama, S., and Klagsbrun, M. (1990). Isolation and characterization of a macrophage-derived heparin binding growth factor. *Cell Regul.* 1: 811–815.
10. Boni-Schnetzler, M. and Pilch, P.F. (1987). Mechanism of EGF receptor autophosphorylation and high affinity binding. *Proc. Natl. Acad. Sci. USA* 84: 7832–7836.
11. Brachmann, R., Lindquist, P.B., Nagashima, M., Kohr, W., Lipari, T., Napier, M., and Derynck, R. (1989). Transmembrane TGF-alpha precursors activate EGF-TGF-alpha receptors. *Cell* 56: 691–700.
12. Brandli, A.W., Adamson, E.D., and Simons, K. (1991). Transcytosis of epidermal growth factor. *J. Biol. Chem.* 266: 8560–8566.
13. Brown, V.I. and Greene, M.I. (1991). Molecular and cellular mechanisms of receptor-mediated endocytosis. *DNA Cell Biol.* 10: 399–409.
14. Burwen, S.J., Barker, M.E., Goldman, I.S., Hradek, G.T., Raper, S.E., and Jones, A.L. (1984). Transport of epidermal growth factor by rat liver: evidence for a nonlysosomal pathway. *J. Cell Biol.* 99: 1259–1265.
15. Buxser, S., Decker, D., and Ruppel, P. (1990). Relationship among types of nerve growth factor receptors on PC12 cells. *J. Biol. Chem.* 265: 12701–12710.
16. Cantley, L.C., Auger, K.R., Carpenter, C., Duckworth, B., Graziani, A., Kapeller, R., and Soltoff, S. (1991). Oncogenes and signal transduction. *Cell* 64: 281–302.
17. Carlin, C.R., Tollefson, A.E., Brady, H.A., Hoffman, B.L., and Wold, W.S.M. (1989). Epidermal growth factor receptor is down-regulated by a 10,400 MW protein encoded by the E3 region of adenovirus. *Cell* 57: 135–144.
18. Carpenter, C.D., Ingraham, H.A., Walton, G.M., Lazar, C.S., Cochet, C., Sowadski, J.M., Rosenfeld, M.G., and Gill, G.N. (1991). Structural analysis of the transmembrane domain of the epidermal growth factor receptor. *J. Biol. Chem.* 266: 5750–5755.
19. Carpenter, G. (1987). Receptors for epidermal growth factor and other polypeptide mitogens. *Ann. Rev. Biochem.* 56: 881–914.
20. Carpenter, G. and Wahl, M.I. (1990). The epidermal growth factor family. In: *Peptide Growth Factors and Their Receptors*, Vol. I, pp. 69–171. (Sporn, M.B. and Roberts, A.B., Eds.). Springer-Verlag, Berlin.
21. Carpenter, G. and Cohen, S. (1979). Epidermal growth factor. *Ann. Rev. Biochem.* 48: 193–216.
22. Carpenter, G., King, Jr., L., and Cohen, S. (1978). Epidermal growth factor stimulates phosphorylation in membrane preparations *in vitro*. *Nature* 276: 409–410.
23. Carpentier, J.L., White, M.F., Orci, L., and Kahn, R.C. (1987). Direct visualization of the phosphorylated epidermal growth factor receptor during its internalization in A-431 cells. *J. Cell. Biol.* 105: 2751–2762.

24. Carraway, K.L. and Cerione, R.A. (1991). Comparison of epidermal growth factor (EGF) receptor-receptor interactions in intact A431 cells and isolated plasma membranes. Large scale receptor micro-aggregation is not detected during EGF-stimulated early events. *J. Biol. Chem.* 266: 8899–8906.
25. Chang, C.-P., Kao, J.P.Y., Lazar, C.S., Walsh, B.J., Wells, A., Wiley, H.S., Gill, G.N., and Rosenfeld, M.G. (1991). Ligand-induced internalization and increased cell calcium are mediated via distinct structural elements in the carboxyl terminus of the epidermal growth factor receptor. *J. Biol. Chem.* 266: 23467–23470.
26. Chen, W.-J., Goldstein, J.L., and Brown, M.S. (1990). NPXY, a sequence often found in cytoplasmic tails, is required for coated pit-mediated internalization of the low density lipoprotein receptor. *J. Biol. Chem.* 265: 3116–3123.
27. Chen, W.S., Lazar, C.S., Lund, K.A., Welsh, J.B., Chang, C.P., Walton, G.M., Der, C.J., Wiley, H.S., Gill, G.N., and Rosenfeld, M.G. (1989). Functional independence of the epidermal growth factor receptor from a domain required for ligand-induced internalization and calcium regulation. *Cell* 59: 33–43.
28. Chen, W.S., Lazar, C.S., Poenie, M., Tsien, R.Y., Gill, G.N., and Rosenfeld, M.G. (1987). Requirement for intrinsic protein tyrosine kinase in the immediate and late actions of the EGF receptor. *Nature* 328: 820–823.
29. Clark, A.J.L., Ishii, S., Richert, N., Merlino, G.T., and Pastan, I. (1985). Epidermal growth factor regulates the expression of its own receptor. *Proc. Natl. Acad. Sci. USA* 82: 8374–8378.
30. Cochet, C., Filhol, O., Payrastre, B., Hunter, T., and Gill, G.N. (1991). Interaction between the epidermal growth factor receptor and phosphoinositide kinases. *J. Biol. Chem.* 266: 637–644.
31. Cochet, C., Kashles, O., Chambaz, E.M., Borello, I., King, C.R., and Schlessinger, J. (1988). Demonstration of epidermal growth factor-induced receptor dimerization in living cells using a chemical cross-linking agent. *J. Biol. Chem.* 263: 3290–3295.
32. Coffey, R.J., Jr., Derynck, R., Wilcox, J.N., Bringman, T.S., Goustin, A.S., Moses, H.L., and Pittelkow, M.R. (1987). Production and auto-induction of transforming growth factor-α in human keratinocytes. *Nature* 328: 817–820.
33. Cohen, S. and Fava, R.A. (1985). Internalization of functional epidermal growth factor: receptor/kinase complexes in A-431 cells. *J. Biol. Chem.* 260: 12351–12358.
34. Cosman, D., Lyman, S.D., Idzerda, R.L., Beckmann, M.P., Park, L.S., Goodwin, R.G., and March, C.J. (1990). A new cytokine receptor superfamily. *Trends Biochem. Sci.* 15: 265–270.
35. Countaway, J.L., Nairn, A.C., and Davis, R.J. (1992). Mechanism of desensitization of the epidermal growth factor receptor protein-tyrosine kinase. *J. Biol. Chem.* 267: 1129–1140.
36. De Diego, J.G., Gorden, P., and Carpentier, J.L. (1991). The relationship of ligand receptor mobility to internalization of polypeptide hormones and growth factors. *Endocrinology* 128: 2136–2140.
37. Decker, S.J. (1990). Epidermal growth factor and transforming growth factor-alpha induce differential processing of the epidermal growth factor receptor. *Biochem. Biophys. Res. Commun.* 166: 615–621.
38. Decker, S.J. (1989). Epidermal growth factor-induced truncation of the epidermal growth factor receptor. *J. Biol. Chem.* 264: 17641–17644.
39. Decker, S.J., Alexander, C., and Habib, T. (1992). Epidermal growth factor (EGF)-stimulated tyrosine phosphorylation and EGF receptor degradation in cells expressing EGF receptors truncated at residue 973. *J. Biol. Chem.* 267: 1104–1108.
40. Derynck, R., Roberts, A.B., Winkler, M.E., Chen, E.Y., and Goeddel, D.V. (1984). Human transforming growth factor-α precursor structure and expression in *E. coli*. *Cell* 28: 287–297.
41. Dickson, R.B. (1990). Stimulatory and inhibitory growth factors and breast cancer. *J. Steroid Biochem. Mol. Biol.* 37: 795–803.
42. Downing, J.R., Roussel, M.F., and Sherr, C.J. (1989). Ligand and protein kinase C downmodulate the colony-stimulating factor-1 receptor by independent mechanisms. *Mol. Cell. Biol.* 9: 2890–2896.

43. Downward, J., Parker, P., and Waterfield, M.D. (1984). Autophosphorylation sites on the epidermal growth factor receptor. *Nature* 311: 483–485.
44. Downward, J., Yarden, Y., Mayes, E., Scrace, G., Totty, N., Stockwell, P., Ullrich, A., Schlessinger, J., and Waterfield, M.D. (1984). Close similarity of epidermal growth factor receptor and v-*erb-B* oncogene protein sequences. *Nature* 307: 521–527.
45. Doxsey, S.J., Brodsky, F.M., Blank, G.S., and A. Helenius. (1987). Inhibition of endocytosis by anti-clathrin antibodies. *Cell* 50: 453–463.
46. Dunn, W.A. and Hubbard, A.L. (1984). Receptor-mediated endocytosis of epidermal growth factor by hepatocytes in the perfused rat liver: ligand and receptor dynamics. *J. Cell. Biol.* 98: 2148–2159.
47. Dunn, W.A., Connolly, T.P., and Hubbard, A.L. (1986). Receptor-mediated endocytosis of epidermal growth factor by rat hepatocytes: receptor pathway. *J. Cell. Biol.* 102: 24–36.
48. Earp, H.S., Austin, K.S., Blaisdell, J., Rubin, R.A., Nelson, K.G., Lee, L.W., and Grisham, J.W. (1986). Epidermal growth factor (EGF) stimulates EGF receptor synthesis. *J. Biol. Chem.* 261: 4777–4780.
49. Ebner, R. and Derynck, R. (1991). Epidermal growth factor and transforming growth factor-α: differential intracellular routing and processing of ligand-receptor complexes. *Cell Regul.* 2: 599–612.
50. Eveleth, D.D. and Bradshaw, R.A. (1988). Internalization and cycling of nerve growth factor in PC12 cells: interconversion of type II (fast) and type I (slow) nerve growth factor receptors. *Neuron.* 1: 929–936.
51. Fanger, B.O., Austin, K.S., Earp, H.S., and Cidlowski, J.A. (1986). Cross-linking of epidermal growth factor receptors in intact cells: detection of initial stages of receptor clustering and determination of molecular weight of high-affinity receptors. *Biochemistry* 25: 6414–6420.
52. Felder, S., Miller, K., Moehren, G., Ullrich, A., Schlessinger, J., and Hopkins, C.R. (1990). Kinase activity controls the sorting of the epidermal growth factor receptor within the multivesicular body. *Cell* 61: 623–634.
53. Fox, C.F. and Das, M. (1979). Internalization and processing of the EGF receptor in the induction of DNA synthesis in cultured fibroblasts: the endocytic activation hypothesis. *J. Supramol. Struct.* 10: 199–214.
54. Geiger, D., Carpentier, J.-L., Gordon, P., and Orci, L. (1989). Down-regulation of insulin receptors is related to internalization. *Exp. Cell. Res.* 185: 33–40.
55. Gilman, A.G. (1987). G proteins: transducers of receptor-generated signals. *Ann. Rev. Biochem.* 56: 615–649.
56. Gladhaug, I.P. and Christoffersen, T. (1987). Kinetics of epidermal growth factor binding and processing in isolated intact rat hepatocytes. *Eur. J. Biochem.* 164: 267–275.
57. Goldschmidt-Clermont, P.J., Kim, J.W., Machesky, L.M., Rhee, S.G., and Pollard, T.D. (1991). Regulation of phospholipase Cγ1 by profilin and tyrosine phosphorylation. *Science* 251: 1231–1233.
58. Goldstein, J.L., Anderson, R.G.W., and Brown, M.S. (1979). Coated pits, coated vesicles, and receptor-mediated endocytosis. *Nature* 279: 679–685.
59. Grenfell, S., Smithers, M., Miller, K., and Solari, R. (1989). Receptor-mediated endocytosis and nuclear transport of human interleukin 1alpha. *Biochem. J.* 264: 813–822.
60. Guilbert, L.J. and Stanley, E.R. (1986). The interaction of ^{125}I-colony-stimulating factor-1 with bone marrow-derived macrophages. *J. Biol. Chem.* 261: 4024–4032.
61. Haigler, H.T., Maxfield, F.R., Willingham, M.C., and Pastan, I. (1980). Dansylcadaverine inhibits internalization of [^{125}I]-epidermal growth factor in BALB 3T3 cells. *J. Biol. Chem.* 255: 1239–1241.
62. Hanks, S.K. and Quinn, A.M. (1991). Protein kinase catalytic domain sequence database: identification of conserved features of primary structure and classification of family members. *Meth. Enzym.* 200: 38–63.

63. Hanover, J.A., Beguinot, L., Willingham, M.C., and Pastan, I.H. (1985). Transit of receptors for epidermal growth factor and transferrin through clathrin-coated pits. Analysis of the kinetics of receptor entry. *J. Biol. Chem.* 260: 15938–15945.
64. Hanover, J.A., Willingham, M.C., and Pastan, I. (1984). Kinetics of transit of transferrin and epidermal growth factor through clathrin-coated membranes. *Cell* 39: 283–293.
65. Helin, K. and Beguinot, L. (1991). Internalization and down-regulation of the human epidermal growth factor receptor are regulated by the carboxyl-terminal tyrosines. *J. Biol. Chem.* 266: 8363–8368.
66. Hepler, J.R., Earp, H.S., and Harden, T.K. (1988). Long-term phorbol ester treatment down-regulates protein kinase C and sensitizes the phosphoinositide signaling pathway to hormone and growth factor stimulation. *J. Biol. Chem.* 263: 7610–7619.
67. Heumann, R., Schwab, M., Merkl, R., and Thoenen, H. (1984). Nerve growth factor-mediated induction of choline acetyltransferase in PC12 cells: evaluation of the site of action of nerve growth factor and the involvement of lysosomal degradation products of nerve growth factor. *J. Neurosci.* 4: 3039–3050.
68. Higashiyama, S., Abraham, J.A., Miller, J., Fiddes, J.C., and Klagsbrun, M. (1991). A heparin-binding growth factor secreted by macrophage-like cells that is related to EGF. *Science* 251: 936–939.
69. Honegger, A., Dull, T.J., Bellot, F., Van, O.E., Szapary, D., Schmidt, A., Ullrich, A., and Schlessinger, J. (1988). Biological activities of EGF-receptor mutants with individually altered autophosphorylation sites. *EMBO. J.* 7: 3045–3052.
70. Honegger, A.M., Dull, T.J., Felder, S., Van, O.E., Bellot, F., Szapary, D., Schmidt, A., Ullrich, A., and Schlessinger, J. (1987). Point mutation at the ATP binding site of EGF receptor abolishes protein-tyrosine kinase activity and alters cellular routing. *Cell* 51: 199–209.
71. Hopkins, C.R., Gibson, A., Shipman, M., and Miller, K. (1990). Movement of internalized ligand-receptor complexes along a continuous endosomal reticulum. *Nature* 346: 335–339.
72. Huckle, W.R., Hepler, J.R., Rhee, S.G., Harden, T.K., and Earp, H.S. (1990). Protein kinase C inhibits epidermal growth factor-dependent tyrosine phosphorylation of phospholipase Cγ and activation of phosphoinositide hydrolysis. *Endocrinology* 127: 1697–1705.
73. Ibbotson, K.J., Harrod, J., Gowen, M., D'Souza, S., Smith, D.D., Winkler, M.E., Derynck, R. and Mundy, G.R. (1986). Human recombinant transforming growth factor alpha stimulates bone resorption and inhibits formation *in vitro*. *Proc. Natl. Acad. Sci. USA* 83: 2228–2232.
74. Imamura, K., Spriggs, D., and Kufe, D. (1987). Expression of tumor necrosis factor receptors on human monocytes and internalization of receptor bound ligand. *J. Immunol.* 139: 2989–2992.
75. Jackle, S., Runquist, E.A., Miranda, B.S., and Havel, R.J. (1991). Trafficking of the epidermal growth factor receptor and transferrin in three hepatocytic endosomal fractions. *J. Biol. Chem.* 266: 1396–1402.
76. Jiang, L.W. and Schindler, M. (1990). Nucleocytoplasmic transport is enhanced concomitant with nuclear accumulation of epidermal growth factor (EGF) binding activity in both 3T3-1 and EGF receptor reconstituted NR-6 fibroblasts. *J. Cell. Biol.* 110: 559–568.
77. Kay, D.G., Lai, W.H., Uchihashi, M., Khan, M.N., Posner, B.I., and Bergeron, J.J.M. (1986). Epidermal growth factor receptor kinase translocation and activation *in vivo*. *J. Biol. Chem.* 261: 8473–8480.
78. Keating, M.T. and Williams, L.T. (1987). Processing of the platelet-derived growth factor receptor. *J. Biol. Chem.* 262: 7932–7937.
79. Keen, J.H. (1990). Clathrin and associated assembly and disassembly proteins. *Ann. Rev. Biochem.* 59: 415–438.
80. Khan, M.N., Baquiran, G., Brule, C., Burgess, J., Foster, B., Bergeron, J.J., and Posner, B.I. (1989). Internalization and activation of the rat liver insulin receptor kinase *in vivo*. *J. Biol. Chem.* 264: 12931–12940.
81. King, A.C. and Cuatrecasas, P. (1981). Peptide hormone-induced receptor mobility, aggregation, and internalization. *N. Engl. J. Med.* 305: 77–88.

82. King, G.L. and Johnson, S.M. (1985). Receptor-mediated transport of insulin across endothelial cells. *Science* 277: 1583–1586.
83. Klagsbrun, M. (1990). The affinity of fibroblast growth factors (FGFs) for heparin; FGF heparan sulfate interactions in cells and extracellular matrix. *Curr. Opin. Cell. Biol.* 2: 857–863.
84. Knauer, D.J., Wiley, H.S., and Cunningham, D.D. (1984). Relationship between epidermal growth factor receptor occupancy and mitogenic response. *J. Biol. Chem.* 259: 5623–5631.
85. Koch, C.A., Anderson, D., Moran, M.F., Ellis, C., and Pawson, T. (1991). SH2 and SH3 domains: elements that control interactions of cytoplasmic signaling proteins. *Science* 252: 668–674.
86. Koland, J.G. and Cerione, R.A. (1988). Growth factor control of epidermal growth factor receptor kinase activity via an intramolecular mechanism. *J. Biol. Chem.* 263: 2230–2237.
87. Korc, M. and Magun, B.E. (1985). Recycling of epidermal growth factor in a human pancreatic carcinoma cell line. *Proc. Natl. Acad. Sci. USA* 82: 6172–6175.
88. Kramer, H., Cagan, R.L., and Zipursky, S.L. (1991). Interaction of bride of sevenless membrane-bound ligand and the sevenless tyrosine-kinase receptor. *Nature* 352: 207–212.
89. Kudlow, J.E., Cheung, C.Y., and Bjorge, J.D. (1986). Epidermal growth factor stimulates the synthesis of its own receptor in a human breast cancer cell line. *J. Biol. Chem.* 261: 4134–4138.
90. Lai, W.H., Guyda, H.J., and Bergeron, J.J.M. (1986). Binding and internalization of epidermal growth factor in human term placental cells in culture. *Endocrinology* 118: 413–423.
91. Lai, W.H., Cameron, P.H., Wada, I., Doherty, J.J., G. Kay, D., Posner, B.I., and Bergeron, J.J. (1989). Ligand-mediated internalization, recycling, and downregulation of the epidermal growth factor receptor *in vivo*. *J. Cell. Biol.* 109: 2741–2749.
92. Lai, W.H., Cameron, P.H., Doherty, J.J., Posner, B.I., and Bergeron, J.J. (1989). Ligand-mediated autophosphorylation activity of the epidermal growth factor receptor during internalization. *J. Cell. Biol.* 109: 2751–2760.
93. Landreth, G.E., Williams, L.K., and Reiser, G.D. (1985). Association of the epidermal growth factor receptor kinase with the detergent-insoluble cytoskeleton of A431 cells. *J. Cell. Biol.* 101: 1341–1350.
94. Larkin, J.M., Brown, M.S., Goldstein, J.L., and Anderson, R.G. (1983). Depletion of intracellular potassium arrests coated pit formation and receptor-mediated endocytosis in fibroblasts. *Cell* 33: 273–285.
95. Lax, I., Mitra, A.K., Ravera, C., Hurwitz, D.R., Rubinstein, M., Ullrich, A., Stroud, R.M., and Schlessinger, J. (1991). Epidermal growth factor (EGF) induces oligomerization of soluble, extracellular, ligand-binding domain of EGF receptor. *J. Biol. Chem.* 266: 13828–13833.
96. Lax, I., Bellot, F., Howk, R., Ullrich, A., Givol, D., and Schlessinger, J. (1989). Functional analysis of the ligand binding site of EGF-receptor utilizing chimeric chicken/human receptor molecules. *EMBO J.* 8: 421–427.
97. Lee, D.C., Rose, T.M., Webb, N.R., and Todaro, G.J. (1985). Cloning and sequence analysis of a cDNA for rat transforming growth factor-α. *Nature* 313: 489–491.
98. Livneh, E., Prywes, R., Kashles, N., Reiss, O., Sasson, I., Mory, Y., Ullrich, A., and Schlessinger, J. (1986). Reconstitution of human epidermal growth factor receptors and its deletion mutants in cultured hamster cells. *J. Biol. Chem.* 261: 12490–12497.
99. Lund, K.A., Opresko, L.K., Starbuck, C., Walsh, B.J., and Wiley, H.S. (1990). Quantitative analysis of the endocytic system involved in hormone-induced receptor internalization. *J. Biol. Chem.* 265: 15713–15723.
100. Lutz, W., Salisbury, J.L., and Kumar, R. (1991). Vasopressin receptor-mediated endocytosis: current view. *Am. J. Physiol.* 261: F1–F13.
101. Maratos-Flier, E., Chiou-Ying, Y.K., Verdin, E.M., and King, G.L. (1987). Receptor-mediated vectorial trancytosis of epidermal growth factor by Madin–Darby canine kidney cells. *J. Cell. Biol.* 105: 1595–1601.
102. Margolis, B.L., Lax, I., Kris, R., Dombalagian, M., Honegger, A.M., Howk, R., Givol, D., Ullrich, A., and Schlessinger, J. All autophosphorylation sites of epidermal growth factor (EGF)

receptor and HER2/*neu* are located in their carboxyl-terminal tails. *J. Biol. Chem.* 264: 10667–10671.

103. Marshall, S. (1985). Kinetics of insulin receptor internalization and recycling in adipocytes. *J. Biol. Chem.* 260: 4136–4144.

104. Martin, P., Vass, W.C., Schiller, J.T., Lowry, D.R., and Velu, T.J. (1989). The bovine papillomavirus E5 transforming protein can stimulate the transforming activity of EGF and CSF-1 receptors. *Cell* 59: 21–32.

105. Massague, J. and Kelly, B. (1987). Internalization of transforming growth factor-β and its receptor in BALB/c 3T3 fibroblasts. *J. Cell. Physiol.* 128: 216–222.

106. Mathews, L.S. and Vale, W.W. (1991). Expression cloning of an activin receptor, a predicted transmembrane serine kinase. *Cell* 65: 973–982.

107. Matrisian, L.M., Rodland, K.D., and Magun, B.E. (1987). Disruption of intracellular processing of epidermal growth factor by methylamine inhibits epidermal growth factor-induced DNA synthesis but not early morphological or transcriptional events. *J. Biol. Chem.* 262: 6908–6913.

108. Matrisian, L.M., Planck, S.R., and Magun, B.E. (1984). Intracellular processing of epidermal growth factor I. Acidification of [^{125}I]-epidermal growth factor in intracellular organelles. *J. Biol. Chem.* 259: 3047–3053.

109. McBride, K., Rhee, S.G., and Jaken, S. (1991). Immunocytochemical localization of phospholipase C-γ in rat embryo fibroblasts. *Proc. Natl. Acad. Sci. USA* 88: 7111–7115.

110. McCrae, C., Maxwill, L., and Ceredig, R. (1988). Intracellular pathway of interleukin-2 receptors studied using immunogold electron microscopy. *Immuno. Cell. Biol.* 66: 185–191.

111. McCune, B. K., Prokop, C.A., and Earp, H.S. (1990). Transient epidermal growth factor (EGF)-dependent suppression of EGF receptor autophosphorylation during internalization. *J. Biol. Chem.* 265: 9715–9721.

112. McCune, B.K. and Earp, H.S. (1989). The epidermal growth factor receptor tyrosine kinase in liver epithelial cells. The effect of ligand-dependent changes in cellular location. *J. Biol. Chem.* 264: 15501–15507.

113. McKanna, J.A., Haigler, H.T., and Cohen, S. (1979). Hormone receptor topology and dynamics: morphological analysis using ferritin-labeled epidermal growth factor. *Proc. Natl. Acad. Sci. USA* 76: 5689–5693.

114. Miller, K., Beardmore, J., Kanety, H., Schlessinger, J., and Hopkins, C.R. (1986). Localization of the epidermal growth factor (EGF) receptor within the endosome of EGF-stimulated epidermoid carcinoma (A431) cells. *J. Cell. Biol.* 102: 500–509.

115. Moolenaar, W.H., Biermann, A.J., Tilly, B.C., Verlaan, I., Defize, L.H.K., Honegger, A.M., Ullrich, A., and Schlessinger, J. (1988). A point mutation at the ATP-binding site of the EGf receptor abolishes signal transduction. *EMBO J.* 8: 707–710.

116. Murthy, U., Basu, M., Sen-Majumdar, A., and Das, M. (1986). Perinuclear location and recycling of epidermal growth factor receptor kinase: immunofluorescent visualization using antibodies directed to kinase and extracellular domains. *J. Cell. Biol.* 103: 333–342.

117. Nesterov, A., Reshetnikova, G., Vinogradova, N., and Nikolsky, N. (1990). Functional state of the epidermal growth factor-receptor complexes during their internalization in A-431 cells. *Mol. Cell. Biol.* 10: 5011–5014.

118. Northwood, I.C. and Davis, R.J. (1988). Activation of the epidermal growth factor receptor tyrosine protein kinase in the absence of receptor oligomerization. *J. Biol. Chem.* 263: 7450–7453.

119. O'Brien, R.M., Soos, M.A., and Siddle, K. (1987). Monoclonal antibodies to the insulin receptor stimulate the intrinsic tyrosine kinase activity by cross-linking receptor molecules. *EMBO J.* 6: 4003–4010.

120. Pearse, B.M. and Crowther, R.A. (1987). Structure and assembly of coated vesicles. *Ann. Rev. Biophys. Biophys. Chem.* 16: 49–68.

121. Planck, S.R., Finch, J.S., and Magun, B.E. (1984). Intracellular processing of epidermal growth factor II. Intracellular cleavage of the COOH-terminal region of [^{125}I]-epidermal growth factor. *J. Biol. Chem.* 259: 3053–3057.
122. Plowman, G.D., Green, J.M., McDonald, V.L., Neubauer, M.G., Disteche, C.M., Todaro, G.J., and Shoyab, M. (1990). The amphiregulin gene encodes a novel epidermal growth factor-related protein with tumor inhibitory activity. *Mol. Cell. Biol.* 10: 1969–1981.
123. Podlecki, D.A., Smith, R.M., Kao, M., Tsai, P., Huecksteadt, T., Brandenburg, D., Lasher, R.S., Jarett, L., and Olefsky, J.M. (1987). Nuclear translocation of the insulin receptor—a possible mediator of insulin's long-term side effects. *J. Biol. Chem.* 262: 3362–3368.
124. Posner, B.I., Khan, M.N., and Bergeron, J.M. (1982). Endocytosis of peptide hormones and other ligands. *Endo. Rev.* 3: 280–298.
125. Powell, P.P. and Klagsbrun, M. (1991). Three forms of rat basic fibroblast growth factor are made from a single mRNA and localize to the nucleus. *J. Cell. Physiol.* 148: 202–210.
126. Rakowicz-Szulczynska, E.M., Otwiaska, D., Rodeck, U., and Koprowski, H. (1989). Epidermal growth factor (EGF) and monoclonal antibody to cell surface EGF receptor bind to the same chromatin receptor. *Arch. Biochem. Biophys.* 268: 456–464.
127. Raper, S.E., Burwen, S.J., Barker, M.E., and Jones, A.L. (1987). Translocation of epidermal growth factor to the hepatocyte nucleus during rat liver regeneration. *Gastroenterology* 92: 1243–1250.
128. Renfrew, C.A. and Hubbard, A.L. (1991). Sequential processing of epidermal growth factor in early and late endosomes of rat liver. *J. Biol. Chem.* 266: 4348–4356.
129. Ristimaki, A. (1989). Transforming growth factor alpha stimulates prostacyclin production by culture human vascular endothelial cells more potently than epidermal growth factor. *Biochem. Biophys. Res. Commun.* 160: 1100–1105.
130. Savion, N., Vlodavsky, I., and Gospodarowicz, D. (1981). Nuclear accumulation of epidermal growth factor in cultured bovine corneal endothelial and granulosa cells. *J. Biol. Chem.* 256: 1149–1154.
131. Sbraccia, P., Wong, K.-Y., Brunettik, A., Rafaeloff, R., Trischitta, V., Hawley, D.M., and Goldfine, I.D. (1990). Insulin down-regulates insulin receptor number and up-regulates insulin receptor affinity in cells expressing a tyrosine kinase-defective insulin receptor. *J. Biol. Chem.* 265: 4902–4907.
132. Schaudies, R.P. and Savage, Jr., C.R. (1986). Intracellular modification of [^{125}I]-labeled epidermal growth factor by normal human foreskin fibroblasts. *Endocrinology* 118: 875–886.
133. Schreiber, A.B., Winkler, M.E., and Derynck, R. (1986). Transforming growth factor-alpha: a more potent angiogenic mediator than epidermal growth factor. *Science* 232: 1250–1253.
134. Schrieber, A.B., Libermann, T.A., Lax, I., Yarden, Y., and Schlessinger, J. (1983). Biological role of epidermal growth factor-receptor clustering. Investigation with monoclonal anti-receptor antibodies. *J. Biol. Chem.* 258: 846–853.
135. Schultz, G.S., White, M., Mitchell, R., Brown, G., Lynch, J., Twardzik, D.R., and Todaro, G.J. (1987). Epithelial wound healing enhanced by transforming growth factor-alpha and vaccinia growth factor. *Science* 235: 350–352.
136. Schwartz, A.L. (1990). Cell biology of intracellular protein trafficking. *Ann. Rev. Immunol.* 8: 195–229.
137. Shepherd, V.L. (1989). Intracellular pathways and mechanisms of sorting in receptor-mediated endocytosis. *Trends Pharmacol. Sci.* 10: 458–462.
138. Shoyab, M., Plowman, G.D., McDonald, V.L., Bradley, J.G., and Todaro, G.J. (1989). Structure and function of human amphiregulin: a member of the epidermal growth factor family. *Science* 243: 1074–1076.
139. Shoyab, M., McDonald, V.L., Bradley, J.G., and Todaro, G.J. (1988). Amphiregulin: a bifunctional growth-modulating glycoprotein produced by the phorbol 12-myristate 13-acetate-treated human breast adenocarcinoma cell line MCF-7. *Proc. Natl. Acad. Sci. USA* 85: 6528–6532.

140. Sorkin, A. and Carpenter, G. (1991). Dimerization of internalized epidermal growth factor receptors. *J. Biol. Chem.* 266: 23453–23460.
141. Sorkin, A., Westermark, B., Heldin, C.H., and Claesson, W.L. (1991). Effect of receptor kinase inactivation on the rate of internalization and degradation of PDGF and the PDGF beta-receptor. *J. Cell. Biol.* 112: 469–478.
142. Sorkin, A., Waters, C., Overholser, K.A., and Carpenter, G. (1991). Multiple autophosphorylation site mutations of the epidermal growth factor receptor. Analysis of kinase activity and endocytosis. *J. Biol. Chem.* 266: 8355–8362.
143. St. Hilaire, R.J., Hradek, G.T., and Jones, A.L. (1983). Hepatic sequestration and biliary secretion of epidermal growth factor: evidence for a high-capacity uptake system. *Proc. Natl. Acad. Sci. USA* 80: 3797–3801.
144. Stoscheck, C.M. and Carpenter, G. (1984). Downregulation of epidermal growth factor receptors: direct demonstration of receptor degradation in human fibroblasts. *J. Cell. Biol.* 98: 1048–1053.
145. Strosberg, A.D. (1991). Structure/function relationship of proteins belonging to the family of receptors coupled to GTP-binding proteins. *Eur. J. Biochem.* 196: 1–10.
146. Sturani, E., Zippel, R., Toschi, L., Morello, L., Comoglio, P.M., and Alberghina, L. (1988). Kinetics and regulation of the tyrosine phosphorylation of epidermal growth factor in intact A431 cells. *Mol. Cell. Biol.* 8: 1345–1351.
147. Todderud, G. and Carpenter, G. (1988). Presence of mannose phosphate on the epidermal growth factor receptor in A-431 cells. *J. Biol. Chem.* 263: 17893–17896.
148. Tollefson, A.E., Stewart, A.R., Yei, S., Saha, S.K., and Wold, W.S.M. (1991). The 10,400- and 14,500-dalton proteins encoded by region E3 of adenovirus form a complex and function together to downregulate the epidermal growth factor receptor. *J. Virol.* 65: 3095–3105.
149. Ullrich, A. and Schlessinger, J. (1990). Signal transduction by receptors with tyrosine kinase activity. *Cell* 61: 203–212.
150. Ullrich, A., Coussens, L., Hayflick, J.S., Dull, T.J., Gray, A., Tam, A.W., Lee, J., Yarden, Y., Libermann, T.A., Schlessinger, J., Downward, J., Mayes, E.L.V., Whittle, N., Waterfield, M.D., and Seeburg, P.H. (1984). Human epidermal growth factor receptor cDNA sequence and aberrant expression of the amplified gene in A431 epidermoid carcinoma cells. *Nature* 309: 418–425.
151. Ushiro, H. and Cohen, S. (1980). Identification of phosphotyrosine as a product of epidermal growth factor-activated protein kinase in A-431 cell membranes. *J. Biol. Chem.* 255: 8363–8365.
152. Van Belzin, N., Rijken, P.J., Hage, W.J., DeLaat, S.W., and Boonstra, J. (1988). Direct visualization and quantitative analysis of epidermal growth factor-induced receptor clustering. *J. Cell. Physiol.* 134: 413–420.
153. van Bergen en Henegouwen, P.M.P., Defize, L.H.K., de Kroon, J., van Damme, H., Verkleij, A.J., and Boonstra, J. (1989). Ligand-induced association of epidermal growth factor receptor to the cytoskeleton of A431 cells. *J. Cell. Biochem.* 39: 455–465.
154. van Deurs, B., Petersen, O.W., Olsnes, S., and Sandvig, K. (1989). The ways of endocytosis. *Int. Rev. Cytol.* 117: 131–177.
155. Wada, I., Lai, W.H., Posner, B.I., and Bergeron, J.J.M. (1992). Association of the tyrosine phosphorylated epidermal growth factor receptor with a 55 kDa tyrosine phosphorylated protein at the cell surface and in endosomes. *J. Cell. Biol.* 116: 321–330.
156. Walker, F., Nice, E., Fabri, L., Moy, F.J., Liu, J.-F., Wu, R., Scheraga, H.A., and Burgess, A.W. (1990). Resistance to receptor-mediated degradation of a murine epidermal growth factor analogue (EGF-Val-47) potentiates its mitogenic activity. *Biochemistry* 29: 10635: 10640.
157. Walton, G.M., Chen, W.S., Rosenfeld, M.G., and Gill, G.N. (1990). Analysis of deletions of the carboxyl terminus of the epidermal growth factor reveals self-phosphorylation at tyrosine 992 and enhanced *in vivo* tyrosine phosphorylation of cell substrates. *J. Biol. Chem.* 265: 1750–1754.
158. Weiner, D.B., Liu, J., Cohen, J.A., Williams, W.V., and Greene., M.I. (1989). A point mutation in the neu oncogene mimics ligand induction of receptor aggregation. *Nature* 339: 230–231.

159. Wells, A., Welsh, J.B., Lazar, C.S., Wiley, H.S., Gill, G.N., and Rosenfeld, M.G. (1990). Ligand-induced transformation by a noninternalizing epidermal growth factor receptor. *Science* 247: 962–964.
160. Welsh, J.B., Gill, G.N., Rosenfeld, M.G., and Wells, A. (1991). A negative feedback loop attenuates EGf-induced morphological changes. *J. Cell. Biol.* 114: 533–543.
161. Wiley, H.S., Herbst, J.J., Walsh, B.J., Lauffenburger, D.A., Rosenfeld, M.G., and Gill, G.N. (1991). The role of tyrosine kinase activity in endocytosis, compartmentation, and downregulation of the epidermal growth factor receptor. *J. Biol. Chem.* 266: 11083–11094.
162. Wiley, H.S., VanNostrand, W., McKinley, D.N., and Cunningham, D.D. (1985). Intracellular processing of epidermal growth factor and its effect on ligand-receptor interactions. *J. Biol. Chem.* 260: 5290–5295.
163. Wong, S.T., Winchell, L.F., McCune, B.K., Earp, H.S., Teixido, J., Massague, J., Herman, B., and Lee, D.C. (1989). The TGF-α precursor expressed on the cell surface binds to the EGF receptor on adjacent cells, leading to signal transduction. *Cell* 56: 495–506.
164. Wu, D., Wang, L., Sato, G.H., West, K.A., Harris, W.R., Crabb, J.W., and Sato, J.D. (1989). Human epidermal growth factor (EGF) receptor sequence recognized by EGF competitive monoclonal antibodies. *J. Biol. Chem.* 264: 17469–17475.
165. Yamashiro, D.J. and Maxfield, F.R. (1987). Kinetics of endosome acidification in mutant and wild-type Chinese-hamster ovary cells. *J. Cell. Biol.* 105: 2713–2721.
166. Yankner, B.A. and Shooter, E.M. (1979). Nerve growth factor in the nucleus: interaction with receptors on the nuclear membrane. *Proc. Natl. Acad. Sci. USA* 76: 1269–1273.
167. Yarden, Y. and Schlessinger, J. (1987). EGF induces rapid reversible aggregation of the purified EGF receptor. *Biochemistry* 26: 1443–1451.
168. Yarden, Y. and Schlessinger, J. (1987). Self-phosphorylation of EGF receptor: evidence for a model of intermolecular allosteric activation. *Biochemistry* 28: 1434–1442.

MEMBRANE TRAFFIC THROUGH THE LATE STAGES OF THE YEAST SECRETORY PATHWAY

Eric A. Whitters, Henry B. Skinner, and Vytas A. Bankaitis

ABSTRACT

A powerful combination of genetic and molecular approaches has been employed to dissect the mechanisms by which yeast execute a variety of protein transport functions through the late stages of the secretory pathway. These various processes include protein transport through and from the yeast Golgi, the fusion of Golgi- derived secretory vesicles to the yeast plasma membrane, the sorting of vacuolar proteins from

Advances in Cell and Molecular Biology of Membranes
Volume 1, pages 307–339

ISBN: 1-55938-362-3

the secretory pathway, and endocytosis. These genetic and molecular approaches have identified proteins that play crucial roles in such protein and membrane trafficking events. In this chapter, we describe some of the advances that have recently been made in these areas.

I. INTRODUCTION

The secretory pathway of eukaryotic cells represents a primary means for generating and maintaining an appropriate level of biochemical compartmentation within the cell. Morphologically, this pathway is defined by a series of distinct membranous organelles that house functions associated with the modification of secretory proteins and the delivery of proteins from the cytoplasm to any one of a number of noncytoplasmic destinations (*51*). The basic nature of the secretory pathway has been conserved across the eukaryotic kingdom. Proteins enter the secretory pathway at the level of insertion through the endoplasmic reticulum (ER) membrane, are subsequently transported to and through the Golgi complex, and are finally delivered to the cell surface as a result of fusion of Golgi-derived secretory vesicles with the plasma membrane. These various stages of intercompartmental protein transport are considered to be catalyzed by the formation and specific consumption of transport vesicles (reviewed in ref. *66*). A great deal of current effort is being focused on the characterization of elegant *in vitro* systems that have reconstituted the transport of proteins from one membrane-enclosed compartment to another (reviewed in ref. *66*).

A detailed analysis of the yeast secretory pathway has been made possible by the now classic work of Novick et al. who isolated a number of conditional-lethal mutations that manifested themselves as stage-specific blocks in the yeast secretory pathway (*49,50*). These mutations identified 23 *SEC* genes and the 21 genes, mutant alleles of which cause a clear stage-specific secretory block, are summarized in Figure 1.

Molecular and biochemical analyses of the nine *SEC* genes, and their products, that function to stimulate protein transport from the ER to the yeast Golgi, have led to penetrating insights into the roles of several of these *SEC* proteins in intracellular protein trafficking. These studies have converged with those employing the *in vitro* mammalian systems to dramatically illustrate the direct mechanistic analogies between the yeast and the mammalian secretory pathways (*8,13,50,85*). Moreover, it is now appreciated that these early acting *SEC* gene products are likely to exhibit multiple execution points, including execution points that occur late in the secretory pathway (*85*). The late-acting *SEC* genes have permitted an analysis of stage-specific functions that are not required for the early stages of the secretory pathway. It is to these late functions that we will restrict our attention in this review.

Whereas much progress has been made in the last decade with respect to an analysis of the secretory pathway in yeast, our understanding of endocytosis in yeast is in its infancy. This is in contrast to the rapid progress that has been made

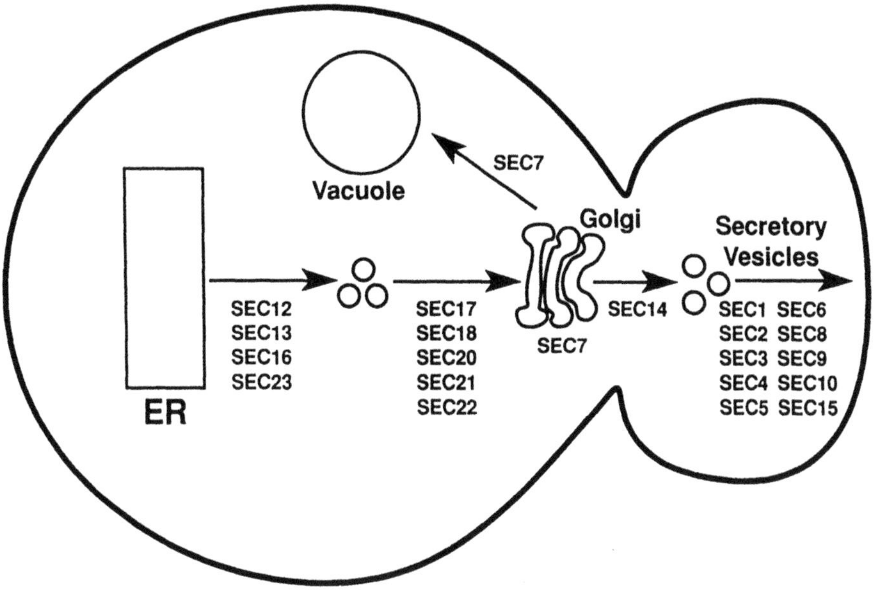

Figure 1. A schematic representation of the yeast secretory pathway with respect to protein transport from the endoplasmic reticulum to the cell surface or the vacuole. The first execution points of 21 of the original 23 *SEC* gene products are indicated (*50*).

in characterizing the endocytic pathway of mammalian cells. It has been determined that plasma membrane receptors are internalized into animal cells by a specific clustering into clathrin-coated pits followed by packaging into endocytic vesicles (*23,57*). These endocytic vesicles are targeted for fusion with endosomal compartments; acidic organelles that effect a disengagement of ligand from many receptors. The ligands are finally delivered to lysosomes, whereas receptors can be recycled back to the cell surface. An elucidation of the mammalian endocytic pathway has benefited greatly from the variety of marker substrates, including a number of enveloped viruses, that have been available to study the process (*26,27*). Only recently have candidate endocytic markers been developed for experimental use in the yeast system. In this review, we will also discuss the progress that has been made with respect to descriptions of the endocytic pathway in yeast. Moreover, we will try to relate what common relationships may exist between secretory pathway and endocytic functions in yeast.

II. GOLGI SECRETORY FUNCTION IN YEAST

Several genes have recently been identified as candidates for encoding polypeptides that are required for protein transport through the Golgi. The *YPT1*, *ARF1*, and

ARF2 gene products have been shown to be ras-like GTP-binding proteins that localize predominantly to the Golgi apparatus (*74,78*). *YPT1* is an essential gene for yeast viability whereas the ARF genes comprise an essential gene family. As these GTP-binding proteins are very likely to follow the SEC4p paradigm for mechanism of GTP-binding protein function in intercompartmental protein transport (discussed at length in Section III), we will not treat the function of these proteins in specific detail here. Those issues, and the recent findings that the YPT1p and ARFp play crucial roles in protein transport from the yeast ER to the Golgi, have been reviewed in detail elsewhere (*13*). Rather, we wish to focus our attention on what has been learned from analysis of the *SEC7* and *SEC14* gene products. These polypeptides have classically been considered to exhibit their first execution point at the level of protein transport through the Golgi (*13,49,50*).

The present data suggest that the SEC7p is involved in several execution points in protein transport through the secretory pathway. First, *sec7ts* mutants challenged with restrictive conditions accumulated incompletely glycosylated forms of the secretory proteins invertase and prepro-a-factor, and the vacuolar pro-carboxypeptidase Y (*19*). In the case of invertase or prepro-a-factor, the intracellular species fell into three forms that were distinguished by: (1) the number of mannose residues that were added to the core-glycosylated forms of these secretory proteins; and (2) the types of glycosyl linkages that were involved in the addition of mannose. The *sec7ts* mutants were able to support early and intermediate N-linked glycosyl modifications on trapped secretory proteins, but failed to mature those N-linked oligosaccharide chains as indicated by the absence of terminal mannose in α–(1→3) linkage on such chains (*19*). These findings have been interpreted to indicate that SEC7p was involved in multiple stages of protein transport through the yeast Golgi, and that, perhaps, the *sec7ts* alleles were somewhat leaky (*19*). This leakiness would permit inefficient protein transport through the early Golgi, and thereby result in the detection of heterogeneously modified pools of trapped secretory glycoproteins. Although the absence of glycosyl maturation of secretory proteins in *sec7ts* mutants argues convincingly for an early first-execution point for SEC7p, these mutants have also been reported to exhibit defects in fluid phase endocytosis (see Section V). These findings provide further support for the notion of multiple execution points for the SEC7p in the yeast secretory and, perhaps, endocytic pathways.

The *SEC7* gene has been shown to encode a 2008 residue polypeptide with an apparent molecular mass of some 230 kDa (*1,21*). SEC7p serves an essential function as *sec7°* alleles represent recessive haploid-lethal mutations. Subcellular fractionation experiments demonstrated that: (1) the SEC7p distributed approximately equally between cytosolic and membrane fractions; (2) that SEC7p association with membranes was sensitive to treatment with urea, but was insensitive to solubilization with nonionic detergents; and (3) that the SEC7p exhibited a significant colocalization with the Golgi marker KEX2p in double-label immunofluorescence experiments (*21*). These collective data identified the SEC7p as a peripheral membrane protein of the yeast Golgi. Franzusoff et al. have also demonstrated that the SEC7p is a phosphoprotein, although at present there is no evidence that the phosphorylation state of the SEC7p correlates with function (*21*).

Some recent and elegant studies have been performed in the laboratories of Franzusoff and Howell that have shed considerable light onto what the role of SEC7p might be in intercompartmental protein transport (*20*). These workers found that inclusion of anti-SEC7p serum to an *in vitro* cell-free assay designed to support transport of prepro-a-factor substrate from the ER to the Golgi specifically inhibited such transport. Moreover, these SEC7p antibodies caused the accumulation of presumed ER-derived transport vesicles in the assay cocktail. Such vesicles could be recovered by immunoisolation with SEC7p antibodies, indicating the presence of SEC7p on the vesicle surface. The collective data have been proposed to indicate that SEC7p represents a constituent of a nonclathrin coat that functions in transport vesicle metabolism at multiple stages in the yeast secretory pathway (*20*). Interestingly, these data also suggest that the first SEC7p execution point is at the level of ER to Golgi transport and not at the level of transport through the Golgi. Analysis of conditional *sec7°* terminal phenotypes should help resolve this issue.

It has been demonstrated that *sec14ts* mutants, challenged with restrictive conditions, accumulated an intracellular form of secretory invertase that had acquired mature N-linked glycosyl chains (*19*), and that the acquisition of these terminal glycosyl modifications occurred with wild-type kinetics (our unpublished data). Furthermore, we have found that the maturation of the vacuolar proteinases carboxypeptidase Y (CPY) and proteinase A (PrA), from a late Golgi zymogen form to the mature vacuolar form, was only kinetically delayed (our unpublished data), even in the face of what seems to be a tight secretory block (*5*). Thus, in contrast to the SEC7p, the SEC14p appears to represent a truly compartment-specific transport factor. At present, there is also no evidence that the SEC14p is required for endocytic pathway function in yeast as *sec14ts* mutants appeared to exhibit normal capacities for fluid phase endocytosis at 37 °C (*62*). The data argue for a specific involvement of the SEC14p in catalyzing the transport of protein from a late Golgi compartment (*5,13*).

The *SEC14* gene encodes a hydrophilic 35-kDa polypeptide that executes an essential function for yeast viability, and the SEC14p is distributed between cytosolic (65%) and membrane (35%) fractions as judged by subcellular fractionation experiments (*5*). Additional studies have shown the SEC14p to cofractionate with the KEX2p in more refined fractionation experiments, and to colocalize with KEX2p in a double-label immunofluorescence regimen (*16*). These data demonstrated a physical association of the SEC14p with the compartment that requires SEC14p function (i.e., the Golgi), and established that the SEC14p is directly involved in stimulating Golgi secretory function in yeast. It has now been established that the SEC14p is the yeast phosphatidylinositol (PI)/phosphatidylcholine (PC) transfer protein, a polypeptide that catalyzes the energy-independent transport of PI or PC as monomers between membrane bilayers *in vitro* (*4*). This discovery represented the first assignment of an *in vivo* function for a phospholipid transfer protein, and has set the stage for a detailed analysis of what these ubiquitous

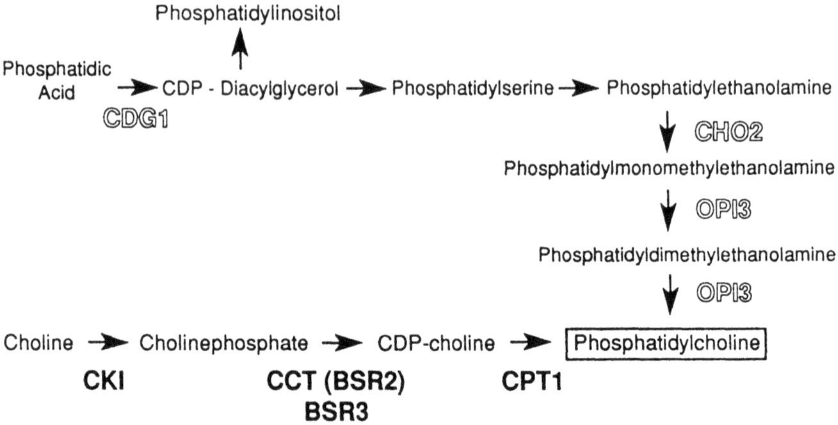

Figure 2. The relationship between the essential SEC14p requirement for yeast Golgi secretory function and the yeast cell's capability to synthesize phosphatidylcholine (PC). Three of the genes, loss of function of which leads to a bypass of SEC14p condition, block PC synthesis via the CDP-choline pathway at the designated stages. These genes are identified by the solid black script. Mutations in genes that are required for PC synthesis via the methylation pathway do not result in bypass of SEC14p, and these genes are identified by the open script. Additional details are presented in the text.

proteins might do in eukaryotic cells (for a further discussion of these points see refs. *4,14,16* and *28*).

Key insights into the *in vivo* role played by the SEC14p in stimulating Golgi secretory function have been gleaned from a molecular analysis of genes, mutant alleles of which suppress the secretory defects associated with *sec14* lesions. Cleves et al. have found that mutations in one of at least six genes was sufficient to generate a condition where the normally essential SEC14p requirement was specifically bypassed in the corresponding mutant yeast (*15,16*). The demonstration that mutations blocking defined aspects of PC biosynthesis by yeast resulted in such a specific bypass of SEC14p has proven particularly informative. Of the six *sec14* suppressor loci identified, three (i.e., *BSR2*, *BSR3*, and *BSR4)* encoded gene products that were required for yeast to be able to synthesize PC via the CDP-choline pathway (*16*; see Fig. 2).

Molecular analysis of *BSR4* revealed that this gene was allelic to *CKI*, the structural gene for the enzyme choline kinase that catalyzes the first reaction in the CDP-choline pathway. Moreover, *cki*° mutations were shown to effect an efficient suppression of haploid-lethal *sec14*° mutations (*16*). Our data now indicate that *BSR2* is allelic to *CCT*, the structural gene for the enzyme cholinephosphate cytidylyltransferase that executes the penultimate step in PC synthesis via the CDP-choline pathway, whereas *BSR3* encodes a function that is required for the stimulation of cytidylyltransferase activity in yeast cells (T.P. McGee, M.K.Y.

Fung, and V.A. Bankaitis, submitted). Finally, *cpt1°* strains, which are defective in the enzyme choline phosphotransferase that catalyzes the ultimate reaction in the CDP-choline PC biosynthetic pathway, were also found to exhibit the bypass of SEC14p condition (*16*). Strikingly, although yeast cells possess two pathways for PC synthesis (Fig. 2), only loss of function mutations that affect PC synthesis via the CDP-choline pathway suppressed *sec14* defects. Yeast strains that were not able to synthesize PC by the alternate route (i.e., the methylation pathway) nonetheless still exhibited the essential requirement for SEC14p function (*16*; Fig. 2).

The finding that null mutations in the structural genes for enzymes of the CDP-choline pathway bypass the normally essential cellular requirement for SEC14p suggests two points. First, the data suggest that PC biosynthesis is inherently toxic to the yeast cell; a rather interesting concept as PC biosynthesis is generally regarded as an essential cellular activity. Second, the data imply a detoxification role for the SEC14p; one that is designed to neutralize the inherently deleterious nature of CDP-choline pathway function. On the basis of these insights, Bankaitis and co-workers have proposed that the SEC14p functions to maintain an appropriately high PI/PC ratio in yeast Golgi membranes (*14,16*). This high PI/PC content is deemed to be critical for the secretory competence of these membranes. Outstanding questions that remain include: (1) does the SEC14p control Golgi membrane composition by regulated transport of PI, or PC, or both PI and PC, or does SEC14p accomplish such a controlling function without catalyzing lipid transport (i.e., by regulating phospholipid biosynthesis in some other fashion); (2) what is the molecular basis for why mutations that block CDP-choline pathway function suppress *sec14°*, whereas lesions affecting methylation pathway function do not; and (3) why is the proposed PI/PC content of Golgi so critical for the secretory competence of Golgi membranes? Resolution of these questions will go a long way towards providing an understanding of two fundamental issues in cell biology. That is, issues relating to the *in vivo* function of this particular phospholipid transfer protein, and to the precise function of a Golgi-specific transport factor.

III. GOLGI-DERIVED SECRETORY VESICLE METABOLISM IN YEAST

Ten genes have been recognized whose products are required for the targeting or fusion of Golgi-derived secretory vesicles to the yeast plasma membrane (*49, 50*). Temperature-sensitive alleles of each of these 10 genes—*SEC1*, *SEC2*, *SEC3*, *SEC4*, *SEC5*, *SEC6*, *SEC8*, *SEC9*, *SEC10*, and *SEC15*—cause the accumulation of secretory vesicles in the corresponding yeast strains challenged with restrictive conditions. Moreover, fluid phase endocytosis in these *sects* yeast strains (as measured by uptake of lucifer yellow; see Section V) is also strongly inhibited at the nonpermissive temperature of 37 °C (*62*). Thus, these particular *SEC* gene products play some role in both the endocytic and late exocytic pathways of membrane traffic. The available evidence suggests an extensive functional interaction between eight of these 10 gene products (with *SEC6* and *SEC9* currently excluded from the interaction), and identifies the *SEC4* gene product as a central element around which these functional interactions revolve.

The work of Novick and co-workers on SEC4p function in yeast has yielded seminal insights into how fusion of transport vesicles to membranes is regulated, and has provided the initial demonstration that a small GTP-binding protein (i.e., the *SEC4* gene product) plays a critical role in vesicle-mediated protein transport, (*49,50*). Subsequently, it has been demonstrated that GTP-binding proteins are involved in many aspects of vesicle-mediated protein transport in both the exocytic and endocytic pathways (*10*). Thus, study of the SEC4p function has established a paradigm for understanding GTP-binding protein function in protein transport. A molecular analysis of *SEC4* revealed that this gene is essential for yeast viability and has the potential to encode a polypeptide with an inferred molecular mass of 23.5 kDa (*71*). The inferred SEC4p primary sequence displayed a 32% homology with the human H-ras protooncogene and identified the SEC4p as a member of the ras family of GTP-binding proteins. Using ligand blot assays, Goud et al. demonstrated that the SEC4p does indeed exhibit GTP-binding activity (24). Moreover, Walworth et al. found that *sec4* alleles producing a mutant Sec4p that was unable to bind GTP scored as nonfunctional *in vivo* (*83*). These collective data demonstrated that GTP binding was required for SEC4p function *in vivo*.

Subcellular fractionation experiments with wild-type yeast strains showed that the SEC4p distributed predominantly to the 10,000xg pellet fraction, with only small amounts from 10,000xg recovered from 10,000xg pellet and cytosolic fractions (*24*). Subsequently, more discriminating sucrose density-gradient fractionations indicated that the pelletable SEC4p cofractionated with a plasma membrane marker. Immunofluorescence localization studies confirmed that, under these conditions, the SEC4p localized largely to the yeast plasma membrane (*24*). By contrast, imposition of the *sec6ts* secretory block, which results in the accumulation of Golgi-derived secretory vesicles, resulted in the redistribution of SEC4p from the 10,000xg pellet fraction to the 100,000xg pellet fraction. Density-gradient fractionation experiments indicated that the SEC4p in the 100,000xg pellet fraction colocalized with secretory vesicles (*24*). Again, further support for the fractionation data was forthcoming from immunofluorescence experiments that indicated a shift of the SEC4p from the plasma membrane to a punctate cytoplasmic organization that was consistent with localization to secretory vesicles (*24*).

The SEC4p exhibits a very rapid rate of membrane association ($t_{1/2}$ < 1 min), irrespective of the finding that the SEC4p primary sequence failed to exhibit significant hydrophobic character (*24,71*). The SEC4p sequence is punctuated by two terminal cysteine residues, a sequence that is similar to the signature CAAX box that defines the C-terminus of RAS (*71*). These terminal cysteine residues are essential for SEC4p function as their deletion resulted in a soluble SEC4p species that exhibited unimpaired GTP-binding activity but, nonetheless, was nonfunctional *in vivo* (*83*). These data indicated that the membrane-binding properties of SEC4p are a function of the SEC4p C-terminus, and that this membrane binding is required for SEC4p function. Goud et al. also found that membrane-associated SEC4p behaved as an integral membrane protein by the criteria of its nonextractability from membranes with alkaline sodium carbonate and its partitioning into the detergent phase of an aqueous Triton X-114 solution (*24*). While there has been no direct demonstration of a posttranslational modification of SEC4p, the finding that SEC4p overproduction specifically elevated the soluble SEC4p pool suggested that SEC4p attachment to membranes was specific and saturable. The demonstration that the BET2 gene product is required for efficient targeting of SEC4p to membranes *in vivo*, coupled with the observation that the BET2p is homologous to the β-subunit of a prenyl (i.e., farnesyl)-transferase, provides further support for the notion that the SEC4p is modified by a prenyl group in a manner similar to RAS and other GTP-binding proteins (*65*).

Walworth et al. have proposed a model for SEC4p function that strives to take in account these various data (*83*). These workers hypothesize that SEC4p function is in principle analogous to that of the *E. coli* translation elongation factor EF-Tu, and the main features of this model are depicted in Figure 3. In this model, soluble SEC4p exchanges GDP for GTP and binds a newly formed secretory vesicle, thereby rendering the secretory vesicle competent for targeting to the plasma membrane. This event likely requires the participation of a guanine nucleotide exchange activity to stimulate the reaction so that it occurs at a rate that is physiologically meaningful (*35*). Docking of the secretory vesicle ensues—an event whose specificity is proposed to be a function of a plasma membrane effector protein that engages the SEC4p—and fusion occurs. Subsequent to fusion, SEC4p is released from effector by virtue of a conformational change induced by GTP hydrolysis. This hydrolysis almost certainly requires the participation of a GTPase-activating protein (GAP), as GTP-bound SEC4p is a stable entity *in vitro* (*35*). In any case, SEC4p release allows both SEC4p and effector to participate in another round of secretory vesicle targeting and fusion (Fig. 3).

An important prediction that arises from this model is that an irreversibly activated SEC4p should serve as a potent inhibitor of this targeting fusion cycle because of its ability to irreversibly occupy the plasma membrane effector. To test this prediction, Walworth et al. constructed a SEC4pIle133 variant that was deduced from the analogous oncogenic ras mutation to result in a SEC4p that remained locked in its active state, even in the absence of GTP binding (*83*). This Sec4p Ile133 exhibited the expected biochemical properties and, at low levels of expression, caused a dominant *ts* secretory block that was similar to that caused by the original *sec4ts* alleles with respect to phenotypic, biochemical, and morphological properties. High level expression of Sec4p Ile133 resulted in a dominant lethal secretory block. The toxic effects of Sec4pIle133 were completely abrogated by deletion of the two C-terminal cysteine residues that are crucial for membrane association of the SEC4p. Taken together, the data suggest that Sec4pIle133 can catalyze the first round of secretory vesicle targeting/fusion, but cannot release effector once fusion has occurred. This leads to a progressive titration of effector and results in a secretory block. Deletion of the terminal cysteine residues elimi-

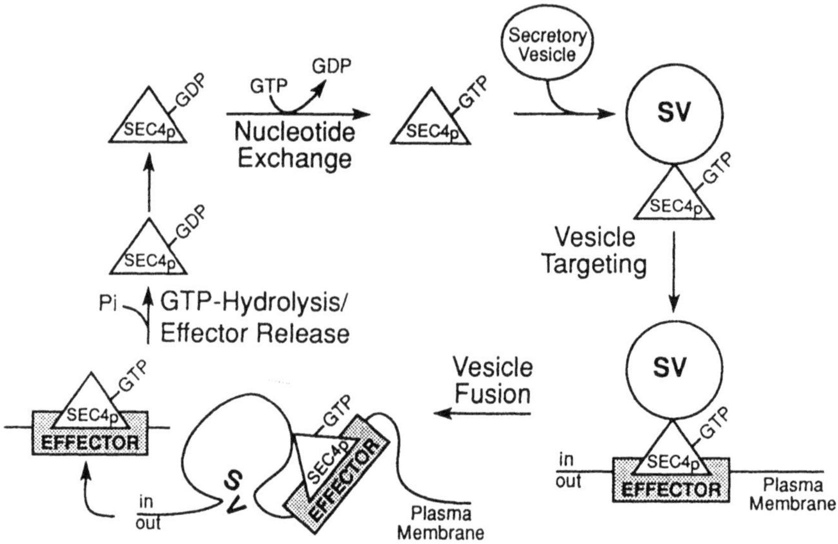

Figure 3. An illustration of how a GTP-binding protein (SEC4p) is proposed to control fusion of secretory vesicles (SV) to the yeast plasma membrane. This model invokes a GTP-hydrolysis dependant cycle of SEC4p engagement and disengagement with secretory vesicle membranes and a plasma membrane associated effector, or docking protein (adapted from ref. 83).

nates the dominant negative effect because the mutant Sec4p can no longer bind secretory vesicle membranes and, as a result, is not able to engage effector.

As indicated above, there are strong reasons for anticipating that SEC4p function will be regulated by auxiliary proteins that stimulate nucleotide exchange by the SEC4p or stimulate SEC4p GTPase activity. Furthermore, the plasma membrane effector that is postulated to interact with SEC4p in a cycle of engagement and release remains to be identified. Genetic studies have shown an extensive pattern of genetic interactions between *sec4ts* alleles and mutations in seven of the remaining nine *SEC* genes whose products function in secretory vesicle fusion to the plasma membrane (*71*). For example, *sec4ts* exhibited synthetic lethal interactions with *sec1ts*, *sec2ts*, *sec3ts*, *sec8ts*, *sec10ts*, and *sec15ts* mutations. Moreover, duplication of *SEC4* clearly suppressed *sec2ts*, *sec8ts*, and *sec15ts* mutations, whereas a weaker suppression of *sec1ts*, *sec5ts*, and *sec10ts* was observed. Thus, there is at present no genetic evidence for functional interaction between SEC4p and SEC6p or SEC9p. The current evidence for functional interaction between SEC4p and SEC1p, SEC3p, SEC5p, or SEC10p is intriguing but not particularly compelling, and powerful genetic evidence for interaction between SEC4p and SEC2p, SEC8p, and SEC15p has been obtained. Because the genetic interactions between *SEC4* and *SEC15* and *SEC2* are particularly notable, these latter two genes have provided an initial focus for study in the hope that direct information concerning some of these SEC4p accessory proteins might be forthcoming.

Molecular analysis of *SEC15* has revealed this gene to be essential for yeast viability, with the potential to encode a 911 amino acid hydrophilic polypeptide with a predicted molecular mass of 116 kDa (*72*). Subcellular fractionation experiments demonstrated that the SEC15p was a peripheral membrane protein that exhibited some cofractionation with a plasma membrane marker, and could also be demonstrated to fractionate as a high molecular mass (>1000 kDa) soluble species (*72*).

Some additional insights into SEC15p function were gleaned from the analysis of the striking consequences associated with SEC15p overproduction in otherwise wild-type yeast cells. Overproduction of SEC15p was found to interfere with secretory vesicle fusion to the plasma membranes, as evidenced by the observation that secretory vesicles aggregated within a specific patch of the growing bud in SEC15p overproducing cells (*72*). This patching reflected, at least in part, the intracellular localization of the SEC15p as the protein was visualized in the vesicle patch by immunofluorescence techniques. Such patch formation was not seen when Sec15pts was overproduced, or when SEC15p was overproduced in *sec2ts* or *sec4ts* strains. These collective data indicated that secretory vesicle patching was a physiologically relevant manifestation of overproduction of wild-type SEC15p function, and suggested that the SEC2p and SEC4p execution points preceded that of the SEC15p. Salminen and Novick suggest that the data are consistent with a model in which SEC15p acts to facilitate the docking of SEC4p-"labeled" secretory vesicles to the plasma membrane (*72*). The observed patching of secretory vesicles would be considered to reflect the inappropriate localization of SEC15p to secretory vesicles, an event that would cause the subsequent, and similarly inappropriate, vesicle-vesicle docking reactions. This model implies a role for SEC15p in effector function.

Molecular analysis of *SEC2* has also revealed this to be an essential gene, one encoding a hydrophilic polypeptide of 759 residues and an apparent molecular mass of 84 kDa (*48*). The SEC2p fractionated predominantly to the cytosolic fraction and was recovered in high molecular mass complexes of 500 to 700 kDa. On the basis of the following observations, it is now clear that the critical functional domain of the SEC2p resides in the amino-terminal one-half of the protein. Two *sec2ts* alleles were demonstrated to represent opal nonsense mutations, and gene disruption experiments revealed that deletion of the last 251 residues of the SEC2p were inconsequential for function whereas removal of the last 368 residues resulted in a thermolabile SEC2p (*48*). Homology searches revealed that the amino-terminal one-third of the SEC2p exhibited a 25% homology to the rod region of myosin heavy chain and to other cytoskeletal proteins. Nair et al. have suggested that this homology identifies an a-helical coiled-coil domain within the SEC2p that may be involved in controlling essential protein-protein interactions that presently remain undefined (*48*).

IV. BIOGENESIS OF THE YEAST VACUOLE

The available data suggest that the yeast vacuole is the terminal destination of endocytosed material (see Section V). As such, an understanding of membrane and protein transport to the vacuole in yeast is an issue that will likely be of direct relevance to an appreciation of the yeast endosomal pathway. Indeed, the yeast vacuole is often considered to represent a lysosomal compartment because it is an acidic compartment that represents the residence of several major hydrolytic activities (i.e., proteinases, RNAse, a-mannosidase, and alkaline phosphatase), and has even been suggested to represent the major degradative compartment in the yeast cell (*40,84*). The problem of vacuole biogenesis can, at a rather superficial level, be considered to involve issues relating to: (1) how proteins are sorted from the secretory pathway to the vacuole; and (2) the machinery that supports delivery of sorted material to the vacuole. Pioneering studies performed by Stevens and Schekman suggested that yeast vacuolar proteins are transported from the ER to the Golgi together with secretory proteins destined for the yeast plasma membrane or secretion into the extracellular milieu (*79*; Fig. 4).

Subsequently, vacuolar proteins are sorted away from the secretory pathway in either a late Golgi compartment, or shortly after exit from the Golgi. Considerable effort has been directed towards the identification of sorting signals that are elaborated by yeast vacuolar proteins. These signals do not follow the mammalian paradigm where the lysosomal targeting signal is defined by addition of mannose 6-phosphate moieties onto N-linked glycosyl chains of lysosomal proteins. Rather, the yeast vacuolar sorting signals appear to reside in the propeptides of soluble vacuolar zymogens (*32,82*), or, perhaps, in the cytosolic tails and/or transmembrane regions of vacuolar membrane proteins (*39*). The vacuolar species that have been analyzed in greatest detail with respect to these issues include the soluble proteinases carboxypeptidase Y (CPY) and proteinase A (PrA), and the vacuolar integral membrane protein alkaline phosphatase (vALP). At present there are no obvious features that are shared between the sorting signals of various vacuolar proteins. The studies from which these basic conclusions were derived have been reviewed in detail elsewhere (*40,68*). For the purposes of this review, however, we wish to restrict our attention to what has been learned concerning the cellular functions that are involved in vacuolar biogenesis and delivery of protein to the yeast vacuole.

Several basic approaches have been employed to genetically dissect vacuolar function/biogenesis in yeast. Studies of mutants obtained by each of these approaches have converged on the problem of protein transport through the vacuolar pathway. One approach that was employed by Jones (*33*) involved a genetic screen designed to identify mutant strains of yeast exhibiting defects in vacuolar proteolytic activities (e.g., CPY). These *pep* mutants fell into 17 complementation groups and exhibited pleiotropic deficiencies in vacuolar proteolytic functions. With one notable exception, all *pep* mutants were defective in sorting of vacuolar proteases to the yeast vacuole (*33,34,40*). Instead, these vacuolar species were mislocalized to the yeast cell surface. These general localization defects resulted in

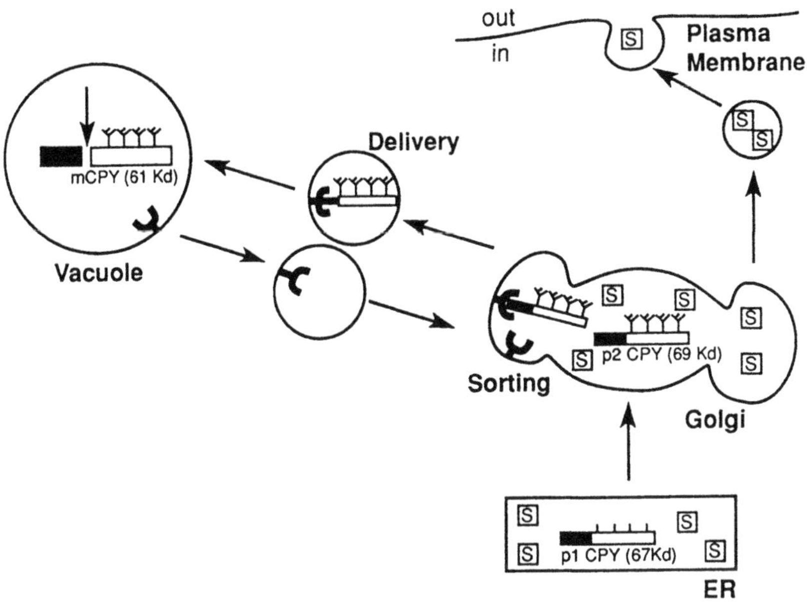

Figure 4. Sorting of soluble vacuolar proteins from the secretory pathway and into the vacuolar pathway. Soluble vacuolar proteins such as carboxypeptidase Y (CPY) traffic from the ER as core-glycosylated zymogens (p1) to the Golgi where remodeling of the core glycosyl chains occurs to yield the p2 form of the zymogen. At the level of a late Golgi compartment, the p2 CPY is sorted from the secretory pathway. This sorting event is proposed to involve a sorting receptor that recognizes a sorting signal that is encoded at the amino-terminal region of the zymogen. Unlike the lysosomal targeting signal in mammalian cells the yeast signal does not involve a carbohydrate component. Delivery of sorted material to the vacuole is thought to occur by a vesicular intermediate and may involve one or more prevacuolar compartments (e.g., endosomes). Proteolytic maturation of the p2 CPY to the enzymatically active mature form (mCPY) occurs either in the vacuole or in a prevacuolar compartment. These issues have been reviewed in detail elsewhere (*40*).

a very inefficient maturation and activation of these protease zymogens. The one exception to the mislocalization mechanism was represented by mutations in *PEP4*, the PrA structural gene, and identified PrA as a critical component of the vacuolar protease activation pathway (*2,86*).

Another approach involved genetic selections designed to directly identify yeast mutants that mislocalized either a genuine vacuolar species (i.e., CPY) to the cell surface, or secreted an invertase hybrid protein that, because of the engineering of vacuolar sorting information onto this normally secreted protein, was typically targeted to the yeast vacuole. The former version of this selection scheme was exploited by Rothman and Stevens who identified the corresponding genes with the vacuolar protein localization (*VPL*) (*67*). The latter version of this selection strategy was exploited by Bankaitis et al. and Banta et al. designated the corresponding genes identified as vacuolar protein targeting (VPT) (*3,6*). At present, the *vpl* and *vpt* mutants have been combined into at least 40 complementation groups that have now been redesignated as vacuolar protein sorting (*VPS*) (reviewed in ref. *40*). There is considerable genetic overlap between and *pep* and *vps* complementation groups, and these overlaps have been summarized elsewhere (*40,64*). As there has been an effort to redesignate all vacuolar biogenesis-relevant genes into the *VPS* nomenclature, we will adopt this general nomenclature in this review.

All told, there presently exist at least 47 unique complementation groups defined by mutations that result in mislocalization of yeast vacuolar proteins. These findings suggest that either protein transport to the yeast vacuole is a complex process that requires the proper involvement of many gene products, or that this is a process that can be easily perturbed by indirect mechanisms, or both. It was originally hoped that a vacuolar-sorting receptor would be identified in analysis of the *vps* mutants. Although such an expectation has not yet been realized, the *VPS* and *PEP* genes have provided the initial focus for a molecular analysis of how proteins are delivered to the vacuole, and have provided the first clues as to the nature of the cellular machinery that insures this delivery.

Biochemical analyses of *vps* yeast mutants have shown that, as expected, these strains exhibited pleiotropic defects in the localization of soluble vacuolar proteinases to the vacuole (*64,68*). Rather, the majority (>80%) of the CPY synthesized by the cell was exported to the cell surface with rapid kinetics (*3*), and this secretion of CPY required the function of the late *SEC* gene products that promote the fusion of Golgi-derived secretory vesicles to the cell surface (*3,67*). On the basis of these data it was concluded that sorting of CPY out of the secretory pathway was an active process, defects in which resulted in secretion of CPY by default. Further support for this view was obtained from studies demonstrating that vacuolar proteins lacking sorting signals were secreted (*32,82*). This inappropriate secretion of CPY was accompanied by a marked processing defect as the secreted CPY was recovered as a Golgi-modified precursor form (i.e., p2 proCPY, see Fig. 4) (*3,64,67*). Moreover, similar processing and mislocalization defects were also observed for other soluble vacuolar proteinases; that is PrA and proteinase B (PrB). In these cases, however, while the severities of the maturation defects were comparable to those of CPY, the magnitude of mislocalization to the cell surface was generally less (*64*). It remains to be determined in which intracellular compartment these precursor forms of PrA and PrB reside.

Interestingly, only a small subset of vps mutant strains exhibited major defects in the maturation of vALP (*40*). These mutants (i.e., *vps11*, *vps16*, *vps18*, and *vps33*) were also unique in that these exhibited vacuolar α-mannosidase activity on the cell surface. α-Mannosidase is a vacuolar membrane associated enzyme whose catalytic domain appears to be disposed to the vacuole lumen. These collective data suggested that vacuolar delivery of membrane proteins overlapped with, but was at least in part independent of, the machinery required for delivery of soluble proteins to the yeast vacuole.

Morphological examination of the vacuoles of *vps* mutant strains has proven itself to be a useful tool for distinguishing unique classes of *vps* mutants. The vacuoles of wild-type cells can be fluorescently labeled with either fluorescein isothiocyanate (FITC; see Section VI), or by an endogenous fluorophore that is loaded into the vacuoles of strains exhibiting specific defects in the yeast adenine biosynthetic pathway (reviewed in ref. *40*). Wild-type cells generally exhibit one to several large vacuoles that can be visualized by such fluorescence-labeling methods. Most *vps* mutants exhibit the wild-type staining profile and these have been designated as class A mutants (*6*). A second class of *vps* mutants exhibit a number (ca. 10) of small vacuoles that may represent fragmented vacuoles, or some type of intermediate structure in vacuole biogenesis. Mutants exhibiting these morphological properties have been designated as class B mutants (*6*). The last *vps* mutant category has been designated class C, and is characterized by the absence of detectable vacuoles (as operationally defined by absence of fluorophore staining) in the mutant yeast cells (*6*). Rather, these mutants exhibit a proliferation of small membrane vesicles, and other membranes of unknown etiology, in the cytoplasm. Interestingly, conjugation of two complementing class C *vps* mutants led to a very rapid assembly of an apparently normal vacuole in the zygote, suggesting that these abnormal membrane structures may represent intermediates in vacuolar biogenesis (*6*). It should also be noted that class C mutants are represented by the *VPS11*, *VPS16*, *VPS18*, and *VPS33* complementation groups; the same groups that were markedly defective for delivery of soluble and membrane proteins to the vacuole (see above). The various biochemical and morphological manifestations of all *vps* mutants, including those of the class C variety, are seen at all temperatures, and the most severe phenotypic consequence yet observed for any *vps* mutant (even *vps°*) is a *ts* growth phenotype. These data suggest that vacuolar functions may only be essential under conditions of environmental stress (see ref. *40*).

Since efficient sorting of lysosomal proteins in mammalian cells requires acidification of compartments in the lysosomal pathway, the ability of yeast *vps* mutants to acidify vacuoles has been investigated. Quinacrine fluorescence of yeast vacuoles was employed as a qualitative measure of vacuolar acidification. Of the 40 unique *vps* mutants, the *vps3*, *vps4*, and *vps6* mutants were remarkably defective in quinacrine labeling of vacuoles, suggesting that vacuole acidification might be important for protein delivery to the yeast vacuole. Indeed, the yeast vacuole harbors an H^+-ATPase that consists of eight subunits and is presumed to mediate acidification of the vacuole (*36*). The pH of the vacuolar lumen has been measured using pH-sensitive vital dyes and determined to be 6.1 to 6.2 (*59,87*). Both *vps3* and *vps6* strains were markedly deficient in vacuolar H^+-ATPase activity assayed in vacuolar fractions prepared from such strains (*68,70*). These findings, coupled with data indicating that challenge of yeast cells with bafilomycin A1 (a specific inhibitor of vacuolar H^+-ATPase) abolished quinacrine staining of vacuoles, and concomitantly effected secretion of newly synthesized vacuolar proteinases to the cell surface (*6*), strongly suggested a role for acidification in vacuolar protein targeting. There is now some controversy, however, with regard to whether vacuolar acidification defects directly affect vacuolar protein targeting or not.

Evidence that argues for an uncoupling of vacuolar acidification and vacuolar protein sorting has been gathered from analyses of mutants that fail to synthesize subunits of the H^+-ATPase, or have otherwise been isolated on the basis of being defective in vacuolar acidification. Both approaches have produced mutants that were manifestly defective in vacuolar acidification, but exhibited much more modest defects in vacuolar protein localization than did the *vps* mutants.

As noted above, the yeast vacuolar H^+-ATPase is composed of eight subunits exhibiting apparent molecular weights of 100, 69, 60, 42, 36, 32, 27, and 17 kDa (*36*). Yamashiro et al. isolated clones of the structural gene for the 60 kDa subunit (i.e., *VAT2*), and determined the consequences of a *vat2* null mutation (*87*). These workers found that *vat2°* mutants were viable, but unable to grow at neutral pH. The *vat2* mutants also exhibited what appeared to be normal vacuoles on the basis of Nomarski interference optics and immunofluorescence labeling with antibodies directed against vALP. Moreover, vacuolar fractions prepared from such strains were devoid of H^+-ATPase activity, and the vacuole lumen in *vat2* mutants exhibited a pH of 7.1. This pH was alkaline with respect to the pH = 6.1 value measured for wild-type cells (*87*). Taken together, these data suggested that: (1) the H^+-ATPase functions to acidify the vacuole; and (2) that the vacuole plays a crucial role in internal pH homeostasis. Interestingly, *vat2°* cells secreted only some 20% of the CPY made by them (*87*). This is to be contrasted with the secretion of >80% of the CPY produced by *vps* mutants (see above).

Preston et al. employed an elegant fluorescence-based screening strategy to identify mutants that were defective in vacuole acidification (*59*). One such mutant was measured to have a pH of 6.9 for the vacuole lumen. Again, this value represented a significant change relative to the wild-type pH of 6.2 that was measured for that compartment. Moreover, *vph1* strains were also quite normal with respect to vacuolar sorting properties (E. W. Jones, personal communication). Taken together, these data suggested that acidification was only required for highly efficient delivery of proteins to the vacuole. Furthermore, these findings have now raised the question of whether vacuolar protein localization defects are the primary result of acidification defects, or vice versa. Clearly, this is an issue that requires further study.

The molecular analysis of *PEP* and *VPS* gene function represents a new frontier in the study of how eukaryotic cells regulate vacuolar protein traffic. There is considerable promise for these studies providing key information concerning the nature of the compartments that make up the yeast vacuolar, and perhaps the endosomal, pathway. Such analyses should be facilitated by the finding that none of the *VPS* genes presently characterized at the molecular level are essential for yeast vegetative growth. Also, it is clear that interesting proteins will be identified. Already, the involvement of a protein kinase (*29*), a GTP-binding protein (*69*), an ATP binding protein (*7*), a putative zinc finger protein (*60,63*), and a putative cytoplasmic motor (*11,64*) in vacuolar localization has been demonstrated. An important challenge for the future is the achievement of a precise understanding of which step of the vacuolar delivery pathway is defective in each *vps* mutant strain. Although the *VPS* mnemonic implies sorting defects, there is no compelling evidence that any of the presently identified *PEP* or *VPS* genes encodes a product that is directly involved in sorting. Post-sorting delivery defects represent equally plausible mechanisms for mislocalization. Clearly, some distinction as to which *PEP* or *VPS* gene products are directly involved in the vacuolar pathway, and which exert only indirect effects, will also have to be made if a coherent understanding is to be extracted from the plethora of mutants already available. Moreover, it seems that the vacuolar sorting receptor(s) remains to be identified and characterized.

V. THE CASE FOR ENDOCYTOSIS IN YEAST

Whereas studies of endocytosis in mammalian cells have benefited from the large number of markers whose internalization into cells can be followed experimentally, the availability of such reagents for study of endocytosis in yeast has been quite restricted. As a result, studies of endocytic pathway function in yeast have lagged well behind those in mammalian systems. Indeed, it was only several years ago that the ability of yeast cells to endocytose was first demonstrated (*62*). Consequently, the early efforts directed at analysis of the yeast endocytic pathway have primarily concentrated on the identification of endocytic substrates and on the development of assays by which endocytosis of such substrates could be reliably measured.

Several different substrates have been used to probe endocytosis by yeast cells and these can be divided into two categories: (1) nonspecific markers; and (2) specific markers for receptor-mediated endocytosis. Nonspecific markers have included two fluorescent compounds [lucifer yellow carbohydrazide (i.e., LY) and fluorescein isothiocyanate-conjugated dextran (FITC-dextran)], mammalian viruses, and a-amylase isolated from a gram-positive bacterium. These reagents have been taken as indicators of nonreceptor-mediated endocytosis, and the superiority of LY to the others is clear (see below). The yeast α-factor pheromone has provided a good candidate probe for receptor-mediated endocytosis. In this section, we wish to briefly review the advances and some of the problems that have been forthcoming from the use of each of these marker systems.

The first demonstration of endocytosis in yeast was provided by Howard Riezman who employed LY as a marker for fluid-phase endocytosis (*62*). The experimental system involved incubating yeast cells with LY, extensively washing the cells to remove uninternalized LY, and quantitating the cell-associated LY by spectrofluorimetry of cell lysates. In this fashion, Riezman was able to determine that yeast cells were able to internalize LY to a significant extent; 180-ng LY/mg cell protein at 37 °C. This wash-resistant association of LY was greatly reduced, but not abolished, by incubation at 0 °C. Shift of cells having bound LY at 0 °C, to a 30 °C temperature, greatly stimulated LY uptake. On the basis of these data, it was proposed that the wash-resistant association of LY with yeast cells involved two steps. The first was a temperature-independent association that reflected trapping of LY in the yeast periplasm, and the second reflected temperature-dependent uptake of LY into the yeast cell. Fluorescence microscopy demonstrated that the internalized LY resided in the yeast vacuole. Moreover, internalization of LY into the yeast vacuole was severely inhibited by challenge of cells with metabolic poisons such as sodium azide and energy uncouplers (*62*). These data indicated that LY uptake was energy-dependent. As might be expected, the temperature-independent association of LY was also energy-independent. Finally, Riezman was able to demonstrate that the amount of LY internalization was a linear function of external LY concentration; indicative of nonsaturability of the LY uptake process. The definitive features of LY uptake (i.e., temperature-dependence, energy-dependence, and nonsaturability) are hallmarks of fluid phase endocytosis. Thus, LY was identified as an appropriate fluid-phase endocytic marker for yeast. Moreover, the vacuole was implicated as the terminal compartment for endocytosed material.

Riezman extended his studies of LY endocytosis in yeast to include conditions where the exocytic pathway for protein secretion was blocked at specific stages (*62*). This was accomplished by measuring LY uptake in *sec* mutants that had been challenged with the appropriate nonpermissive conditions. As indicated above, these *sec* mutants comprise a collection of strains which exhibit defined conditional-lethal defects in executing stage-specific functions in the yeast secretory pathway (see Fig. 1). First, Riezman observed that all 10 representative *sec* lesions blocking fusion of Golgi-derived secretory vesicles with the plasma membrane resulted in concomitant defects in fluid phase endocytosis (*62*). These data strongly suggested a functional overlap between the endocytic pathway and the latest stage of the exocytic pathway, and two interpretations of such data were offered. Either the endocytic and late exocytic pathways are: (1) obligatorily linked in a cycle of membrane traffic; or (2) the pathways are independent but the late-acting *SEC* gene products exhibit essential execution points in each. Second, the data also showed that certain defects in the ER to Golgi leg of the secretory pathway (represented by the *sec53ts*, *sec59ts*, *sec20ts*, and *sec21ts* mutations) did not reduce the efficacy with which LY was internalized into the yeast vacuole, suggesting that efficient execution of the early stages of the secretory pathway was not required for efficient endocytosis. However, other defects in the same general stage of secretory pathway function (i.e., *sec18ts* and *sec23ts*) exhibited an accompanying block in fluid-phase endocytosis. This apparent paradox can be explained by invoking execution points for the SEC18p and SEC23p in both early and late stages of the secretory pathway. Indeed, recent work has demonstrated a rather global requirement for the SEC18p at multiple stages in the eukaryotic secretory and endocytic pathways. The yeast SEC18p is structurally and functionally homologous to the mammalian N-ethylmaleimide-sensitive factor (NSF) that is required for transport of proteins from the ER to the Golgi, through the Golgi stack, and for fusion of endocytic vesicles in an *in vitro* system (*8,17,66,85*). Finally, one of the Golgi-blocked mutants (*sec7ts*) was defective in LY uptake whereas the other *(secl4ts)* was not. Since the SEC14p gene product is required for late Golgi secretory function (see Section II), it seems likely that a significant interface between the exocytic and endocytic pathways in yeast will be limited to post-Golgi events.

As stated above, LY uptake in yeast fulfills the criteria for fluid phase endocytosis and has identified LY as a reliable endocytic probe. In this regard, it is worth examining briefly the behavior of other nonspecific probes for endocytosis that have been employed. The data obtained with those systems have to some extent corroborated the principal conclusions of the LY work. On the other hand, these systems exhibit significant potential for artifact and, in at least one case, a convincing argument for artifact has been made.

Makarow and co-workers have described the uptake of three different macromolecular probes by yeast cells: animal viruses, bacterial α-amylase, and 70-kDa FITC-dextran (*43,44,45*). The experiments of interest were designed along the basic lines already discussed above for the LY work but the detection for internalization was in some cases different. When radiolabeled virus (e.g., vesicular stomatitis virus or Semliki Forest virus) was used as probe, virions were incubated with yeast spheroplasts and noninternalized virus was destroyed by a proteinase K challenge. In experiments using α-amylase or FITC dextran as probe, whole yeast cells and spheroplasts were used. No internalized α-amylase was liberated from spheroplasts by digestion with exogenous protease and subsequently washed away. FITCC-Dextran uptake was qualitatively assayed by fluorescence microscopy. In all cases, uptake of these probes was found to exhibit time-, temperature-, and concentration-dependence (*43,44,45*). Moreover, the intracellular destination of the internalized material was suggested to be the vacuole on the basis of two lines of evidence. First, incubation of cells with 70-kDa FITC-dextran led to a fluorescence signal that was found to reside in the vacuole by direct microscopic visualization (*45*). Second, subcellular fractionation of spheroplasts that had taken up radiolabeled VSV or Semliki Forest virus showed a cofractionation of a minor percentage of label with fractions enriched in vacuoles (*43*). Finally, the *sec1ts* secretory vesicle block effectively abolished amylase uptake or FITC-dextran delivery to the vacuole upon imposition of restrictive conditions. Thus, the major features of yeast endocytosis (as measured by nonspecific probes) were not controversial. However, the inherent difficulties associated with several aspects of the macromolecular probes employed by Makarow require mention. First, the use of fusogenic viruses, such as VSV and Semliki Forest virus, in conjunction with yeast spheroplasts in an uptake assay, is fundamentally problematic. Indeed, in those studies Makarow did not report the morphology of spheroplasts that had been challenged with virus. A large component of viral uptake could have resulted from multiple fusion events that generated cell-cell syncytia, a form of internalization that is fundamentally distinct from endocytosis. It is noteworthy that in the spheroplast fractionation experiments, the bulk of the label (i.e., virus) cofractionated with plasma membrane. Virus in vacuolar fractions represented only a very minor percentage (<10%) of the cell-associated virus (*43*). Moreover, only some 2% of input virus ever achieved a cell-associated protease-resistant state under standard conditions in the first place. The inefficiency of, and general difficulty with these assays, effectively disqualify animal viruses as reliable endocytic probes for yeast. Uptake of α-amylase also appeared to be inefficient and generally problematic as the protein is itself not easily proteolyzed, therefore complicating uptake measurements (*44*).

Finally, the suitability of 70-kDa FITC-dextran as an endocytic probe must be addressed. The operative assumption in the Makarow experiments was that FITC-dextran was taken up as a macromolecule and delivered to the lysosome via some endocytic route. More recent experiments in the laboratory of Jones (*34*), however, suggest that this apparent uptake of 70-kDa FITC-dextran by yeast actually reflected uptake of low molecular weight fluorescent contaminants (*58*). Preston, Jones, and Murphy found that commercial preparations of FITC-dextran were heavily contaminated with at least three low molecular weight fluorescent compounds which could not be removed by dialysis. Serial steps of gel filtration chromatography, followed by ethanol extraction, resoled these contaminants from the 70-kDa FITC-dextran fraction and revealed that 10 mg of commercial FTC-dextran contained 4 nmoles of FITC, 400 nmoles of a FITC-dextran degradation product, and 100 nmoles of fluorescein amine. Moreover, the composite fraction of these fluorescent contaminants was capable of labeling yeast vacuoles when presented to cells in the concentration expected to be present in a 100-mg/ml solution of commercial FITC-dextran (the standard labeling concentration). Finally, purified FITC-dextran failed to label yeast vacuoles, and the fluorescent material found in the vacuoles of yeast cells incubated with commercial FITC-dextran was isolated and determined to be FITC. Taken together, these data identified FITC, and not FITC-dextran, as the material taken up by yeast cells. The finding that FITC uptake was pH-dependent, in a manner suggesting that the neutral form of FITC was not efficiently taken up, raised the possibility that FITC uptake was by diffusion and not by endocytosis at all (*58*). These uncertainties limit the usefulness of FITC-dextran as an endocytic probe in yeast.

Although the LY system has provided insight into the yeast endocytic process, the immediate future for determinations of endocytic activity in yeast cells resides in the use of what is likely a natural substrate for receptor-mediated endocytosis in yeast, the α-factor peptide pheromone. The evidence for endocytosis of α-factor by yeast is circumstantial but, nonetheless, quite appealing. *S. cerevisiae* exists in three cell types. Two of these represent haploid states with different mating types, *a* and α. Each haploid type produces an extracellular peptide pheromone that stimulates a complex mating response by the other haploid type. Thus, *a*-cells elaborate an α-factor receptor on the surface of their plasma membrane and vice versa. Although the *a*-factor and α-factor peptides are entirely unrelated, the response of the appropriate target cells to each is very similar. One aspect of this mating response is a transient, cell-cycle stage specific growth arrest. Successful mating between pairs of stimulated haploids of opposite mating-type yields the asexual and pheromone nonresponse *a*/α diploid. Further details concerning yeast mating processes can be found in several excellent reviews (*22,38,77*). For the purposes of this review, however, it is useful to consider mating pheromone action on yeast cells as being directly analogous to peptide hormone action in animal cells. A common feature of peptide hormone receptors in mammals is their internalization into the cell by endocytic means, often an event triggered by binding of receptor to ligand. This ligand induced "downregulation" also applies to the yeast α-factor receptor and has provided an important piece of evidence supporting the notion that α-factor may enter yeast cells via a receptor mediated endocytic pathway.

The down regulation of α-factor receptor in yeast was first described by Jenness and Spatrick (*30*). Using radiolabeled α-factor as probe Jenness et al. (*31*) had established that *a* cells exhibited some 8000 receptors on their surface, on a per cell basis, and that these receptors bound α-factor with high affinity ($Kd = 6 \times 10^{-9}M$). Challenge of *a* cells with α-factor led to an immediate reduction in α-factor binding capacity that was followed by a rather rapid recovery of α-factor binding activity (*30*). The loss of binding activity was completely abolished by addition of energy poisons, and regeneration of binding activity required ongoing protein synthesis. Loss of binding activity did not require active protein synthesis, however. Moreover, correlative measurements suggested that loss of binding activity was directly proportional to α-factor internalization. At high α-factor concentrations the data were consistent with internalization of α-factor/receptor complexes. At low α-factor concentrations it was quite clear that unoccupied receptors were also internalized at a stimulated rate relative to the untreated condition (*30*). Taken together, the data supported a receptor-mediated uptake of α-factor pheromone by yeast cells.

These conclusions were independently corroborated by Chvatchko et al. (*12*) who directly monitored radiolabeled α-factor uptake by yeast cells of the *a*-mating type. The critical aspect of this assay was the distinction of internalized α-factor from bound, but not internalized, α-factor. The latter form was released by a pH-2 wash of yeast cells. Using this assay, it was determined that α-factor uptake occurred in a time-, temperature-, and energy-dependent manner. Whereas binding of α-factor to the yeast cell surface did not compromise α-factor structural integrity, internalization resulted in rapid degradation of the radiolabeled pheromone. Indeed, this degradation was compromised by defects in vacuolar proteases, suggesting that the vacuole was the ultimate destination of internalized α-factor (*12*). It is interesting to note that synchronized uptake of prebound α-factor exhibited an initial lag period of some 3 minutes, a feature also observed for receptor-mediated endocytosis in mammalian cells. In the mammalian system this lag presumably reflects receptor clustering, receptor recruitment into coated pits, and ultimate closure of such pits into coated vesicles (*23,57*). While it is tempting to draw a direct analogy to this scenario in the yeast system, there is at present no solid evidence that the mammalian paradigm for receptor-mediated endocytosis is strictly recapitulated in yeast. Indeed, recent studies on clathrin function in yeast raise the possibility that there may be fundamental differences between the endocytic pathways of fungal and mammalian systems (see Section VI).

Recent genetic studies of endocytosis in yeast suggest that fluid phase endocytosis and receptor-mediated uptake of mating pheromone may occur by at least partially overlapping pathways. This conclusion stems from the identification of yeast mutants that are defective in LY accumulation in the vacuole, and the demonstration that such mutants exhibit concomitant defects in α-factor degradation (*12*). Chvatchko et al. (1986) screened a library of random *ts* mutants for the inability to accumulate LY in vacuoles at 37 °C. Two such mutants were recognized and determined to identify two genes, *END1* and *END2*. Interestingly, both *end1* and *end2* mutants were defective for LY accumulation, not only at 37 °C but at 24 °C as well. This temperature-independent defect also extended to pheromone responsiveness at 24 °C. *end2* mutants were some 10-fold less sensitive to exogenous α-factor than were wild-type cells, while *end1* mutants were at least three-orders of magnitude less sensitive than wild-type strains. The *end1* mutants exhibited near wild-type capacities for binding α-factor but were initially, and it seems incorrectly (see below), reported to be some 90% defective in α-factor uptake relative to wild-type (*12*).

Ultrastructural examination of the *end* mutants showed clear structural abnormalities. The *end1* mutants grown at 24 °C exhibited the absence of a large vacuole (normally a dominant ultrastructural feature of wild-type yeast), the proliferation of vesicles in the cytoplasm, and an elevated number of plasma membrane invaginations at the cell surface (*12*). *end2* mutants grown at 24 °C did not exhibit any of the *end1* ultrastructural abnormalities. Rather, these strains appeared to exaggerate a membrane bound organelle of unknown identity that was usually located in the vicinity of the nucleus. It was suggested that this organelle could represent an endocytic compartment (*12*). There is, at present, no direct evidence to support this hypothesis, however.

What is the nature of the proposed endocytic defects in the *end* mutants? As described above, the phenotypic screen by which these mutants were originally recognized was based on the absence of accumulated endocytic material in the mutant strains. Such a phenotype could be realized by: (1) a defect in uptake of endocytic material; and (2) a defect in retention of internalized endocytic material. The former mechanism is directly relevant to endocytosis whereas the latter is of uncertain interest. The *end1* mutants have been characterized in the greatest detail. Dulic and Riezman isolated *END1* clones, discovered that *END1* is not an essential gene at temperatures below 37 °C, and found that *end1* null (*end1°*) mutants exhibited all of the phenotypic and ultrastructural properties associated with the original *end1* allele (*18*). These *end1°* strains exhibited rather mild defects in α-factor uptake. However, α-factor was not degraded in these mutants. These data identified the *END1* gene product as being involved is some aspect of α-factor metabolism other than its uptake, and corrected the earlier claim by Chvatchko et al. that *end1* strains were specifically defective in α-factor internalization (*12,18*). Penetrating clues as to why *end1* mutants are defective in α-factor degradation were forthcoming from the observations that *end1* mutants fail to form a recognizable vacuole, and that, by biochemical, genetic, and morphological criteria, these mutants fall into the class C category for *vps* mutants exhibiting wholesale defects in vacuolar biogenesis (*6,18*; see Section V). Indeed, *END1* was found to be allelic to *VPS11* (*18*). Thus, it seems that *end1* mutants are primarily defective in retention of material in and/or delivery of endocytosed material to, the ultimate endocytic compartment (i.e., the vacuole). Although the role of the *END2* gene product in endocytosis remains to be determined, it is clear that more discriminating isolation schemes for yielding yeast endocytic mutants are required, and that a comprehensive genetic analysis of endocytosis in yeast still lies in the future.

The demonstration that radiolabeled α-factor is taken up by yeast cells and, presumably, targeted to the vacuole for degradation, has made available another approach with which to probe the nature of the yeast endocytic pathway: that is, to isolate and characterize intracellular compartments that are involved in the trafficking of internalized pheromone from the plasma membrane to the vacuole. Singer and Riezman observed that, by creating a synchronous round of labeled α-factor uptake by yeast cells at 15 °C, uptake occurred with rather normal kinetics, whereas delivery to the vacuole was markedly delayed (*76*). Warming of the cells to 30 °C resulted in a chase of internalized pheromone into the vacuole. Challenge of the chase reaction with metabolic poisons completely inhibited delivery of internalized pheromone to the vacuole in a manner that was reversible and independent of perturbations of vacuolar pH. These data suggested that the labeled α-factor was trapped in an intracellular compartment, passage through which was temperature- and energy-dependent. Subcellular fractionation experiments indicated that the trapped α-factor sedimented at a density that was clearly distinct from plasma membrane and vacuole, and that the α-factor in this fraction was resistant to exogenous protease challenge in the absence, but not the presence, of detergent (*76*). Taken together, the data are consistent with the notion that an intracellular compartment was identified that exhibited the dual properties of accepting internalized α-factor from the cell surface and subsequently donating that α-factor to the vacuole. As such, this compartment fulfills some of the basic criteria of an endosome. Although a definitive demonstration that this compartment represents an endocytic organelle has not yet been provided, a detailed characterization of the resident proteins of this compartment, and their cognate structural genes, is warranted.

VI. THE FUNCTION OF CLATHRIN IN YEAST

The studies that best exemplify the application of "wreck and check" strategies to the problem of endocytosis have involved genetic analyses of clathrin function in yeast. The basic clathrin unit is the triskelion, a structure that involves the noncovalent association of three identical heavy chains (M_r = ca. 180 kDa) with three light chains (M_r = ca. 25 kDa) (*37,81*). Mammalian cells express two varieties of clathrin light chain whereas yeast express only one light chain type (*9,75*). Clathrin triskelions have the capacity to either polymerize into extensive polyhedral lattices that serve as cytoplasmic coats for specific classes of intracellular transport vesicles, or to polymerize into other types of membrane associated patches (*9,52*). Initial implications of clathrin function in endocytosis were derived from cytological studies that localized clathrin polymers to plasma membrane invaginations (i.e., pits) and intracellular vesicles in cells that were undergoing various types of endocytosis (reviewed in refs. *23* and *57*). These and other studies have led to the thought that clathrin functions in driving the formation of specific transport vesicles, facilitates the clustering of specific cell-surface receptors, and may be involved in protein-sorting events in both the endocytic and exocytic pathways (*9,23,52*). As such, clathrin presented an appealing target for functional studies *in vivo*.

Detailed molecular and genetic analyses of clathrin heavy- and light-chain function in yeast have been performed in several laboratories. A key breakthrough that made these studies possible was provided by Mueller and Branton who purified a coated vesicle fraction from yeast and demonstrated that the coat material could be disassembled into triskelions (*46*). These findings, coupled with the discoveries that these triskelions could be resolved into subunits of 190 kDa and 36 kDa, and that these triskelions could spontaneously repolymerize into empty coatlike baskets, led to the purification of yeast clathrin heavy and light chains (*41,46,53,75*). Specific antisera were raised against each chain-type, and used to screen the appropriate expression libraries. In this fashion, the yeast clathrin heavy-chain (*CHC1*) and light-chain (*CLC1*) structural genes were isolated (*41,53,75*). Nucleotide sequence analysis of *CHC1* revealed a 50% identity between the yeast and rat clathrin heavy-chain sequences (*41a*). The yeast and mammalian light chains exhibited considerably more divergence; only an 18% identity at the primary sequence level (*75*). However, regions of significant homology were observed nonetheless. First, an amino-terminal region spanning residues 28–49 of the yeast protein shared a 45% identity with a sequence that is conserved in all mammalian light chains analyzed. A second domain, spanning residues 193–213 in the yeast light chain, exhibited a 55% primary sequence identity to an insert sequence that is specifically found in clathrin light chains prepared from mammalian brain (*75*). Finally, yeast light chains also resembled mammalian light chains in their acidic and heat-stable character, and their ability to bind Ca^{2+} and calmodulin. Thus, the physical similarities between yeast and mammalian clathrin subunits are clear. In order to determine the consequences of clathrin dysfunction in yeast, null mutations in *CHC1* and *CLC1* have been constructed, and the phenotypic properties of such null mutants examined. We will first discuss the studies relating to disruption of *CHC1* since that information is at present more complete. It also seems that the general concepts gleaned from those studies will be largely applicable to *CLC1* as well (see below).

Two groups have reported on the disruption of *CHC1* in yeast. Two different results were obtained, and this has sparked considerable debate in the field as to which results represented the genuine item. Payne and Schekman found that haploid strains bearing a deletion of *CHC1* grew rather slowly, but were, nonetheless, viable (*53*). These findings suggested that the clanthrin heavy chain did not fulfill some essential function in the yeast cell. The viability of Δ*chc1* null mutants could not be easily explained by the existence of clathrin heavy-chain homologs as low-stringency DNA hybridization and antibody cross-reactivity experiments failed to detect any obvious potential candidates (*53*). Lemmon and Jones, however, found Δ*chc1* to be a haploid-lethal mutation in other genetic backgrounds, and described an unlinked modifier locus (*SCD1*) whose allelic state determined whether Δ*chc1* was a lethal mutation or not (*41*). Each set of these results was reproducible and without a doubt correct. The question that remained was which of these results gave the most accurate picture of the true Δ*chc1* phenotype. Resolution of this issue demands precise information concerning the relationship of the *CHC1* and *SCD1* gene products. Although such information is still unavailable, recent work has provided some preview to what the final judgement may be.

In principle, two interpretations for the modifier *SCD1* locus could be envisioned First, the *SCD1* locus could be defined by suppressor mutations that alleviate the normally lethal consequences of Δ*chc1* lesions. In this model, *SCD1* might well encode a product whose function is relevant to clathrin heavy-chain function in yeast. Indeed, the *SCD1* mnemonic (for suppressor of clathrin deficiency) implicitly favored this notion. An alternative explanation for the *SCD1* modifier involves a less exciting scenario; one in which Δ*chc1* strains are viable but unable to tolerate other mutations that would further debilitate an already sickly strain. In that latter model, *SCD1* (i.e., wild-type), strains would tolerate Δ*chc1*, whereas *scd1* alleles would be recessive to *SCD1* and lethal when combined with Δ*chc1*. That is, *scd1* and Δ*chc1* would exhibit synthetic lethality. The weight of the evidence collected to date appears to favor the latter hypothesis. First, a rather broad survey of yeast strains showed that all but one could tolerate Δ*chc1* (*55*). Second, Munn et al. found that mutations in a large number of genes (likely greater than 47) could affect the viability of Δ*chc1* cells in a manner similar to that seen for *SCD1* (*47*). These mutations, including *scd1*, were almost always recessive to wild-type; suggestive of loss of function events. These findings, coupled with the demonstration that cells were hypersensitive to various conditions of environmental stress (e.g., low pH, elevated temperatures, and challenge with low levels of amino acid analogues; *47*), strongly suggested a pattern of nonspecific genetic interactions. On this basis, Munn et al. have concluded that *CHC1* is genuinely a nonessential yeast gene (*47*). Although this result was likely not the one anticipated at the outset of the clathrin studies, the availability of viable Δ*chc1* strains permitted a direct analysis of clathrin function in yeast. Furthermore, these studies showed that clathrin plays an important role in the yeast physiological scheme as Δ*chc1* mutants are sickly (see above). Indeed, Δ*chc1* strains have the propensity to undergo an increase in ploidy; perhaps due to some growth advantage permitted by their increased genomic content (*41,42*).

The impressive evidence linking clathrin function to intracellular membrane traffic in eukaryotic cells prompted Payne et al. to survey the efficiencies at which dΔ*chc1* yeast strains execute various aspects of intracellular protein and membrane traffic (*55,56*). In particular, the abilities of such mutants to undergo what is likely to represent the yeast version of receptor-mediated endocytosis, to secrete protein, and to sort the appropriate proteins to the yeast vacuole were examined. Surprisingly, none of these processes were remarkably defective in Δ*chc1* strains.

Analysis of uptake of ^{35}S-radiolabeled α-factor was used as a measure for receptor-mediated endocytosis in Δ*chc1* versus *CHC1* strains (*56*). This assay was initially described by Chvatchko et al. (*12*) and is discussed in more detail in Section V (see above). Payne and coworkers found that α-factor uptake was reduced only some two- to threefold in the appropriate Δ*chc1* strains relative to wild-type strains (*56*). Significantly, this reduction in uptake could not be accounted for by a reduction in cell-surface α-factor receptor molecules in Δ*chc1* yeast, and this reduction in uptake did not desensitize Δ*chc1* strains from the appropriate α-factor-induced cell-cycle arrest. These data indicated that the yeast endocytic pathway remained largely functional in Δ*chc1* strains. Pulse-chase experiments indicated that the rate of secretory invertase and α-factor transport from the ER to the Golgi of Δ*chc1* strains occurred with wild-type kinetics, whereas further transport of secretory proteins from the Golgi to the yeast cell surface took approximately twice as long for a Δ*chc1* strain relative to an isogenic *CHC1* strain (*54,55*). Similar pulse-chase experiments demonstrated that the biogenesis of vacuolar proteinase precursors was also quite normal in Δ*chc1* versus *CHC1* strains (*56*). Moreover, these vacuolar proteinases were properly delivered to morphologically recognizable vacuoles in Δ*chc1* strains. It should be noted, however, that Δ*chc1* cells exhibited a clear and abnormal exaggeration of intracellular membranes. For example, the vacuoles of such cells appeared unusual in that these exhibited a fragmented and multivesicular character (*55,56*). Furthermore, the cytoplasm of Δ*chc1* cells was littered with abnormal membrane enclosed structures that have been correlated with defective Golgi bodies. Taken together, these studies suggested that clathrin was not essential for endocytosis, protein secretion, or vacuolar protein localization in yeast. However, data collected from analyses of the latter two processes implicated the yeast Golgi complex as the organelle for which clathrin function was most likely relevant.

The most penetrating insights into the role of clathrin function in yeast were reported by Payne and Schekman who discovered that Δ*chc1* cells were remarkably defective in the proteolytic maturation of the precursor form of the secreted α-factor pheromone (*54*). The secreted precursor form had failed to experience the initial processing event that is normally catalyzed by the KEX2p, an integral membrane endoprotease that has been operationally assigned to reside in yeast Golgi membranes (*22,73*). Payne and Schekman demonstrated that the inability of KEX2p to cleave the pro-α-factor was correlated with the failure of Δ*chc1* cells to retain KEX2p in the Golgi. Rather, some 75% of the total cellular KEX2p was mislocalized to the cell surface at steady-state (*54*). This was in marked contrast to wild-type cells which did not exhibit significant amounts of KEX2p at the cell surface. In light of these findings, several proposals for clathrin function have been offered (*54*). These are diagrammed in Figure 5.

First, clathrin may serve to retain certain proteins in the Golgi by excluding their packaging into Golgi-derived transport vesicles. This could be considered a mecha-

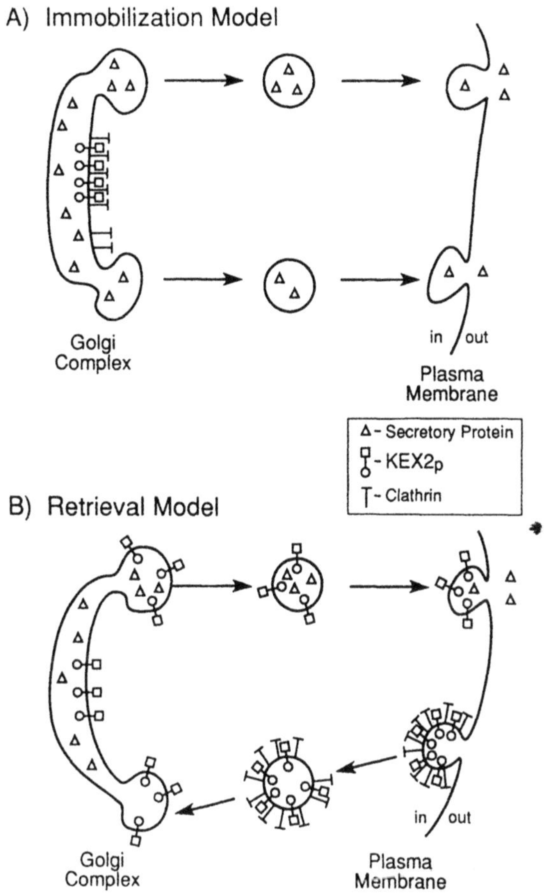

Figure 5. Proposed functions for the role of clathrin in net retention of an integral membrane protein (KEX2p) in the yeast Golgi. **A.** Retention by immobilization where a clathrin network in the Golgi traps the KEX2p. **B.** Net retention by retrieval where clathrin is a component of a very rapid and efficient scavenging pathway that retrieves escaped KEX2p from the cell surface and recycles that KEX2p population back to the Golgi system. This figure was adapted from Payne (*52*).

nism of retention by immobilization (Fig. 5A). Second, clathrin may function in an efficient recapture mechanism that retrieves escaped Golgi resident proteins from the plasma membrane; perhaps via some specialized endocytic route. Such a retrieval model is diagrammed in Figure 5B. In any event, clathrin appears to play an important role in the maintenance of Golgi structure in yeast. More precise models await the localization of clathrin to specific intracellular membranes in yeast, and the test of whether other integral membrane proteins of the Golgi are similarly misdirected to the cell surface of Δ*chc1* strains.

As alluded to above, deletion of the clathrin light-chain structural gene (*CLC1*) in yeast yielded phenotypes that were very similar to those seen for Δ*chc1* strains. Specifically, Δ*chc1* strains grew poorly and failed to proteolytically process the α-factor precursor, by analogy to Δ*chc1* strains, a phenotype suggestive of KEX2p mislocalization to the cell surface (*75*). Interestingly, Δ*chc1* and Δ*clc1* lesions were not equivalent in at least one potentially important respect. Whereas Δ*chc1*, *scd1* double mutants were inviable, Δ*clc1*, *scd1* mutants exhibited the slow growth phenotype characteristic of Δ*clc1* strains (*75*). These data suggested that residual clathrin heavy-chain function may have persisted in strains that were devoid of light-chain function. This hypothesis is in line with studies of mammalian clathrin that have shown light-chain function is not required for polymerization of triskelions into cages *in vitro* (*9*). As such, in at least one functional aspect, the light chain can be viewed in broad terms as a regulatory subunit that modulates clathrin heavy-chain function. The observation that yeast double mutants bearing both Δ*clc1* and Δ*chc1* lesions are viable, and phenotypically indistinguishable from each of the corresponding single mutants, is consistent with the notion that light-chain function in yeast is exerted through modulation of heavy-chain function (*75*). It will be of interest to determine whether coated vesicles can be recovered from Δ*clc1* yeast, and if these coats are composed of heavy chain triskelion polymers.

The finding that the most obvious pathways for protein transport appear not to be remarkably dysfunctional in clathrin-deficient yeast strains is not immediately consistent with the models formulated from studies of clathrin function in mammalian cells. How can this apparent difference be reconciled? Perhaps it cannot. The real possibility exists that clathrin fulfills a different function in yeast than it does in mammalian cells. We note, however, that a real gap in our understanding of clathrin function in yeast still exists, and is revealed by the lack of data pertaining to the localization of clathrin in the yeast cell. It is not yet known whether clathrin is restricted to the yeast Golgi, whether it decorates the cytoplasmic face of the plasma membrane, and so forth. The fact that clathrin-coated vesicles can be purified from yeast suggests an association of clathrin with some type of transport vesicle (*46*). However, the origin and destination of such vesicles is presently a mystery. The current data gleaned from Δ*chc1* mutants cannot be placed into an appropriate physiological context without these various pieces of information. Also, the consideration that yeast exhibit a redundant pathway that can absorb the loss of a clathrin-requiring pathway remains a viable alternative. It seems that a reasonable strategy to pursue might be the generation of dominant-interfering forms of mutant clathrin heavy chains. Mutant yeast expressing such "poison" forms of clathrin may exhibit different secretory or endocytic properties than do the null strains. Nevertheless, as pointed out by Payne (*52*), it is also important to remember that the considerable body of evidence implicating clathrin in lysosomal protein delivery and endocytosis is itself somewhat circumstantial. Moreover, this evidence is heavily dependent on an unambiguous assignment of the compartments that clathrin is associated with in mammalian cells. Given the current debate concerning the relationship between early and late endosomes in mammalian cells (i.e., whether these are two stable and distinct compartments or, more simply, derivatives of each other, see ref. *25* and the contribution of Murphy, this volume), it seems that making such assignments is not a trivial matter. The acid test of mammalian clathrin function *in vivo* has not yet been performed but, with the advent of gene disruption technology in mammalian cells (*80*), the tools for such an analysis are at hand. It will be those genetic studies in mammalian systems that will go a long way towards resolving whether clathrin function in yeast and mammals is similar, or not.

VII. SUMMARY

In this chapter, we have attempted to describe the recent progress that has been made in the areas of late secretory pathway function and endocytosis in yeast. The yeast system has yielded fundamental insights into the mechanisms for how GTP-binding proteins stimulate fusion of secretory vesicles to their target membrane, how a phospholipid transfer protein functions to drive late Golgi secretory function, and how proteins are sorted from the secretory pathway and delivered to the vacuole. Moreover, it now appears that reliable methods for studying endocytic processes in yeast have been developed, and one can anticipate rapid progress in this area in the future. Clearly, the role of clathrin in the yeast intracellular protein traffic program remains a question that must be resolved. Nonetheless, it is certain that further exploitation of the power afforded by the genetic and biochemical tractability of the yeast experimental system will yield even more fundamental insights into how proteins traffic into and through eukaryotic cells.

ACKNOWLEDGMENTS

Work in the authors' laboratory was supported by grants from the NIH (GM44530) and NSF (DCB9003750). We thank Sandy Henson for expert preparation of the manuscript.

REFERENCES

1. Achstetter, T., Franzusoff, A., Field, C. and Schekman, R. (1988). *SEC7* encodes an unusual, high molecular weight protein required for membrane traffic from the yeast Golgi apparatus. *J. Biol. Chem.* 263: 11711–11717.
2. Ammerer, G., Hunter, C., Rothman, J.H., Saari, G.C., Walls, L.A., and Stevens, T.H. (1986). *PEP4* gene of *Saccharomyces cerevisiae* encodes proteinase A, a vacuolar enzyme required for processing of vacuolar precursors. *Mol. Cell. Biol.* 6: 2490–2499.
3. Bankaitis, V.A., Johnson, L.M., and Emr, S.D. (1986). Isolation of yeast mutants defective in protein targeting to the vacuole. *Proc. Natl. Acad. Sci. USA* 83: 9075–9079.
4. Bankaitis, V.A., Aitken, J.F., Cleves, A.E., and Dowhan, W. (1990). An essential role for a phospholipid transfer protein in yeast Golgi function. *Nature* (Lond.) 347: 561–562.
5. Bankaitis, V.A., Malehorn, D.E., Emr, S.D., and Greene, R. (1989). The *Saccharomyces cerevisiae SEC14* gene encodes a cytosolic factor that is required for transport of secretory proteins from the yeast Golgi complex. *J. Cell Biol.* 108: 1271–1281.

6. Banta, L.M., Robinson, J.S., Klionsky, D.J., and Emr, S.D. (1988). Organelle assembly in yeast characterization of yeast mutants defective in vacuolar biogenesis and protein sorting. *J. Cell Biol.* 107: 1369–1383.
7. Banta, L.M., Vida, T.A., Herman, P.K., and Emr, S.D. (1990). Characterization of yeast Vps33p, a protein required for vacuolar protein sorting and vacuole biogenesis. *Mol. Cell Biol.* 10: 4638–4649.
8. Beckers, C.J.M., Block, M.R., Glick, B.S., Rothman, J.E., and Blach, W.E. (1989). Vesicular transport between the endoplasmic reticulum and the Golgi stack requires the NEM-sensitive fusion protein. *Nature* (Lond.) 339: 397–398.
9. Brodsky, F.M. (1988). Living with clathrin: its role in intracellular membrane traffic. *Science* (Wash. DC). 242: 1396–1402.
10. Chavrier, P., Vingron, M., Sander, C., Simons, K., and Zerial, M. (1990). Localization of low molecular weight GTP binding proteins to exocytic and endocytic compartments. *Cell* 62: 317–329.
11. Chen, M.S., Obar, R.A., Schroeder, C.C., Austin, T.W., Poodry, C.A., Wadsworth, S.C., and Vallee, R.B. (1991). Multiple forms of dynamin are encoded by *shibire*, a *Drosophilia* gene involved in endocytosis. *Nature* (Lond.). 351: 583–586.
12. Chvatchko, Y., Howard, I., and Riezman, H. (1986). Two yeast mutants defective in endocytosis are defective in pheromone response. *Cell* 46: 355–364.
13. Cleves, A.E. and Bankaitis, V.A. (1992). Secretory pathway function in *Saccharomyces cerevisiae*. In: *Advances in Microbial Physiology*, Vol. 33, pp. 73–144 (Rose, A.H., and Tempest, D.W., Eds.). Academic Press, London.
14. Cleves, A.E., McGee, T.P., and Bankaitis, V.A. (1991). Phospholipid transfer proteins: a biological debut. *Trends in Cell Biol.* 1: 30–34.
15. Cleves, A.E., Novick, P.J., and Bankaitis, V.A. (1989). Mutations in the *SAC1* gene suppress defects in yeast Golgi and yeast actin function. *J. Cell Biol.* 109: 2939–2950.
16. Cleves, A.E., McGee, T.P., Whitters, E.A., Champion, K.M., Aitken, J.R., Dowhan, W., Goebl, M., and Bankaitis, V.A. (1991). Mutations in the CDP-choline pathway for phospholipid biosynthesis bypass the requirement for an essential phospholipid transfer protein. *Cell* 64: 789–800.
17. Diaz, R., Mayorga, L.S., Weidman, P.J., Rothman, J.E., and Stahl, P.D. (1989). Vesicle fusion following receptor-mediated endocytosis requires a protein active in Golgi transport. *Nature* (Lond.). 339: 398–400.
18. Dulic, V. and Reizman, H. (1989). Characterization of the *END1* gene required for vacuole biogenesis and gluconeogenic growth of budding yeast. *EMBO J.* 8: 1349–1359.
19. Franzusoff, A. and Schekman, R. (1989). Functional compartments of the yeast Golgi apparatus are defined by the *sec7* mutation. *EMBO J.* 8: 2695–2702.
20. Franzusoff, A., Lauzè, E. and Howell, K.E. (1992). Immunnoisolation of Sec7p-coated transport vesicles from the yeast secretory pathway. *Nature* (Lond.) 355: 173–175.
21. Franzusoff, A., Redding, K., Crosby, J., Fuller, R.S., and Schekman, R. (1991). Localization of components involved in protein transport through the yeast Golgi apparatus. *J. Cell Biol.* 112:27–37.
22. Fuller, R.S., Sterne, R.E., and Thorner, J. (1988). Enzymes required for yeast prohormone processing. *Annu. Rev. Physiol.* 50: 345–362.
23. Goldstein, J.L., Brown, M.S., Anderson, R.G.W., Russell, D.W. and Schneider, W.J. (1985). Receptor-mediated endocytosis: concepts emerging from the LDL receptor system. *Annu. Rev. Cell Biol.* 1: 1–39.
24. Goud, B., Salminen, A., Walworth, N.C., and Novick, P.J. (1988). A GTP-binding protein required for secretion rapidly associates with secretory vesicles and the plasma membrane in yeast. *Cell* 53:753–768.
25. Griffiths, G. and Gruenberg, J. (1991). The arguments for pre-existing early and late endosomes. *Trends in Cell Biol.* 1: 5–9.

26. Helenius, A. (1984). Semliki Forest virus penetration from endosomes: a morphological study. *Biol. Cell* 51: 181–186.
27. Helenius, A., Kartenbeck, J., Simons, K., and Fries, E. (1980). On the entry of Semliki Forest virus into BHK-21 cells. *J. Cell Biol.* 84: 404–420.
28. Helmkamp, G.M., Jr. (1986). Phospholipid transfer proteins: mechanisms of action. *J. Bioenerg. Biomembr.* 18: 71–90.
29. Herman, P.K., Stack, J.H., DeModena, J.A., and Emr, S.D. (1991). A novel protein kinase homolog essential for protein sorting to the yeast lysosome-like vacuole. *Cell* 64: 425–437.
30. Jenness, D.D. and Spatrick, P. (1986). Down regulation of the α-factor pheromone receptor in *S. cerevisiae*. *Cell* 46: 345–353.
31. Jenness, D.D., Burkholder, A.C., and Hartwell, L.H. (1986). Binding of α-factor pheromone to *Saccharomyces cerevisiae* a cells: dissociation constant and number of binding sites. *Mol. Cell. Biol.* 6: 318–320.
32. Johnson, L.M., Bankaitis, V.A., and Enr, S.D. (1987). Distinct sequence determinants direct intracellular sorting and modification of a yeast vacuolar protein. *Cell* 48: 875–885.
33. Jones, E.W. (1977). Proteinase mutants of *Saccharomyces cerevisiae*. *Genetics* 85: 23–33.
34. Jones, E.W. (1984). The synthesis and function of proteases in *Saccharomyces cervisiae*: genetic approaches. *Annu. Rev. Genet.* 18: 233–270.
35. Kabcenell, A.K., Goud, B., Northrup, J.K., and Novick, P. (1990). Binding and hydrolysis of guanine nucleotides by Sec4p, a yeast protein involved in the regulation of vesicular traffic. *J. Biol. Chem.* 265: 9366–9372.
36. Kane, P.M., Yamashiro, C.T., and Stevens, T.H. (1989). Biochemical characterization of the yeast vacuolar H^+-ATPase. *J. Biol. Chem.* 264: 19236–19244.
37. Kirchhausen, T. and Harrison, S.C. (1981). Protein organization in clathrin trimers. *Cell* 23:755–761.
38. Klar, A.J.S., Strathern, J.N., and Hicks, J.B. (1984). Developmental pathways in yeast. In: *Microbial Development*, pp. 151–195. (Losick, R., Ed.). Cold Spring Harbor Laboratory, Cold Spring Harbor, NY.
39. Klionsky, D.J. and Emr, S.D. (1990). A new class of lysosomal/vacuolar protein sorting signals. *J. Biol. Chem.* 265: 5349–5352.
40. Klionsky, D.J., Herman, P.K., and Emr, S.D. (1990). The fungal vacuole: composition, function, and biogenesis. *Microbiol. Rev.* 54: 266–292.
41. Lemmon, S.K. and Jones, E.W. (1987). Clathrin requirement for normal growth of yeast. *Science* (Wash. DC).238: 504–509.
41a. Lemmon, S.K., Pellicena-Palle, A., Conley, K., and Freund, C.L. (1991). Sequence of the clathrin heavy chain from *Saccharomyces cerevisiae* and requirement of the COOH terminus for clathrin function. *J. Cell Biol.* 112: 65–80.
42. Lemmon, S.K., Freund, C., Conley, K., and Jones, E.W. (1990). Genetic instability of clathrin-deficient strains of *Saccharomyces cerevisiae*. *Genetics* 124: 27–38.
43. Makarow, M. (1985). Endocytosis in *Saccharomyces cerevisiae*: internalization of enveloped viruses into spheroplasts. *EMBO J.* 4:1855–1860.
44. Makarow, M. (1985). Endocytosis in *Saccharomyces cerevisiae*: internalization of α-amylase and fluorescent dextran into cells. *EMBO J.* 4: 1861–1866.
45. Makarow, M. and Nevalainen, L.T. (1987). Transport of a fluorescent macromolecule via endosomes to the vacuole in *Saccharomyces cerevisiae*. *J. Cell Biol.* 104:67–75.
46. Mueller, S.C. and Branton, D. (1984). Identification of coated vesicles in *Saccharomyces cerevisiae*. *J. Cell Biol.* 98: 341–346.
47. Munn, A.L., Silveira, L., Elgort, M., and Payne, G.S. (1991). Viability of clathrin heavy-chain-deficient *Saccharomyces cerevisiae* is compromised by mutations at numerous loci: implications for the suppression hypothesis. *Mol. Cell. Biol.* 11: 3868–3878.

48. Nair, J., Müller, H., Peterson, M., and Novick, P.J. (1990). Sec2 protein contains a coiled-coil domain essential for vesicular transport and a dispensible carboxy terminal domain. *J. Cell Biol.* 1897–1909.
49. Novick, P., Ferro, S., and Schekman, R. (1981). Order of events in the yeast secretory pathway. *Cell* 25: 461–469.
50. Novick, P., Field, C., and Schekman, R. (1980). Identification of 23 complementation groups required for posttranslational events in the yeast secretory pathway. *Cell* 21: 205–215.
51. Palade, G. (1975). Intracellular aspects of the process of protein secretion. *Science* (Wash. DC) 189: 347–358.
52. Payne, G.S. (1990). Genetic analysis of clathrin function in yeast. *J. Membr. Biol.* 116: 93–105.
53. Payne, G.S. and Schekman, R. (1985). A test of clathrin function in protein secretion and cell growth. *Science* (Wash. DC). 230: 1009–1014.
54. Payne, G.S. and Schekman, R. (1989). Clathrin: a role in the intracellular retention of a Golgi membrane protein. *Science* (Wash. DC). 245: 1358–1365.
55. Payne, G.S., Hasson, T.B., Hasson, M.S., and Schekman, R. (1987). Genetic and biochemical characterization of clathrin-deficient *Saccharomyces cerevisiae. Mol. Cell. Biol.*7: 3888–3898.
56. Payne, G.S., Baker, D., van Tuinen, E., and Schekman, R. (1988). Protein transport to the vacuole and receptor-mediated endocytosis by clathrin heavy chain-deficient yeast. *J. Cell Biol.* 106: 1453–1461.
57. Pearse, B.M.F. and Bretscher, M. (1981). Membrane recycling by coated vesicles. *Annu. Rev. Biochem.* 50: 85–101.
58. Preston, R.A., Murphy, R.F., and Jones, E.W. (1987). Apparent endocytosis of fluorescein isothiocyanate-conjugated dextran by *Saccharomyces cerevisiae* reflects uptake of low molecular weight impurities, not dextran. *J. Cell Biol.* 105: 1981–1987.
59. Preston, R.A., Murphy, R.F., and Jones, E.W. (1987). Assay of vacuolar pH in yeast and identification of acidification-defective mutants. *Proc. Natl. Acad. Sci. USA* 86:7207–7031.
60. Preston, R.A., Manolson, M.F., Becherer, K., Weidenhammer, E., Kirkpatrick, D., Wright, R., and Jones, E.W. (1991). Isolation and characterization of *PEP3,* a gene required for vacuolar biogenesis in *Saccharomyces cerevisiae. Mol. Cell. Biol.* 11: 5801–5812.
61. Raymond, C.K., O'Hara, P.J., Eichinger, G., Rothman, J.H., and Stevens, T.H. (1990). Molecular analysis of the yeast *VPS3* gene and the role of its product in vacuolar protein sorting and vacuolar segregation during the cell cycle. *J. Cell Biol.* 111: 877–892.
62. Riezman, H. (1985). Endocytosis in yeast: several of the yeast secretory mutants are defective in endocytosis. *Cell* 40: 1001–1009.
63. Robinson, J.S., Graham, T.R., and Emr, S.D. (1991). A putative zinc finger protein, *Saccharomyces cervisiae* Vps18p, affects late Golgi functions required for vacuolar protein sorting and efficient α-factor prohormone maturation. *Mol. Cell. Biol.* 11: 5813–5824.
64. Robinson, J.S., Klionsky, D.J., Banta, L.M., and Emr, S.D. (1988). Protein sorting in *Saccharomyces cerevisiae*: isolation of mutants defective in the delivery and processing of multiple vacuolar hydrolases. *Mol. Cell. Biol.* 8: 4936–4948.
65. Rossi, G., Jiang, Y., Newman, A.P., and Ferro-Novick, S. (1991). Dependence of Yptl and Sec4 membrane attachment on Bet2. *Nature* (Lond.). 351: 158–161.
66. Rothman, J.E. and Orci, L. (1990). Movement of proteins through the Golgi stack: a molecular dissection of vesicular transport *FASEB J.* 4: 1460–1468.
67. Rothman, J.H. and Stevens, T.H. (1986). Protein sorting in yeast: mutants defective in vacuole biogenesis mislocalize vacuolar proteins into the late secretory pathway. *Cell* 47: 1041–1051.
68. Rothman, J.H., Yamashiro, C.T., Kane, P.M., and Stevens, T.H. (1989). Protein targeting to the yeast vacuole. *Trends in Biochem. Sci.* 14: 347–350.

69. Rothman, J.H., Raymond, C.K., Gilbert, T., O'Hara, P.J., and Stevens, T.H. (1990). A putative GTP binding protein homologous to interferon-inducible Mx proteins performs an essential function in yeast protein sorting. *Cell* 61: 1063–1074.
70. Rothman, J.H., Yamashiro, C.T., Raymond, C.K., Kane, P.M., and Stevens, T.H. (1989). Acidification of the lysosome-like vacuole and the vacuolar H^+-ATPase are deficient in two yeast mutants that fail to sort vacuolar proteins. *J. Cell Biol.* 109: 93–100.
71. Salminen, A. and Novick, P.J. (1987). A *ras*-like protein is required for a post-Golgi event in yeast secretion. *Cell* 49: 527–538.
72. Salminen, A. and Novick, P. (1989). The Sec15 protein responds to the function of the GTP binding protein, Sec4, to control vesicular traffic in yeast. *J. Cell Biol.* 109:1023–1036.
73. Schekman, R. (1985). Protein localization and membrane traffic in yeast. *Annu. Rev. Cell Biol.* 1: 115–143.
74. Segev, N., Mulholland, J., and Botstein, D. (1988). The yeast GTP-binding YPT1 protein and a mammalian counterpart are associated with the secretion machinery. *Cell* 52: 915–924.
75. Silveira, L.A., Wong, D.H., Masiarz, F.R., and Schekman, R. (1990). Yeast clathrin has a distinctive light chain that is important for cell growth. *J. Cell Biol.* 111: 1437–1449.
76. Singer, B. and Riezman, H. (1990). Detection of an intermediate compartment involved in transport of α-factor from the plasma membrane to the vacuole in yeast. *J. Cell Biol.* 110: 1911–1922.
77. Sprague, G.F. Jr., Blair, L.C., and Thorner, J. (1983). Cell interactions and regulation of cell type in the yeast *Saccharomyces cerevisiae*. *Annu. Rev. Microbiol.* 37: 623–660.
78. Stearns, T., Willingham, M.C., Botstein, D., and Kahn, R.A. (1990). ADP-ribosylation factor is functionally and physically associated with the Golgi complex. *Proc. Natl. Acad. Sci. USA* 87: 1238–1242.
79. Stevens, T., Esmon, B., Schekman, R. (1982). Early stages in the yeast secretory pathway are required for transport of carboxypeptidase Y to the vacuole. *Cell* 30: 439–448.
80. Thomas, K.R. and Capecchi, M.R. (1987). Site-directed mutagenesis by gene targeting in mouse embryo-derived stem cells. *Cell* 51: 503–512.
81. Ungewickell, E. and Branton, D. (1981). Assembly units of clathrin coats. *Nature* (Lond.). 282: 420–422.
82. Valls, L.A., Hunter, C.P., Rothman, J.H. and Stevens, T.H. (1987). Protein sorting in yeast the localization determinant of yeast vacuolar carboxypeptidase Y resides in the propeptide. *Cell* 48: 887–897.
83. Walworth, N.C., Goud, B., Kabcenell, A.K., and Novick, P. (1989). Mutational analysis of *SEC4* suggests a cyclical mechanism for the regulation of vesicular traffic. *EMBO J.* 8: 1685–1693.
84. Wiemken, A., Schellenberg, M., and Urech, K. (1979). Vacuoles: the sole compartments of digestive enzymes in yeast (*Saccharomyces cerevisiae*). *Arch. Microbiol.* 123: 23–25.
85. Wilson, D.W., Wilcox, C.A., Flynn, G.C., Chen, E., Kuang, W-J., Henzel, W.J., Block, M.R., Ullrich, A., and Rothman, J.E. (1989). A fusion protein required for vesicle-mediated transport in both mammalian cells and yeast. *Nature* (Lond.). 339: 355–359.
86. Woolford, C.A., Daniels, L.G., Park, F.J., Jones, E.W., van Arsdell, J.N., and Innis, M.A. (1986). The *PEP4* gene encodes an aspartyl protease implicated in the posttranslational regulation of *Saccharomyces cerevisiae* vacuolar hydrolases. *Mol. Cell. Biol.* 6: 2500–2510.
87. Yamashiro, C.T., Kane, P.M., Wolczyk, D.F., Preston, R.A., and Stevens, T.H. (1990). Role of vacuolar acidification in protein sorting and zymogen activation: a genetic analysis of the yeast vacuolar proton-translocating ATPase. *Mol. Cell. Biol.* 10: 3737–3749.

REGULATION OF LYSOSOMAL TRAFFICKING AND FUNCTION DURING GROWTH AND DEVELOPMENT OF *DICTYOSTELIUM DISCOIDEUM*

James A. Cardelli

Advances in Cell and Molecular Biology of Membranes
Volume 1, pages 341–390

ISBN: 1-55938-362-3

ABSTRACT

The mechanisms regulating the trafficking of proteins in eukaryotic cells are not completely understood, although this process is fundamentally important to ensure the proper organization and function of all intracellular organelles. The targeting of lysosomal enzymes in human fibroblasts and other higher eukaryotic cells is currently one of the better described processes. In these cells, lysosomal enzymes are phosphorylated on mannose residues in N-linked oligosaccharides, which act as lysosomal targeting signals. However, many animal cells as well as plants, yeast, and *Dictyostelium discoideum* cells sort lysosomal enzymes by an alternative phosphomannosyl independent route. *D. discoideum* has proven to be an excellent system to study alternative sorting mechanisms because of the relative ease of application of genetics, cell biology, and biochemistry. An added advantage of the organism is its relatively simple developmental cycle during which the synthesis and modification of lysosomal enzymes are regulated and the lysosomal system undergoes functional and morphological changes. These studies have led to the following conclusions: (1) The enzymes are synthesized as glycosylated membrane associated precursor polypeptides which are cleaved to mature soluble forms; (2) sulfate and phosphate on N-linked oligosaccharides do not play a role in sorting or transport; (3) acidic pH compartments are not required for correct sorting of hydrolases but are required for complete proteolytic processing; (4) proteolytic processing of precursors may be required for sorting to lysosomes; (5) starvation of cells induces the secretion of lysosomal enzymes; and (6) during development genes coding for lysosomal enzymes are transcriptionally regulated, and lysosomes are altered to function more like classical secretory granules.

I. INTRODUCTION

The biosynthesis and targeting of lysosomal enzymes have been well characterized in a few mammalian systems such as human fibroblasts (reviewed in ref. *117*). Enzymes in these cells are synthesized on membrane-bound polysomes as precursor (prepro) polypeptides which are glycosylated and proteolytically cleaved (to remove the signal or "pre"-peptide) following translocation into the lumen of the ER. Following transport of these proteins to a pre-Golgi or *cis*-Golgi compartment, their N-linked oligosaccharide side-chains are phosphorylated on mannose residues in a reaction critically dependent on protein conformation. These phosphomannosyl residues act as sorting signals and are recognized by either the cation-dependent or cation-independent mannose 6-phosphate receptor (CD-MPR or CI-MPR, respectively). These receptors mediate transport and sorting of the phosphorylated hydrolases to prelysosomal compartments where acidic pH conditions facilitate the release of the ligands from the receptors allowing recycling of these receptors back to the Golgi. The enzymes are subsequently packaged into lysosomes and proteolytically processed to mature forms. Fibroblasts from patients suffering from the disease mucolipidosis-II (I-cell disease) are missing the enzyme responsible for phosphorylation, and as a result, many of the lysosomal enzymes are missorted and secreted in precursor form. This observation and variety of other results strongly support the model described above.

It has also become clear that alternative (MPR-independent) mechanisms exist for lysosomal enzyme targeting. For instance, not all lysosomal enzymes in I-cell fibroblasts are missorted. Some of these enzymes (β-glucocerebrosidase and acid phosphatase) are membrane-associated during much of their existence (*53,208*) and are correctly localized (*122,152,201*). Furthermore, other cells in I-cell patients such as hepatocytes and leukocytes contain normal intracellular levels of those soluble lysosomal enzymes which are missorted in I-cell fibroblasts (*116,161,207*). Unlike the MPR-dependent pathway, the MPR-independent targeting mechanisms operating in these cells has yet to be defined. Furthermore, an examination of lysosomal enzyme targeting in plants (*30*), and lower eukaryotes such as *Saccharomyces cerevisiae* (*111*), and *Dictyostelium discoideum* (*24*) has indicated a mechanism for targeting of hydrolases that does not rely on phosphorylated carbohydrate moieties. Thus, alternative targeting mechanisms (of which we know very little) may play a more universal role in lysosomal trafficking than previously thought.

The simple eukaryote, *Dictyostelium discoideum* (which contains no detectable MPRs but has a secretory pathway similar to higher eukaryotic cells), has proven to be a particularly useful system to investigate the mechanisms regulating lysosomal enzyme synthesis, processing, and sorting because of the relative ease of application of genetic and biochemical methods (reviewed in refs. *24,27*). Furthermore, these same methods are being applied to this organism to investigate other more universal processes pertaining to the lysosomal system including endocytosis, autophagocytosis, and secretion. *D. discoideum* also has a relatively simple developmental cycle (culminating in the formation of two differentiated cell types), and many of the genes coding for lysosomal proteins are developmentally regulated (*133,136*). Finally, the amoeboid cells more closely resemble animal cells than do yeast and plant cells because the cellular slime mold lacks a cell wall and contains lysosomes that appear similar to mammalian lysosomes in morphology. Thus, *D. discoideum* is yielding useful information concerning a wide variety of different aspects of the lysosomal system.

In this chapter, I will review what is currently known about the synthesis, transport, processing, localization, and secretion of *D. discoideum* lysosomal enzymes during growth and development of the organism. In addition to this general description of the lysosomal transport pathways, I will also discuss the role of N-linked oligosaccharides, vesicular pH, proteolytic processing, and precursor membrane association in the transport, sorting and secretion of hydrolases. Finally, I will describe the changes that occur to the lysosomal system during development. This last section will include a review of three topics: (1) the regulation of genes coding for lysosomal enzymes; (2) the changes in posttranslation modifications; and (3) the restructuring of lysosomes late in development.

II. FACTORS REGULATING THE INTRACELLULAR TRANSPORT AND SORTING OF LYSOSOMAL ENZYMES

In its natural habitat, *Dictyostelium discoideum* (also referred to as the cellular slime mold because of the mucopolysaccharides it secretes during development) exists as solitary amoebae which feed by phagocytosis. In the laboratory, the wild-type strains (such as NC-4) grow on plates or in suspension using bacteria as a food source, while mutant derivatives (axenic strains Ax-2 and Ax-3) are capable of growth in liquid media because they have acquired the ability to pinocytose nutrients.

A. Description of the Lysosomal and Endosomal Transport Pathways

Synthesis and Transport of Lysosomal Hydrolases

The three best characterized lysosomal enzymes in *D. discoideum* (acid phosphatase, α-mannosidase, and β-glucosidase) are all synthesized in the RER as N-glycosylated precursor polypeptides analogous to what has been observed for most mammalian lysosomal enzymes (Table 1). *In vitro* translation studies (*22,23*) have identified the primary translation products for two of these enzymes, α-mannosidase (120 kDa) and β-glucosidase (94 kDa). Furthermore, the co-translational addition of dog pancreas microsomes results in the formation of higher molecular weight precursor polypeptides (140 kDa for α-mannosidase and 105 kDa for β-glucosidase) identical in molecular weight to the forms observed *in vivo* following a brief pulse with radiolabeled amino acids. Membranes added posttranslationally have no effect. These polypeptides synthesized *in vitro* reside in the lumen of the ER and contain the same number of N-linked oligosaccharides as their cellular counterparts. Polypeptides similar in molecular weight to the primary translation products are produced when the *in vivo* pulse labeled and *in vitro* modified precursors are digested with *endo*-β-acetylglucosaminidase H (*endo* H), an enzyme which removes the high mannose form of N-linked oligosaccharides. Although these data suggest the signal sequence is not removed during translocation of these proteins into the lumen of the ER, current studies (Section II-B) support the hypothesis that signal sequence cleavage does occur. Both direct amino acid terminal sequence analysis and more indirect approaches indicate that the signal sequence is cleaved from lysosomal enzymes in animal cells following translocation into the ER (*52,172,180*). In summary, the initial events in the synthesis of *D. discoideum* lysosomal enzymes are similar to what has been observed for the synthesis of mammalian lysosomal enzymes: proteins are synthesized as glycosylated preproforms in the ER followed by the removal of the signal sequence to generate the proforms.

A unique feature of the three best characterized lysosomal enzyme precursors in *D. discoideum* is that immediately following translocation into the lumen of the ER the proteins tightly associate with membranes (*12,23,148*). This interaction is

Table 1. Properties of *Dictyostelium discoideum* Lysosomal Enzymes and Genes

Enzyme	Actual or Predicted (p) Molecular Mass (kDa)			Actual or Predicted # of N-linked Oligosacch.	Half-time (mins) for:		Molec.[a] Studies	Dev. Induction[b] Pattern	Ref.
	Holoenzyme	Proform	Mature		Transport ER to the Golgi	Conversion Proform to Mature			
α-man	220–280	140	60+58	9–10	13	35	cDNA/gene	VE	1
β-glu	220	105	100	5–6	4	22	cDNA	E/L	2
acid phos	115–135	56	55	5–6	4	10			3
β-hexos	140–168		66	2–3			cDNA	VE	4
α-glucos	150		84						5
α-fucos		50(p)	50	5–6(p)			cDNA/gene	E	6
Protein I	38								7
Cys Prot-I		37(p)					cDNA/gene	M	8
Cys Prot-II		40(p)		3(p)			cDNA/gene	M	9

Notes: [a]The cDNA and/or the structural gene has been cloned and characterized.

[b]VE = very early; induction in gene expression during preaggregation within the first 4 hours of development; E = early, induction in gene expression during the aggregation phase from 4–12 hours; M = middle, induction during the slug stage from 10–16 hours; L = late, induction during culmination from 18–24 hours.

1. α-Mannosidase: Free and Loomis (1974); Free et al. (1976); Livi et al. (1985a); Cardelli et al. (1986b); Mierendorf et al. (1983, 1985); Pannell et al. (1982); Schatzle et al. (1991a, 1991b).
2. β-Glucosidase: Dimond and Loomis (1976); Loomis (1980b); Cardelli et al. (1986b); Golumbeski and Dimond (1987); Richardson (1991).
3. Acid phosphatase: Loomis and Kuspa (1984); Bennett and Dimond (1986); Bush and Cardelli (1989); Bush et al. (1990).
4. β-Hexosaminidase: Dimond et al. (1973); Loomis (1978); Judelson et al. (1987); Graham et al. (1988).
5. α-Glucosidase: Every and Ashworth (1973); Borts and Dimond (1981a,b).
6. α-Fucosidase: Muller-Taubenberger et al. (1989); May et al. (1989).
7. Proteinase I/B: North (1985); North and Whyte (1984); Gustafson and Thon (1979).
8. Cysteine Proteinase-I: Williams et al. (1985).
9. Cysteine Proteinase-II: Pears et al. (1985); Pears and Williams (1987); Datta and Firtel (1987).

maintained until proteins are proteolytically processed in a late Golgi/endosomal compartment (explored at greater length in Section II-B). The membrane-associated precursor polypeptides are transported at different rates to the Golgi complex (Table 1) where the N-linked oligosaccharides are phosphorylated and sulfated on mannose residues. The Golgi complex is undefined morphologically in growing *D. discoideum* cells compared to animal cells; however, the lectin WGA used to visualize the Golgi in animal cells also labels a perinuclear region in the slime mold (*104*). Interestingly, developing *D. discoideum* cells do have prominent Golgi stacks (*186*). However, some of the carbohydrate processing enzymes found in mammalian Golgi are missing (*71*), although a low-density membrane fraction can sulfate and phosphorylate mannose residues *in vitro* similar to what is observed for the mammalian Golgi (*73,101,119*). Finally, soluble factors prepared from the slime mold can mediate transport of proteins through the Golgi stacks in Chinese Hamster Ovary cells *in vitro* (*166*), suggesting that the *D. discoideum* Golgi is functionally similar to Golgi membranes in higher eukaryotes.

The different rates of transport of *D. discoideum* lysosomal enzymes are consistent with the observations of others, indicating that individual members within the secretory, membrane, and lysosomal protein groups in animal cells are transported from the ER to the Golgi with different kinetics (*5,57,77,131*). Transport of proteins from the ER most likely does not require a receptor, suggesting different exit rates may be the result of inherently different rates of folding or oligomerization of proteins to achieve a transport competent state (*103,169*). Changes in the conformation of *D. discoideum* lysosomal enzymes do, in fact, have a dramatic effect on transport kinetics. For instance, the α-mannosidase structural gene mutant, *HMW437*, synthesizes an apparently normal size α-mannosidase precursor that remains in the ER (*215*). Sedimentation velocity experiments and protease analysis *in vitro* have indicated the precursor is altered in conformation which may account for why it is retained in the ER following synthesis. In fact, many proteins genetically altered in amino acid sequence are retained in the ER presumably as a result of improper folding or the inability to oligomerize (*103,169*). In many cells, improperly folded proteins may be retained in the ER due to association with the ER localized protein BiP (a member of the heat shock protein family); however, this has not been examined in the slime mold.

N-linked Glycosylation and Posttranslational Modifications

The glycosylation of lysosomal enzymes in the lumen of the ER of *D. discoideum* occurs in a manner very similar to that observed for mammalian cells (*74*). The N-linked oligosaccharides are derived from a lipid-linked precursor oligosaccharide composed of three glucose residues, nine mannose residues, and two N-acetylglucosamine residues which are bound to dolichol pyrophosphate (*76*). Following transfer of the oligosaccharide to asparagine residues, the glucose residues are removed by two different α-glucosidases residing in the ER. One or two mannose residues are then removed by a neutral α-mannosidase which is found

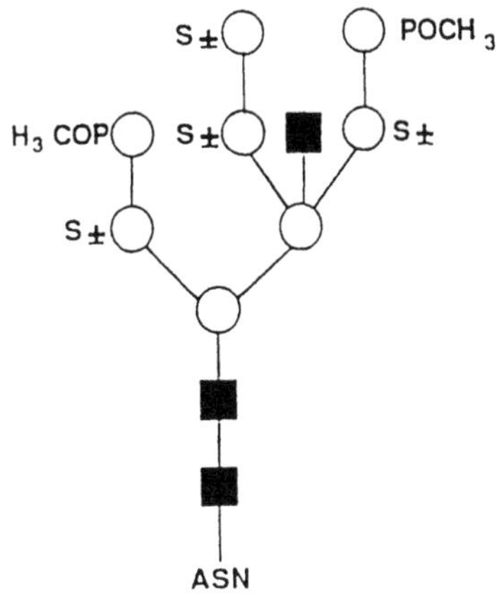

Figure 1. Structure of N-linked oligosaccharides found in *Dictyostelium.* ●, Man; ■, GlcNAc; and S, SO_4. See Freeze (1986) and Freeze et al (1990) for furthermore details.

in the ER and/or the Golgi (*188*). In mammalian cells, the N-linked oligosacccharide is often extensively processed resulting in the removal of additional mannose residues and the addition of other peripheral sugars such as fucose, galactose, and sialic acid (*115*). These modifications do not occur to glycoproteins in growing *D. discoideum* cells but the oligosaccharides are subjected to other forms of posttranslational modifications as described below (Fig. 1).

Phosphorylation. D. discoideum lysosomal enzymes contain mannose 6-phosphate on their N-linked oligosaccharides (Fig. 1) which is potentially important because this modification acts as a targeting signal for many lysosomal enzymes in mammalian cells (*117*). In fact, slime mold enzymes are excellent ligands for the mammalian CI-MPR (*66,68*) and have been used in affinity column chromatography to purify the receptor (*56,81*); however, no detectable MPRs have been found in this organism.

In mammalian cells, GlcNac-1-phosphate is first added to the 6 position of mannose to generate an acid-sensitive phosphodiester bond which is cleaved in the Golgi by α-N-acetyl glucosaminidase (*116*). A similar GlcNac-1-phosphate transferase has been identified in *D. discoideum* (*35,120*), but there is no evidence for the presence of the uncapping enzyme α-N-acetylglucosaminidase. Such an enzyme most likely exists since mannose 6-phosphate residues in *D. discoideum* are found as acid-stable methylated phosphodiesters in contrast to the acid labile esters in animal cells (*36,80*). The methyl group is derived from S-adenosylmethionine (*73*) but the biochemical nature of the reaction remains unknown, although an *in vitro* assay has recently been developed (Freeze, personal communication) and should prove useful in the description of this process.

Sulfation. The lysosomal enzyme precursor polypeptides are sulfated immediately upon arrival in Golgi membranes (Fig. 1; 148). Freeze et al. (64) have shown that the sulfate esters occur as mannose 6-sulfate, and a cell free system has been developed which can mimic the *in vivo* reaction (*101,119*). On average, each N-linked oligosaccharide contains three sulfate groups (and two phosphate groups) which accounts for the extremely anionic nature of the lysosomal enzymes (*64,67,72,113*). These mannose 6-sulfate residues are also very immunogenic in rabbits and mice yielding polyclonal antibodies which cross-react with most lysosomal enzymes (*42,108,112*). The epitope consisting of a specific arrangement of mannose 6-sulfate on the α-1,6 branch (*70,108*) has been termed Common Antigen 1 (CA-1), and is found on all lysosomal enzymes and 3 to 5% of total cellular proteins (*112,113*). It is expressed preferentially on secreted, acidic lysosomal glycoproteins although nonlysosomal enzymes, such as the plasma membrane localized cyclic nucleotide phosphodiesterase, also contain this epitope.

Sulfated glycoproteins are not unique to slime molds. They have also been identified as members of the group of bovine pituitary polypeptide hormones (*89*), in sea urchin embryos (*97*), and as members of rat kidney-basement membrane proteins (*123*). Others have also reported that mammalian lysosomal enzymes can be sulfated on mannose and tyrosine residues although the importance of this modification remains unknown (*206*).

Other modifications. A novel "intersecting" GlcNac residue (Fig. 1) on N-linked oligosaccharides in *D. discoideum* has been identified and the enzyme responsible for this modification has been partially characterized (*36,189*). Furthermore, peripheral and core fucose residues have been reported (*98,105*), and Gustafson and Milner (*92*) have reported that lysosomal enzymes may be phosphorylated on serine residues. Therefore, lysosomal enzymes in *D. discoideum* undergo a variety of postranslational modifications some of which are unique and others which are conserved among a variety of organisms.

Proteolytic Processing of Enzymes

The three best characterized lysosomal enzymes in *D. discoideum* are all synthesized as glycosylated precursors that are cleaved through intermediate forms to mature soluble enzymes localized to lysosomes (*24,27*). α-Mannosidase is synthesized as a 140-kDa precursor that is processed with a half-life of 25 to 30 minutes to a mature form of 58 kDa, and multiple intermediate forms of 80 to 84 kDa (*147,148,165*). The 60-kDa mature polypeptide is cleaved (half-time of formation 35 min) from this population of multiple intermediate polypeptides (*21*). β-Glucosidase is processed from a precursor form of 105 kDa through a 103-kDa intermediate (half-time of processing 10 min) to a mature form of 100 kDa (half-time of formation 20 min; *21*). Finally, acid phosphatase is synthesized as a 56-kDa precursor that is processed through an intermediate form to a 55-kDa mature form (half-time of formation of 10 min; *17*). The decrease in molecular mass for all these proteins is the result of proteolytic processing and is not the result of modifications to N-linked oligosaccharides. Newly synthesized acid phosphatase and β-glucosidase are cleaved faster then α-mannosidase because they are transported more quickly from the ER to the compartment where cleavage occurs (Table 1). In contrast, β-hexosaminidase is apparently not cleaved before or after arrival in lysosomes (*88*). Most lysosomal enzymes in animal cells, plant cells, and yeast are also proteolytically processed; however, the rate of processing of enzymes in animal cells is 10- to 100-fold slower than in *D. discoideum* (*95,190*).

The kinetics and pathways of processing of *Dictyostelium* and animal cell lysosomal enzymes are well described; however, less is known about the biochemistry of the proteinases involved. Treatment of *D. discoideum* cells with cysteine proteinase inhibitors results in the intracellular accumulation of intermediate forms of α-mannosidase and β-glucosidase (*174*), a result also observed for the mammalian enzymes cathepsin D, α-glucosidase, α-N-acetylglucosaminidase, and aryl sulfatase (*99,203*). In addition, raising the pH of endosomes and lysosomes (Section II-2) with ammonium chloride also results in the intracellular accumulation of intermediate forms of *D. discoideum* lysosomal hydrolases (*25*), and under both conditions, the intermediate forms accumulate in vesicles of density equal to that of the vesicles containing lysosomal enzyme activity. Furthermore, in untreated cells these same dense hydrolase-containing vesicles appear to be the location where radiolabeled mature forms first appear which, together with the above results, suggests the final proteolytic processing event to generate mature enzymes is catalyzed by cysteine proteinases residing in mature lysosomes.

Treatment of purified *D. discoideum* and mammalian lysosomal enzyme precursors and intermediate forms with trypsin or with lysosomal extracts results in the generation of mature enzymes (*78,83,215*, unpublished results). The result observed using trypsin is not unexpected because cysteine proteinases also cleave at lysine and arginine residues. In fact, the amino terminal sequence of many mature lysosomal enzymes is usually preceded by monobasic or dibasic amino acids in the precursor polypeptide (Section II-B). In all of the cases examined, the final processing events using lysosomal extracts is prevented by the addition of sulhydryl reagents supporting the role of lysosomally localized cysteine proteinases.

Multiple acid proteinases have been identified in *D. discoideum* (*154*) and two different cysteine proteinases have been purified from growing cells (*93,155*). Furthermore, the genes coding for the cysteine-like proteinases (proteinase-I and proteinase-II) have been cloned (*167*), and the mechanisms regulating their expression have been analyzed; however, the role of these enzymes in lysosomal enzyme processing remains to be determined. In Section II-B, I will discuss recent data concerning the biochemical properties and intracellular location of the proteinases which are involved in processing precursor polypeptides; an event which may be important in sorting.

Two Transport Pathways for Secretion of Enzymes

Lysosomal enzymes can be secreted from *D. discoideum* cells by two different pathways; one appears to be constitutive, while the other is regulated by a variety of genetic, developmental, and nutritional factors (*21,27,148*). In the constitutive pathway, a small percentage (5–10%) of newly synthesized precursor polypeptides bypass lysosomes and are rapidly secreted from cells. A variable portion of lysosomal enzyme precursors in human fibroblasts (*94*), smooth muscle cells (*96*), hepatocytes (*180*), and endothelial cells (*96*) are also secreted. The most likely explanation for this phenomena is that sorting to lysosomes is not 100% efficient, and a variable portion of newly synthesized precursors exit cells following a "default" constitutive secretion pathway. Inefficiency in sorting may be the result of the targeting machinery (perhaps receptors) being saturated, as revealed by the missorting of yeast hydrolases that are overexpressed (*111*).

Mature forms of lysosomal enzymes are also secreted in a regulated fashion from *D. discoideum* cells (*3,16,45,148,165*) and a variety of higher eukaryotic cells (including mast cells, neutrophils, macrophages, kidney cells; reviewed in ref. *106*) and lower eukaryotic cells [including *Acanthamoebae castellannii* (*102*) and *Tetrahymena pyriformis* (*7*)]. During growth of *D. discoideum*, newly processed mature enzymes begin to be secreted 2 to 4 hours after synthesis (*148,213*). The extracellular concentration of enzyme activity can approach 70% of the total enzyme activity in exponentially growing cells, and cells secrete greater than 90% of their total enzymes when they reach stationary phase (*3,16*). Newly synthesized mature forms of the enzymes apparently pass through a nonsecretory compartment (perhaps newly formed lysosomes) before reaching the lysosomal compartment from which they are secreted (*213*).

Secretion of mature lysosomal enzymes in axenically growing cultures of *D. discoideum* is a relatively slow linear process (*27,213*) that can be stimulated fivefold when cells are removed from their nutrient source (*38,45,181*) and resuspended in a buffered salt solution (starvation conditions). The presence of external amino acids inhibits secretion, the process is energy-dependent, and it occurs in the absence of cell lysis (*45*). Acid phosphatase is secreted with linear kinetics (5–10% per hr) while a group of glycosidase enzymes including α-mannosidase, β-glucosidase, β-galactosidase, and β-hexosaminidase are secreted with biphasic kinetics and to a greater extent (50–80% of the glycosidases are extracellular within 2 hr of starvation). Furthermore, cycloheximide preferentially inhibits the secretion of acid phosphatase (*45*). Together, these results suggest mature acid phosphatase may be localized in a physically distinct compartment from the glycosidases.

Two lines of evidence, however, support the hypothesis that lysosomal enzymes in *D. discoideum* reside in the same intracellular compartments. First, acid phosphatase and the glycosidases show the same distribution when organelles are fractionated on self-forming Percoll gradients (*17*). Second, indirect immunofluorescence studies using antibodies to acid phosphatase, β-glucosidase, and α-mannosidase indicate all three enzymes are colocalized in greater than 95% of the punctate vesicles observed in growing cells (*17*). In *Acanthamoebae*, lysosomal enzymes secreted at different rates are also colocalized to the same compartments (*102*), and in Section III, I will discuss some of factors which may regulate the differential secretion of hydrolases during starvation.

Endocytosis

Genetic and biochemical studies of animal cells, yeast, and the slime mold have indicated the intracellular lysosomal enzyme transport pathway merges with the endocytic pathway (*4,14,48,90,176,179*). The axenic strains of *D. discoideum* endocytose nutrients by both phagocytic and pinocytic routes, while under normal growth conditions, wild-type strains are only capable of growth by phagocytosis of bacteria (*32*). Furthermore, the requirements for growth in axenic cultures are fulfilled on the basis of pinocytosis alone which is a very dynamic process in growing cells and involves a significant flow of membranes. In fact, *D. discoideum* can internalize the equivalent of its entire plasma membrane every 45 minutes (*40,199*). The endocytic compartment also exists in an equilibrium with the extracellular space; fluid phase markers accumulate linearly for 1 to 2 hours until a plateau is reached (*48,109*). The steady-state is maintained by equivalent rates of influx and efflux of fluid phase markers, and the intracellularly accumulated fluid phase markers can be completely washed out of cells during the chase period (*109*). Together, these results suggest that fluid phase markers accumulate in *D. discoideum* cells in a single functionally connected compartment (or multiple compartments indistinguishable on a kinetic basis) which is in equilibrium with the extracellular medium; dead-end compartments are not observed.

Sustained growth occurs in axenic cultures of the *D. discoideum* strains Ax2 and Ax3 suggesting that the pinocytosed nutrients must reach the digestive lysosomal compartments; two lines of evidence exist to support this hypothesis. First, Percoll gradient fractionation of organelles prepared from cells pulse-labeled with ^{3}H-dextran and chased for various times indicates that the fluid-phase marker first accumulates in low density vesicles (*48*). During the chase period, compartments containing the fluid-phase marker increase very rapidly in density (half-time of 5 min) reaching values equivalent to the density of lysosomal organelles. This is consistent with the maturation of (or transport of vesicles from) light endosomes to dense secondary lysosomes, analogous to the process described for mammalian cells (*146*). The second line of evidence comes from experiments using density shift in which we have determined that greater than 90% of the lysosomal enzyme-containing compartments are functionally connected to the endosomal system by the ability of these organelles to receive material added to the extracellular medium (*27*). Thus, in *D. discoideum* and *Acanthamoebae* most lysosomal enzymes appear to reside in secondary lysosomal compartments which can release fluid phase markers and lysosomal enzymes to the extracellular medium.

Genetic studies have also indicated that a functional endosomal pathway may be important for lysosomal trafficking. For instance, the *D. discoideum* mutant

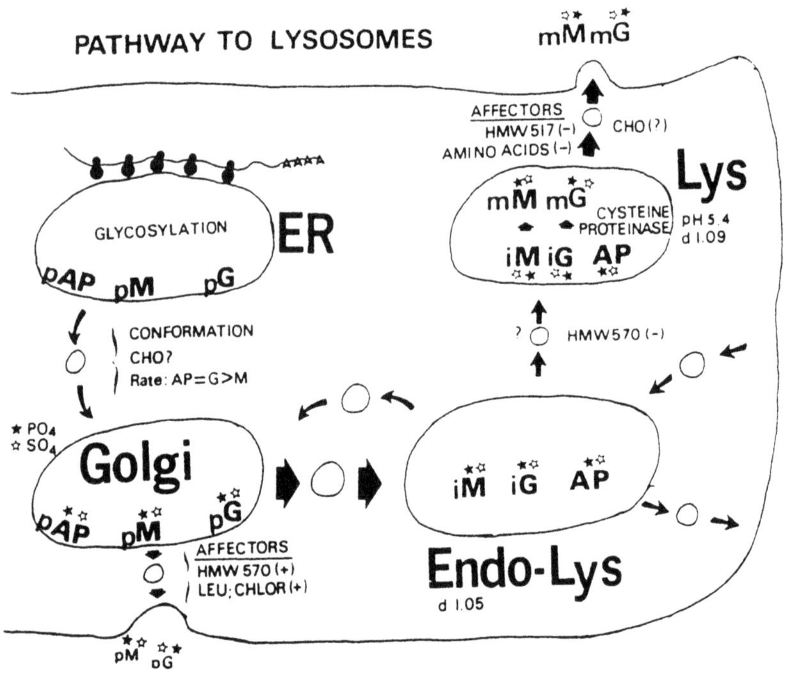

Figure 2. Transport, targeting and secretion pathways for *Dictyostelium* lysosomal enzymes. The figure represents the various transport, processing and secretory pathways followed by precursor forms (p), intermediate forms (i), and mature forms (m) of acid phosphatase (AP), α-mannosidase (M), and β-glucosidase (G). HMW570 refers to a lysosomal enzyme sorting mutant (Ebert et al, 1989); leu and chlor refer to leupeptin and chloroquine respectively; endo/lys refers to endosomal/lysosomal compartments.

HMW570 missorts and oversecretes unprocessed lysosomal enzymes, and this strain is defective in the formation of dense secondary lysosomes (*48*); pinocytosed fluid phase markers remain in low-density compartments (Section II-B). A similar yeast mutant has also been described (*31*). Furthermore, O'Halloran and Anderson (*157*) have used antisense mutagenesis to create *D. discoideum* cell lines that do not produce the clathrin heavy chain which results in strains that grow slowly, do not pinocytose fluid phase markers, contain abnormal large vesicles, and do not develop. Finally genetic studies (and biochemical studies) have implicated a number of yeast and animal cell GTP- binding proteins in the regulation of vesicular traffic along the secretory and endocytic pathways (*10,87,187*). Two *D. discoideum* genes (SAS-1 and SAS-2) have also been characterized (*183*) which code for proteins having significant homology to the yeast proteins, Ypt1 and Sec4. In yeast, Ypt1 and Sec4 control membrane flow between the ER and the Golgi, and between the Golgi and the cell surface, respectively. The function of the Sas proteins in vesicular transport and trafficking remains to be determined. More recently, we have cloned and characterized 18 cDNAs coding for proteins belonging to the ras superfamily. Included in this group are potential homologues to human Rac-1 (a rho member) and homologues to mammalian Rab 1, Rab 2, Rab 4, Rab 7, and Rab 11 (J. Bush and J. Cardelli, unpublished results).

Figure 2 summarizes the discussion in Section A concerning the lysosomal pathways in *D. discoideum*. At least three lysosomal enzymes, acid phosphatase, α-mannosidase, and β-glucosidase are synthesized as precursor polypeptides that are extensively N-glycosylated in the lumen of the ER with typical mannose-rich N-linked oligosaccharides. The precursors are tightly associated with ER membranes (see Section II-B for additional information) and are transported at different rates to the Golgi compartment where they are sulfated and phosphorylated on mannose residues. The majority of the hydrolases are proteolytically processed through intermediate forms to mature soluble forms and deposited in dense secondary lysosomes while a smaller percentage of the precursors bypass lysosomes and follow the constitutive secretory pathway which leads to their export from cells. The lysosomally-localized mature enzymes are secreted at low rates from growing cells and at fivefold higher rates when cells are starved in a buffered salt solution. Although the enzymes are released with different kinetics, they are colocalized to the same compartments. In the next section, I will review current information concerning the factors which regulate the transport and sorting of *D. discoideum* lysosomal hydrolases.

B. Identification of Factors Important in Transport and Sorting of Hydrolases

The Role of N-linked Oligosaccharides and Posttranslational Modifications

Mannose 6-phosphate residues on N-linked oligosaccharides of lysosomal enzymes in many animal cells act as sorting signals for transport of these proteins to lysosomes (*117*). Although *D. discoideum* lysosomal enzymes were known to be extensively phosphorylated and sulfated, little was known prior to 1989 about the role of these modifications in transport and sorting of enzymes. Even though these cells apparently lack MPRs (*24*), it is still possible that these modifications or other unidentified modifications play an important role in the transport and/or sorting of hydrolases. Unfortunately, no newly synthesized α-mannosidase, β-glucosidase, or acid phosphatase is detected following immunoprecipitation of proteins from cells pulse-labeled with ^{35}S-met in the presence of tunicamycin (unpublished observations). This suggests the nonglycosylated precursors are either very unstable or glycosylation is important in their synthesis. This is not an unusual observation, and others have reported that tunicamycin can effect both synthesis and stability of glycoproteins (*159*). This problem has, therefore, been approached by biochemically analyzing mutants that are altered in the construction and modification of N-linked oligosaccharides (Fig. 3).

Description of mutants: α-glucosidase II deficient mutant. The mutant strain M31 was originally isolated as an α-mannosidase deficient strain (*60*); the

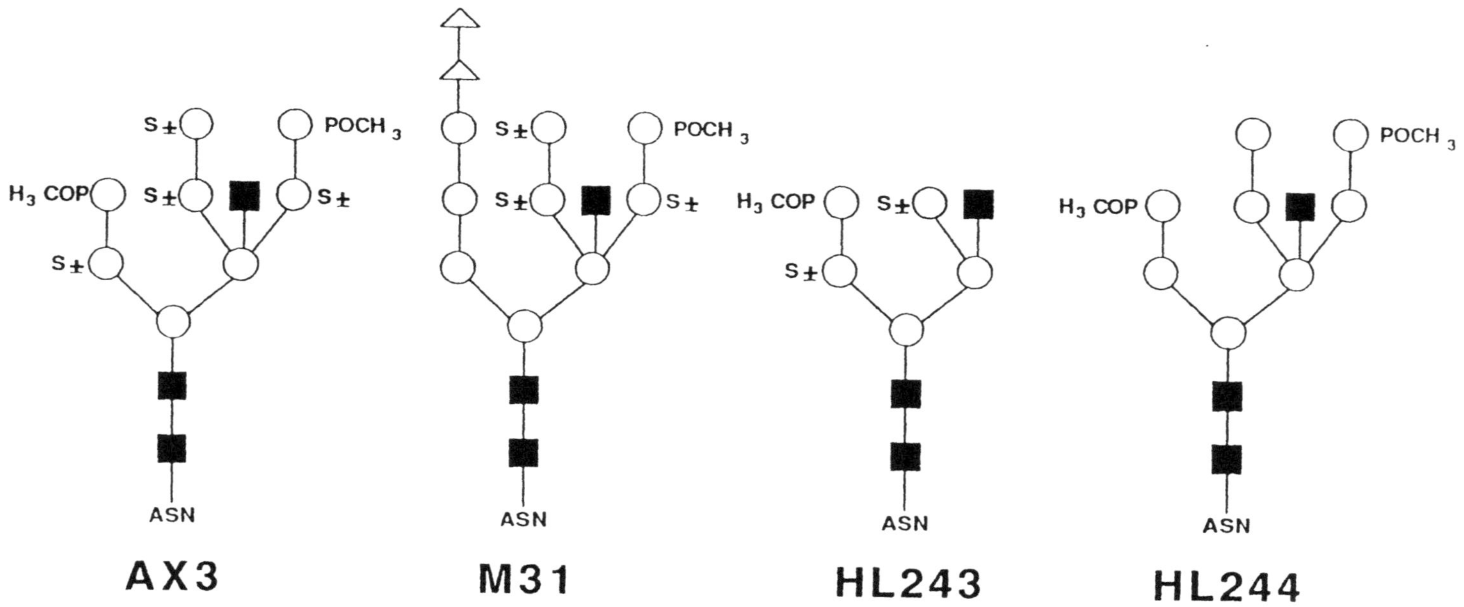

Figure 3. Structure of N-linked oligosaccharides derived from wild-type (Ax3) and mutant strains. Strain M31 lacks processing α-glucosidase II and cannot remove the innermost two glucose residues. HL243 and HL244 do not express the epitope CA-1 which consists of mannose 6-sulfate expressed in specific locations on the α-1,6 branch.

mutation *modA* maps to linkage group I (*63*) and decreases the total charge on several lysosomal enzymes. N-linked oligosaccharides prepared from three lysosomal enzymes contain on average threefold less sulfate and one instead of the two phosphate groups normally found for wild-type (*65,67*). Subsequently, it was reported that lysosomal enzymes secreted from this strain retained the two innermost α-1,3-linked glucose residues, and that this strain was deficient in ER-localized α-1,3 glucosidase activity (*69*). The structure of N-linked oligosaccharides isolated from this strain are indicated in Figure 3.

Mutants defective in the synthesis of Common Antigen 1. As described in Section II-A, a specific arrangement of sulfated sugars on N-linked oligosaccharides is required to define the shared lysosomal epitope CA-1. Using monoclonal antibodies to CA-1, a mutagenized population of cells was screened for loss of expression of the determinant, and five strains were isolated (*113*). Strains HL240-HL243 are reduced 30 to 50% in the amount of sulfated macromolecules, while strain HL244 lacks greater than 90% protein sulfation. Subsequent biochemical analysis has revealed that N-linked oligosaccharides from HL244 lysosomal enzymes are identical in size and composition to wild-type but were less negatively charged (*26*). The remaining negative charge can be accounted for by the presence of 1-3 phosphodiesters in HL244; no sulfate esters are detected. HL244 does not appear to be defective in sulfotransferase activity (*119*). Instead, the mutant apparently lacks a transporter for the high energy sulfate donor 5′ phosphoadenosine 5′ phosphosulfate, and therefore cells cannot accumulate this compound in the Golgi (H.H. Freeze, personal communication).

HL241 and HL243 each synthesize nonglucosylated, truncated lipid linked oligosaccharides with the structure $Man_6GlcNac_2$ (*76*). N-linked oligosaccharides prepared from the mutant lysosomal enzymes resemble each other but are smaller and less negatively charged than wild-type. In wild-type cells, 20% of the phosphorylated N-linked oligosaccharides contain one residue of mannose 6-phosphate, while 80% have greater than two such residues (*75*). In HL241 and HL243, about 80% have only one mannose 6-phosphate, and 20% have two. The structures of N-linked oligosaccharides from HL243 and HL244 are indicated in Figure 3.

Analysis of mutants: strain HL244. Radiolabeled pulse-chase, subcellular fractionation, and indirect immunofluorescence experiments have shown that the complete absence of mannose 6-sulfate does not affect the correct proteolytic processing, rate of transport, or sorting of three lysosomal enzymes (*18,26,74*). However, the mutant secretes mature enzymes at a twofold slower rate than wild-type. Together, these results suggest mannose 6-sulfate on N-linked oligosaccharides does not act as a transport, processing, or sorting signal for *D. discoideum* lysosomal enzymes, although these moieties may play a role in the regulated secretion of enzymes from lysosomes (Section III).

Strains HL241 and HL243. Pulse-chase radiolabeling experiments indicate that newly synthesized lysosomal enzymes are transported from the ER to the Golgi at two- to threefold slower rates in the mutants compared to wild-type (*18,74,75*). Precursor forms of the enzymes are not oversecreted, and fractionation of cellular organelles on Percoll gradients indicates that the processed enzymes are targeted to dense lysosomes. Therefore, a reduction in the size of N-linked oligosaccharides, and a decrease in phosphorylation and sulfation does not influence correct sorting of lysosomal enzymes, although these changes do affect the exit rate of precursors from the ER. This slower rate of transport of enzymes may be the result of a change in the conformation (*47,214*) or folding rate of the protein, or it may be due to alteration in glycoproteins which normally function as transport components. Finally, lysosomal enzymes are secreted two- to threefold more slowly from the mutants than from wild-type cells suggesting again that carbohydrate may regulate secretion of enzymes.

Strain M31. Subcellular fractionation and radiolabel pulse-chase experiments indicate lysosomal enzymes are transported 2 to 3 times slower from the ER to the Golgi in the mutant M31 than wild-type enzymes (*18,47*). The mutant forms of the enzymes leaving the ER are still glucosylated (*69*) which suggests that removal of the glucose residues is not necessary for transport out of the ER. In contrast, glucosylated mammalian lysosomal enzymes (*121*) and some secretory proteins (*91,130*) remain in the ER until glucose is removed. Slower transport of enzymes in M31 may be the result of improper folding since the mutant α-mannosidase precursor is more sensitive to protease digestion *in vitro* than the wild-type (*47*). Interestingly, acid phosphatase, but not the glycosidases, appears to also be delayed in transport from the Golgi to dense lysosomes (*18*). Both indirect immunofluorescence and Percoll gradient fractionation studies reveal, however, that all lysosomal enzymes examined are correctly targeted to lysosomes (*18*). Finally, consistent with the other mutants described above, all enzymes are secreted with slower kinetics from M31 compared to wild-type.

Together, these studies indicate that the retention of glucose residues, the absence of mannose 6-sulfate, a reduction in mannose 6-phosphate (including its absence from the glucosylated branch), and the absence of the four outer mannose residues on the nonglucosylated branches, do not affect the accuracy of proteolytic processing or targeting of these lysosomal enzymes. These studies do not eliminate the possibility that either phosphate or other uncharacterized modifications to N-linked oligosaccharides act as lysosomal sorting signals. In fact, of the 20 mutant strains isolated producing lysosomal enzymes that are less negatively charged on N-linked oligosaccharides, none are completely lacking mannose 6-phosphate (H.J. Judelson, unpublished results) suggesting that this defect may be lethal. However, as described in Section II-B, it seems more probable that the information for sorting resides is in the amino acid sequence of the hydrolases and not in the attached carbohydrate side-chains.

The Role of Acidic pH Compartments

Endosomes and lysosomes are acidic. Lysosomes and endosomes are acidic compartments in mammalian cells with pH values ranging from 4.8 to 6.4 (*144*); however, these are not the only compartments along the secretory pathway that are acidic. Coated vesicles, the *trans*-Golgi stack, and condensing vacuoles also have intravesicular pH values less than 7.0. The low pH in endosomes of fibroblasts and other mammalian cells is necessary to release lysosomal enzymes from the MPRs allowing these receptors to return to the Golgi compartment to bind newly arriving phosphorylated enzymes (*144,202*). Weak bases like ammonium chloride accumulate in these acidic compartments and effectively raise the pH (*158*) which results in the inability of receptors to release their ligands. As a result, newly synthesized precursors to lysosomal enzymes are missorted and secreted (*144,202*).

Spectrofluorometric techniques have indicated that lysosomes and endosomes have an average intravesicular pH of 5.4 in axenically growing *D. discoideum* cells (*25*). Steck and his colleagues have investigated the intracellular distribution of vacuolar proton pumps and the mechanism(s) responsible for acidification of these compartments in *D. discoideum*. An initial report indicated that the prominent H^+-ATPase activity (presumed to be the vacuolar proton pump) codistributed on sucrose gradients with endosomal vacuoles, while vesicles containing lysosomal enzymes contained only minor amounts of the ATPase (*163*). In a more recent study (*164*), these authors report that when vesicles are prepared in the absence of Mg^{2+}, vacuoles containing the proton pumps can be separated from both endosomes (labeled with the fluid phase marker dextran) and lysosomes (acid phosphatase activity). Furthermore, coupling can be achieved *in vitro* in the presence of Mg^2 and cell cytosol, suggesting that the endocytic circuit is acidified by a discreet and separable proton pumping organelle. These proton pumping stations may function similarly to the acidosomes of *Paramecium* (*58*).

Maintenance of acidic compartments is not required for proper sorting. Exposure of growing cells to low concentrations of ammonium chloride raises the endosomal/lysosomal pH from 5.4 to 6.1 and slightly slows the rate of proteolytic processing, but increases in pH do not prevent the correct localization of newly synthesized lysosomal enzymes (*25*). Raising the intravacuolar pH to 6.4 results in the accumulation of intermediate forms of the enzymes that are slowly transported to lysosomes and very inefficiently processed to mature forms. Wood and Kaplan (*212*) have also reported that monensin, a carboxylic proton ionophore which raises the pH in distal Golgi compartments (*197*), delays the transport (in agreement with our results), but not the processing of α-mannosidase. More importantly, the percentage of newly synthesized precursors that are missorted and secreted does not increase significantly under conditions where the vacuolar pH is 6.4 (*17,25*).

These results suggest acidic vacuolar compartments in *D. discoideum* are important for complete processing of lysosomal enzymes and, perhaps, for efficient transport of processed enzymes to lysosomes, but low pH is *not* necessary for correct localization of the enzymes. This later result is in contrast to what is observed in cells that utilize MPRs to target lysosomal enzymes (*144,203*). In these cells, raising the pH results in the missorting of newly synthesized lysosomal enzymes, presumably because the targeting receptors remain associated with the hydrolases while low pH facilitates release of ligands generating ligand-free receptors.

Similar conclusions concerning the role of vacuolar pH have been reached by others studying cell lines that do not rely on the CI-MPR for targeting of enzymes. For instance, von Figura and co-workers have determined that cathepsin C and cathepsin D in Morris hepatoma 7777 cell lines (*140*) and U937 monocytes (*11*) respectively, appear to be localized correctly by a pH-independent mechanism. Others have also found that sorting of myeloperoxidase to the azurophil granule in neutrophils occurs in the presence of ammonium chloride (*95,195*). Finally, yeast appears to sort hydrolases in the absence of an acidic pH gradient. The gene coding for a 60-kDa subunit of H^+-ATPase (VAT2) was cloned and used to generate null cell lines. The vacuolar pH in this mutant is 7.1 compared to 6.1 in wild-type vacuoles, and the vacuolar acidification defective mutant displayed only a minor defect in sorting (*216*). In contrast, yeast cells treated with weak bases to neutralize the pH (*182*), or with the drug bafilomycin A (which inhibits vacuolar ATPases; 4), are defective in the acidification of the lysosome like vacuole *and* missort a variable percentage of newly synthesized hydrolases, suggesting that the drugs may indirectly influence sorting.

Although additional work needs to be done, the results to date suggest hydrolase targeting mechanisms that do not rely on the CI-MPRs may not require an acidified lysosomal/endosomal network in order to function efficiently. This hypothesis does not eliminate the need for potential receptors (although none have been identified). However, if such targeting receptors exist, they would need to be capable of releasing ligands under only slightly acidic or even neutral pH conditions in order to return to the ligand binding compartment free of cargo.

Interestingly, *D. discoideum* cells exposed to the weak base, chloroquine, missort (and oversecrete) a large percentage of newly synthesized lysosomal enzymes even though this amine only slightly raises the endosomal/lysosomal pH (*25*). Enzyme precursors pass through the Golgi complex (they become *endo* H-resistant) but they do not reach lysosomes, and they are not proteolytically processed prior to export from cells. This indicates chloroquine may induce missorting by a pH-independent mechanism. Experiments involving the timed addition of chloroquine also reveal that this amine acts on a distal Golgi or prelysosomal compartment where precursors are first cleaved to induce missorting of newly synthesized enzymes. Furthermore, chloroquine has been shown to be an inhibitor of cysteine and other *D. discoideum* proteinases (*154*), and, together with the above data, suggests this amine may induce missorting of lysosomal enzymes by preventing processing of precursor polypeptides. In fact, as described in the next section precursor processing does appear to be an important event in the sorting process.

The Role of Proteolytic Processing of Precursors

As discussed in Section II-A, most lysosomal enzymes in *D. discoideum*, yeast and animal cells are synthesized as precursor (prepro) polypeptides that are proteolytically processed to mature forms. This has been demonstrated by pulse chase radiolabeling experiments in which the radiolabeled precursors (proforms) disappear over time and are replaced by lower molecular weight mature polypeptide chains. The signal peptide ("pre"-piece) is removed cotranslationally and therefore the first detectable radiolabeled form is the proform. Additionally, cleavage has been confirmed by comparing the N-terminal amino acid sequence of the mature subunits to the inferred amino acid sequence of the cloned genes coding for the enzyme (*139,185*). In many cases, precursors are first cleaved to intermediate forms which in turn are processed to mature forms of the enzymes by lysosomally-localized cysteine proteinases (Section II-A). In the following section, I will describe current information regarding the intracellular locations where precursors are cleaved, the nature of the cleavage sites, the proteinases involved, and the role of precursor processing. Figure 4 summarizes this information for the α-mannosidase polypeptide.

Intracellular compartments where precursor cleavage occurs. Only precursor forms of α-mannosidase and β-glucosidase are observed in the ER and Golgi membranes following fractionation of radiolabeled cellular organelles on sucrose gradients (*21,148*). This was originally interpreted to indicate that precursors are cleaved after exit from the Golgi and perhaps upon arrival in lysosomes. In contrast, Wood and Kaplan (*212*) concluded (based on Percoll gradient fractionations) the precursor form of α-mannosidase is cleaved in the Golgi compartment. More recently, using Percoll gradient fractionation of cell extracts, we have determined (*174*) that intermediate forms of newly synthesized α-mannosidase and β-glucosidase first appear in a cellular compartment intermediate in density between the Golgi complex (sulfotransferase used as a marker) and dense lysosomes (acid phosphatase used as a marker).

A density shift approach (*37*) was used to determine if precursors are cleaved in a late Golgi compartment (not functionally connected to the cell surface), or in a compartment connected to the endosomal pathway (this compartment may include some portion of the late Golgi). The basis of this procedure is that the equilibrium density of endosomal vesicles loaded with horseradish peroxidase (HRP) increases dramatically on Percoll gradients when these organelles are exposed to the substrate diaminobenzidene in the presence of hydrogen peroxide. Accordingly, a protein or enzyme is considered to reside in an endosomal compartment if it also shows a shift in density. Using this approach we observe that essentially all radiolabeled intermediate and mature forms of α-mannosidase and β-glucosidase are localized to compartments functionally connected to the endosomal network (*27*). In contrast, very little of the precursor polypeptides shift in density, suggesting that these proteins are cleaved immediately upon arrival in prelysosomal/endosomal compartments, or that transport of intermediate forms to these compartments occurs

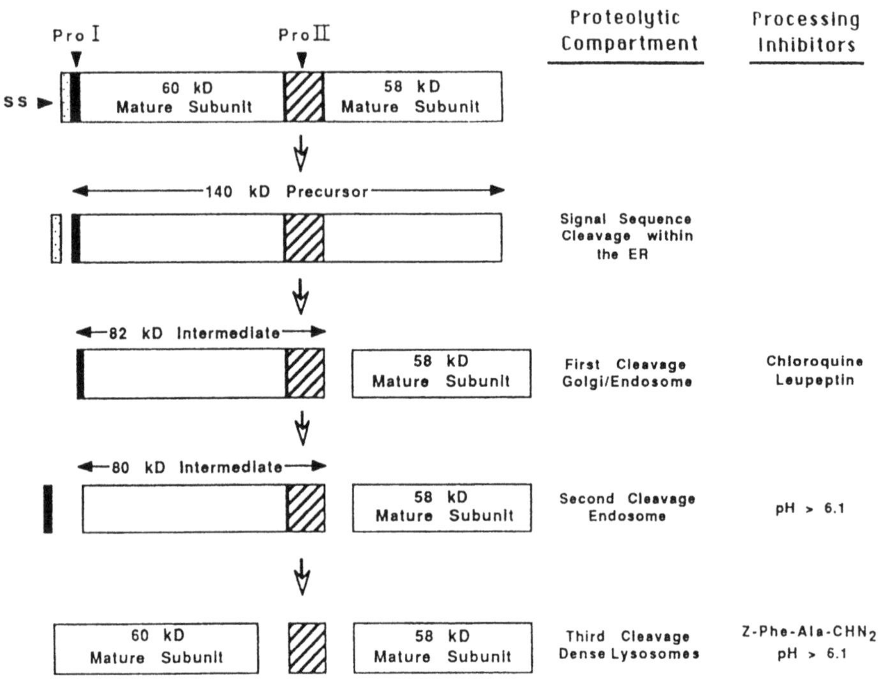

Figure 4. Predicted processing pathway for the α-mannosidase precursor. This model is based on results from the following publications: Mierendorf et al. (1983,1985); Cardelli et al. (1986b;1989;1990b); Richardson et al. (1988); Pannell et al. (1982); Schatzle et al. (1992).

concomitantly with processing. Based on the observation that most protease inhibitors can only enter cells by endocytosis, we favor the hypothesis that cleavage of precursors occurs in compartments functionally connected with endosomes. These compartments may still include portions of the Golgi complex because others have found that the *trans*-Golgi network compartments are accessible to the endosomal pathway in certain cells including the slime mold (*117*).

Gieselmann et al. reported (*82*) that processing of precursors to mature cathepsin D occurs in a prelysosomal compartment in mammalian cells. In contrast, others have determined that cleavage of β-glucuronidase and β-galactosidase occurs just prior to, or concomitant with, arrival in lysosomes (*12*). Processing of the precursor forms of yeast vacuolar hydrolases also appears to occur upon transport of these proteins to the lysosome-like vacuole (*111*).

Nature of the cleavage sites. Trypsin has been found to cleave mammalian and *D. discoideum* lysosomal enzyme precursor polypeptides at the locations along the polypeptide backbone recognized by the cellular-processing enzymes, suggesting these endogenous enzymes are trypsin-like in activity (*78,83,214*). As determined by sequence analysis, the predicted cleavage sites for mouse cathepsin L (*171*), the α-chain of human β-hexosaminidase (*139*), human cathepsin B (*28*), and *D. discoideum* α-mannosidase (*185*) consist of monobasic or dibasic amino acids (lysines and arginines) which appear (based on computer assisted analysis) to be embedded in an unstructured hydrophilic region preceeded by a β-turn (unpublished results). Cleavage at these basic amino acids is often followed by exoproteolytic (perhaps amino peptidase catalyzed?) trimming to generate the authentic amino terminus on the mature polypeptides. For instance, the α-mannosidase 140-kDa precursor is cleaved in a prelysosomal compartment to yield 80 to 84-kDa intermediate polypeptides and 58-kDa mature subunit (Fig. 4). Two noncontiguous lysines are found at positions -3 and -7 in a stretch of seven amino acids preceding the first amino acid (tyrosine position +1) of the mature 58-kDa subunit (*185*). The lysine at position -7 is found in a β-turn, and both lysines are predicted to reside in a hydrophilic stretch of 10 amino acids. Interestingly, cleavage of the 80 to 84-kDa intermediate forms of α-mannosidase to release the 60-kDa mature subunit and the "pro-II" polypeptide is also predicted to occur in a similar region containing noncontiguous arginines (*185*). Other uncleaved regions of the polypeptide contain mono- and dibasic amino acids, but very few of these amino acids (a total of 65) are found in regions predicted to have both a β-turn and a hydrophilic stretch. These results suggest that lysosomal enzyme precursors and (perhaps intermediate forms) are cleaved by enzymes that recognize overall structure in addition to specific amino acids. Cohen and co-workers (*173*) have also proposed that cleavage of precursors to polypeptide hormones occurs at basic amino acids which are located inside or immediately adjacent to regions with high probability of β-turn formation.

Biochemical nature of proteolytic enzymes. To further define the nature of precursor-cleaving proteinases, radiolabeled *D. discoideum* cells were chased in growth medium in the presence of a variety of protease inhibitors followed by immunoprecipitation of α-mannosidase and β-glucosidase, and SDS-PAGE (Fig. 5). Only two inhibitors, leupeptin and antipain (cysteine and serine protease inhibitors), prevent processing of precursors to α-mannosidase, β-glucosidase, and acid phosphatase (*17,174*). Another inhibitor Z-phe-ala-CHN2 (cysteine protease inhibitor) prevents processing of intermediate forms, (but not precursor forms) to mature forms, thus identifying a cysteine proteinase as responsible for the final maturation step (*174*; Fig. 5). Furthermore, unlike the formation of mature enzymes, processing of precursor polypeptides does not require an acidic pH compartment (*25*). Leupeptin and antipain can inhibit the activity of enzymes belonging to both the serine and cysteine class of proteinases; however, specific serine proteinase inhibitors (TLCK and PMSF) do not prevent processing of precursor polypeptides. Finally, the addition of EDTA to developing cells (the experiment has not been done with growing cells) results in the inhibition of processing of the α-mannosidase precursor (*184*); however metalloproteinase inhibitors do not prevent processing (*174*), suggesting EDTA is not functioning by chelating heavy

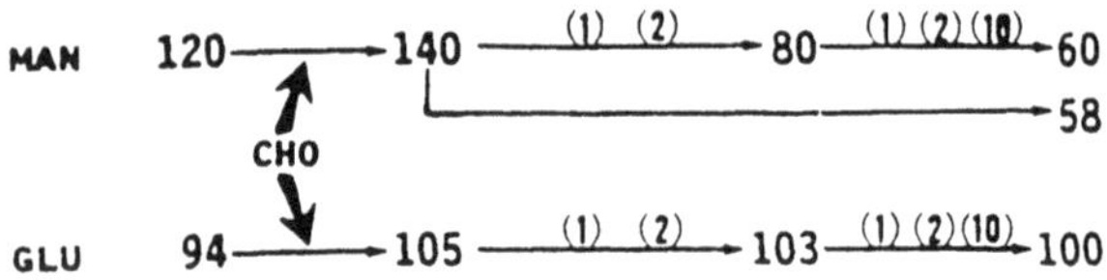

INHIBITOR	CLASS
1 LEUPEPTIN	SER, CYS
2 ANTIPAIN	SER, CYS
3 PMSF	SER, CYS
4 E-64	CYS
5 EPNP	ASP
6 DAN	ASP
7 TPCK	SER, CYS
8 PHENANTHROLINE	METALLO
9 PEPSTATIN	ASP
10 Z-PHE-ALACHN$_2$	CYS
11 TLCK	SER, CYS
12 Z-GLY-PHENH$_2$	METALLO

Figure 5. The effect of 12 proteinase inhibitors on proteolytic processing of α-mannosidase and β-glucosidase. The cotranslational addition of carbohydrate and subsequent processing of the enzymes are depicted in the top of the figure. The numbers correspond to the inhibitors listed below, and depict which inhibitors prevent processing as well as the step during proteolysis which is inhibited. Abbreviations used: TLCK, N-p-tosyl-L-lysine chloromethyl ketone; A-Phe-AlaCHN2, benzyloxycarbonyl-L-phenylalanyl-l-l-alanine-diazomethyl ketone; A-Gly-PheNH$_2$, carbobenzoxy-L-glycine-phenylalanine-amide; TPCK, L-1-tosylamide 2-phenylethyl chloromethyl ketone. PMSF, phenylmethylsulfonyl fluoride; DNA, diazoacetyl-DL-norleucine methyl ester; EPNP; 1,2-epoxy-3-(p-nitrophenoxy) propane.

metals. Therefore, the precursor processing enzyme(s) appears to differ in both intracellular location and biochemical properties from the lysosomally-localized cysteine proteinases (Section II-A) responsible for generating mature subunits, although both enzymes appear to be trypsin-like and cleave at basic amino acid residues.

A similar observation has been reported for the processing of cathepsin D in human fibroblasts (*82,83*); the precursor polypeptides are processed in prelysosomal compartments, the reaction does not require an acidic pH environment, and thiol proteinase inhibitors do not prevent cleavage while EGTA does. Together, these results suggest that the lysosomal precursor processing enzyme(s) in at least two organisms may be a Ca^{2+}-dependent neutral serine-like proteinase resembling the yeast Kex-2 proteinase (*79*) which is responsible for cleaving the precursor to the secreted pheromone, α-factor a. Kex-2 also cleaves at lysine/arginine residues, is localized in the Golgi, and is inhibited by EDTA and leupeptin. Furthermore, the enzyme correctly cleaves mammalian prohormones and neuropeptides when expressed in mammalian cells suggesting it may be a prototype for processing enzymes (*200*).

Importance of proteolytic processing. Surprisingly, the addition of leupeptin or antipain to cultures of growing cells (or EDTA to developing cells) not only prevents processing of precursor polypeptides but also induces the rapid secretion of the uncleaved precursors; the inhibitors do not induce the secretion of mature forms (*17,174,184*). Chloroquine also induces the missorting of unprocessed precursors (*25*) and may be acting as a protease inhibitor (Section II-B). This result, which has been observed for three different lysosomal enzymes, suggests the initial cleavage event may be important in the proper localization of some *D. discoideum* lysosomal enzymes. In contrast, conditions which prevent processing of intermediate forms (e.g., neutral endosomal/lysosomal pH, the presence of thiol proteinase inhibitors) do not induce missorting of enzymes; instead, the intermediate forms accumulate in dense lysosomes (*25,174*). Cleavage is apparently not an absolute requirement for sorting in the slime mold because lysosomal β-hexosaminidase is probably not cleaved (*88*).

To my knowledge, *D. discoideum* is the only identified organism in which proteolytic processing of lysosomal enzymes may be important in the localization process. In contrast, proteolytic processing is not essential for the sorting of hydrolases to the yeast vacuole (*111*) or the plant vacuole (*30*), and the generation of mannose 6-phosphate on N-linked oligosaccharides of fibroblast enzymes (the sorting signal for cells using the MPR) occurs prior to processing of precursors (*117*).

A number of testable models can account for the observations described above. For instance, cleavage of precursors in a late Golgi or endosomal compartment may be necessary to generate or expose a lysosomal sorting signal embedded in the intermediate form of the enzyme; uncleaved precursors with buried sorting signals would exit the cell along the constitutive secretory pathway. Alternatively, proteolytic processing may be responsible for releasing the membrane-associated precursor polypeptides which may be necessary to allow potential sorting receptors to return to the sorting compartment to bind newly arrived precursors. In the absence of these receptors, uncleaved precursors would be secreted. Finally, the inhibitors may prevent proper sorting of hydrolases through some indirect effect; for example, processing of membrane proteins may be necessary for vesicular transport and fusion.

There is precedence for each of these possibilities. Proteolytic processing may be necessary to release a conformational restraint on inactive proteins to generate biologically active ones which commonly results in a gain in enzymatic function (*153*). Alternatively, cleaved proteins may acquire other structural properties as observed for uvomorulin which gains cell adhesive function upon cleavage (*162*).

In addition, upon cleavage the new amino terminal of the thrombin receptor functions as a ligand to activate the receptor (*205*).

Proteolysis has also been implicated in protein localization in other systems. As an example, trypsin treatment of testicular sperm *in vitro* is sufficient to correctly cleave specific plasma membrane proteins (such as PH-20 and PH-30) which induces their redistribution in the plasma membrane to the posterior or anterior head domain (*170*). The redistribution of proteins coincides with, and may facilitate, the maturation of the sperm cells. Furthermore, Scheller and colleagues (*210*) have demonstrated that bioactive peptides are cleaved from a precursor polypeptide in a late Golgi compartment in bag cell neurons of *Aplysia*, an event which may trigger the condensation and sorting of processing intermediates to different dense core vesicles.

Proteolysis has also been demonstrated to be important in certain membrane fusion events. For instance, in many enveloped viruses (including HIV) endoproteolytic cleavage of viral proteins occurring in the Golgi (*192*) is necessary to release an internal hydrophobic domain with fusogenic potential required for full expression of viral infectivity (*143*). In addition, metalloproteinases have been demonstrated to be important in endocytosis and the secretion of newly synthesized proteins in HepG2 cells (*196*); the fusion of sea urchin gametes (*178*); and in exocytosis in mast cells and adrenal chromaffin cells (*154*). Finally, Opresko and Karpf (*160*) have reported that proteolysis of endocytosed vitellogenin may be important in endosomal targeting in *Xenopus* oocytes. They propose that cleavage of vitellogenin may remove the protein from a receptor, and that the absence of occupied receptors allows endosomes to fuse with platelets.

The Nature of Precursor Polypeptide Membrane Association

The *D. discoideum* lysosomal enzymes are tightly associated with ER and Golgi membranes while in transit to lysosomes, while proteolytically processed mature forms are soluble in the lumen of high-density lysosomes (*21,23,148*). This is a strong association based on the inability to release precursors following treatment of opened membranes with high salt, mannose 6-phosphate, chelating agents (EDTA), chaotropic agents (KI), and at low pH (*23*). The molecular nature of this interaction remains to be determined; however, analysis of the complete predicted amino acid sequence encoded in the cloned α-mannosidase structural gene has allowed us to eliminate certain possibilities (*185*). First, the amino acid sequence contains no predicted transmembrane regions that could to anchor the precursor, consistent with the observation that secreted precursors have similar (or slightly higher molecular weights) compared to intracellular precursor forms. Second, the carboxy terminal region is not hydrophobic, suggesting the absence of a signal for the attachment of a glycolipid anchor. Third, the amino acid sequence does not contain signals for the attachment of either palmitic or myristic acid. Finally, amphipathic sequences have been shown to be important in the binding of some

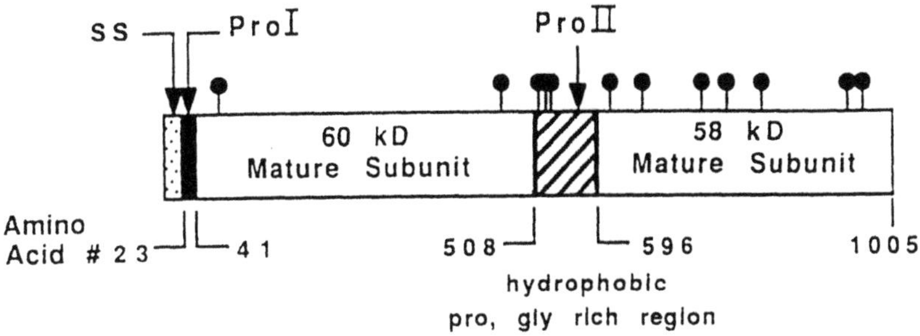

Figure 6. Schematic diagram of the α-mannosidase precursor polypeptide. The "pro" regions represent amino acids not found in the purified enzyme, which consists of a heterotetramer of 58 and 60 kDa subunits. The lollipop structures represent potential sites for N-glycosylation. The figure is adapted from Schatzle et al. (1992).

proteins to membranes; however, based on the computer program of Eisenberg et al. (*50*), this structure is not apparent in the α-mannosidase precursor.

Conceivably, the α-mannosidase protein may be anchored to membranes by the uncleaved N-terminal signal sequence. In support of this, *in vitro* translation experiments in the presence and absence of dog pancreas microsomes indicate that the primary translation product and the ER-localized deglycosylated polypeptide are identical in molecular weight (*22,23*). However, given the large size of the precursor (120 kDa) and the inaccuracy of SDS-PAGE in predicting molecular weight for some polypeptides, this conclusion may be unwarranted. In fact, the predicted signal sequence is probably cleaved because it is of the same size and general composition of other known cleavable signal peptides of lysosomal, secretory, and plasma membrane proteins (*185*).

Analysis of the α-mannosidase amino acid sequence using the programs of Kyte and Doolittle (*118*) and Chou and Fasman (*29*) indicates the presence of eight regions (each greater than nine amino acids long) distributed throughout the proteins predicted to be hydrophobic β-sheets. Interestingly, six of these hydrophobic β-sheets are found in the central portion of the precursor including three in the pro-II region (Fig. 6) (which separates the 60-kDa and 58-kDa mature polypeptide chains). Conceivably these hydrophobic sheets may in part account for the association of only the α-mannosidase precursor with membranes (mature forms are soluble and contain few predicted hydrophobic stretches). Alternatively, the precursors may be tightly associated with a targeting receptor or other tightly associated membrane proteins which may account for association with membranes.

Precursors to lysosomal enzymes are also membrane-associated in animal cells that do not rely exclusively on mannose 6-phosphate signals for sorting. For instance, Diment et al. (*41*) reported that cathepsin D is membrane-associated in endosomes of rabbit alveolar macrophages, and this association is not disrupted by high pH or mannose 6-phosphate, suggesting it is not mediated by MPRs. More recently, Rijnboutt et al. (*175*) demonstrated the membrane association of lysosomal cathepsin D, glucocerebrosidase, and the sphingolipid-activating protein in HepG2 cells. Membrane association of cathepsin D is not dependent on N-linked glycosylation, and the enzyme is released from membranes following proteolysis. These proteins, unlike *D. discoideum* enzymes, apparently associate with membranes upon arrival in Golgi compartments. Finally, Erickson and co-workers have also found that the proenzyme forms of porcine cathepsin L and cathepsin D are capable of binding to membranes at acidic pH in the presence of 10-mM mannose 6-phosphate and chelating agents (*145*). Furthermore, binding of the proforms *in vitro* is reduced following treatment of membranes with trypsin, suggesting that the association is protein-protein in nature. This prompted the authors to suggest that a "lysosomal proenzyme receptor" may bind lysosomal proenzymes in a prelysosomal compartment to direct their transport to lysosomes.

Interestingly, both human glucocerebrosidase and cathepsin D precursors are membrane-associated and also contain a significant number of long hydrophobic β-sheets. Cathepsin D has three such structures, all longer than 13 amino acids, and glucocerebrosidase has four, two longer than 14 amino acids in length (unpublished results).

Additional work is needed to determine if proteolysis is required for the release of membrane-associated *D. discoideum* lysosomal enzymes targeted to lysosomes, although the secretion of precursors identical in molecular weight to the intracellular forms suggests major processing is not required for detachment. It will also be important to determine if membrane-association is required for the localization of enzymes, although as described in the next section, this interaction is apparently not sufficient.

Genetic and Molecular Approaches to Characterize the Sorting Process

Organisms like *S. cerevisiae* and *D. discoideum* can be analyzed biochemically and genetically, making them very useful systems to answer important biological problems. For instance, in yeast, two independent genetic approaches have led to the isolation of a large number of mutants defective in the ability to sort hydrolases (*111*). These mutants define over 40 complementation groups; many of these mutants exhibit extreme defects in sorting, and apparently do not synthesize a vacuole. These studies suggest that sorting in yeast is a complicated process requiring a large number of genes, and that the proper transport and sorting of lysosomal enzymes involves a number of distinct steps which are regulated by different proteins. These proteins may function in two ways: (1) recognition and sorting of hydrolases away from other secretory proteins; or (2) packaging of hydrolases into transport vesicles that must be directed to fuse with the appropriate acceptor compartment and recycle receptors from prelysosomal compartments back to the Golgi complex.

Complementation of temperature-sensitive mutations have led to the cloning and characterization of a number of yeast genes important in sorting (*111*). For instance, Herman et al. (*100*) report that the gene *VPS15* encodes a novel protein kinase essential for correct delivery of hydrolases to the vacuole. The authors propose that protein phosphorylation may act as a molecular "switch" to divert proteins from the default secretory pathway to a pathway directed towards the vacuole. These genetic approaches, combined with *in vitro* assays (capable of reconstituting vacuolar protein transport and sorting) will help elucidate the molecular mechanisms regulating sorting.

Genetic approaches have also been applied to *D. discoideum* and have resulted in the independent isolation of more than 70 mutants altered in lysosomal enzyme secretion (*49*). The vast majority of these mutants undersecrete enzymes during growth and starvation, and they will be discussed in greater detail in Section III-B. We have, however, biochemically characterized a mutant that oversecretes lysosomal enzymes and appears to be defective in sorting. The mutant strain *HMW570* oversecretes more than 85% of the newly synthesized α-mannosidase and β-glucosidase precursors as uncleaved polypeptides; these proteins pass through the Golgi but do not reach lysosomes, suggesting the mutant is defective in localization (*48*). The precursors in the mutant are membrane-associated, suggesting association is not sufficient for localization to lysosomes.

Interestingly, the mutant is also defective in endocytosis. First, subcellular fractionation studies indicate the majority of a fluid-phase marker (accumulated to steady state) resides in low density compartments characteristic of endosomes, while in contrast, in wild-type cells most of the fluid-phase marker is delivered to dense secondary lysosomes. Second, the mutant accumulates fluid-phase markers at only one-half the rate, and to only one-half the extent compared to uptake in wild-type cells.

Phenotypically, the *D. discoideum* mutant *HMW570* is similar to some of the higher eukaryotic endocytic mutants described by others which also missort and oversecrete lysosomal enzymes (reviewed in ref. *177*). However, missorting of the enzymes in these mammalian cell mutants may be the result of the inability of cells to acidify endosomes, a process necessary to allow MPRs to return from endosomes to the Golgi in an unoccupied state. *HMW570*, however, appears normal in acidification of endosomes and lysosomes, and acidification is not required for the localization of *D. discoideum* enzymes in wild-type cells (Section II-B).

The mutant *HMW570* more closely resembles the yeast endocytic mutant *end1* (*31*) which is in the same complementation group as the vacuolar lysosomal enzyme sorting mutant vps11 (*111*). These mutations may define genes whose products function in both the endocytic and the lysosomal transport pathways. For instance, in *D. discoideum* these proteins may be necessary for the formation of mature lysosomes which are obviously required for the correct sorting of hydrolases. Alternatively, *HMW570* may also be defective in the precursor proteolysis step, a process which may be important in the fusion of endosomes with lysosomes in *Xenopus* oocytes (*160*). Precursors would, therefore, reside in endosomes longer than normal resulting in a greater probability of incorporation into secretion vesicles recycling from endosomes to the cell surface.

Molecular approaches. Recombinant DNA techniques are the most sophisticated approaches currently available to determine the molecular basis of organellar targeting signals on proteins. Application of these methods has helped elucidate the sorting mechanisms for a wide variety of proteins including those targeted to the nucleus, mitochondria, chloroplasts, peroxisomes, the ER, the Golgi complex, and lysosomes. In general, mutagenized versions of the genes coding for the protein under investigation are expressed in the appropriate cells (or *in vitro*), and the processing and localization of the protein is examined. This approach identifies regions of the protein necessary for sorting. In addition, chimeric proteins are constructed which contain portions of the protein under investigation fused to a reporter protein which under normal conditions is not sorted to the compartment being examined. This approach is useful in identifying regions of the protein sufficient for localization.

The application of these approaches first requires the molecular cloning of genes coding for the targeted proteins. A number of *D. discoideum* genes have been cloned coding for authentic (the encoded product has been shown to reside in lysosomes) and putative (the encoded product resembles other known lysosomal enzymes based on amino acid homology) lysosomal enzymes (see Table 1). The cloned authentic lysosomal genes include α-mannosidase (*185*), β-glucosidase (J.R. Richardson and J.A. Cardelli, unpublished results), β-hexosaminidase (*88*), while the putative lysosomal genes include α-fucosidase (*150*), cysteine proteinase I (*211*) and cysteine proteinase II (*167*). The encoded proteins are all characterized as having typical N-terminal signal peptides (*204*). Although all of these signal peptides conform in structure to those known to be cleaved in the ER in other cells, no direct evidence exists for cotranslational processing of these peptides in *D. discoideum*.

Kaplan and colleagues have tentatively identified a region of *D. discoideum* β-hexosaminidase that appears to be sufficient for localization of the enzymes to lysosomes (personal communication). Portions of the gene (encoding a complete ORF of 532 amino acids) were cloned in front of, and in frame with, the yeast gene coding for invertase, an enzyme which does not normally enter the secretory pathway and is easily assayed making it a useful marker enzyme to monitor localization to lysosomes. Fusion proteins containing only the first 22 amino acids of β-hexosaminidase (the predicted signal peptide) are secreted when the chimeric gene is introduced into cells, suggesting the signal peptide contains no lysosomal localization information while fusions containing more than 314 of the N-terminal amino acids are localized to lysosomes. Interestingly, a fusion protein containing only 97 amino acids appears to accumulate in a prelysosomal or late Golgi compartment, suggesting that a signal for lysosomal targeting may be located between amino acids 97 and 314.

Comparable studies have been initiated concerning the structural gene coding for α-mannosidase. The α-mannosidase precursor polypeptide consists of 1005 amino acids divided into five domains (Fig. 6): (1) a 22-amino acid signal sequence at the amino terminus; (2) the highly polar "pro-I" region (positions 23–40) containing seven lysines; (3) the 60-kDa subunit (positions 41–507) containing two possible glycosylation sites; (4) a second "pro-II" region (positions 508–595) containing three potential glycosylation sites; and (5) the 58-kDa subunit (positions 596–1005) containing seven potential glycosylation sites. Interestingly, the three N-linked oligosaccharides in the very hydrophobic "pro-II "region are very closed spaced at amino acid positions 520, 528, and 539 in the precursor. Of particular interest will be the determination of the targeting role of the two "pro" regions (stretches of amino acids not conserved in the mature protein) found in the 140-kDa precursor. I also predict that the α-mannosidase precursors engineered to be missing the cleavage sites just upstream of the 58-kDa sequence (positions 589–593) will be missorted when the mutated gene is expressed *in vivo* (Section II-B).

In yeast, molecular studies have indicated that the "pro" regions of at least two hydrolases are sufficient and necessary for the localization of these proteins to the lysosome-like vacuole (*111*). The N-terminal, 50 amino acids of preprocarboxypeptidase Y, is sufficient to direct an invertase fusion protein into the vacuole, and the essential sorting region has recently been narrowed down to four contiguous amino acids. In addition, the N-terminal, 76 amino acids of preproproteinase A, is sufficient for localization of this protein to the vacuole. These two "pro" regions do not share any sequence homology suggesting that either these proteins are sorted by two different recognition systems, or that a region similar in secondary or tertiary structure is recognized as common.

The "pro" regions of polypeptide chains appear to play an important role in the localization of proteins to the lysosome like vacuole in plants. For instance, the vacuolar protein sporamin is synthesized as a "prepro" polypeptide containing no N-linked glycans, which is consistent with data indicating that carbohydrate modifications are not required for sorting to the vacuole (*141*). Expression of a mutated version of the protein with a deleted "pro" region leads to the missorting and secretion of sporamin suggesting that the "pro" peptide is required for correct targeting. The "pro" peptide in sporamin is similar to the "pro-I" region in the α-mannosidase polypeptide in that it is very small (16 amino acids), and very hydrophilic (containing five charged and three polar amino acids). In comparison, the 18-amino acid-long "pro-I" region of α-mannosidase contains seven charged and five polar amino acids.

III. DEFINING THE MECHANISMS REGULATING SECRETION OF MATURE LYSOSOMALLY LOCALIZED ENZYMES

As discussed in Section II-A, most lower and higher eukaryotic cells secrete lysosomal enzymes in precursor form by a pathway which bypasses lysosomes. Since the percentage of secreted precursor polypeptides can be increased under certain experimental conditions which prevent proper sorting, this suggests this export pathway probably represents the constitutive default secretion pathway. A

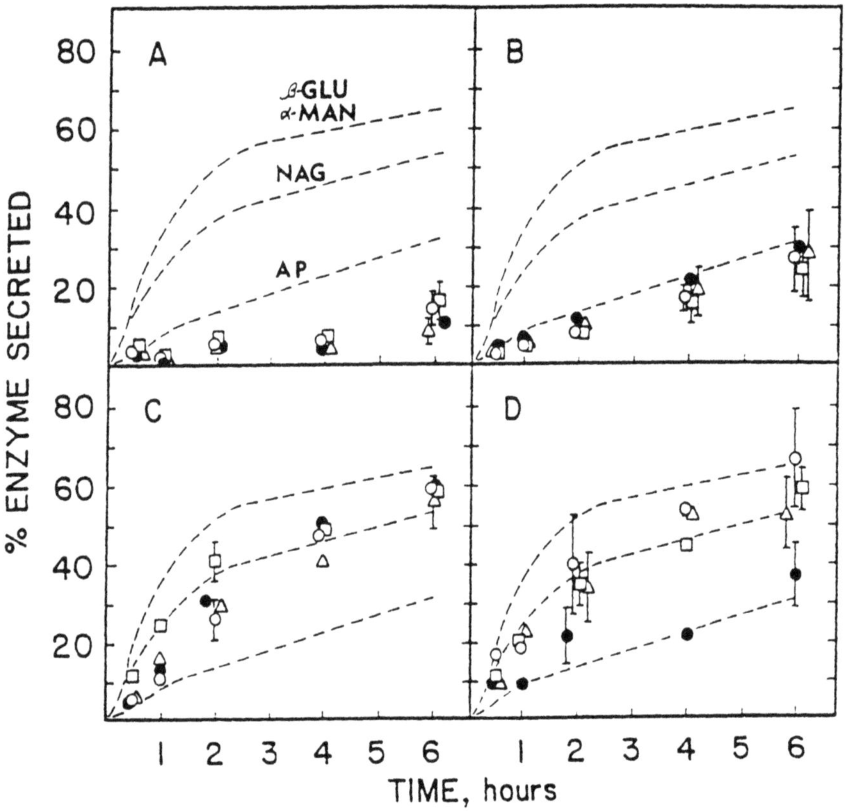

Figure 7. Percent of enzymes secreted by the mutants under standard secretion conditions. The standard error for each point is represented by vertical bars. If there is no bar, then the standard error is within the dimensions of the symbol. Dotted lines represent wild-type secretion for α-mannosidase, β-glucosidase, acid phosphatase and N-acetylglucosaminidase. (A) Class I strain, HMW500 (all enzymes affected); (B) Class II, HMW522 (glycosidase only affected); (C) Class III, HMW540 (acid phosphatase only affected); (D) Class IV HMW568 (secretion of enzymes affected only during growth). N-acetylglucosaminidase (○); α-mannosidase (□); β-glucosidase (△); acid phosphatase (●).

second route of export is available, and involves the regulated release of lysosomally localized mature enzymes. This pathway operates in many cells including a number of higher eukarytic cells (reviewed in ref. *106*), such as mast cells, kidney cells, and phagocytic cells (macrophages and neutrophils), as well as lower eukaryotic cells such as *Acanthamoebae castellannii* (*102*), *Tetrahymena pyriformis* (*7*), *Leishmani donovani* (*86*), and *Dictyostelium discoideum* (*45*). Because of the relative ease of application of genetics and biochemistry, the slime mold has proven to be a useful system to investigate regulated secretion of lysosomal enzymes.

As detailed in Section II-A, suspension of growing cells in a nonnutrient buffer (standard secretion conditions) initiates development and results in a fivefold increase in the rate of secretion of acid phosphatase and most of the glycosidases (*16,45,165*). The glycosidases are secreted very efficiently (50–80% of the enzymes are released in a few hours with biphasic kinetics), while acid phosphatase is secreted to a lesser extent (10–20% secreted by 3 hr) with linear secretion kinetics (Panel A, Fig. 7). Interestingly, all of these enzymes are colocalized to the same intracellular compartments (*17*), suggesting that the lower secretion rate for acid phosphatase may be the result of preferential retention of this enzyme in lysosomal compartments compared to the glycosidases. In this section, I will discuss recent information concerning biochemical and genetic approaches to identify the molecular mechanisms regulating the differential secretion of lysosomal enzymes.

A. Biochemical Approaches

Induced secretion of lysosomal enzymes may result from an increase in the rate at which mature lysosomes fuse to the plasma membrane and release their contents, or it may occur due to an increase in the rate at which enzymes are incorporated into vesicles which bud from the surface of lysosomal compartments to shuttle contents to the cell surface. An increased incorporation rate of enzymes into vesicles may be due to either the formation of vesicles with greater lumenal volumes (containing more enzymes), or a greater rate of formation of the same size vesicles during starvation, or to a more efficient packaging of enzymes into these vesicles recycling at a constant rate. Earlier studies have suggested that mature lysosomes do not fuse with the plasma membrane during starvation (*45*), although the percentage of total cellular lysosomal enzymes in secondary lysosomes was not determined. More recently, we have demonstrated (*27*; using the density shift protocol of Courtoy, *37*) that greater than 90% of acid phosphatase and the glycosidases reside in shared compartments considered to be secondary lysosomes (these compartments can be loaded with a fluid-phase marker added to cultures of growing cells). This suggests that exocytic fusion of lysosomes with the plasma membrane leading to the release of enzymes is an infrequent event.

Starvation induces only a modest increase in the rate of release of ^{3}H-dextran loaded into secondary lysosomes and endosomes compared to release of this fluid marker during growth (unpublished results), suggesting that the increase in secretion of lysosomal enzymes is probably not due to an increase in the size (volume) or number of recycling vesicles returning to the cell surface from lysosomal compartments. Instead, it appears more likely that the enzymes are more efficiently packaged into vesicles returning to the cell surface. In fact, α-mannosidase behaves kinetically exactly like a fluid-phase marker during starvation-induced secretion. Others studying the induced release of lysosomal enzymes from *Acanthamoebae* have reached similar conclusions concerning secretory mechanisms (*102*).

These studies suggest the differential rate of release of enzymes from *D. discoideum* lysosomes (and perhaps lysosomes in other cells) may be controlled by regulating the association of these proteins with a nonsecretory component in the lumen of lysosomes. In a recent review, Storrie (*194*) summarizes three hypotheses to explain preferential retention of hydrolases in lysosomes. First, Hohman and Bowers (*102*) propose that only molecules that can bind to the lysosomal membrane at low pH can be exocytosed via small vesicles to the plasma membrane where they are released due to neutral pH conditions. The second proposal states that lysosomal enzymes interact with an organellar matrix. The rate of secretion of enzymes via recycling vesicles would be governed in part by the equilibrium constant for association of these enzymes with each other and this insoluble "matrix". In fact, the interiors of lysosomes in many cells appear electron-dense suggesting the presence of a dense matrix. Finally, differences in diffusion rates based on molecular size may dictate how efficiently enzymes are packaged into small secretory vesicles.

The identification of the chemical or nutritional agents that influence secretion will help experimentally to better distinguish the above models, and will facilitate identification of the mechanisms involved. For instance, low concentrations of sucrose and other nonmetabolizable sugars stimulate secretion of lysosomal enzymes from growing cells *D. discoideum*, and further stimulate the already induced secretion of enzymes from starved cells (unpublished results; *38*). In addition, these sugars induce the formation of large vacuoles distributed throughout the cytoplasm, while metabolizable sugars like glucose have no effect on secretion or giant vacuole formation. This observation has also been observed with the mammalian cell line PyS-2 in the presence of sucrose (*209*). It is possible that vacuolation and/or the sugars themselves disrupt the lysosomal "matrix", resulting in the secretion of the hydrolases and fluid phase markers at the same rate. Interestingly, these results suggest that the "nonsecretory" compartment containing mature lysosomal enzymes can be rendered secretory-competent under certain conditions.

Caffeine, vanadate, ammonium chloride, and high salt are some of the substances known to inhibit secretion of *D. discoideum* lysosomal enzymes (*85,110*; Cardelli, unpublished results). Caffeine reversibly inhibits axenic growth, fluid phase pinocytosis, and the secretion of lysosomal enzymes induced by starvation in the presence of sucrose. Caffeine also increases the cytosolic concentration of calcium which is consistent with inhibition of secretion observed with lathium, a calcium transport inhibitor. Changes in intracellular calcium levels may influence vesicular traffic, and the reduction in secretion of lysosomal enzymes may simply reflect reduced membrane flow. Changes in intracellular Ca^{2+} concentrations have been shown to be important in the secretion of lysosomal enzymes from some (*191*), but not all (*209*) animal cells. Ammonium chloride also reduces secretion rates of mature enzymes from *D. discoideum* (*17*; unpublished results), and raises intravesicular pH (*25*); its effect may also be accounted for by reduced flow of membranes (Cardelli, unpublished results). In contrast, Swank and co-workers (*13*) have reported that treatment of mouse macrophages with ammonium chloride stimulates the secretion of both precursor and mature forms of lysosomal enzymes. Finally, vanadate, which inhibits ATPases and phosphatases in other systems, also reversibly inhibits growth, pinocytosis, and secretion of slime mold lysosomal enzymes (*110*), but efflux of fluid phase markers is not effected. This suggests that phosphorylation events may be important in the packaging of lysosomal enzymes into recycling vesicles.

Current studies are underway to define both the signals responsible for induced secretion, as well as the mechanism by which this signal(s) is recognized and transduced. As revealed mainly from studies of macrophages and neutrophils (*106*), these secretion signals may consist in part of changes in the concentration of calcium, and increased phosphorylation and phosphoinositide metabolism.

B. Genetic Approaches

Genetic and biochemical approaches have proven to be a powerful combination in the analysis of the secretory and vacuolar targeting pathways in yeast. Similar approaches are also being used to study *D. discoideum*, and over 70 independently isolated *D. discoideum* mutants have been characterized which are defective in lysosomal enzyme secretion (*49*). The majority of the mutants can be classified into one of four defined secretory classes (Fig. 7) based on their behavior under standard secretion assay conditions (Section II-A). The first class contains 20 members and undersecretes both the glycosidases and acid phosphatase, while the second class (16 members) undersecretes only glycosidases with normal acid phosphatase secretion. Nine strains (class III) oversecrete only acid phosphatase while the fourth class of mutants is normal in secretion of enzymes under standard starvation conditions, but secretion is defective during growth. These results indicate that there are shared genes that regulate the secretion of both classes of enzymes, as well as separate groups of genes influencing secretion of the glycosidases and acid phosphatase. Altered secretion of enzymes may be the result of mutations which: (1) prevent proper localization of lysosomal enzymes (see Section II-B); (2) affect the production, movement, and fusion of vesicles; (3) alter the packaging efficiency of mature enzymes into recycling vesicles; or (4) alter regulation of these processes.

To help define the signals and cellular components responsible for secretion of mature enzymes from lysosomes, we are biochemically analyzing those secretion mutants that correctly localize and process mature enzymes. For instance, *HMW517* releases hydrolases at less than one-fifth the rate of release of enzymes from wild-type cells, although subcellular fractionation studies and radiolabeled pulse-chase experiments have indicated that processed forms of the enzymes are targeted to dense secondary lysosomes in the mutant (*27*). The mutant is also normal in influx and efflux rates of a fluid phase markers, and cells contain a functional secretory pathway capable of transporting macromolecules to the cell surface. The mutation in *HMW517*, therefore, does not prevent processing and transport of hydrolases, nor does it affect the secretory and endocytic pathways. The mutation may, however, inactivate a gene which functions to facilitate packaging of mature enzymes into secretory vesicles, or alternatively, the gene product may function to recognize or generate the signal responsible for induced secretion.

Native gel and nonequilibrium gel electrophoresis indicates that lysosomal enzymes in *HMW517* are less negatively charged than wild-type enzymes (*49*). In fact, many of the secretion mutants are altered in posttranslational modifications of lysosomal enzymes, while many of the mutants selected for defects in posttranslational modification are themselves defective in secretion (*46*). This suggests that a potential functional relationship exists between secretion and modifications of N-linked oligosaccharides. For instance, as described in Section II-B, a reduction in phosphorylation, the retention of glucose residues and the absence of mannose 6-sulfate, or the outer mannose residues on the nonglucosylated branches of N-linked oligosaccharides does not prevent sorting, but does alter the secretion kinetics of the hydrolases during growth and starvation.

Secretion of lysosomal enzymes from animal cells is also under genetic control (*106*). In response to acidotropic amines, normal macrophages secrete large amounts of mature lysosomal enzymes, while in contrast, macrophages from congenic-pale ear mice fail to secrete lysosomally localized enzymes although they have an efficient constitutive secretory pathway (*156*).

IV. DEVELOPMENTAL REGULATION OF THE LYSOSOMAL SYSTEM

Compared to the extensive studies investigating hydrolase transport and sorting in higher eukayotes, significantly less work has been done in the area of regulation of lysosome assembly including expression of the luminal and membrane proteins that make up the organelle. A distinct advantage to studying *D. discoideum* is that it undergoes a well-defined developmental cycle, and the expression of many of the genes coding for lysosomal enzymes is developmentally regulated (reviewed in *133,136,137*). In this section, I will describe current attempts to define the mechanisms operating during development to regulate both the synthesis and modification of lysosomal hydrolases, as well as the functional and physical changes occurring to the lysosomal system.

A. Description of the Developmental Cycle

Upon starvation, *D. discoideum* undergoes a developmental process which results in the genesis of two major differentiated cell-types, spore cells, and stalk cells, which make up the fruiting body. This process is characterized by defined morphological changes accompanied by stage-specific (temporal) changes in the expression of many developmentally regulated genes (*1,2,20*), including the genes coding for lysosomal enzymes (*137*). The developmental stages include: (1) the aggregation stage when single cells communicate with each other and form a mass (aggregate) of 100,000 cells (0–10 hr); (2) the slug stage (10–18 hr) when the organized mass of prestalk and prespore cells moves on a solid surface in response to light; and (3) the culmination stage (18–24 hr) when the organism undergoes a morphological change to generate a fruiting body consisting of mature spore cells on top of dead vacuolated stalk cells.

B. Developmental Regulation of Genes Coding for Lysosomal Enzymes

Induction of α-mannosidase activity is one of the earliest developmental events in *D. discoideum* (*132*). The cellular specific activity of the enzyme is very low in early log-phase amoebae (3×10^5 cells/ml) using bacteria as a food source, and begins to increase immediately upon starvation, accumulating linearly for the next 20 hours and plateauing as cells undergo the terminal stages of differentiation by 24 to 26 hours of development (*128,132*). The α-mannosidase structural gene mutants develop normally suggesting the enzyme is not essential for this process (*60,127*). Starvation is sufficient for induction of α-mannosidase gene expression, but it is not necessary; late log-phase cultures of amoebae (5×10^6 cells/ml) using bacteria as a food source express the α-mannosidase gene (*184*). This same regulatory pattern is observed (*16*) in cells growing in liquid broth (axenic cultures), although the gene is induced at lower cell titers (10^5 cells/ml). Furthermore, α-mannosidase levels in cells at titers greater than 5×10^5 cells/ml are already induced to near peak developmental levels, with a further increase in enzyme level occurring as cells enter stationary phase.

The accumulation of α-mannosidase activity is the result of an increase in the relative rate of synthesis of the α-mannosidase precursor (*126,128*) which can be accounted for by a parallel increase in mRNA concentration (*128,184*). As determined by Northern blot analysis, mRNA levels reach a maximum by 8 hours of development, and then begin to decline by 14 to 22 hours. Nuclear run on assays also indicate that the gene is regulated at the level of transcription (*184*). Finally, cell-cell contacts and exogenous pulses of cAMP do not significantly influence expression of the α-mannosidase gene, although these same conditions do modulate expression of other genes during early development. Interestingly, EDTA, which disrupts cell-cell interaction during early development (and has been used to examine the role of these contacts in gene expression), prevents the processing and induces the missorting of the α-mannosidase precursor.

Buffer conditioned by cells growing at high densities can induce expression of α-mannosidase (*184*) in cells growing at low density (growth conditions where the gene is normally silent). To account for this, Clarke has proposed (*32,33*) that during axenic growth or growth using bacteria as a food-source cells produce and secrete a protein, termed the prestarvation factor (PSF), which accumulates in the culture medium. Axenically growing cells apparently respond to lower concentrations of PSF (at lower cell titers) than cells using bacteria as a food source because bacteria may compete with PSF for binding sites on the cell surface. When the concentration of PSF is high relative to the concentration of the bacterial food source, cells detect this factor (perhaps via a receptor) and respond by increasing the transcription of the α-mannosidase gene and a number of other genes which are also induced during early development. Interestingly, although starvation is not required for gene induction, it is sufficient, and the mechanism by which it works may differ from induction of gene expression during growth.

The control of α-mannosidase gene expression has also been analyzed genetically. Direct screening has identified five aggregation minus-mutants which accumulate negligible amounts of α-mannosidase activity during axenic growth and development (*127*). Genetic complementation studies reveal that all of these mutations map outside the α-mannosidase structural gene, which is consistent with some of these lesions defining regulatory genes. One mutation, *mad-404*, results in the complete lack of expression of the α-mannosidase gene (*126,128*), and causes the misregulation of one-half of the genes normally expressed early in development in wild-type cells (*19,20*). Another mutation, *mad-426*, prevents α-mannosidase precursor processing (accounting in part for the lack of accumulation of enzyme activity), and the precursor molecules are slowly transported, missorted, and secreted from cells during development (*127,215*). The mutations *mad-414* and *mad-417* slow the processing of abnormally modified precursors (*127*). In summary, there exists a small number of developmentally essential genes which also regulate the synthesis, modification, and processing of α-mannosidase.

β-Glucosidase activity increases prior to aggregation (for cells previously grown using bacteria as a food source), decreases during the slug stage, and increases again during culmination (*34,84*). These changes in enzyme activity are paralleled by changes in the concentration of mRNA and in the relative rate of synthesis of the β-glucosidase precursor (*84*, J.R. Richardson and J.A. Cardelli, unpublished results).

Other lysosomal enzymes are developmentally regulated including acid phosphatase, β-hexosaminidase, and α-glucosidase-I (see Table 1); however much less is known about the mechanisms regulating expression of these genes. A good deal more is known about the regulation of expression of genes coding for proteins that may be localized to lysosomes. These genes which include α-fucosidase, cysteine proteinase-I, and cysteine proteinase-II (Table 1) are transcriptionally activated at later times in development than the α-mannosidase gene, and cAMP has been shown to act as a modulating agent; however a further description of these genes is beyond the scope of this review.

C. Changes in Posttranslational Modifications of Lysosomal Enzymes

Multiple lysosomal enzymes undergo a developmentally regulated change in posttranslational modifications to their N-linked oligosaccharides (*114,149*). Less acidic late forms of lysosomal enzymes (as revealed by isoelectric focussing gel electrophoresis) begin to appear at 6 hours of development, and by 14 hours these new forms represent over 50% of the total enzyme activity. The less acidic forms contain much less mannose 6-phosphate and mannose 6-sulfate, and, not surprisingly, Golgi-localized sulfotransferase activity decreases as development proceeds (*98*). The early (more negatively charged forms) and late forms of α-mannosidase differ in susceptibility to proteolytic inactivation and thermal stability; the late form is more stable to both conditions (*60*). Interestingly, the late form of α-mannosidase, some of which may be in precursor form, is preferentially secreted late in development (*149*). This suggests that late in development the machinery for targeting of lysosomal enzymes functions less efficiently, or that newly synthesized enzymes are missing a modification required for sorting. Finally, the late form of α-mannosidase arises by *de novo* synthesis and is not the result of the early form being desulfated and dephosphorylated (*129*).

The generation of the late forms of the enzymes is developmentally controlled as revealed by the existence of many independently isolated aggregation minus developmental mutants which, unlike the wild-type strains, do not alter posttranslational modification of enzymes during starvation. As an example, the mutants continue to sulfate their carbohydrate side-chains throughout the developmental period unlike wild-type strains which lose sulfotransferase activity. Interestingly, the absence of changes in posttranslational modifications to α-mannosidase results in an equilibrium between the rate of synthesis and the rate of inactivation of the enzyme (the specific activity of the enzyme remains constant over time), although the protein subunits are not degraded (*129*). This change in posttranslational modification therefore appears to be required for the accumulation of α-mannosidase activity in wild-type cells during later times in development because the vegetative forms and early developmental forms of the enzyme are inactivated during this period. Thus, although the enzyme is synthesized and processed in the aggregation minus developmental mutants, without the alterations in posttranslational modifications which occurs in wild-type cells, the mutant forms of the enzymes are inactivated.

Other changes in carbohydrate modifications have been reported to occur during development (*98,105*). For instance, the N-linked oligosaccharides on many non-lysosomal and lysosomal proteins are processed from $GlcNac_2Man_8$ species (vegetative forms) to $GlcNac_2Man_{5-7}$ species, an event that corresponds in time to the appearance of neutral α-mannosidase-2 activity. Golgi-localized galactosyltransferase activity also accumulates during this stage of development while fucosylation of core GlcNacs on N-linked oligosaccharides decreases.

These changes may be of little functional significance and may instead simply reflect the developmental changes that occur in the modification machinery. Alternatively, these alterations in modifications to lysosomal enzymes may play an important physiological role. For instance, the late forms of the enzymes may be more stable, and therefore, would be expected to remain functional in spores that may remain dormant for long periods of time. Also, the preferential secretion of the late forms of the enzymes may be necessary for enzymatic processing of the cell surface of prespore, prestalk, and/or mature spore cells. Together these results suggest that changes in the posttranslational modification of lysosomal enzymes may be a necessary response to a changing lysosomal and extracellular environment. In the next section, I will review recent evidence indicating that lysosomes undergo morphological and functional changes during development.

D. Changes in the Structure and Function of Lysosomes

The lysosomal vacuolar system also changes during development. Early in development, cells previously grown using bacteria as a food source decrease their phagocytic activity while concurrently increasing autophagocytic capacity (*133*). During this change from a phagocytic to autophagocytic mode of digestion, lysosomal enzymes are efficiently secreted while later in development secretion of enzymes decreases (*16*).

In addition, prespore cells produce vesicles during development, termed prespore vesicles (PSVs), which contain components used to construct the spore coat (*51*). These components, which include six different glycoproteins and a mucopolysaccharide, are exocytosed during culmination while the spore cells climb the stalk prior to encapsidation at about 20 hours of development. Lenhard et al. (*124*) have hypothesized that the PSVs resemble secretory granules based on their ability to concentrate and store material to be released in a regulated fashion during spore formation. PSVs can be separated from lysosomes on the basis of density, but surprisingly, they appear to contain newly synthesized as well as preexisting lysosomal hydrolases (*125*). This suggests that PSVs may actually represent modified lysosomes, a finding that is not without precedence. For instance, rennin containing secretory granules of juxtaglomerular cells also contain lysosomal enzymes (*59,198*). Furthermore, cytolytic lymphocytes contain specialized granules (which are important in cell mediated cytolysis) which contain secretory proteins localized in dense cores and lysosomal proteins in multivesicular domains (*15*). These lytic granules have an acidic pH, and they take up exogenous fluid phase markers which is consistent with their being modified lysosomes. Finally, azurophilic granules in neutrophils contain hydrolytic enzymes which can be secreted in response to the appropriate stimuli (*54*).

There are a number of possibilities to account for the origin of PSVs. For instance vesicles might transport hydrolases from lysosomes directly to PSVs, or during development certain lysosomes may be reconstructed into PSVs. Alternatively, newly formed PSVs lacking hydrolases may fuse with preformed lysosomes. Regardless of the mechanism involved, *D. discoideum* will be a useful system to investigate the relationship between lysosomes and secretory like granules which may be more closely related to each other then previously thought.

V. SUMMARY AND FUTURE PERSPECTIVES

In this review I have emphasized by way of comparative analysis the importance of studying systems like *D. discoideum* (which is amenable to genetic and biochemical approaches) in order to increase our knowledge about alternative (MPR-independent) lysosomal enzyme targeting pathways. Although the MPR-dependent transport pathway is still the best characterized, it is clear a significant number of mammalian cell types, as well as organisms such as yeast, plants, and the slime mold, rely on alternative (MPR-independent) sorting mechanisms. In many ways, the processing and modification of *D. discoideum* lysosomal enzymes are similar to the lysosomal biosynthesis pathways in MPR-dependent cells (Fig. 2). For instance, the enzymes are: proteolytically processed to mature forms; phosphorylated and sulfated on N-linked oligosaccharides; and transported at different rates to dense lysosomes. However, it is clear that the lysosomal trafficking pathway in *D. discoideum* and other cells using alternative (MPR-independent) pathways differ in many important respects from the localization process in fibroblasts. For instance, the maintenance of acidic-pH vacuolar compartments is not required for correct sorting of hydrolases in *D. discoideum*, as well as in yeast and some other mammalian cell lines. Also, N-linked oligosaccharide side-chains (including modifications like sulfate and phosphate) may not play a direct role in sorting. Furthermore, inhibition of processing of precursor polypeptides in the slime mold results in missorting and hypersecretion of uncleaved precursors, suggesting that processing and sorting may be coupled. Also precursor forms of lysosomal enzymes in the slime mold and many other MPR-independent cell types are tightly associated with membranes until formation of mature enzymes. What is becoming clear from these studies is that *D. discoideum* and other MPR-independent cells may share common mechanisms operating to localize hydrolases to lysosomes.

In *D. discoideum*, the expression of many genes coding for lysosomal enzymes is developmentally regulated; the mature lysosomal enzymes undergo changes in posttranslational modifications, and they are secreted in a regulated manner. Finally lysosomes themselves are morphologically and functionally altered during development to more closely resemble secretory granules. *D. discoideum* is well-suited for the studies listed above because classical genetic (cells are normally haploid) and molecular genetic approaches can be employed. Gene disruption techniques and antisense mutagenesis (both creating null mutants) are being used to dissect the lysosomal targeting and biogenesis pathways. Furthermore, unlike yeast and plants, this organism does not have a cell wall and contains a lysosomal network more similar to the vacuolar system of mammalian cells. Current studies are being directed towards understanding the molecular nature of the sorting signals on lysosomal hydrolases and the biogenesis and function of lysosomes in growing and developing cells.

ACKNOWLEDGMENTS

Research by the author was supported by a grant from the National Institutes of Health (DK 39232) and by funding from the LSUMC-S Center for Excellence in Arthritis Research and the Center for Excellence in Cancer Research. The author would like to thank members of his laboratory, John Bush, John Schatzle and especially Tracey Ruscetti, for critical reading of the manuscript, and Cynthia Harrison for help in the preparation of this document.

REFERENCES

1. Alton, T.A. and Lodish, H.F. (1977a). Developmental changes in mRNAs and protein synthesis in *Dictyostelium discoideum*. *Dev. Biol.* 60: 180–206.
2. Alton, T.A. and Lodish, H.F. (1977b). Translational control of protein synthesis during early stages of differentiation of the slime mold, *Dictyostelium discoideum*. *Cell* 12: 301–310.
3. Ashworth, J.M., and Quance, J. (1972). Enzyme synthesis in myxamoebae of the cellular slime mold, *Dictyostelium discoideum*. *Biochem. J.* 126: 601–608.
4. Banta, L.M., Robinson, J.S., Klionsky, D.J., and Emr, S.D. (1988). Organelle biogenesis in yeast: Characterization of yeast mutants defective in vacuolar biogenesis and protein sorting. *J. Cell Biol.* 107: 1369–1383.
5. Barriocanal, J.G., Bonifacino, J.S., Yuan, L., and Sandoval, I.V. (1986). Biosynthesis, glycosylation, movement through the Golgi system by an N-linked carbohydrate independent mechanism of three lysosomal integral membrane proteins. *J. Biol. Chem.* 261: 16755–16763.
6. Bennett, V.D., and Dimond, R.L. (1986). Biosynthesis of two developmentally distinct acid phosphatase isozymes in *Dictyostelium discoideum*. *J. Biol. Chem.* 261: 5355–5362.
7. Blum, J.J. and Rothstein, T.L. (1975). Lysosomes in *Tetrahymena*. In: *Lysosomes in Biology and Pathology*, Vol. 4, pp. 33–45. (Dingle, J.T. and Dean, R.T., Eds.). Elsevier, New York.
8. Borts, R.H. and Dimond, R.L. (1981a). The α-glucosidases of *Dictyostelium discoideum*. I. Identification and characterization. *Dev. Biol.* 87: 176–184.
9. Borts, R.H. and Dimond, R.L. (1981b). The α-glucosidases of *Dictyostelium discoideum*. II. Developmental regulation and cellular localization. *Dev. Biol.* 87: 185–192.
10. Bourne, H.R., Sanders, D.A., and McCormick, F (1990) The GTPase superfamily a conserved switch for diverse cell functions. *Nature* 348: 125–132.
11. Braulke, T., Geuze, H.J., Slot, J.W., Hasilik, A., and von Figura, K. (1987). On the effects of weak bases and monensin on sorting and processing of lysosomal enzymes in human cells. *Eur. J. Cell Biol.* 43: 316–321.
12. Brown, J.A. and Swank, R.T. (1983). Subcellular redistribution of newly synthesized macrophage lysosomal enzymes. *J. Biol. Chem.* 258: 15323–15328.
13. Brown, J.A., Novak, E.K., and Swank, R.T. (1985). Effects of ammonia on processing and secretion of precursor and mature lysosomal enzyme from macrophages of normal and pale ear mice: Evidence for two distinct pathways. *J. Cell Biol.* 100: 1894–1904.
14. Brown, W.J., Goodhouse, J. and Farquhar, M.G. (1988). Mannose-6-phosphate receptors for lysosomal enzymes cycle between the Golgi complex and endosomes. *J. Cell Biol.* 103: 1235–1247.
15. Burkhardt, J.K., Hester, S., Lapham, C.K., and Argon, Y. (1990). The lytic granules of natural killer cells are dual function organelles combining secretory and pre-lysosomal compartments. *J. Cell. Biol.* 111: 2327–2340.
16. Burns, R.A., Livi, G.P., and Dimond, R.L. (1981). Regulation and secretion of early developmentally controlled enzymes during axenic growth in *Dictyostelium discoideum*. *Dev. Biol.* 84: 407–416.
17. Bush, J.M. and Cardelli, J.A. (1989). Processing, transport, and secretion of Acid Phosphatase in *Dictyostelium discoideum*. *J. Biol. Chem.* 264: 7630–7636.
18. Bush, J.M. and Cardelli, J.A. (1990). Alterations to N-linked oligosaccharide which affect intracellular transport and regulated secretion but not sorting of lysosomal acid phosphatase in *Dictyostelium discoideum*. *Arch. Bioch. Biophy.* 283: 158–166.
19. Cardelli, R.A., Livi, G.P., and Mierendorf, R.C., Knecht, D.A., and Dimond, R.L. (1984). Gene regulation during early development of the cellular slime mold *Dictyostelium discoideum*. In: *Molecular Biology of Development (UCLA) Symposia on Molecular and Cellular Biology*, Vol. 19, pp. 427–435. (Davidson, E.H. and Firtel, R.A., Eds.). Alan R. Liss, New York.

20. Cardelli, J.A., Knecht, D.A., Wunderlich, R., and Dimond, R.L. (1985). Major changes in gene expression occur during at least four stages of development of *Dictyostelium discoideum. Dev. Biol.* 110: 147–156.
21. Cardelli, J.A., Golumbeski, G.S., and Dimond, R.L. (1986a). Lysosomal enzymes in *Dictyostelium discoideum* are transported at distinctly different rates. *J. Cell Biol.* 102: 1264–1270.
22. Cardelli, J., Mierendorf, R.C., and Dimond, R.L. (1986b). Initial events involved in the synthesis of the lysosomal enzyme α-mannosidase in *Dictyostelium discoideum. Arch. Biochem. Biophys.* 244: 338–345.
23. Cardelli, J.A., Golumbeski, G.S., Woychick, N.A., Ebert, D.L., Mierendorf, R.C., and Dimond, R.L. (1987). Defining the intracellular localization pathways followed by lysosomal enzymes in *Dictyostelium discoideum*. In: *Methods in Cell Biology*, Vol. 28, pp. 139–155 (Spudich, J., Ed.), Academic Press, Orlando.
24. Cardelli, J.A. and Dimond, R.L. (1988). Transport and targeting of lysosomal enzymes in the cellular slime mold, *Dicytostelium discoideum*. In: *Protein Transfer and Organelle Biogenesis*, pp. 363–399. (Ras, R.C. and Robbins, P.W., Eds.). Academic Press, New York.
25. Cardelli, J.A., Richardson, J.R., and Miears, D. (1989). Role of acidic intracellular compartments in the biosynthesis of *Dictyostelium* lysosomal enzymes: the weak bases ammonium chloride and chloroquine differentially affect proteolytic processing and sorting. *J. Biol. Chem.* 264: 3454–3463.
26. Cardelli, J.A., Bush, J.M., and Freeze, H.H. (1990a). Sulfate on N-linked oligosaccharides affect secretion but are not essential for the transport, proteolytic processing, and sorting of lysosomal enzymes in *Dictyostelium discoideum. J. Biol. Chem.* 265: 8847–8853.
27. Cardelli, J.A., Schatzle, J., Bush, J.M., Richardson, J. Ebert, D.L., and Freeze, H.H. (1990b). Biochemical and genetic analysis of the biosynthesis, sorting and secretion of *Dictyostelium* lysosomal enzymes. *Dev. Genet.* 11: 454–462.
28. Chan, S.J., McCormic, M.B., and Steiner, D.F. (1988). Nucleotide and predicted amino acid sequences of cloned human and mouse preprocathepsin B cDNAs. *Proc. Natl. Acad. Sci. USA* 83: 7721–7725.
29. Chou, P.Y. and Fasman, G.D. (1978). Empirical predictions of protein conformations. *Ann. Rev. Bioch.* 47: 251–276.
30. Chrispeels, M.J. and Tague, B.W. (1991). Protein sorting in the secretory system of plant cells. *Inter. Rev. Cyt.* 125: 1–45.
31. Chvatchko, Y., Howard, I., and Roizman, H. (1986). Two yeast mutants defective in endocytosis are defective in pheromone response. *Cell* 46: 355–364.
32. Clarke, M., Rilen, K., and Kayman, S.C. (1987). Density dependence induction of discoidin I: Synthesis in exponentially growing cells of *Dictyostelium discoideum. Differentiation* 34: 79–87.
33. Clarke, M., Kayman, S.C., and Yang, J. (1988). Analysis of the prestarvation response in growing cells of *Dictyoselium discoideum. Dev. Genet.* 9: 315–326.
34. Coston, M.B. and Loomis, W.F. (1969). Isozymes of β-glucosidase in *Dictyostelium discoideum. J. Bacteriol.* 100: 1208–1217.
35. Couso, R., Lang, L., Roberts, R.M., and Kornfeld, S. (1986). Phosphorylation of the oligosaccharide of uteroferrin by UDP-GlcNAc: Glycoprotein N-acetyl glucosamine-1-phosphotransferases from rat liver, *Acanthamoeba castellani* and *Dictyostelium discoideum* requires α-1,2-linked mannose residues. *J. Biol. Chem.* 261: 6326–6331.
36. Cousò, R., van Halbeek, H., Reinhold, V., and Kornfeld, S. (1987). The high mannose of oligosaccharides of *Dictyostelium discoideum* glycoproteins contain a novel intersecting N-acetyl glucosamine residue. *J. Biol. Chem.* 262: 4521–4523.
37. Courtoy, P.J., Quintart, J., and Bandhuin, P. (1984). Shift of equilibrium density induced by 3,3′ diaminobenzidene cytochemistry: A new procedure for the analysis and purification of peroxidase containing organelles. *J. Cell Biol.* 98: 870–876.

38. Crean, E.V. and Rossomando, E.F. (1979). Effects of sugars on glycosidase secretion in *Dictyostelium discoideum*. *J. Gen. Microbiol.* 110: 315–322.
39. Datta, S. and Firtel, R.A. (1987). Identification of the sequences controlling cAMP regulation and cell-type specific expression of a prestalk-specific gene in *Dictyostelium discoideum*. *Mol. Cell Biol.* 7: 149–159.
40. De Chastellier, C., Ryter, A., and Thilo, L. (1983). Membrane shuttle between plasma membrane, phagosomes, and pinosomes in *Dictyostelium discoideum* amoeboid cells. *Eur. J. Cell Biol.* 30: 233–243.
41. Diment, S. Leech, M.S., and Stahl, P. (1988) Cathepsin D is membrane associated in macrophage endosomes. *J. Biol. Chem.* 263: 6901–6907.
42. Dimond, R.L., and Loomis, W.F. (1976). Structure and function of β-glucosidases in *Dictyostelium discoideum*. *J. Biol. Chem.* 251: 2680–2687.
43. Dimond, R. L., Brenner, M., and Loomis, W.F. (1973). Mutations affecting N-acetylglucosaminidase in *Dictyostelium discoideum*. *Proc. Natl. Acad. Sci. USA* 70: 3356–3360.
44. Dimond, R.L., Mayer, M., and Loomis, W.F. (1976). Characterization and developmental regulation of β-galactosidase isozymes in *Dictyostelium discoideum*. *Dev. Biol.* 52: 74–82.
45. Dimond, R.L., Burns, R.A., and Jordan, K.B. (1981). Secretion of Lysosomal enzymes in the cellular slime mold, *Dictyostelium discoideum*. *J. Biol. Chem.* 256: 6565–6572.
46. Dimond, R.L., Knecht, D.A., Jordan, K.B., Burns, R.A., and Livi, G.P. (1983). Secretory mutants in cellular slime mold, *Dictyostelium discoideum*. *Methods Enzymol.* 96: 815–828.
47. Ebert, D.L., Bush, J.M, Dimond, R.L., and Cardelli, J.A. (1989a). Biogenesis of lysosomal enzymes in the α-glucosidase II-deficient modA mutant of *Dictyostelium discoideum*: retention of α-1,3-linked glucose on N-linked oligosaccharides delays intracellular transport but does not alter sorting of α-mannosidase or β-glucosidase *Arch. Bioch. Bioph.* 273: 479–490.
48. Ebert, D.L., Freeze, H.H., Richardson, J.M., Dimond, R.L., and Cardelli, J.A. (1989b). A *Dictyostelium discoideum* mutant that missorts (and oversecretes) lysosomal enzyme precursors is defective in endocytosis. *J. Cell Biol.* 109: 1445–1456.
49. Ebert, D.L., Jordan, K.B., and Dimond, R.L. (1990). Lysosomal enzyme secretory mutants of *Dictyostelium discoideum*. *J. Cell Sci.* 96: 491–500.
50. Eisenberg, D., Weiss, R.M., and Terwilliger, T.C. (1984). The hydrophobic moments detects periodicity in protein hydrophobicity. *Proc. Natl. Acad. Sci. USA* 81: 140–144.
51. Erdos, G.W. and West, C.M. (1989). Formation and organization of the spore coat proteins of *Dictyostelium discoideum*. *Exp. Mycol.* 13: 169–182.
52. Erickson, A.H., Conner, G.E., and Blobel, G. (1981). Biosynthesis of a lysosomal enzyme. Partial structure of two transient and functionally distinct NH2-terminal sequences in cathepsin D. *J. Biol. Chem.* 256: 11224–11231.
53. Erickson, A.H., Ginns, E.I., Barranger, J.A. (1985). Biosynthesis of the lysosomal enzyme, glucocerebrosidase. *J. Biol. Chem.* 260: 14319–14324.
54. Estensen, R.D., White, J.G., and Holmes, B. (1974). Specific degranulation of human polymorphonuclear leukocytes. *Nature* 248: 347–348.
55. Every, D. and Ashworth, J.M. (1973). The purification and properties of the extracellular glycosidases of the cellular slime mold *Dictyostelium discoideum*. *Biochem. J.* 133: 37–47.
56. Fischer, H.D., Creek, D.E., and Sly, W.S. (1982). Binding of phosphorylated oligosaccharides to immobilized phosphomannosyl receptors. *J. Biol. Chem.* 257: 9938–9943.
57. Fitting, T. and Kabat, D. (1982). Evidence for a glycoprotein "signal" involved in transport between subcellular organelles. *J. Biol. Chem.* 257: 14011–14017.
58. Fok, A.K. and Allen, R.D. (1990). The phagosome-lysosome membrane system and its regulation in *Paramecium*. *Inter. Rev. Cyto.* 123: 61–94.
59. Fransen, J.A.M. (1987). Lysosomal characteristics of human renin-containing granules. *Lab. Invest.* 56: 124.
60. Free, S.J. and Loomis, W.F. (1974). Isolation of mutations in *Dictyostelium discoideum* affecting α-mannosidase. *Biochimie* 84: 159–174.

61. Free, S.J. and Schimke, R.T. (1978). Effects of a posttranslational modification mutation on different developmentally regulated glycosidases in *Dictyostelium*. *J. Biol. Chem.* 253: 4107–4111.
62. Free, S.J., Schimke, R.T., and Loomis, W.F. (1976). The structural gene for α-mannosidase in *Dictyostelium discoideum*. *Genetics* 84: 159–174.
63. Free, S., Schimke, R.T., Freeze, H., and Loomis, W.F. (1978). Characterization and genetic mapping of *modA*. A mutation in the posttranslational modification of the glycosidases of *Dictyostelium discoideum*. *J. Biol. Chem.* 253: 4102–4106.
64. Freeze, H.H. (1985). Mannose 6-sulfate is present in the N-linked oligosaccharides of lysosomal enzymes of *Dictyostelium*. *Arch. Biochem. Biophys.* 243: 690–693.
65. Freeze, H. and Miller, A.L. (1980). *modA*: A posttranslational mutation affecting phosphorylated and sulfated glycoproteins in *Dictyostelium discoideum*. *J. Biol. Chem.* 261: 135–141.
66. Freeze, H., Miller, A.L., and Kaplan, A. (1980). Acid hydrolases from *Dictyostelium discoideum* contain phosphomannosyl recognition markers. *J. Biol. Chem.* 255: 11081–11084.
67. Freeze, H.H., Yek, R., Miller, A.L., and Kornfeld, S. (1983a). Structural analysis of the asparagine-linked oligosaccharides from three lysosomal enzymes of *Dictyostelium discoideum*. *J. Biol. Chem.* 258: 14874–14879.
68. Freeze, H.H., Yeh, R.Y., and Miller, A.L. (1983b). Uptake of α-D-mannosidase and β-D-glucosidase from *Dictyostelium discoideum* via phosphohexosyl receptor on normal human fibroblasts. *J. Biol. Chem.* 258: 8928–8933.
69. Freeze, H.H., Yeh, R., Miller, A.L., and Kornfeld, S. (1983c). The *modA* mutant of *Dictyostelium discoideum* is missing the α-1,3-glucosidase involved in asparagine linked oligosaccharide processing. *J. Biol. Chem.* 258: 14880–14884.
70. Freeze, H.H., Mierendorf, R.C., Wunderlich, R., and Dimond, R.L. (1984). Sulfated oligosaccharides block antibodies to many *Dictyostelium discoideum* acid hydrolases. *J. Biol. Chem.* 259: 10641–10643.
71. Freeze, H.H. (1986). Modifications of lysosomal enzymes in *Dictyostelium discoideum*. *Mol. Cell Biochem.* 72: 47–65.
72. Freeze,, H.H. and Wolgast, D. (1986b). Structural analysis of N-linked oligosaccharides from glycoproteins secreted by *Dictyostelium discoideum*. *J. Biol. Chem.* 261: 127–134.
73. Freeze, H.H. and Wolgast, D. (1986a). Biosynthesis of methyl phosphomannosyl residues in the oligosaccharides of *Dictyostelium discoideum*. *J. Biol. Chem.* 261: 135–141.
74. Freeze, H.H., Bush, J.M., and Cardelli, J.A. (1990) Biochemical and genetic analysis of an antigenic determinant found on N-linked oligosaccharides in *Dictyostelium*. *Dev. Genetics* 11: 463–472.
75. Freeze, H.H., Koza-Taylor, P., Saunders, A., and Cardelli, J.A. (1989b). The effect of altered N-linked oligosaccharide structures on maturation and targeting of lysosomal enzymes in Dictyostelium. *J. Biol. Chem.* 264: 19278–19286.
76. Freeze, H.H., Willies, L., Hamilton, S., and Koza-Taylor, P. (1989a). Two mutants of *Dictyostelium discoideum* that lack a sulfated carbohydrate antigenic determinant synthesize a truncated lipid-linked precursor of N-linked oligosaccharides. *J. Biol. Chem.* 264: 5653–5659.
77. Fries, E., Gustafsson, L., and Peterson, P.A. (1984). Four secretory proteins synthesized by hepatocytes are transported from endoplasmic reticulum to Golgi complex at different rates. *EMBO J.* 3: 147–152.
78. Frisch, A., and Neufeld, E.F. (1981). Limited proteolysis of the β-hexosaminidase precursor in a cell-free system. *J. Biol. Chem.* 256: 8242–8246.
79. Fuller, R.S., Brake, A., and Thorner, J. (1989). Yeast prohormone processing enzyme (Kex 2 gene product) is a Ca^{2+}-dependent serine protease. *Proc. Natl. Acad. Sci. USA* 86: 1434–1438.
80. Gabel, C.A., Costello, C.E., Reinhold, V.N., Kurz, L., and Kornfeld, S. (1984). Identification of methylphosphomannosyl residues as components of the high mannose oligosaccharides of *Dictyostelium discoideum* glycoproteins. *J. Biol. Chem.* 259: 13762–13769.

81. Geuze, H.J., Slot, J. W., Strous, G. J. A. M., Hasilik, A., and von Figura, K. (1984). Ultrastructural localization of the mannose 6-phosphate receptor in rat liver. *J. Cell Biol.* 98: 2047–2054.
82. Gieselmann, V., Pohlmann, R., Hasilik, A., and von Figura, K. (1983). Biosynthesis and transport of cathepsin D in cultured human fibroblasts. *J. Cell Biol.* 97: 1–5.
83. Gieselmann, V., Hasilik, A., and von Figura, K. (1985). Processing of human cathepsin D in lysosomes *in vitro*. *J. Biol.* 260: 3215–3220.
84. Golumbeski, G.S., and Dimond, R.L. (1987). Developmentally regulated expression of temporally distinct β-glucosidase isozymes in *Dictyostelium discoideum*. *Dev. Biol.* 123: 494–499.
85. Gonzalez, C., Klein, G., and Satre, M. (1990). Caffeine, an inhibition of endocytosis in *Dictyostelium discoideum* amoebae. *J. Cell Phys.* 144: 408–415.
86. Gottlieb, M. and Dweyer, D.M. (1981). Protozoan parasite of humans: surface membrane with externally disposed acid phosphatase. *Science* 212: 939–940.
87. Goud, B., Salminen, A., Walworth, N.C., and Novick, P.J. (1988) A GTP-binding protein required for secretion rapidly associates with secretory vesicles and the plasma membrane in yeast. *Cell* 53: 753–768.
88. Graham, T.R., Zassenhaus, H.P., and Kaplan, A. (1988). Molecular cloning of the cDNA which encodes β-N-acetylhexosaminidase A from *Dictyostelium discoideum*. *J. Biol. Chem.* 263: 16823–16829.
89. Green, E. D., Gruenebaum, J., Bielinska, M., Baenziger, J.V., and Boim, I. (1984). Sulfation of leutropin oligosaccharides with a cell-free system. *Proc. Natl. Acad. Sci. USA* 81: 5320–5324.
90. Griffiths, G., Hoflack, B., Simons, K., Mellman, I., and Kornfeld, S. (1988). The mannose 6-phosphate receptor and the biogenesis of lysosomes. *Cell* 52: 329–341.
91. Gross, V., Andrus, T., Tran-Thi, T.-A., Schwarz, R.T., Decker, K., and Heinrich, P.C. (1983). 1-Deoxynorjirimycin impairs oligosaccharide processing of α-1-proteinase inhibitors and inhibits its secretion in primary cultures of rat hepatocytes. *J. Biol. Chem.* 258: 12203–12209.
92. Gustafson, G.L. and Milner, L.A. (1980). Occurrence of N-acetylglucosamine 1-phosphate in proteinase I from *Dictyostelium discoideum*. *J. Biol. Chem.* 255: 7208–7210.
93. Gustafson, G.L. and Thon, L.A. (1979). Purification and characterization of a proteinase from *Dictyostelium discoideum*. *J. Biol. Chem.* 254: 12471–12478.
94. Hasilik, A. and Neufeld, E.F. (1980). Biosynthesis of lysosomal enzymes in fibroblasts. *J. Biol. Chem.* 255: 4937–4945.
95. Hasilik, A. and von Figura, K. (1984). Processing of lysosomal enzymes in fibroblasts. In: *Lysosomes in Biology and Pathology*, pp. 3–16 (Dingle, J.T., Dean, R.T., and Sly, W., Eds.). Elsevier, Amsterdam.
96. Hasilik, A., Voss, B., and von Figura, K. (1981). Transport and processing of lysosomal enzymes by smooth muscle cells and endothelial cells. *Exp. Cell Res.* 133: 23–30.
97. Heifetz, A. and Lennarz, W.J. (1979). Biosynthesis of N-glycosidically linked glycoproteins during gastrulation of sea urchin embryos. *J. Biol. Chem.* 254: 6119–6127.
98. Henderson, E.J. (1984). The role of glycoproteins in the life cycle of the cellular slime mold *Dictyostelium discoideum*. In: *The Biology of Glycoproteins*, pp. 371–443 (Ivatt, R., Ed.). Plenum, New York.
99. Hentze, M., Hasilik, A., and von Figura, K. (1984). Enhanced degradation of cathepsin D synthesized in the presence of the threonine analog β-hydroxynorvaline. *Arch. Biochem. Biophy.* 230: 375–382.
100. Herman, P.K., Stack, J.H., DeModena, J.A., and Emr, S.D. (1991). A novel protein kinase homolog essential for protein sorting to the yeast lysosome-like vacuole. *Cell* 64: 425–437.
101. Hohmann, H., Gerish, G., Lee, R., and Huttner, W.B. (1985). Cell-free sulfation of the contact site A glycoprotein of *Dictyostelium discoideum* and of a partially glycosylated precursor. *J. Biol. Chem.* 260: 13869–13878.
102. Hohman, T.C. and Bowers, B. (1984). Hydrolase secretion is a consequence of membrane recycling. *J. Cell Biol.* 98: 246–252.

103. Hurtley, M. and Helenius, A. (1989) Protein oligomerization in the endoplasmic reticulum. *Ann. Rev. Cell Biology* 5: 277–308.
104. Ingalls, H.M., Barcelo, G., Wuestehube, L.J., and Cuna, E.J. (1989). *Differentiation* 41: 87–98.
105. Ivatt, R.L., Das, O.P., Henderson, E.J., Robbins, P.W. (1984). Glycoprotein biosynthesis in *Dictyostelium discoideum*: developmental regulation of the protein-linked glycans. *Cell* 38: 561–567.
106. Jessup, W., Leoni, P., and Dean, R.T. (1985). Constitutive and triggered lysosomal enzyme secretion. In: *Developments in Cell Biology I. Secretory Processes*, pp. 38–57. (Dean, R.T. and Stahl, P., Eds.). Butterworths, London.
107. Judelson, H.S., Freeze, H.H., and Dimond, R.L. (1987). Characterization and distribution of multiple antigens on N-linked oligosaccharides of *Dictyostelium discoideum* proteins. *Arch. Bioch. Bioph.* 253: 305–314.
108. Judelson, H.S., Burns, R.A., and Dimond, R.L. (1987). A locus regulating N-acetylglucosaminidase synthesis during development in *Dictyostelium*. *Dev. Biol.* 120: 170–176.
109. Klein, G. and Satre, M. (1986). Kinetics of fluid phase pinocytosis in *Dictyostelium discoideum* amoebae. *Bioch. Bioph. Res. Comm.* 138: 1146–1152.
110. Klein, G., Cotter, D.A., Martin, J.B., and Satre, M. (1989). Vanadate, an inhibitor of growth, development, and endocytosis in *Dictyostelium discoideum* amoebae. *J. Cell Sci.* 94: 127–134.
111. Klionsky, D.J., Herman, P.K., and Emr, S.D. (1990). The fungal vacuole composition, function and biogenesis. *Micro. Reviews* 54: 266–292.
112. Knecht, D.A. and Dimond, R.L. (1981). Lysosomal enzymes possess a common antigenic determinant in the cellular slime mold: *Dictyostelium discoideum*. *J. Biol. Chem.* 256: 3564–3575.
113. Knecht, D.A., and Dimond, R.L. Wheeler, S., and Loomis, W.F. (1984). Antigenic determinants shared by lysosomal proteins of *Dictyostelium discoideum*. *J. Biol. Chem.* 259: 10633–10640.
114. Knecht, D.A., Green, E.D., Loomis, W.F., and Dimond, R.L. (1985). Developmental changes in the modification of lysosomal enzymes in *Dictyostelium discoideum*. *Dev. Biol.* 107: 490–502.
115. Kornfeld, R. and Kornfeld, S. (1985). Assembly of asparagine-linked oligosaccharides. *Annu. Rev. Biochem.* 54: 631–634.
116. Kornfeld, S. (1986). Trafficking of lysosomal enzymes in normal and disease states. *J. Clin. Invest.* 77: 1–6.
117. Kornfeld, S. and Mellman, I. (1989). The biogenesis of lysosomes. *Annu. Rev. Bioch.* 5: 483–525.
118. Kyte, J. and Doolittle, R.F. (1982). A simple method for displaying the hydropathic character of a protein. *J. Mol. Biol.* 157: 105–132.
119. LaCoste, C.H., Freeze, H.H., Jones, J.A., and Kaplan, A. (1989). Characteristics of the sulfation of N-linked oligosaccharides in vesicles from *Dictyostelium discoideum*: *in vitro* sulfation of lysosomal enzymes. *Arch Biochem. Bioph.* 273: 505–515.
120. Lang, L., Couso, R., and Kornfeld, S. (1986). Glycoprotein phosphorylation in simple eukaryotic organisms. Identification of UDP-GlcNAc: glycoprotein N-acetylglucosamine-1-phosphotransferase activity and analysis of substrate specificity. *J. Biol. Chem.* 261: 6320–6325.
121. Lemansky, P., Gieselmann, V., Hasilik, A., and von Figura, K. (1984). Cathepsin D and β-hexosaminidase synthesized in the presence of 1-deoxynorjirimycin accumulate in the endoplasmic reticulum. *J. Biol. Chem.* 259: 10129–10135.
122. Lemansky, P., Gieselmann, V., Hasilik, A., von Figura, K. (1985). Synthesis and transport of lysosomal acid phosphatase in normal and I-cell fibroblasts. *J. Biol. Chem.* 260: 9023–30.
123. Lemkin, M. and Farquhar, M.G. (1981). Sulfated and nonsulfated glycosaminoglycans and glycopeptides are synthesized by kidney *in vivo* and incorporated into glomerular basement membranes. *Proc. Natl. Acad. Sci. USA* 78: 1726–1730.
124. Lenhard, J.M., Siegel, A., and Free, S.J. (1989a). Developing *Dictyostelium* cells contain the lysosomal enzyme α-mannosidase in a secretory granule. *J. Cell Biol.* 109: 2761–2769.
125. Lenhard, J.M., Kasperek, E., Moore, B.R., and Free, S.J. (1989). Developing *Dictyostelium* cells contain two distinct acid hydrolase containing vesicles. *Exp. Cell Res.* 182: 142–155.

126. Livi, G.P., Cardelli, J.A., Mierendorf, R.C., and Dimond, R.L. (1984). Developmental control of α-mannosidase-1 synthesis in *Dictyostelium discoideum*. In: *Molecular Biology of Development*, UCLA Symposia on Molecular and Cellular Biology, Vol. 19, pp. 447–457. (Davidson, E.H. and Firtel, R.A., Eds.). Alan R. Liss, New York.
127. Livi, G.P., Cardelli, J.A., and Dimond, R.L. (1985a). α-Mannosidase-1 mutants of *Dictyostelium discoideum* and early aggregation-essential genes regulate enzyme precursor synthesis, modification, and processing. *Differentiation* 29: 207–215.
128. Livi, G.P., Cardelli, J.A., Mierendorf, R.C., and Dimond, R.L. (1985b). Regulation of lysosomal α-mannosidase-1 synthesis during development in *Dictyostelium discoideum*. *Dev. Biol.* 110: 514–520.
129. Livi, G.P., Woychick, N.A., and Dimond, R.L. (1985c). Lysosomal enzyme inactivation associated with defects in posttranslational modification during development in *Dictyostelium discoideum*. *Differentiation* 30: 83–91.
130. Lodish, H.F. and Kong, N. (1984). Glucose removal from N-linked oligosaccharides is required for efficient maturation of certain secretory glycoproteins from the rough endoplasmic reticulum to the Golgi complex. *J. Cell Biol.* 98: 1720–1729.
131. Lodish, H.F., Kong, N., Snider, M., and Strous, G.J.A.M. (1983). Hepatoma secretory proteins migrate from rough endoplasmic reticulum to Golgi at characteristic rates. *Nature (London)* 304: 80–83.
132. Loomis, W.F. (1970). Developmental regulation of α-mannosidase in *Dictyostelium discoideum*. *J. Bacteriol.* 103: 375–381.
133. Loomis, W.F. (1975). *Dictyostelium discoideum: A Developmental System*. Academic Press, New York.
134. Loomis, W.F. (1978). Genetic analysis of the gene for N-acetylglucosaminidase in *Dictyostelium discoideum*. *Genetics* 88: 277–284.
135. Loomis, W.F. (1980b). A β-glucosidase gene of *Dictyostelium discoideum*. *Dev. Genet.* 1: 241–246.
136. Loomis, W.F. (1982). *The Development of Dictyostelium discoideum*. Academic Press, New York.
137. Loomis, W.F. (1989). Essential genes for development of *Dictyostelium discoideum*. *Progress in Mol. and Cell. Biol.* 11: 159–183.
138. Loomis, W.F. and Kuspa, A. (1984). Biochemical and genetic analysis of pre-stalk specific acid phosphatase in *Dictyostelium*. *Dev. Biol.* 102: 498–503.
139. Mahuran, D.J., Neote, K., Klavins, M.H., Leung, A., and Gravel, R.A. (1988). Proteolytic processing of pro-α and pro-β precursors from human β-hexosaminidase. *J. Biol. Chem.* 263: 4612–4618.
140. Mainferme, F., Wattiaux, R., and von Figura, K. (1985). Synthesis, transport, and processing of Cathepsin C in Morris hepatoma 7777 cells and rat hepatocytes. *Eur. J. Bioch.* 153: 211–216.
141. Matsuoka, K. and Nakamura, K. (1991). Propeptide of a precursor to a plant vacuolar protein required for vacuolar targeting. *Proc. Natl. Acad. Sci. USA* 88: 834–838.
142. May, T., Kern, H., Muller-Taubenberger, and Nellen, W.F. (1989). Identification of a *cis*-acting element controlling induction of early gene expression in *Dictyostelium discoideum*. *Mol. Cell. Biol.* 9: 4653–4659.
143. McCune, J.M., Rabin, L.G., Feinberg, M.B., Lieberman, M., Kosek, J.C., Reyes, G.R., and Weissman, I.L. (1988). Endoproteolytic cleavage of gp 160 is required for the activation of human immunodeficiency virus. *Cell* 53: 55–67.
144. Mellman, I., Fuchs, R., and Helenius, A. (1986) Acidification of the endocytic and exocytic pathway. *Annu. Rev. Biochem.* 55: 663–700.
145. McIntyre, G.F. and Erickson, A.H. (1991). Procathepsins L and D are membrane-bound in acidic microsomal vesicles. *J. Biol. Chem.* (In Press).
146. Merion, M. and Sly, W.S. (1983). The role of intermediate vesicles in the adsorptive endocytosis and transport of ligand to lysosomes by human fibroblasts. *J. Cell Biol.* 96: 644–650.

147. Mierendorf, R.C., Cardelli, J.A., Livi, G.P., and Dimond, R.L. (1983). Synthesis of related forms of the lysosomal enzyme α-mannosidase in *Dictyostelium discoideum*. *J. Biol. Chem.* 258: 5878–5884.
148. Mierendorf, R.C., Cardelli, J.A., and Dimond, R.L. (1985). Pathways involved in targeting and secretion of a lysosomal enzyme in *Dictyostelium discoideum*. *J. Cell Biol.* 100: 1777–1787.
149. Moore, B.R., Gossels, J.D., and Free, S.J. (1985). Changes in the posttranslational modification of lysosomal enzymes during development of *Dictyostelium*. *Differentiation* 29: 7–13.
150. Muller-Taubenberger, A., Westphal, M., Noegel, A., and Gerisch, G. (1989). A developmentally regulated gene product from *Dictyostelium discoideum* shows high homology to human α-L-fucosidase. *FEBS Lett.* 246: 1855–192.
151. Mundy, D.I. and Strittmatter, W.J. (1985). Requirement for metallo-endoprotease in exocytosis: Evidence in mast cells and adrenal chromaffin cells. *Cell* 40: 645–656.
152. Neufeld, E.F. and McKusick, V.A. (1983). Disorders of lysosomal enzyme synthesis and localization: I-Cell disease and pseudo-Hurler polydistrophy. In: *Metabolic Basis of Inherited Disease*, 5th ed., pp. 778–787 (Stanbury, J.B., Wyngaarden, J.B., Fredrickson, D.S., Goldstein, J.L., and Brown, M.S., Eds.). McGraw-Hill, New York.
153. Neurath, H. (1989). Proteolytic processing and physiological regulation. *Trends Biochem. Sci.* 14: 268–271.
154. North, M.J. (1985). Cysteine proteinases of cellular slime molds. *Biochem. Soc. Trans.* 13: 288–290.
155. North, M.J. and Whyte, A. (1984). Purification and characterization of two acid proteinases from *Dictyostelium discoideum*. *J. Gen. Microbiol.* 130: 123–134.
156. Novak, J.E., Stoehr, S.J., and Boxer, L.A. (1984). Platelet storage pool deficiency in mouse pigment mutations associated with seven distinct genetic loci. *Blood* 63: 536–544.
157. O'Halloran, T.J. and Anderson, R.G.W. (1990). Creation of clathrin deficient *Dictyostelium discoideum* cells. *J. Cell. Biol.* 111: 447a.
158. Ohkuma, S. and Poole, B. (1978). Fluorescence probe measurement of intralysosomal pH in living cells and the perturbation of pH by various agents. *Proc. Natl. Acad. Sci. USA* 75: 3327–3331.
159. Olden, K., Bernard, B.A., Humphries, M.J., Yeo, K-T., White, S.L., Newton, S.A., Bauer, H.C., and Parent, J.B. (1985). Function of glycoprotein glycans. *Trends Biochem. Sci.* 10: 76–82.
160. Opresko, L.K. and Karpf, R.A. (1987). Specific proteolysis regulates fusion between endocytic compartments in Xenopus oocytes. *Cell* 51: 557–568.
161. Owada, M. and Neufeld, E.F. (1982). Is there a mechanism for introducing acid hydrolases into liver lysosomes that is independent of mannose 6-phosphate recognition? Evidence from I-cell disease. *Biochem. Biophys. Res. Commun.* 105: 814–820.
162. Ozawa, M. and Kemler, R. (1990). Correct proteolytic cleavage is required for the cell adhesive function of uvomovalin. *J. Cell Biol.* 111: 1645–1650.
163. Padh, H., Lavasa, M., and Steck, T.L. (1989). Prelysosomal acidic vacuoles in *Dictyostelium*. *J. Cell Biol.* 108: 865–874.
164. Padh, H., Lavasa, M., and Steck, T.L. (1991). Endosomes are acidified by association with discrete proton-pumping vacuoles in Dictyostelium *J. Biol. Chem.* 266: 5514–5520.
165. Pannell, R., Wood, L., and Kaplan, A. (1982). Processing and secretion of α-mannosidase forms by *Dictyostelium discoideum*. *J. Biol. Chem.* 257: 9861–9865.
166. Paquet, M.R., Pfeffer, S.R., Burczak, J.D., Glick, B.S., and Rothman, J.E. (1986). Components responsible for transport between successive Golgi cisternae are highly conserved in evolution. *J. Biol. Chem.* 261: 4367–4370.
167. Pears, C.J., Mahbugani, H.M., and Williams, J.G. (1985). Characterization of two highly diverged but developmentally coregulated cysteine proteinase genes in *Dictyostelium discoideum*. *Nucleic Acids Res.* 13: 8853–8866.

168. Pears, C.J. and Williams, J.G. (1987). Identification of a DNA sequence element required for efficient expression of a developmentally regulated and cAMP–inducible gene of *Dictyostelium discoideum*. *EMBO J.* 6: 195–200.
169. Pelham, H.R.B. (1989). Control of protein exit for the endoplasmic reticulum. *Ann. Rev. Cell Biology* 5: 1–24.
170. Phelps, B.M., Koppel, D.E., Primakoff, P., and Myles, D.G. (1990). Evidence that proteolysis is an initial step in the mechanism of formation of sperm cell domains. *J. Cell Biol.* 111: 1839–1846.
171. Portnoy, D.A., Erickson, A.H., Kochan, J, Ravetch, J.V., and Unkeless, J.C. (1986) Cloning and characterization of a mouse cysteine proteinase. *J. Biol. Chem.* 261: 14697–14703.
172. Proia, R.L. and Neufeld, E.F. (1982). Synthesis of α-hexosaminidase in cell-free translation and in intact fibroblasts: An insoluble precursor chain in a rare form of Tay-Sachs disease. *Proc. Natl. Acad. Sci. USA* 79: 6360–6364.
173. Rholam, M., Nicolas, P., and Cohen, P. (1986). Precursors for peptide hormones share common secondary structures forming features at the proteolytic processing site. *FEBS Lett.* 207: 1–6.
174. Richardson, J.R., Woychik, N.A., Ebert, D., Dimond, R.L., and Cardelli, J.A. (1988). Inhibition of early but not late proteolytic processing events induces the missorting and oversecretion of precursor forms of lysosomal enzymes in *Dictyostelium discoideum*. *J. Cell Biol.* 107: 2097–2107.
175. Rijnboutt, S., Aerts H.M., Geuze, H.J. Tager, J, and Strous, G.J. (1991). Mannose 6-phosphate-independent membrane association of Cathespin D, Glucocerebrosidase, and sphingolipid-activating protein in HepG2 cells. *J. Biol. Chem.* 266: 4862–4868.
176. Robbins, A.R., Oliver, C., Bateman, J.L., Krag, S.S., Galloway, C.J., and Mellman, I. (1984). A single mutation in Chinese hamster ovary cells impairs both Golgi and endosomal functions. *J. Cell Biol.* 99: 1296–1308.
177. Robbins, A.R. (1988). Endocytosis and compartmentalization of lysosomal enzymes in normal and mutant mammalian cells: mannose 6-phosphate dependent pathways. In: *Protein Transfer and Organellar Biogenesis*, pp. 463–520. (Das, R. and Robbins, P., Eds.). Academic Press, New York.
178. Roe, J.L., Farach, H.A., Strittmatter, W.J., and Lennarz, W.J. (1988). Evidence for involvement of metalloendoproteases in a step in sea urchin gamete fusion *J. Cell Biol.* 107: 539–544.
179. Roff, C.F., Fuchs, R., Mellman, I., and Robbins, A.R. (1988). Chinese hamster ovary cell mutants with temperature sensitive defects in endocytosis. I. Loss of function on shifting to the nonpermissive temperature. *J. Cell Biol.* 103: 2282–2297.
180. Rosenfeld, M.G., Kreibich, G., Popov, D., Kato, K., and Sabatini, D.D. (1982). Biosynthesis of lysosomal hydrolases: Their synthesis in bound polysomes and the role of co- and posttranslational processing in determining their subcellular distribution. *J. Cell Biol.* 93: 135–143.
181. Rossomando, E.F., Maldonado, B., Crean, E.V., and Kollar, E.J. (1978). Protease secretion during onset of development in *Dictyostelium discoideum*. *J. Cell Sci.* 30: 305–318.
182. Rothman, J.H., Yamashiro, C.T., Raymond, C.K. Dane, P.M., and Stevens, T.H. (1989). Acidification of the lysosome-like vacuole and the vacuolar H-ATPase are deficient in the two yeast mutants that fail to sort vacuolar proteins *J. Cell Biol.* 109: 93–100.
183. Saxe, S.S. and Kimmell, A.R. (1990). SAS1 and SAS2, GTP-binding protein genes in *Dictyostelium discoideum* with sequence similarities to essential genes in *Saccharomyces cerevisiae*. *Mol. Cell. Biol.* 10: 2367–2378.
184. Schatzle, J., Rothi, A., Clarke, M., and Cardelli, J.A. (1991). Developmental regulation of the α-mannosidase gene in *Dictyostelium discoideum*; control is at the level of transcription and is affected by cell density. *Mol. Cell. Biol.* 11: 3339–3347.
185. Schatzle, J., Bush, J.M., and Cardelli, J.A. (1992). Molecular cloning and characterization of the developmentally regulated gene coding for lysosomal α-mannosidase in *Dictyostelium discoideum*. *J. Biol. Chem.* 267: 4000–4007.

186. Schwarz, H. and Thilo, L. (1983). Membrane traffic in *Dictyostelium discoideum*: Plasma membrane glycoconjugates internalized and recycled during fluid phase pinocytosis enter the Golgi complex. *Eur. J. Cell Biol.* 31: 212–219.
187. Segev, N., Mulholland, J., and Botstein, D. (1988). The yeast GTP-binding Ypt1 protein and a mammalian counterpart are associated with the secretion machinery. *Cell* 52: 915–924.
188. Sharkey, D.J. and Kornfeld, S. (1985). Evidence for processing α-mannosidases that appear in developing *Dictyostelium discoideum*. In: *Glycoconjugates, Proceedings of the Eight International Symposium* Vol. 2, pp. 462–463.
189. Sharkey, D.J. and Kornfeld. (1989). Identification of an N-acetyl glucosaminyltransferase in *Dictyostelium discoideum* that transfers an "interesting" N-acetylglucosamine to high mannose oligosaccharides. *J. Biol. Chem.* 204: 10411–10419.
190. Skudlarek, M.D., Novak, E.K., and Swank, R.T. (1984). Processing of lysosomal enzymes in macrophages and kidney. In: *Lysosomes in Biology and Pathology*, pp. 17–43. (Dingle, J.T., Dean, R.T., and Sly, W., Eds.). Elsevier, Amsterdam.
191. Smolen, J.E, Stoehr, S.J, and Buxer, L.A. (1986). Human neutrophils permeabilzed of digitonin respond with lysosomal enzyme release when exposed to micromolar levels of free calcium. *Bioch. Biophys. Acta* 886: 1–17.
192. Stein, B.S. and Engleman, E.G. (1990). Intracellular processing of the gp 160 HIV-1 envelope precursor. *J. Biol. Chem.* 265: 2640–2649.
193. Stevens, T.H., Rothman, J.H., Payne, G.S., and Schekman, R. (1986). Gene dosage dependent secretion of yeast vacuolar carboxypeptidase Y. *J. Cell Biol.* 102: 1551–1557.
194. Storrie, B. (1988). Assembly of lysosomes: perspectives from comparative molecular cell biology. *Int. Rev. Cyto.* 111: 53–105.
195. Stromberg, K., Persson, A.M., and Olsson, I. (1985). The processing and intracellular transport of myeloperoxidase. Modulation by lysosomotropic agents and monensin. *Eur. J. Cell Biol.* 39: 424–431.
196. Strous, G.J., Van Kerkhof, P., Dekken, J., and Schwartz, A.L. (1988). Metalloprotease inhibitors block protein synthesis intracellular transport, and endocytosis in hepatoma cells. *J. Biol. Chem.* 263: 18197–18204.
197. Tartakoff, A.M. (1983). Perturbation of vesicular traffic with the carboxylic ionophore monensin. *Cell* 32: 1026–1028.
198. Taugner, R., Whalley, A., Angermuller, S., Buhrle, C.P., and Hackenthal, E. (1985). Are the renin containing granules of juxtaglomerular epitheloid cells modified lysosomes. *Cell Tissue Res.* 239: 575.
199. Thilo, C. and Vogel, G. (1980). Kinetics of membrane internalization and recycling during pinocytosis in *Dictyostelium discoideum*. *Proc. Natl. Acad. Sci. USA* 77: 1015–1019.
200. Thomas, G, Thorne, B.A., Thomas, L., Allen, R.G., Hruby, D.E., Fuller, R. and Thorner, J. (1988) Yeast Kes-2 endopeptidase correctly cleaves a neuroendocrine prohormone in mammalian cells. *Science* 241: 226–230.
201. van Dongen, J.M., Barneveld, R.A., Geuze, H.J., Galjaard, H. (1984). Immunocytochemistry of lysosomal hydrolases and their precursor forms in normal and mutant human cells. *Histochem. J.* 16: 941–54.
202. Von Figura, K. and Hasilik, A. (1986). Lysosomal enzymes and their receptors. *Annu. Rev. Biochem.* 55: 167–194.
203. Von Figura, K., Steckel, F., and Hasilik, A. (1983). Juvenile and adult metachromatic leukodystrophy: Partial restoration of arylsulfatase A (cerebroside) activity by inhibition of thiol proteinases. *Proc. Natl. Acad. Sci. USA* 80: 6066–6070.
204. Von Heijne, G. (1980). Towards a comparative anatomy of N-terminal topogenic protein sequences. *J. Mol. Biol.* 189: 239–242.
205. Vu, T., Hung, O.T., and Coughlin, S.R. (1991). Molecular cloning of a functional thrombin receptor reveals a novel proteolytic mechanism of receptor activation. *Cell* 64: 1057–1068.

206. Waheed, A. and Van Etten, R.L. (1985). Phosphorylation and sulfation of arylsulfatase-A accompanies biosynthesis of the enzyme in normal and carcinoma cell-lines. *Biochem. Biophys. Acta* 847: 53–61.

207. Waheed, A., Pohlmann, R., Hasilik, A., von Figura, K., van Elsen, A., and Leroy, J.G. (1982). Deficiency of UDP-N-acetylglucosamine: lysosomal enzyme N-acetylglucosamine-1-phosphotransferase in organs of I-cell patients. *Biochem. Biophys. Res. Commun.* 105: 1052–1058.

208. Waheed, A., Gottschalk, S., Hille, A., Krentler, C., Pohlmann, R., Braulke, T., Hauser, H., Geuze, H., and von Figura, K. (1988). Human lysosomal acid phosphatase is transported as a transmembrane protein to lysosomes in transfected baby hamster cells. *EMBO J.* 7: 2351–58.

209. Warren, L. (1989). Stimulated secretion of lysosomal enzymes by cells in culture. *J. Biol. Chem.* 264: 8835–8842.

210. Wayne, S.S., Fisher, J.M., and Scheller, R.H. (1990) Sorting within the regulated secretary pathway occurs in the trans. Golgi network. *J. Cell Biol.* 110: 1–12.

211. Williams, J.G., North, M.J., and Mahbubani, H. (19855). A developmentally regulated cysteine proteinase in *Dictyostelium discoideum. EMBO J.* 4: 999–1006.

212. Wood, L. and Kaplan, A. (1985). Transit of α-mannosidase during its maturation in *Dictyostelium discoideum. J. Cell Biol.* 101: 2063–2069.

213. Wood, L., Pannell, R.N., and Kaplan, A. (1983). Linked pools of processed α-mannosidase in *Dictyostelium discoideum. J. Biol. Chem.* 258: 9426–9430.

214. Woychik, N.A., Cardelli, J.A., and Dimond, R.L. (1986). A conformationally altered precursor to the lysosomal enzyme α-mannosidase accumulates in the endoplasmic reticulum in a mutant strain of *Dictyostelium discoideum. J. Biol. Chem.* 261: 9595–9602.

215. Woychik, N.A. and Dimond, R.L. (1987). A single mutation prevents normal intracellular transport of multiple lysosomal proteins from the endoplasmic reticulum. *J. Biol. Chem.* 262: 10008–10014.

216. Yamashiro, C.T., Kane, P.M., Wolczyk, D.F., Preston, R.A., and Stevens, T.H. (1990). Role of vacuolar acidification in protein sorting and zymogen activation: A genetic analysis of the yeast vacuolar proton-translocating ATPase. *Mol. Cell. Biol.* 10: 3737–3749.

TOWARDS AN UNDERSTANDING OF THE INHERITANCE OF MAMMALIAN LYSOSOMES AND YEAST VACUOLES

Brian Storrie

ABSTRACT

Recent work strongly indicates that both the mammalian lysosome and the fungal vacuole contain substantial contributions from preexisting lysosomes and vacuoles.

Advances in Cell and Molecular Biology of Membranes
Volume 1, pages 391–406

ISBN: 1-55938-362-3

Much of this is a result of intercommunication within the organelle population. In polykaryons between fibroblastic or epithelial cell lines or in yeast zygotes, rapid bidirectional intermixing of lysosomal or vacuolar contents is observed. Organelle intercommuncation appears to be an importance principle in the inheritance of both lysosomes and vacuoles. From the standpoint of organelle inheritance, mammalian lysosomes, although existing often as many individual units at the time of cell division, are inherited as a single functional entity. Daughter vacuoles in the budding yeast *Saccharomyces cerevisiae* appear to be seeded from parental vacuoles by a tubulovesicular extension into the bud. It is likely that both lysosomes and vacuoles are derived from preexisting organelles. However, the fundamental question of whether lysosomes or vacuoles must derive from preexisting lysosomes and vacuoles or may form *de novo* still remains a point of contention. Research on the identification and characterization of molecules responsible for lysosomal and vacuolar morphology, dynamics, and inheritance is at its very beginning stages. Our understanding of the relationship between the lysosome and the vacuole and preexisting organelles is likely to hinge upon our understanding of the molecules involved in the process.

I. INTRODUCTION

In Chapter 1 of this volume Robert Murphy has summarized competing models for lysosomal and vacuolar biogenesis in mammalian, plant, and fungal cells. This chapter is concerned with a question which, at first glance, centers on later stages of the lysosomal life cycle. Namely, how are lysosomes inherited or, in molecular terms, what is the distribution of preexisting molecules between daughter lysosomes in vertebrates and the corresponding vacuoles in plants and fungi as cells grow and divide. In its simplest sense, the question can be distilled to: "Do all lysosomes and vacuoles inherit molecules from preexisting lysosomes and vacuoles?"

In summarizing work relevant to this question, I shall review data on (1) mammalian lysosomal heterogeneity, (2) lysosomal intercommunication and the plasticity of lysosomal and vacuolar morphology, (3) the formation and inheritance of yeast vacuoles, an extensively studied vacuolar example, and (4) probable molecules involved in regulating lysosomal dynamics and morphology. In the end, answers to the points explored in this chapter are also important to deciding whether or not lysosomes are formed *de novo*, as proposed by the maturation hypothesis, or by expansion and division of preexisting structures, as proposed by the vesicle-shuttle model for lysosome formation. This chapter will concentrate on the mammalian lysosome and will be presented in essay format.

Before beginning to discuss answers to these questions, a working definition of a lysosome is in order. In a simplistic sense, a lysosome may be defined as a degradative, membrane-limited organelle, composed of resident proteins and lipids, which is a terminal compartment in an endocytic and biosynthetic pathway (see, Fig. 1). In theory, a lysosome may be defined by absolute differences in array of resident molecules. Operationally, however, this has to date been a difficult task.

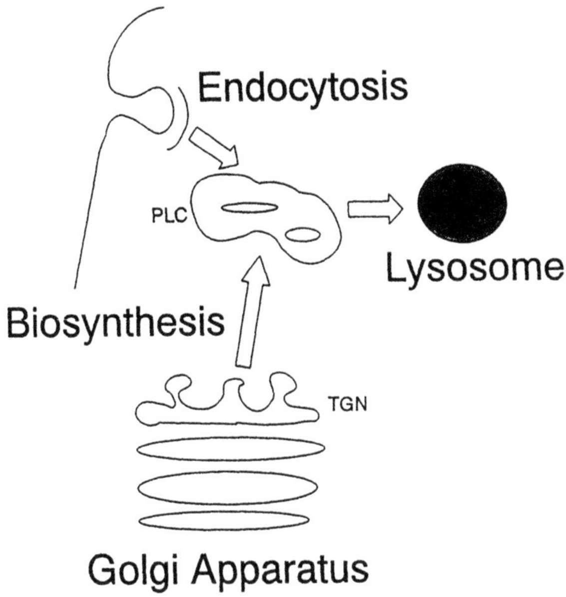

Figure 1. Schematic drawing of the relationship of the mammalian lysosome to the endocytic and biosynthetic pathways. PLC is an abbreviation for prelysosome compartment and is an intermediate structure(s) which is an intersection point for the two pathways. It may be viewed as the first commited step in a pathway leading to lysosomal biogenesis. TGN is an abbreviation for *trans*-Golgi network.

The definition of a lysosome by its array of resident molecules is complicated by the biosynthetic passage of the molecules through endosomes and their possible physiological role in cycling between various compartments. For example, this last aspect may be illustrated by reference to the cation-independent mannose 6-phosphate receptor (CI-MPR). It would be considered, at first glance, a resident protein of the lysosomally related prelysosomal compartment (*17,18*) if its role as an accessory molecule in protein transport between the Golgi apparatus and the prelysosomal compartment was not known (see Chapter 3 by Hoflack and Lobel). At least trace amounts of all lysosomal components, by virtue of their biosynthetic origin, are found in the endoplasmic reticulum and Golgi apparatus. Moreover, at least some lysosomal membrane proteins as first shown by Lippincott-Schwartz and Fambrough (*29*) cycle between the lysosome and the cell surface. This may be true of all lysosomal membrane proteins containing a tyrosine-positive endocytic signaling motif (see Chapter 8 by Roth). Moreover, as shown originally by Peters et al. (*42*) for acid phosphatase and Williams and Fukuda (*64*) for hLAMP-1, all such membrane proteins containing an endocytosis signal may be routed to the lysosome via the cell surface (see references 15a and 56a for reviews). All lysosomal membrane proteins containing a tyrosine-positive endocytic sorting motif are present in trace or greater amounts in the plasma membrane and various endosomal structures. All of these proteins appear to be abundant in the prelysosomal compartment. In striking contrast, other lysosomal membrane proteins, such as rat LIMP II, are devoid of a tyrosine-positive endocytic signaling motif (*55,56*). These proteins may be routed directly to the lysosome (*55*), and may constitute "true" resident proteins. Soluble lysosomal hydrolases such as cathepsin D and ß-glucuronidase contain mannose 6-phosphate, and are targeted to the lysosome as a consequence of binding to the CI-M6PR. These hydrolases are also abundant in the prelysosomal compartment together with the CI-M6PR. The lysosome as a terminal compartment appears to be devoid of CI-M6PR.

Perhaps, a lysosome may be best defined by what it is *not*. A lysosome is the *non-recycling* membrane limited organelle which is the terminus of endocytic and biosynthetic pathways; it does *not* contain recycling protein targeting receptors, either of biosynthetic or endocytic origin.

II. AN OVERVIEW OF ORGANELLE GROWTH, DIVISION, AND SEGREGATION DURING THE CELL CYCLE

Membranous organelles within the eukaryotic cell may be placed into two general classes based on the biosynthetic targeting and processing of the proteins embedded within the organelle membrane(s). In one class are the nucleus, endoplasmic reticulum, Golgi complex, and the various components of the vacuolar apparatus; namely, endosomes, phagosomes, autophagosomes, and lysosomes. The proteins of this organelle class typically possess, at least transiently, signal peptides which target the *cotranslational* insertion of the polypeptide into the membrane of the rough endoplasmic reticulum (ER). In most cases, the signal sequence is then cleaved by signal peptidase within the lumen of the ER once it has served its targeting function. In the other class are mitochondria, chloroplasts, and peroxisomes/glyoxyosomes. The proteins of this second organelle class are all inserted *posttranslationally* into the organelle following the completion of the synthesis of the polypeptide within the cytosol. The members of this class are all thought to have originated through endosymbiotic infection of primitive eukaryotic cells.

The members of the ER related class of organelles are ultimately derived, albeit often indirectly, from the growth and expansion of preexisting ER membrane. Likewise, it has been long recognized that mitochondria and chloroplasts are derived from the growth and division of preexisting mitochondria and chloroplasts. For the mitochondria, for example, radioactive choline and thymidine labels in *Neurospora crassa* (*31–34*) and *Tetrahymena* (*40*), respectively, are diluted on a per organelle basis, in a manner consistent with the formation of new mitochondria from preexisting mitochondria; the average grain count per mitochondrion is diluted as a direct function of cell multiplication in label-free medium, and organelle density varies *en masse* with variations in the choline level in the medium for a choline requiring mutant. Similarly, for human mitochondria, the level of the oxidative enzyme, cytochrome c oxidase, per mitochondrion also varies *en masse* and as a direct function of cell multiplication in the absence of mitochondrial protein synthesis (*47*). The occurrence in electron micrographs of mitochondrial partition figures (e.g., *28,52*) suggests that mitochondria divide. It should be emphasized that the *en masse* variation in mitochondrial lipids and enzymes is also consistent with rapid exchange of molecules within the organelle population; time-lapse observations have long indicated that mitochondria are dynamic structures engaged in rapid fusion and fission processes (*15*). Evidence for rapid exchange of molecules within members of the vacuolar apparatus of both mammalian and yeast cells will be discussed later in this review.

The expansion of organellar membranes is a continuous process in interphase cells. In mammalian cells, protein synthesis occurs throughout G_1, S, and the G_2 phase of the cell cycle. Newly synthesized proteins destined for organelle formation and for secretion are inserted into the rough endoplasmic reticulum throughout interphase, and both the reticular network of the ER and the perinuclear Golgi complex appears to expand throughout interphase. Vesicular traffic, be it biosynthetic or endocytic, occurs at a high rate throughout interphase (for review, see *57*). Similarly for organelles of endosymbiotic origin, such as the mitochondrion, membrane growth is continuous (*43*). For the mitochondrion, membrane growth is accompanied by organelle division; the volume of individual organelle units stays constant, suggesting that growth must be accompanied by organelle division (*43*).

During mitosis, organelles must be segregated between daughter cells. For the nucleus, this is accompanied by reversible vesicularization of the nuclear envelope early in mitosis, most probably as a consequence of the phosphorylation and depolymerization of the laminin intermediate filaments underlying the envelope (for review, see *16*). Likewise, the reticular network of both the ER and Golgi complex fragments during mitosis and later reforms. This process has been proposed by Warren (*57,58*) to be an outcome of the cessation of membrane trafficking that is one of the landmarks of mitosis. As proposed by Warren, vesiculation continues, but fusion is blocked; a complete cycle of membrane trafficking fails to occur and the organelles involved fragment. The actual cessation of both biosynthetic and endocytic membrane trafficking has been speculated to be the consequence of cyclin-modulated reversible phosphorylation of small GTP binding proteins needed for vesicle fusion (*58*). This appears to be the case for at least one fusion event in endocytosis (*54*). For organelles distributed throughout the cytoplasm in many copies per cell, cytokinesis has long been assumed to produce a random segregation of the organelles to the daughter cells. Typically, lysosomes are present in many copies per cell, whereas the vacuole is present in a single copy to a small number per cell.

III. RESIDUAL BODIES AND CONTENT HETEROGENEITY OF LYSOSOMES

The composition of the individual lysosomal units within a cell is the outcome of biosynthetic, endocytic, and autophagic processes. To be a lysosome, each unit must, by definition, possess a complement of lysosomally targeted proteins. In fact, to be fully functional in degradative processes, each must possess the full complement of lysosomal proteins. Hence all lysosomes are likely to be qualitatively identical in their spectrum of endogenous proteins (see also, next section). However, this does not at all mean that lysosomes are identical in composition. With the exception of specialized storage forms of the organelle (e.g., in the polymorphonuclear leukocyte) much of the contents of lysosomes are of endocytic or autophagic origin. In molecular terms, there is absolutely no evidence for the formation of lysosomes or vacuoles as "virgin" structures (primary lysosomes, which bud full-formed from the Golgi apparatus as has been proposed earlier; for review, see *23*). By the very nature of their biogenesis, be it by endosomal maturation or by expansion and division of preexisting structures, lysosomes and vacuoles contain molecules of endocytic and autophagic origin. Hence, lysosomal heterogeneity in content is to be expected. This is an expectation abundantly supported by years of electron microscopic observations (for an overview, see *24*).

Content indigestibility can, and does lead to the accumulation of a residium within individual lysosomes. Classically such lysosomes have been referred to as residual bodies, "a type of . . . lysosome in which much of the content is residual material not susceptible to, or only slowly susceptible to further degradation" (*23*, p. 2). In electron micrographs, these lysosomes appear to be inert, dead-end structures filled with indigestible material. However, physiologically there is no evidence that such lysosomes are nonfunctional. In fact, the physiological evidence strongly indicates that these structures are involved in a number of different transport pathways. This can be best illustrated by examples. In humans, various single enzyme defects which limit the digestive capacity of lysosomes result in the accumulation of residuum and a consequent expansion of individual lysosomal size (*35*). Ultimately, this leads to premature death of the affected individual in adolescence (*35*). In cell culture, lysosomal storage diseases are correctable by the addition of the missing enzyme to the medium. The added enzyme is internalized and transferred to preexisting lysosomes (residual bodies) where its presence leads to the degradation of the residuum. In culture, the equivalent of a single enzyme defect can be induced by incubating fibroblasts or macrophages in the presence of dilute sucrose. These cells can not digest sucrose since they lack invertase. The sugar accumulates in lysosomes and the individual lysosomes swell (*6,7,13*). The addition of invertase to the culture media leads to enzyme internalization and the subsequent hydrolysis of the sucrose to membrane permeant monomers. The size of the entire lysosomal population shrinks and becomes normal again (*6,7*). In both cases, these apparent "residual" bodies have proven to be active in the endocytic pathway. Similarly, sucrose swollen lysosomes are active in molecular exchanges with other lysosomes (*9,13,48*).

Content heterogeneity within the mammalian lysosomal population should be transient. Most endocytic and autophagic material is digested. In many cell types, cell growth and division leads to dilution of undigestible material. Moreover, mammalian lysosomes appear to be in constant communication with each other. In cell-fusion experiments between homologous and heterologous pairings of mammalian cells, lysosomal contents are exchanged rapidly within the organelle population with a $t_{1/2}$ of approximately 20–30 minutes (*9,13,48*). Similarly lysosomal membrane proteins, scored by immunofluorescent or immunogold labeling with species-specific monoclonal antibodies, intermix rapidly throughout the entire organelle population (*9,10*). By immunofluorescent staining, lysosomal contents and membrane proteins appear to exchange with similar kinetics (*10,13*) suggesting that both are exchanged as part of the same process. Lysosomes may repeatedly collide with nearest neighbors leading to transient fusion of organelles (*48*).

IV. ARE THERE DIFFERENT KINDS OF LYSOSOMES?

The assessment of lysosomal heterogeneity requires not only consideration of exogenous molecules delivered to the lysosome by endocytosis and autophagy, but also consideration of the distribution of endogenous molecules. Are endogenous molecules; that is, those synthesized by the cell itself, distributed heterogeneously within the lysosomal population of mammalian cells in such a manner as to give rise to multiple forms of lysosomes? In discussing this question, the reader should be forewarned that, in part, any answer is an exercise in semantics.

Two major kinds of arguments have been presented for the existence of different kinds of lysosomes. These are first heterogeneity in enzyme activity, and second heterogeneity in endocytic delivery. Both arguments are best illustrated by a discussion of specific examples. Enzyme activity at the individual organelle level has been assayed for a very limited number of enzymes by electron microscopic cytochemistry. As exemplified by the work of Båscy (*1*), Oliver (*37,38*), and Oliver and Yuasa (*39*), cells are fixed, sectioned, and lead or barium precipitates formed by a capture reaction typically with the phosphate or sulfate-degradation product of cytidine monophosphate or aryl sulfate. The distribution of reaction product is then scored relative to organelle profiles thought to be lysosomes. The net conclusion of such work is that lysosomal enzyme activities are distributed nonuniformly in individual secretory cells. The firm interpretation of these results is, however, difficult. The substrates are relatively nonspecific with respect to molecules. Moreover, no evidence is provided regarding where the structures might fall along an endocytic/biosynthetic pathway with respect to the lysosome. To conclude more than that there is compartment diversity within secretory cells is unwarranted; the structures may not in reality be different kinds of lysosomes, but rather intermediates in the endocytic/biosynthetic pathway. Similar data have been obtained biochemically in cell fractionation experiments in which the distribution of lysosomal enzyme activities is plotted versus density. In such experiments, distribution differences may be observed between different enzyme activities. In my experience, these differences vanish if care is taken in the enzyme assays to achieve full activity recovery of (E. Madden, M. Sachdeva, J. Wirt, and B. Storrie, unpublished data). The sugar solutions common for centrifugation media often inhibit the activity of lysosomal saccharidases.

The other major line of reasoning for endogenous lysosomal heterogeneity has come from evidence for the delivery of endocytosed material to an apparent "light" or "early" lysosome population prior to its accumulation in "late" lysosomes. In these experiments, as exemplified by the work of Berg et al. (*4*) with hepatocytes, and Einstein and Gabel (*11,12*) and Storrie et al. (*49*) with cultured fibroblasts, delivery of endocytized molecules into an early lysosomal population appears to be rapid (approximately 10–20 minutes). The early compartment is intermediate in density between endosomes and the total lysosomal population as assessed by the distribution of endocytized marker and ß-hexosaminidase, an endogenous lysosomal enzyme (*4,49*). By enzyme distribution, the compartment contains a small portion of lysosomal enzyme activity (Nycodenz gradient, *4*; Percoll gradient, *49*). Based on gradient position, the compartment is devoid of CI-M6PR (*12,17,36*). The compartment is active in protein degradation (*4,49*). This compartment displays many of the properties characteristic of lysosomes; a relatively high-density, degradative capacity, and the absence of CI-M6PR. The compartment is clearly not a terminal compartment in an endocytic/biosynthetic pathway since it is not the end-site of accumulation of endocytized molecules. Hence, it can not be considered to be a "mature" lysosome. Whether it should be considered to be an "early" or "immature" lysosome is a matter of interpretation. Presumably, although not presently known, the compartment is devoid of recycling endocytic receptors such as the transferrin receptor. If so, the compartment is likely not to be a recycling endocytic compartment. Interestingly, transport between early and late lysosomes appears to be regulated by serum factors, likely insulin-like growth factor II (*11,12*).

Whether or not there are different kinds of lysosomes in single mammalian cells thus remains an open question. "Are early lysosomes transitional structures to a mature lysosomal population or preexisting compartments?" This question can only be answered by further experimentation.

V. LYSOSOMAL INTERCOMMUNICATION AND THE PLASTICITY OF LYSOSOMAL AND VACUOLAR MORPHOLOGY

Mammalian lysosomes intercommunicate (see above). In polykaryons between fibroblastic or epithelial cell lines, essentially complete intermixing of endogenous membrane proteins and exogenous contents within the lysosomal population are observed over a time span of 2 to 3 hours (*9,10,13,48*). Presumably the same process is continuously ongoing in single mammalian cells. Hence, from the standpoint of organelle inheritance, mammalian lysosomes, although existing as many individual units at the time of cell division, are inherited as a single functional entity.

Lysosomes typically are thought of as small spherical organelles (2). However, in some mammalian cells such as macrophages (*27,50*), lysosomes may well be a single entity physically. As with endosomes, which may be tubular and interconnected (*25,53*), lysosomes in these cells appear as a tubuloreticular network (*27,50*). Lysosomes in these experiments were defined as sites of accumulation of endocytic tracer characterized by the presence of the molecule, cathepsin L, a lysosomal enzyme (in other words, mature lysosomes). The formation of tubular lysosomes is dependent on phagocytic load (*27*), microtubules (*50*), kinesin (22), a motor protein, and cytosolic pH (*20*). Tubular lysosomes may be found in fibroblasts and other cell types (*20,48*). The formation of tubular lysosomes is promoted by phorbol esters (*50*; Deng and Storrie, unpublished observations). The formation of tubular lysosomes is also promoted by brefeldin A, a putative disrupter of coat protein mediated intracellular transport processes (*30*).

The formation of tubular versus spherical lysosomes may represent a balance between motor protein and coat protein binding to lysosomes (*41*). Brefeldin A disruption of Golgi coats leads to "premature" vesicle fusion and the formation of microtubule-dependent connections between fused Golgi cisternae and the endoplasmic reticulum. As speculated by Pelham (*41*), a yet to be characterized lysosome-specific coat protein complex may be an important controlling factor in determining lysosome morphology in animal cells. This plasticity of lysosomal morphology is likely to be important in lysosomal inheritance. Fragmented tubular lysosomal elements may then be passively segregated during cytokinesis to daughter cells.

While mammalian cells typically have multiple lysosomes, interphase plant and fungal cells typically have only one or a few vacuoles. This structure is generally large and occupies much of the cell volume. The vacuole of plant and fungal cells is also a microtubule-dependent structure. For example, treatment of yeast *Saccharomyces cerevisiae* cells with nocodazole leads to the fragmentation of the vacuole into numerous smaller structures (*19*). However, vacuolar segregation structures (see below) are nocodazole insensitive, and hence unlikely to be microtubule-dependent (*8*).

Yeast genetics indicates that the morphology of the organelle is dependent on multiple vacuolar-specific gene products (for review, see *26*). Defects in some of these result in fragmentation of the vacuole. Defects in other gene products result in cells which accumulate many vesicles and appear devoid of a wild-type vacuole. Many of these mutants may be in soluble, accessory proteins needed for vacuolar assembly rather than in components which are integral membranes of the vacuole. For example, the mutant *vps33* which results in the accumulation of small vesicles codes for a soluble protein that possesses an ATP binding domain and is presumably an ATPase (*3*). Yeast genetics and the possibility of cloning genes by their ability to correct (i.e., complement) mutations holds great promise for the characterization of proteins responsible for maintaining vacuolar morphology. In the following section, I shall turn to the topic of how the yeast vacuole is inherited and the importance of organelle intercommunication in this process.

Curiously enough in mammals, lysosomal gigantism is associated with pigmentation defects and in humans with early death (for review, see *65*). The Chediak-Higashi syndrome in humans is characterized by giant lysosomes in leukocytes which fail to degranulate, giant melanosomes in melanocytes and keratinocytes, pigment dilution, and an increased susceptibility to infection. Children who survive to their late teens generally die of a terminal lymphoreticular malignancy. The molecular basis of the Chediak-Higashi syndrome is unknown.

VI. INHERITANCE OF YEAST VACUOLES

The yeast *Saccharomyces cerevisiae* has been a favorite experimental subject of eukaryotic geneticists. The biochemical and molecular characterization of mutations in the secretory pathway of yeast has contributed greatly to our understanding of this pathway in both yeast, plants, and mammals. The identification and characterization of yeast mutations in vacuolar assembly and inheritance may also provide important clues concerning the assembly and inheritance of the mammalian lysosome. However, because cell division in *Saccharomyces cerevisiae* and plants and animals differ in several aspects, not all that is true for the *Saccharomyces cerevisiae* vacuole may be true of the mammalian lysosome or the more related plant vacuole. *Saccharomyces cerevisiae* reproduces by budding; cell surface growth is restricted to the bud (*14*). The yeast nuclear envelope during mitosis stretches across the division septum and divides (*5*) rather than fragments during mitosis as the mammalian nucleus does. Hence, certain steps in yeast vacuolar inheritance, such as the initial seeding of the bud with a vacuole, may well be unique to yeast, while many other steps related to more general processes of organelle biogenesis and inheritance may well be true for cells separated by wide phylogenetic distances.

The inheritance of the yeast vacuole at the cellular level has been studied by two different experimental approaches. These have led to the development of contrasting models. In the first the morphology of the vacuole in synchronous cell populations has been characterized (*45,61,63*). In these studies where different approaches to synchronize the cell population were employed, the vacuole was observed to fragment as bud emergence begins during the S phase of the cell cycle. Both parent and daughter cells were later observed to each contain a large vacuole, and it was surmised that vacuole fragments which had distributed between the parent cell and bud had coalesced. Based on these experiments, Severs et al. (*45*) and Wiemken et al. (*63*) have suggested that vacuole inheritance is a matter of regulated organelle fragmentation and coalescence.

In the second approach the morphology of the vacuole and distribution of vacuolar components has been characterized in exponentially growing asynchronous cultures relative to bud emergence and growth. In these studies (*8,44,61*), the parental vacuole was found to continuously maintain its morphology with a separate vacuole being found in intermediate-sized buds. The size of the daughter vacuole was found to increase as the bud enlarged. Morphologically the daughter vacuole appeared to derive from a tubulovesicular vacuole segregation structure connected to the parental vacuole (*8,44*). Consistent with this apparent physical continuity, the daughter cell vacuole contained preexisting molecules inherited from the parental vacuole (*60,61*). Based on these experiments, Weisman et al. (*61*) have proposed that vacuole segregation and inheritance is the result of vesicular or tubular traffic from the parental cell to the bud. These authors attribute the vacuole fragmentation in synchronized cell experiments to artefacts of synchronization (*61*).

Both asynchronous and synchronized cell studies indicate that the daughter vacuole inherits substantial amounts of preexisting molecules from the parental vacuole. Such a pattern of vacuolar inheritance has also been substantiated by genetic experiments in which a pronounced phenotypic lag in the expression of the *PEP4* vacuolar protease mutation in newly sporulated *Saccharomyces cerevisiae* cells has been observed (*66*). Tubulovesicular trafficking between parental and daughter vacuole has been shown to be cell-cycle dependent (*8,60*), bidirectional (*60*), energy dependent (*60*), and insensitive to nocodazole (*8*), a microtubule depolymerizer.

Based on recent genetic studies, vacuolar inheritance appears to be a separate process from vacuolar protein targeting (*46*). The *vac2-1* mutation isolated by Shaw and Wickner provides the first example of a mutation which affects only vacuolar inheritance. At elevated temperature, mutant *vac2-1* cells are defective in the partitioning of molecules from the parental vacuole to the bud and hence fail to form a detectable vacuole. These cells do, however, transport carboxypeptidase Y and proteinase A normally to the vacuole where they are processed to mature forms. Likewise, the cells process and secrete invertase normally at both 23 and 37 °C and nuclei and mitochondria are inherited normally in these cells. The *vac2* gene product has been sequenced (W.T. Wickner, personal communication). Its further characterization should provide important information on the role of individual molecules in vacuolar inheritance.

As reviewed by Klionsky et al. (*26*), a portion of yeast mutants unable to target proteins from the Golgi apparatus to the vacuole also are defective in vacuole morphology and inheritance. For example, *vac1-1* is one such mutation that affects both protein targeting and vacuole inheritance (*62*). In mutant *vac1-1* cells, carboxypeptidase Y, a vacuolar protease, is secreted from the cells rather than transported from the Golgi to the vacuole, and the normal transfer of preexisting molecules from the parental vacuole to the bud is blocked. Cells are frequently formed without a detectable vacuole. In this mutation, as in most of the others affecting vacuolar inheritance, the effect of the mutation may be either direct or indirect. The gene product itself may be directly required for vacuolar inheritance, or it may exert an indirect effect because its normal function is required for the function of another gene product.

The essentiality of three protein products which affect both vacuolar protein targeting and vacuole inheritance has been investigated (*21,44,60*). In all three cases, the approach has been the same: cloning of the corresponding nucleotide sequence by complementation with the mutant and then construction of a yeast cell in which the gene is deleted. The three deletion mutants, all in different genes, exhibited similar phenotypes. The cells grew, albeit slowly. In each case, although the daughter cell acquires little (if any) detectable vacuolar material, the cells do develop a normal size vacuole before producing a new bud. Each of these three deletions produces only a partial block in vacuolar inheritance, suggesting that vacuolar inheritance is a complicated process in which no given gene product is necessarily essential. Moreover, the results suggest either that mutant parental cell must contribute an undetected vacuole compartment to the bud which then nucleates subsequent organelle growth, or that there are multiple pathways for vacuolar inheritance.

VII. PROBABLE MOLECULES INVOLVED IN REGULATING LYSOSOMAL DYNAMICS AND MORPHOLOGY

What are the molecules involved in directing lysosomal or vacuolar assembly, morphology, and inheritance? At present only a mere handful of molecules have been identified. The mannose 6-phosphate receptors clearly have a role as accessory proteins in the transport of soluble, lumenal proteins to the mammalian lysosome. Lysosomes and vacuoles almost certainly are capable of interacting with the cytoskeleton. How this occurs is presently vague. The motor protein, kinesin (*22*), plays a role in lysosome movement and in the formation of tubular lysosomes. How this is regulated is unknown. Similarly, the mechanism underlying lysosome-lysosome intercommunication is unknown. Some of the cloned yeast proteins affecting vacuolar morphology and inheritance may interact with both the vacuole and cytoskeleton. This is a speculation presented by both Raymond et al. (*44*) and Weisman and Wickner (*60*). However, there is no direct evidence for this. What regulates the formation of a vacuole in the bud of *Saccharomyces cerevisiae* is for now unknown.

It is likely that lysosomal and vacuolar morphology and inheritance are regulated by cell cycle-dependent protein kinases. Membrane trafficking, in general, stops during mitosis. The cell division cycle protein, cdc2 protein kinase, may be important. Recently, it has been shown to regulate endosome-endosome fusion (*54*).

VIII. CONCLUSIONS

The evidence summarized in this chapter strongly indicates that both the mammalian lysosome and the fungal vacuole normally contain substantial contributions from preexisting lysosomes and vacuoles. Intercommunication within the organelle population appears to be an important principle in the inheritance of both lysosomes and vacuoles. In concept, it is likely that both lysosomes and vacuoles are derived from preexisting organelles. However, the fundamental point of whether lysosomes or vacuoles must derive from preexisting lysosomes and vacuoles or may form *de novo* still remains a point of contention. Raymond et al. (*44*) have speculated recently that yeast vacuoles may form *de novo*. Although there is general agreement that lysosomes and vacuoles normally contain substantial amounts of preexisting molecules, there is yet to be general agreement on how these organelles are formed.

Much remains to be learned about the inheritance of lysosomes and vacuoles. As emphasized in the previous sections, what controls lysosomal or vacuolar morphology is unknown. The identification and characterization of molecules responsible for lysosomal and vacuolar dynamics is likely to be an important future area of inquiry. Research in this area is at its very beginning stages. An understanding of the relationship between the lysosome and the vacuole and preexisting organelles is likely to hinge upon our understanding of the molecules involved in the process.

ACKNOWLEDGMENTS

The author wishes to express his appreciation to Dr. R. Alscher, Virginia Polytechnic Institute and State University, for critical comments on the manuscript and to Dr. L Weisman, University of California, Los Angeles, for helpful discussions on the yeast vacuole. Any errors in this section of the chapter are the responsibility of the author. Work in the author's laboratory was supported by National Institutes of Health research grant GM 36988.

REFERENCES

1. Båcsy, E. (1982). Enzymatic heterogeneity of adrenocortical lysosomes: an X-ray microanalytical study. *Histochem. J.* 14:99–112.
2. Bainton, D.F. (1981). The discovery of lysosomes. *J. Cell Biol.* 91:66s–76s.
3. Banta, L.M., Vida, T.A., Herman, P.K., and Emr, S.D. (1990). Characterization of yeast vps33p, a protein required for vacuolar protein sorting and vacuole biogenesis. *Mol. Cell. Biol.* 10: 4638–4649.
4. Berg, T., Ford, T., Kindberg, G., Blomhoff, R., and Drevon, C. (1985). Intracellular degradation of asialoglycoproteins in hepatocytes starts in a subgroup of lysosomes. *Exp. Cell Res.* 156: 570–574.
5. Byers, B. (1981). Cytology of the yeast life cycle. In: *The Molecular Biology of the Yeast Saccharomyces: Life Cycle and Inheritance*. pp. 59–96 (Strathern, J., Jones, E., Broach, J., Eds.) Cold Spring Harbor Laboratory, Cold Spring, New York.
6. Cohn, Z.A. and Ehrenreich, B.A. (1969). The uptake, storage, and intracellular hydrolysis of carbohydrates by macrophages. *J. Exp. Med.* 129: 201–225.

7. DeCourcy, K. and Storrie, B. (1991). Osmotic swelling of endocytic compartments induced by internalized sucrose is restricted to mature lysosomes in cultured mammalian cells. *Exp. Cell Res* 192: 52–60.
8. De Mesquita, D.S.G., ten Hoopen, R., and Woldringh, C.L. (1991). Vacuolar segregation to the bud of *Saccharomyces cerevisiae:* an analysis of morphology and timing in the cell cycle. *J. Gen. Microbiol.* 137: 2447–2545.
9. Deng, Y. and Storrie, B. (1988). Animal cell lysosomes rapidly exchange membrane proteins. *Proc. Natl. Acad. Sci. USA* 85: 3860–3864.
10. Deng, Y., Griffiths, G., and Storrie, B. (1991). Comparative behavior of lysosomes and the pre-lysosome compartment (PLC) in *in vivo* cell fusion experiments. *J. Cell Sci.* 99: 571–582.
11. Einstein, R. and Gabel, C.A. (1989). Serum factors alter the extent of dephosphorylation of ligands endocytosed via the mannose 6-phosphate/insulin-like growth factor II receptor. *J. Cell Biol.* 109: 1037–1046.
12. Einstein, R. and Gabel, C.A. (1991). Cell- and ligand-specific dephosphorylation of acid hydrolases: evidence that the mannose 6-phosphatase is controlled by compartmentalization. J. Cell Biol. 112: 81–94.
13. Ferris, A.L., Brown, J.C., Park, R.D., and Storrie, B. (1987). Chinese hamster ovary cell lysosomes rapidly exchange contents. *J. Cell Biol.* 105: 2703–2712.
14. Field, C. and Schekman, R. (1980). Localized secretion of acid phosphatase reflects the pattern of cell surface growth in *Saccharomyces cerevisiae. J. Cell Biol.* 86: 123–128.
15. Frederic, J. (1958). Recherches sur les chondriosomes du cellules vivantes par la microscopie et la microcinématographie en contraste de phase. *Arch. Biol. Liege* 69: 167–349.
15a. Fukuda, M. (1991). Lysosomal membrane glycoproteins. Structure, biosynthesis and intracellular trafficking. *J. Biol. Chem.* 266: 21327–21330.
16. Gerace, L. and Burke, B. (1988). Functional organization of the nuclear envelope. *Ann. Rev. Cell Biol.* 4: 335–374.
17. Griffiths, G., Matteoni, R., Back, R., and Hoflack, B. (1990). Characterization of the cation-independent mannose 6-phosphate receptor enriched prelysosomal compartment in NRK cells. *J. Cell Sci.* 95: 441–461.
18. Griffiths, G., Hoflack, B., Simons, K., Mellman, I., and Kornfeld, S. (1988). The mannose 6-phosphate receptor and the biogenesis of lysosomes. *Cell* 52: 329–341.
19. Guthrie, B.A. and Wickner, W. (1988). Yeast vacuoles fragment when microtubules are disrupted. *J. Cell Biol.* 107: 115–120.
20. Heuser, J. (1989). Changes in lysosome shape and distribution correlated with changes in cytoplasmic pH. *J. Cell Biol.* 108: 855–864.
21. Herman, P.K. and Emr, S.D. (1990). Characterization of *VPS34,* a gene required for vacuolar protein sorting and vacuole segregation in *Saccharomyces cerevisiae. Mol. Cell. Biol.* 10: 6742–6754.
22. Hollenbeck, P.J. and Swanson,, J.A. (1990). Radial extension of macrophage tubular lysosomes supported by kinensin. *Nature* 346: 864–866.
23. Holtzman, E. (1976). *Lysosomes: A Survey.* Springer-Verlag, Vienna and New York. p. 2.
24. Holtzman, E. (1989). Lysosomes. Plenum, New York.
25. Hopkins, C.R., Gibson, A., Shipman, M., and Miller, K. (1990). Movement of internalized ligand-receptor complexes along a continuous endosomal reticulum. *Nature* 346: 335–339.
26. Klionsky, D. J., Herman, P.K., and Emr, S.D. (1990). The fungal vacuole: composition, function, and biogenesis. *Microbiol. Rev.* 54: 266–292.
27. Knapp, P.E. and Swanson, J.A. (1990). Plasticity of the tubular lysosomal compartment in macrophages. *J. Cell Sci.* 95: 433–439.
28. Larsen, W.J. (1970). Genesis of mitochondria in insect fat body. *J. Cell Biol.* 47: 373–383.
29. Lippincott-Schwartz, J. and Fambrough, D.M. (1987). Lysosomal membrane dynamics: structure and interorganelle movement of a major lysosomal membrane glycoprotein. *J. Cell Biol.* 102: 1593–1605.

30. Lippincott-Schwartz, J., Yuan, L., Tipper, C., Amherdt, M., Orci, L., and Klausner, R.D. (1991). Brefeldin A's effects on endosomes, lysosomes, and the TGN suggest a general mechanism for regulating organelle structure and membrane traffic. *Cell* 67: 601–616.
31. Luck, D.J.L. (1963a). Genesis of mitochondria in *Neurospora crassa. Proc. Natl. Acad. Sci. USA* 49: 233–240.
32. Luck, D.J.L (1963b). Formation of mitochondria in *Neurospora crassa.* A quantitative radioautographic study. *J. Cell Biol.* 16: 483–499.
33. Luck, D.J.L (1965a). The influence of precursor size on mitochondrial composition in *Neurospora crassa. J. Cell Biol.* 24: 445–460.
34. Luck, D.J.L (1965b). Formation of mitochondria in *Neorospora crassa.* A study based on mitochondrial density changes. *J. Cell Biol.* 24: 461–470.
35. McKusick, V.A. and Neufeld, E.F. (1983). The mucopolysaccharide storage diseases. In: *The Metabolic Basis of Inherited Disease,* 5th ed. pp. 301–346 (Stanbury, J.B., Wyngaarden, J.R., Fredrickson, D.S., Goldstein, J.L., and Brown, M.S., Eds.). McGraw Hill, New York.
36. Messner, D.J., Griffiths, G., and Kornfeld, S. (1989). Isolation and characterization of membranes from bovine liver which are highly enriched in mannose 6-phosphate receptors. *J. Cell Biol.* 108: 2149–2162.
37. Oliver, C. (1983a). Lysosomal heterogeneity in exocrine acinar cells. *J. Histochem. Cytochem.* 31: 222–223.
38. Oliver, C. (1983b). Characterization of basal lysosomes in exocrine acinar cells. *J. Histochem. Cytochem.* 31: 1209–1216.
39. Oliver, C. and Yuasa, Y. (1987). Distribution of basal lysosomes in exocrine acinar cells. *J. Histochem. Cytochem.* 35: 565–570.
40. Parsons, J.A. (1965). Mitochondrial incorporation of tritiated thymidine in *Tetrahymena pyriformis. J. Cell Biol.* 25: 641–645.
41. Pelham, H.R.B (1991). Multiple targets for brefeldin A. *Cell* 67: 449–451.
42. Peters, C., Braun, M., Weber, B., Wendland, M., Schmidt, B., Pohlman, R., Waheed, A., and von Figura, K. (1990). Targeting of lysosomal membrane protein: a tyrosine containing endocytosis signal in the cytoplasmic tail of lysosomal acid phosphatase is necessary and sufficient for targeting to lysosomes. *EMBO J.* 9: 3497–3506.
43. Posakony, J.W., England, J.M., and Attardi, G. (1977). Mitochondrial growth and division during the cell cycle in HeLa cells. With an Appendix by J.N. Franklin on Estimation of three-dimensional cytological parameters from two-dimensional measurements made on cell sections. *J. Cell Biol.* 74: 468–491.
44. Raymond, C.K., O'Hara, P.J., Eichinger, G., Rothman, J.H., and Stevens, T.H. (1990). Molecular analysis of the yeast *VPS3* gene and the role of its product in vacuolar sorting and vacuolar segregation during the cell cycle. *J. Cell Biol.* 111: 877–892.
45. Severs, N.J., Jordan, E.G., and Williamson, D.H. (1976). Nuclear pore absence from areas of close assocaiton between the nucleus and vacuole in synchronous yeasst cultures. *J. Ultrastruct. Res.* 54: 374–387.
46. Shaw, J.M. and Wickner, W. (1991). *Vac2*: a yeast mutant which distinguishes vacuole segregation from Golgi-to-vacuole protein targeting. *EMBO J.* 10: 1741–1748.
47. Storrie, B. and Attardi, G. (1973). Mode of mitochondrial formation in HeLa cells. *J. Cell Biol.* 56: 833–838.
48. Storrie, B. and Deng, Y. (1992). Protein exchange within the lysosome and pre-lysosome compartment: a mechanism for maintaining organelle functionality. *NATO Symp.* H62: 85–90.
49. Storrie, B., Pool, Jr., R.R., Sachdeva, M., Maurey, K.M., and Oliver, C. (1984). Evidence for both prelysosomal and lysosomal intermediates in endocytic pathways. *J. Cell Biol.* 98: 108–115.
50. Swanson, J.A., Bushnell, A., and Silverstein, S.C. (1987). Tubular lysosome morphology and distribution within macrophages depend on the integrity of cytoplasmic microtubules. *Proc. Natl. Acad. Sci.* USA 84: 1921–1925.

51. Swanson, J.A., Yirinec, B.D., and Silverstein, S.C. (1985). Phorbol esters and horseradish peroxidase stimulate pinocytosis and redirect the flow of pinocytosed fluid in macrophages. *J. Cell Biol.* 100: 851–859.
52. Tandler, B., Erlendson, R.A., Smith, A.L., and Wynder, E.L. (1969). Riboflavin and mouse hepatic structure and function II. division of mitochondria during simple recovery from deficiency. *J. Cell Biol.* 41: 477–493.
53. Tooze, J. and Hollinshead, M. (1991). Tubular early endosomal networks in AtT20 and other cells. *J. Cell Biol.* 115: 635–653.
54. Tuomikoski, T., Felix, M.-A., Dorée, M., and Gruenberg, J. (1989). Inhibition of endocytic vesicle fusion *in vitro* by the cell-cycle control protein kinase cdc2. *Nature* 342: 942–945.
55. Vega, M.A., Rodriguez, F., Segui, B., Calés, C., Alcaldo, J., and Sandoval, I.V. (1991a). Targeting of lysosomal integral membrane protein LIMP II. The tyrosine-lacking carboxyl cytoplasmic tail of LIMP II is sufficient for direct targeting to lysosomes. *J. Biol. Chem.* 266: 16269–16272.
56. Vega, M., Segui-Real, B., Alcalde Garcia, J., Calés, C., Rodriguez, F., Vanderkerckhove, J., and Sandoval, I.V. (1991b). Cloning, sequencing, and expression of a cDNA encoding rat LIMP II, a novel 74-kDa lysosomal membrane protein related to the surface adhesion protein CD36. *J. Biol. Chem.* 266: 16818–16824.
56a. von Figura, K. (1991). Molecular recognition and targeting of lysosomal proteins. *Cur. Op. Cell Biol.* 3: 642–646.
57. Warren, G. (1985). Membrane traffic and organelle division. *Trends. Biochem. Sci.* 10: 439–443.
58. Warren, G. (1989). Mitosis and membranes. *Nature* 342: 857–858.
59. Weisman, L.S. and Wickner, W. (1988). Intervacuole exchange in the yeast zygote: a new pathway in organelle communication. *Science* 214: 589–591.
60. Weisman, L.S. and Wickner, W. (1992). Molecular characterization of *VAC1*, a gene required for vacuole inheritance and vacuole protein sorting. *J. Biol. Chem.* 267: 618–623.
61. Weisman, L.S., Bacallao, R., and Wickner, W. (1987). Multiple methods of visualizing the yeast vacuole permit evaluation of its morphology and inheritance during the cell cycle. *J. Cell Biol.* 105: 1539–1547.
62. Weisman, L.S., Emr, S.D., and Wickner, W.T. (1990). Mutants of *Saccharomyces cerevisiae* that block intervacuole vesicular traffic and vacuole division and segregation. *Proc. Natl. Acad. Sci. USA* 87: 1076–1080.
63. Wiemken, A., Matile, P., and Moor, H. (1970). Vacuolar dynamics in synchronously budding yeast. *Arch. Mikrobiol.* 70: 89–103.
64. Williams, M.A. and Fukuda, M. (1990). Accumulation of membrane glycoproteins in lysosomes requires a tyrosine residue at a particular position in the cytoplasmic tail. *J. Cell Biol.* 111: 955–966.
65. Witkop, Jr., C.J., Quevedo, Jr., W.C., and Fitzpatrick, T.B. (1983). Albinism and other disorders of pigment metabolism. In: *The Metabolic Basis of Inherited Disease.* 5th ed. (Stanbury, J.B., Wyngaarden, J.R., Fredrickson, D.S., Goldstein, J.L., and Brown, M.S., Eds.). McGraw Hill, New York. 301–346.
66. Zubenko, G.S., Park, F.J., and Jones, E.W. (1982). Genetic properties of mutations at the PEP4 locus in *Saccharomyces cerevisiae. Genetics* 102: 679–690.

INDEX

www.ingramcontent.com/pod-product-compliance
Ingram Content Group UK Ltd.
Pitfield, Milton Keynes, MK11 3LW, UK
UKHW020228290926
15470UKWH00001B/421